Lecture Notes in Artificial Intelligence 6743

Edited by R. Goebel, J. Siekmann, and W. Wahlster

Subseries of Lecture Notes in Computer Science

Sergei O. Kuznetsov Dominik Ślęzak
Daryl H. Hepting Boris G. Mirkin (Eds.)

Rough Sets, Fuzzy Sets, Data Mining and Granular Computing

13th International Conference, RSFDGrC 2011
Moscow, Russia, June 25-27, 2011
Proceedings

Springer

Volume Editors

Sergei O. Kuznetsov
National Research University Higher School of Economics
11 Pokrovski Boulevard, 109028 Moscow, Russia
E-mail: skuznetsov@hse.ru

Dominik Ślęzak
University of Warsaw, ul. Banacha 2, 02-097 Warsaw, Poland
E-mail: d.slezak@mimuw.edu.pl

Daryl H. Hepting
University of Regina, 3737 Wascana Parkway, Regina, SK, S4S 0A2, Canada
E-mail: hepting@cs.uregina.ca

Boris G. Mirkin
National Research University Higher School of Economics
11 Pokrovski Boulevard, 109028 Moscow, Russia
E-mail: bmirkin@hse.ru
and Birbeck University of London, Malet Street, London, WC1E 7HX, UK
E-mail: mirkin@dcs.bbk.ac.uk

ISSN 0302-9743 e-ISSN 1611-3349
ISBN 978-3-642-21880-4 ISBN 978-3-642-21881-1 (eBook)
DOI 10.1007/978-3-642-21881-1
Springer Heidelberg Dordrecht London New York

Library of Congress Control Number: 2011929500

CR Subject Classification (1998): I.2, H.2.8, H.2.4, H.3, F.4.1, F.1, I.5, H.4

LNCS Sublibrary: SL 7 – Artificial Intelligence

Typesetting: Camera-ready by author, data conversion by Scientific Publishing Services, Chennai, India

Printed on acid-free paper

Springer is part of Springer Science+Business Media (www.springer.com)

Preface

This volume contains papers presented at the 13th International Conference on Rough Sets, Fuzzy Sets and Granular Computing (RSFDGrC) held during June 25–27, 2011, at the National Research University Higher School of Economics (NRU HSE) in Moscow, Russia. RSFDGrC is a series of scientific events spanning the last 15 years. It investigates the meeting points among the four major disciplines outlined in its title, with respect to both foundations and applications. In 2011, RSFDGrC was co-organized with the 4th International Conference on Pattern Recognition and Machine Intelligence (PReMI), providing a great opportunity for multi-faceted interaction between scientists and practitioners.

There were 83 paper submissions from over 20 countries. Each submission was reviewed by at least three Chairs or PC members. We accepted 34 regular papers (41%). In order to stimulate the exchange of research ideas, we also accepted 15 short papers. All 49 papers are distributed among 10 thematic sections of this volume. The conference program featured five invited talks given by Jiawei Han, Vladik Kreinovich, Guoyin Wang, Radim Belohlavek, and C.A. Murthy, as well as two tutorials given by Marcin Szczuka and Richard Jensen. Their corresponding papers and abstracts are gathered in the first two sections of this volume.

We would like to thank all authors and reviewers for their work and excellent contributions. We express our gratitude to Lotfi A. Zadeh, who suggested many talented scientists to serve as PC members. The success of the whole undertaking would be impossible without collaboration with the Chairs of PReMI-2011, as well as the Chairs of workshops co-organized with the main conference. We also acknowledge the following organizations and sponsoring institutions: National Research University Higher School of Economics (Moscow), Laboratoire Poncelet (UMI 2615 du CNRS, Moscow), International Rough Set Society, International Fuzzy Systems Association, Russian Foundation for Basic Research, ABBYY Software House, Yandex (Moscow), and Springer. Last but not least, we are grateful to all Chairs and organizers of RSFDGrC-2011, especially to Dmitry I. Ignatov, whose endless energy saved us in the most critical stages of conference preparation.

April 2011

Sergei O. Kuznetsov
Dominik Ślęzak
Daryl H. Hepting
Boris G. Mirkin

Organization

General Chair	Boris G. Mirkin, Russia
Conference Chair	Sergei O. Kuznetsov, Russia
Program Co-chairs	Dominik Ślęzak, Poland
	Daryl H. Hepting, Canada
Organizing Chair	Dmitry I. Ignatov, Russia
Tutorial Co-chairs	Chris Cornelis, Belgium
	Sanghamitra Bandyopadhyay, India
Publicity Co-chairs	Jimmy Huang, Canada
	Wei-Zhi Wu, China

Program Committee

Alexey N. Averkin, Russia
Mohua Banerjee, India
Alan Barton, Canada
Ildar Batyrshyn, Russia
Mihir K. Chakraborty, India
Ashok Deshpande, India
Lipika Dey, India
Anna Gomolińska, Poland
Vladimir Gorodetsky, Russia
Aboul E. Hassanien, Egypt
Qinghua Hu, China
M. Gordon Hunter, Canada
Dmitry I. Ignatov, Russia
Masahiro Inuiguchi, Japan
Ryszard Janicki, Canada
Manish Joshi, India
Michiro Kondo, Japan
Rudolf Kruse, Germany
Yasuo Kudo, Japan
Tianrui Li, China
Pawan Lingras, Canada
Ju-Sheng Mi, China
Michinori Nakata, Japan
Hung Son Nguyen, Poland
Sergey Nikolenko, Russia
Vilem Novak, Czech Republic
Witold Pedrycz, Canada
Georg Peters, Germany
Sheela Ramanna, Canada
Hiroshi Sakai, Japan
Gerald Schaefer, UK
Kun She, China
Qiang Shen, UK
Marek Sikora, Poland
Vasily Sinuk, Russia
Andrzej Skowron, Poland
Roman Słowiński, Poland
Jarosław Stepaniuk, Poland
Zbigniew Suraj, Poland
Piotr Synak, Poland
Andrzej Szałas, Poland
Marcin Szczuka, Poland
Noboru Takagi, Japan
Domenico Talia, Italy
Valery Tarasov, Russia
Alexander Tulupiev, Russia
Xizhao Wang, China
Junzo Watada, Japan
Yanping Xiang, China
JingTao Yao, Canada
Nadezhda Yarushkina, Russia
Alexander Yazenin, Russia
Alla Zaboleeva-Zotova, Russia
William Zhu, China
Leonid E. Zhukov, Russia
Wojciech Ziarko, Canada

Additional Reviewers

Andrzej Chmielewski, Poland
Si Yuan Jing, China
Sharmistha Mitra, India
Vsevolod Oparin, Russia
Yulia Orlova, Russia
Herald S. Plesnevich, Russia
Jonas Poelmans, Belgium
Julia Preusse, Germany
Georg Ruß, Germany
Alexander Sirotkin, Russia
Matthias Steinbrecher, Germany
Rustam Tagiew, Germany

Table of Contents

Invited Papers

Tutorial Papers

Rough Sets and Approximations

Coverings and Granules

Fuzzy Set Models

Fuzzy Set Applications

Compound Values

Feature Selection and Reduction

Clusters and Concepts

Rules and Trees

Construction and Analysis of Web-Based Computer Science Information Networks

Jiawei Han

Department of Computer Science
University of Illinois at Urbana-Champaign
hanj@cs.uiuc.edu

With the rapid development of the Web, huge amounts of information are available on the Web in the form of Web documents, structures, and links. It has been a dream of the database and Web communities to harvest information exhibited on the Web and reconcile the unstructured nature of the Web with the semi-structured schemas of the database paradigm. This is a challenging task. Even though databases are currently used to generate Web content in some sites, the schemas of these databases are rarely consistent across a domain. However, with the recent research in Web structure mining and information network analysis, major progress has been made at discovering Web hidden structures, constructing heterogeneous information networks by integration of information from structured databases and Web contents, and performing in-depth analysis for systematic harvesting of such rich information on the Web.

Based on our recent research, we have been developing an innovative Web-based information network analysis system, called WINACS (Web-based Information Network Analysis for Computer Science) [6], which incorporates many recent, exciting developments in data sciences to construct a Web-based computer science information network, and discover, retrieve, rank, cluster, and analyze such an information network. Taking computer science as a dedicated domain, WINACS first discovers Web entity structures, integrates the contents in the DBLP database with that on the Web to construct a heterogeneous computer science information network. With this structure in hand, WINACS is able to rank, cluster and analyze this network and support intelligent and analytical queries. In this talk, we will discuss the principles of information network-based Web mining, show multiple salient features of WINACS and demonstrate how computer science Web pages and DBLP can be nicely integrated to support queries and mining in highly friendly and intelligent ways. We envision the methodologies can be extended to handle many other exciting information networks extracted from the Web, such as general academia, governments, sports and so on.

The WINACS system is being developed at the Data Mining Research Group in Computer Science, Univ. of Illinois, based on our recent research on Web structure mining, such as [8,7], and information network analysis, such as [4,3,2,1,5].

Acknowledgements. The work was supported in part by the U.S. National Science Foundation grants IIS-09-05215, the Network Science Collaborative Technology Alliance Program (NS-CTA) of U.S. Army Research Lab (ARL) under

S.O. Kuznetsov et al. (Eds.): RSFDGrC 2011, LNAI 6743, pp. 1–2, 2011.

contract number W911NF-09-2-0053, and the Air Force Office of Scientific Research MURI award FA9550-08-1-0265. The author would like to express his sincere thanks to all the WINACS project group and the Ph.D. students in the Data Mining Group of CS, UIUC for their dedication and contribution.

References

1. Ji, M., Sun, Y., Danilevsky, M., Han, J., Gao, J.: Graph regularized transductive classification on heterogeneous information networks. In: Proc. 2010 European Conf. on Machine Learning and Principles and Practice of Knowledge Discovery in Databases (ECMLPKDD 2010), Barcelona, Spain (September 2010)
2. Sun, Y., Han, J., Yan, X., Yu, P.S., Wu, T.: PathSim: Meta path-based top-k similarity search in heterogeneous information networks. In: Proc. 2011 Int. Conf. on Very Large Data Based (VLDB 2011), Seattle, WA (August 2011)
3. Sun, Y., Han, J., Zhao, P., Yin, Z., Cheng, H., Wu, T.: RankClus: Integrating clustering with ranking for heterogeneous information network analysis. In: Proc. 2009 Int. Conf. on Extending Data Base Technology (EDBT 2009), Saint-Petersburg, Russia (March 2009)
4. Sun, Y., Yu, Y., Han, J.: Ranking-based clustering of heterogeneous information networks with star network schema. In: Proc. 2009 ACM SIGKDD Int. Conf. on Knowledge Discovery and Data Mining (KDD 2009), Paris, France (June 2009)
5. Wang, C., Han, J., Jia, Y., Tang, J., Zhang, D., Yu, Y., Guo, J.: Mining advisor-advisee relationships from research publication networks. In: Proc. 2010 ACM SIGKDD Conf. on Knowledge Discovery and Data Mining (KDD 2010), Washington D.C (July 2010)
6. Weninger, T., Danilevsky, M., Fumarola, F., Hailpern, J., Han, J., Ji, M., Johnston, T.J., Kallumadi, S., Kim, H., Li, Z., McCloskey, D., Sun, Y., TeGrotenhuis, N.E., Wang, C., Yu, X.: Winacs: Construction and analysis of web-based computer science information networks. In: Proc. of 2011 ACM SIGMOD Int. Conf. on Management of Data (SIGMOD 2011) (system demo), Athens, Greece (June 2011)
7. Weninger, T., Fumarola, F., Han, J., Malerba, D.: Mapping web pages to database records via link paths. In: Proc. 2010 ACM Int. Conf. on Information and Knowledge Management (CIKM 2010), Toronto, Canada (October 2010)
8. Weninger, T., Fumarola, F., Lin, C.X., Barber, R., Han, J., Malerba, D.: Growing parallel paths for entity-page discovery. In: Proc. of 2011 Int. World Wide Web Conf (WWW 2011), Hyderabad, India (March 2011)

Towards Faster Estimation of Statistics and ODEs Under Interval, P-Box, and Fuzzy Uncertainty: From Interval Computations to Rough Set-Related Computations

Vladik Kreinovich

University of Texas at El Paso, El Paso, TX 79968, USA
vladik@utep.edu

Abstract. Interval computations estimate the uncertainty of the result of data processing in situations in which we only know the upper bounds Δ on the measurement errors. In interval computations, at each intermediate stage of the computation, we have intervals of possible values of the corresponding quantities. As a result, we often have bounds with excess width. In this paper, we show that one way to remedy this problem is to extend interval technique to *rough-set computations*, where at each stage, in addition to intervals of possible values of the quantities, we also keep rough sets representing possible values of pairs (triples, etc.).

The paper's outline is as follows: we formulate the main problem (Section 1), briefly overview interval computations techniques solve this problem (Section 2), and then explain how the main ideas behind interval computation techniques can be extended to computations with rough sets (Section 3).

Keywords: interval computations, interval uncertainty, rough sets, statistics under interval uncertainty.

1 Formulation of the Problem

Need for interval computations. In many real-life situations, we need to process data, i.e., to apply an algorithm $f(x_1, \ldots, x_n)$ to measurement results $x_1, \ldots, x_n$.

Measurements are never 100% accurate, so in reality, the actual value x_i of i-th measured quantity can differ from the measurement result $\widetilde{x}_i$. Because of these *measurement errors* $\Delta x_i \stackrel{\text{def}}{=} \widetilde{x}_i - x_i$, the result $\widetilde{y} = f(\widetilde{x}_1, \ldots, \widetilde{x}_n)$ of data processing is, in general, different from the actual value $y = f(x_1, \ldots, x_n)$ of the desired quantity y.

In many practical situations, we only know the upper bound Δ_i on the (absolute value of) the measurement errors Δx_i. In such situations, the only information that we have about the (unknown) actual value of $y = f(x_1, \ldots, x_n)$ is that y belongs to the range $\mathbf{y} = [\underline{y}, \overline{y}]$ of the function f over the box $\mathbf{x}_1 \times \ldots \times \mathbf{x}_n$:

$$\mathbf{y} = [\underline{y}, \overline{y}] = f(\mathbf{x}_1, \ldots, \mathbf{x}_n) \stackrel{\text{def}}{=} \{f(x_1, \ldots, x_n) \,|\, x_1 \in \mathbf{x}_1, \ldots, x_n \in \mathbf{x}_n\}.$$

S.O. Kuznetsov et al. (Eds.): RSFDGrC 2011, LNAI 6743, pp. 3–10, 2011.

The process of computing this interval range based on the input intervals $\mathbf{x}_i$ is called *interval computations*; see, e.g., [4].

Case of fuzzy uncertainty and its reduction to interval uncertainty. In addition to bounds, we can also have expert estimates on Δx_i. An expert usually describes his/her uncertainty by using words from a natural language, like "most probably, the value of the quantity is between 3 and 4". To formalize this knowledge, it is natural to use *fuzzy set theory*, a formalism specifically designed for describing this type of informal ("fuzzy") knowledge; see, e.g., [5].

In fuzzy set theory, the expert's uncertainty about x_i is described by a fuzzy set, i.e., by a function $\mu_i(x_i)$ which assigns, to each possible value x_i of the i-th quantity, the expert's degree of certainty that x_i is a possible value. A fuzzy set can also be described as a nested family of α-cuts $\mathbf{x}_i(\alpha) \stackrel{\text{def}}{=} \{x_i \mid \mu_i(x_i) \geq \alpha\}$.

Zadeh's extension principle can be used to transform the fuzzy sets for x_i into a fuzzy set for y. It is known that for continuous functions f on a bounded domain this principle is equivalent to saying that, for every α,

$$\mathbf{y}(\alpha) = f(\mathbf{x}_1(\alpha), \ldots, \mathbf{x}_n(\alpha)).$$

In other words, fuzzy data processing can be implemented as layer-by-layer interval computations. In view of this reduction, in the following text, we will mainly concentrate on interval computations.

2 Interval Computations: Brief Reminder

Interval computations: main idea. Historically the first method for computing the enclosure for the range is the method which is sometimes called "straightforward" interval computations. This method is based on the fact that inside the computer, every algorithm consists of elementary operations (arithmetic operations, min, max, etc.). For each elementary operation $f(a, b)$, if we know the intervals $\mathbf{a}$ and $\mathbf{b}$ for a and b, we can compute the exact range $f(\mathbf{a}, \mathbf{b})$. The corresponding formulas form the so-called *interval arithmetic*:

$$[\underline{a}, \overline{a}] + [\underline{b}, \overline{b}] = [\underline{a} + \underline{b}, \overline{a} + \overline{b}]; \quad [\underline{a}, \overline{a}] - [\underline{b}, \overline{b}] = [\underline{a} - \overline{b}, \overline{a} - \underline{b}];$$

$$[\underline{a}, \overline{a}] \cdot [\underline{b}, \overline{b}] = [\min(\underline{a} \cdot \underline{b}, \underline{a} \cdot \overline{b}, \overline{a} \cdot \underline{b}, \overline{a} \cdot \overline{b}), \max(\underline{a} \cdot \underline{b}, \underline{a} \cdot \overline{b}, \overline{a} \cdot \underline{b}, \overline{a} \cdot \overline{b})];$$

$$1/[\underline{a}, \overline{a}] = [1/\overline{a}, 1/\underline{a}] \text{ if } 0 \notin [\underline{a}, \overline{a}]; \quad [\underline{a}, \overline{a}]/[\underline{b}, \overline{b}] = [\underline{a}, \overline{a}] \cdot (1/[\underline{b}, \overline{b}]).$$

In straightforward interval computations, we repeat the computations forming the program f step-by-step, replacing each operation with real numbers by the corresponding operation of interval arithmetic. It is known that, as a result, we get an enclosure $\mathbf{Y} \supseteq \mathbf{y}$ for the desired range.

From main idea to actual computer implementation. Not every real number can be exactly implemented in a computer; thus, e.g., after implementing an operation of interval arithmetic, we must enclose the result $[r^-, r^+]$ in a computer-representable interval: namely, we must round-off r^- to a smaller computer-representable value $\underline{r}$, and round-off r^+ to a larger computer-representable value $\overline{r}$.

Sometimes, we get excess width. In some cases, the resulting enclosure is exact; in other cases, the enclosure has excess width. The excess width is inevitable since straightforward interval computations increase the computation time by at most a factor of 4, while computing the exact range is, in general, NP-hard (see, e.g., [6]), even for computing the population variance $V = \frac{1}{n}\cdot\sum_{i=1}^{n}(x_i - \overline{x})^2$, where $\overline{x} = \frac{1}{n}\cdot\sum_{i=1}^{n} x_i$ (see [3]). If we get excess width, then we can use techniques such as centered form, bisection, etc., to get a better estimate; see, e.g., [4].

Reason for excess width. The main reason for excess width is that intermediate results are dependent on each other, and straightforward interval computations ignore this dependence. For example, the actual range of $f(x_1) = x_1 - x_1^2$ over $\mathbf{x}_1 = [0,1]$ is $\mathbf{y} = [0, 0.25]$. Computing this f means that we first compute $x_2 := x_1^2$ and then subtract x_2 from x_1. According to straightforward interval computations, we compute $\mathbf{r} = [0,1]^2 = [0,1]$ and then $\mathbf{x}_1 - \mathbf{x}_2 = [0,1] - [0,1] = [-1,1]$. This excess width comes from the fact that the formula for interval subtraction implicitly assumes that both a and b can take arbitrary values within the corresponding intervals $\mathbf{a}$ and $\mathbf{b}$, while in this case, the values of x_1 and x_2 are clearly not independent: x_2 is uniquely determined by x_1, as $x_2 = x_1^2$.

3 Rough Set Computations

Main idea. The idea behind (rough) set computations (see, e.g., [1,7,8]) is to remedy the above reason why interval computations lead to excess width. Specifically, at every stage of the computations, in addition to keeping the *intervals* $\mathbf{x}_i$ of possible values of all intermediate quantities x_i, we also keep *sets*:

- sets $\mathbf{x}_{ij}$ of possible values of pairs (x_i, x_j);
- if needed, sets $\mathbf{x}_{ijk}$ of possible values of triples (x_i, x_j, x_k); etc.

In the above example, instead of just keeping two intervals $\mathbf{x}_1 = \mathbf{x}_2 = [0,1]$, we would then also generate and keep the set $\mathbf{x}_{12} = \{(x_1, x_1^2) \,|\, x_1 \in [0,1]\}$. Then, the desired range is computed as the range of $x_1 - x_2$ over this set – which is exactly $[0, 0.25]$.

How can we propagate this set uncertainty via arithmetic operations? Let us describe this on the example of addition, when, in the computation of f, we use two previously computed values x_i and x_j to compute a new value $x_k := x_i + x_j$. In this case, we set $\mathbf{x}_{ik} = \{(x_i, x_i + x_j) \,|\, (x_i, x_j) \in \mathbf{x}_{ij}\}$, $\mathbf{x}_{jk} = \{(x_j, x_i + x_j) \,|\, (x_i, x_j) \in \mathbf{x}_{ij}\}$, and for every $l \neq i, j$, we take

$$\mathbf{x}_{kl} = \{(x_i + x_j, x_l) \,|\, (x_i, x_j) \in \mathbf{x}_{ij}, (x_i, x_l) \in \mathbf{x}_{il}, (x_j, x_l) \in \mathbf{x}_{jl}\}.$$

From main idea to actual computer implementation. In interval computations, we cannot represent an arbitrary interval inside the computer, we need an enclosure. Similarly, we cannot represent an arbitrary set inside a computer, we need an enclosure.

To describe such enclosures, we fix the number C of granules (e.g., $C = 10$). We divide each interval $\mathbf{x}_i$ into C equal parts $\mathbf{X}_i$; thus each box $\mathbf{x}_i \times \mathbf{x}_j$ is divided into C^2 subboxes $\mathbf{X}_i \times \mathbf{X}_j$. We then describe each set $\mathbf{x}_{ij}$ by listing all subboxes $\mathbf{X}_i \times \mathbf{X}_j$ which have common elements with $\mathbf{x}_{ij}$; the union of such subboxes is an enclosure for the desired set $\mathbf{x}_{ij}$. This enclosure is a P-upper approximation to the desired set.

This enables us to implement all above arithmetic operations. For example, to implement $\mathbf{x}_{ik} = \{(x_i, x_i + x_j) \mid (x_i, x_j) \in \mathbf{x}_{ij}\}$, we take all the subboxes $\mathbf{X}_i \times \mathbf{X}_j$ that form the set $\mathbf{x}_{ij}$; for each of these subboxes, we enclosure the corresponding set of pairs $\{(x_i, x_i + x_j) \mid (x_i, x_j) \in \mathbf{X}_i \times \mathbf{X}_j\}$ into a set $\mathbf{X}_i \times (\mathbf{X}_i + \mathbf{X}_j)$. This set may have non-empty intersection with several subboxes $\mathbf{X}_i \times \mathbf{X}_k$; all these subboxes are added to the computed enclosure for $\mathbf{x}_{ik}$. One can easily see that if we start with the exact range $\mathbf{x}_{ij}$, then the resulting enclosure for $\mathbf{x}_{ik}$ is an $(1/C)$-approximation to the actual set – and so when C increases, we get more and more accurate representations of the desired set.

Similarly, to find an enclosure for

$$\mathbf{x}_{kl} = \{(x_i + x_j, x_l) \mid (x_i, x_j) \in \mathbf{x}_{ij}, (x_i, x_l) \in \mathbf{x}_{il}, (x_j, x_l) \in \mathbf{x}_{jl}\},$$

we consider all the triples of subintervals $(\mathbf{X}_i, \mathbf{X}_j, \mathbf{X}_l)$ for which $\mathbf{X}_i \times \mathbf{X}_j \subseteq \mathbf{x}_{ij}$, $\mathbf{X}_i \times \mathbf{X}_l \subseteq \mathbf{x}_{il}$, and $\mathbf{X}_j \times \mathbf{X}_l \subseteq \mathbf{x}_{jl}$; for each such triple, we compute the box $(\mathbf{X}_i + \mathbf{X}_j) \times \mathbf{X}_l$; then, we add subboxes $\mathbf{X}_k \times \mathbf{X}_l$ which intersect with this box to the enclosure for $\mathbf{x}_{kl}$.

Toy example: computing the range of $x - x^2$. In straightforward interval computations, we have $r_1 = x$ with the exact interval range $\mathbf{r}_1 = [0, 1]$, and $r_2 = x^2$ with the exact interval range $\mathbf{x}_2 = [0, 1]$. The variables r_1 and r_2 are dependent, but we ignore this dependence and estimate $\mathbf{r}_3$ as $[0, 1] - [0, 1] = [-1, 1]$.

In the new approach: we have $\mathbf{r}_1 = \mathbf{r}_2 = [0, 1]$, and we also have $\mathbf{r}_{12}$. First, we divide the range $[0, 1]$ into 5 equal subintervals $\mathbf{R}_1$. The union of the ranges $\mathbf{R}_1^2$ corresponding to 5 subintervals $\mathbf{R}_1$ is $[0, 1]$, so $\mathbf{r}_2 = [0, 1]$. We divide $\mathbf{r}_2$ into 5 equal subintervals $[0, 0.2]$, $[0.2, 0.4]$, etc. We now compute $\mathbf{r}_{12}$ as follows:

- for $\mathbf{R}_1 = [0, 0.2]$, we have $\mathbf{R}_1^2 = [0, 0.04]$, so only subinterval $[0, 0.2]$ of the interval $\mathbf{r}_2$ is affected;
- for $\mathbf{R}_1 = [0.2, 0.4]$, we have $\mathbf{R}_1^2 = [0.04, 0.16]$, so also only subinterval $[0, 0.2]$ is affected;
- for $\mathbf{R}_1 = [0.4, 0.6]$, we have $\mathbf{R}_1^2 = [0.16, 0.36]$, so two subintervals $[0, 0.2]$ and $[0.2, 0.4]$ are affected, etc.

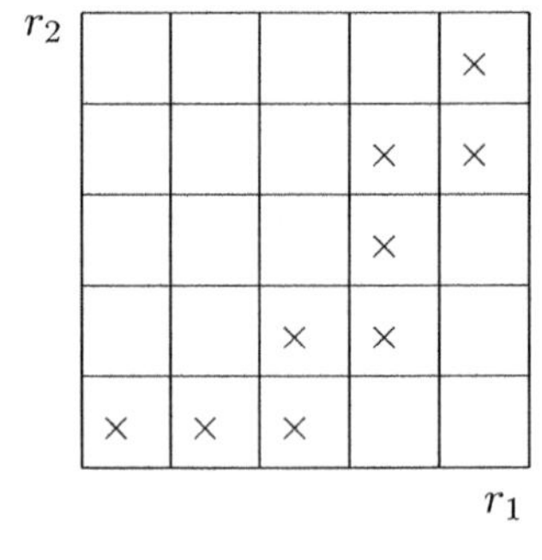

For each possible pair of small boxes $\mathbf{R}_1 \times \mathbf{R}_2$, we have $\mathbf{R}_1 - \mathbf{R}_2 = [-0.2, 0.2]$, $[0, 0.4]$, or $[0.2, 0.6]$, so the union of $\mathbf{R}_1 - \mathbf{R}_2$ is $\mathbf{r}_3 = [-0.2, 0.6]$.

If we divide into more and more pieces, we get the enclosure which is closer and closer to the exact range $[0, 0.25]$.

How to Compute $\mathbf{r}_{ik}$. The above example is a good case to illustrate how we compute the range $\mathbf{r}_{13}$ for $r_3 = r_1 - r_2$. Indeed, since $\mathbf{r}_3 = [-0.2, 0.6]$, we divide this range into 5 subintervals $[-0.2, -0.04]$, $[-0.04, 0.12]$, $[0.12, 0.28]$, $[0.28, 0.44]$, $[0.44, 0.6]$.

- For $\mathbf{R}_1 = [0, 0.2]$, the only possible $\mathbf{R}_2$ is $[0, 0.2]$, so $\mathbf{R}_1 - \mathbf{R}_2 = [-0.2, 0.2]$. This covers $[-0.2, -0.04]$, $[-0.04, 0.12]$, and $[0.12, 0.28]$.
- For $\mathbf{R}_1 = [0.2, 0.4]$, the only possible $\mathbf{R}_2$ is $[0, 0.2]$, so $\mathbf{R}_1 - \mathbf{R}_2 = [0, 0.4]$. This interval covers $[-0.04, 0.12]$, $[0.12, 0.28]$, and $[0.28, 0.44]$.
- For $\mathbf{R}_1 = [0.4, 0.6]$, we have two possible $\mathbf{R}_2$:
  - for $\mathbf{R}_2 = [0, 0.2]$, we have $\mathbf{R}_1 - \mathbf{R}_2 = [0.2, 0.6]$; this covers $[0.12, 0.28]$, $[0.28, 0.44]$, and $[0.44, 0.6]$;
  - for $\mathbf{R}_2 = [0.2, 0.4]$, we have $\mathbf{R}_1 - \mathbf{R}_2 = [0, 0.4]$; this covers $[-0.04, 0.12]$, $[0.12, 0.28]$, and $[0.28, 0.44]$.
- For $\mathbf{R}_1 = [0.6, 0.8]$, we have $\mathbf{R}_1^2 = [0.36, 0.64]$, so three possible $\mathbf{R}_2$: $[0.2, 0.4]$, $[0.4, 0.6]$, and $[0.6, 0.8]$, to the total of $[0.2, 0.8]$. Here, $[0.6, 0.8] - [0.2, 0.8] = [-0.2, 0.6]$, so all 5 subintervals are affected.
- Finally, for $\mathbf{R}_1 = [0.8, 1.0]$, we have $\mathbf{R}_1^2 = [0.64, 1.0]$, so two possible $\mathbf{R}_2$: $[0.6, 0.8]$ and $[0.8, 1.0]$, to the total of $[0.6, 1.0]$. Here, $[0.8, 1.0] - [0.6, 1.0] = [-0.2, 0.4]$, so the first 4 subintervals are affected.

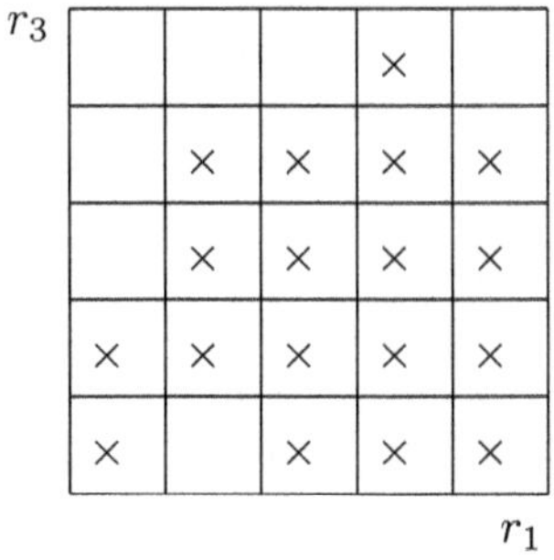

Limitations of this approach. The main limitation of this approach is that when we need an accuracy ε, we must use $\sim 1/\varepsilon$ granules; so, if we want to compute the result with k digits of accuracy, i.e., with accuracy $\varepsilon = 10^{-k}$, we must consider exponentially many boxes ($\sim 10^k$). In plain words, this method is only applicable when we want to know the desired quantity with a given accuracy (e.g., 10%).

Cases when this approach is applicable. In practice, there are many problems when it is sufficient to compute a quantity with a given accuracy: e.g., when we detect an outlier, we usually do not need to know the variance with a high accuracy – an accuracy of 10% is more than enough.

Let us describe the case when interval computations do not lead to the exact range, but set computations do – of course, the range is "exact" modulo accuracy of the actual computer implementations of these sets.

Example: estimating variance under interval uncertainty. Suppose that we know the intervals $\mathbf{x}_1, \ldots, \mathbf{x}_n$ of possible values of $x_1, \ldots, x_n$, and we need to compute the range of the variance $V = \frac{1}{n} \cdot M - \frac{1}{n^2} \cdot E^2$, where $M \stackrel{\text{def}}{=} \sum_{i=1}^n x_i^2$ and $E \stackrel{\text{def}}{=} \sum_{i=1}^n x_i$.

A natural way to compute V is to compute the intermediate sums $M_k \stackrel{\text{def}}{=} \sum_{i=1}^k x_i^2$ and $E_k \stackrel{\text{def}}{=} \sum_{i=1}^k x_i$. We start with $M_0 = E_0 = 0$; once we know the pair (M_k, E_k), we compute $(M_{k+1}, E_{k+1}) = (M_k + x_{k+1}^2, E_k + x_{k+1})$. Since the values of M_k and E_k only depend on $x_1, \ldots, x_k$ and do not depend on x_{k+1}, we can conclude that if (M_k, E_k) is a possible value of the pair and x_{k+1} is a possible value of this variable, then $(M_k + x_{k+1}^2, E_k + x_{k+1})$ is a possible value of (M_{k+1}, E_{k+1}). So, the set $\mathbf{p}_0$ of possible values of (M_0, E_0) is the single point $(0, 0)$, and once we know the set $\mathbf{p}_k$ of possible values of (M_k, E_k), we can compute $\mathbf{p}_{k+1}$ as

$$\{(M_k + x^2, E_k + x) \mid (M_k, E_k) \in \mathbf{p}_k, x \in \mathbf{x}_{k+1}\}.$$

For $k = n$, we will get the set $\mathbf{p}_n$ of possible values of (M, E). Based on this set, we can then find the exact range of the variance $V = \frac{1}{n} \cdot M - \frac{1}{n^2} \cdot E^2$.

What C should we choose to get the results with an accuracy $\varepsilon \cdot \overline{V}$? On each step, we add the uncertainty of $1/C$. So, after n steps, we add the inaccuracy of n/C. Thus, to get the accuracy $n/C \approx \varepsilon$, we must choose $C = n/\varepsilon$.

What is the running time of the resulting algorithm? We have n steps; at each step, we need to analyze C^3 combinations of subintervals for E_k, M_k, and x_{k+1}. Thus, overall, we need $n \cdot C^3$ steps, i.e., n^4/ε^3 steps. For fixed accuracy $C \sim n$, we need $O(n^4)$ steps – a polynomial time, and for $\varepsilon = 1/10$, the coefficient at n^4 is still 10^3 – quite feasible.

For example, for $n = 10$ values and for the desired accuracy $\varepsilon = 0.1$, we need $10^3 \cdot n^4 \approx 10^7$ computational steps – "nothing" for a Gigaherz (10^9 operations per second) processor on a usual PC. For $n = 100$ values and the same desired accuracy, we need $10^4 \cdot n^4 \approx 10^{12}$ computational steps, i.e., 10^3 seconds (15 minutes) on a Gigaherz processor. For $n = 1000$, we need 10^{15} steps, i.e., 10^6 seconds – 12 days on a single processor or a few hours on a multi-processor machine.

In comparison, the exponential time 2^n needed in the worst case for the exact computation of the variance under interval uncertainty, is doable ($2^{10} \approx 10^3$ steps) for $n = 10$, but becomes unrealistically astronomical ($2^{100} \approx 10^{30}$ steps) already for $n = 100$.

Comment. When the accuracy increases to $\varepsilon = 10^{-k}$, we get an exponential increase in running time – but this is OK since, as we have mentioned, the problem of computing variance under interval uncertainty is, in general, NP-hard.

Other statistical characteristics. Similar algorithms can be presented for computing many other statistical characteristics [1].

Systems of ordinary differential equations (ODEs) under interval uncertainty. A general system of ODEs has the form $\dot{x}_i = f_i(x_1, \ldots, x_m, t)$, $1 \le i \le m$. Interval uncertainty usually means that the exact functions f_i are unknown, we only know the expressions of f_i in terms of parameters, and we have interval bounds on these parameters.

There are two types of interval uncertainty: we may have global parameters whose values are the same for all moments t, and we may have parameters whose values may differ at different moments of time – but always within given intervals. In general, we have a system of the type

$$\dot{x}_i = f_i(x_1, \ldots, x_m, t, a_1, \ldots, a_k, b_1(t), \ldots, b_l(t)),$$

where f_i is a known function, and we know the intervals $\mathbf{a}_j$ and $\mathbf{b}_j(t)$ of possible values of a_i and $b_j(t)$.

For the general system of ODEs, Euler's equations take the form

$$x_i(t + \Delta t) = x_i(t) + \Delta t \cdot f_i(x_1(t), \ldots, x_m(t), t, a_1, \ldots, a_k, b_1(t), \ldots, b_l(t)).$$

Thus, if for every t we keep the set of all possible values of a tuple

$$(x_1(t), \ldots, x_m(t), a_1, \ldots, a_k),$$

then we can use the Euler's equations to get the exact set of possible values of this tuple at the next moment of time.

The reason for exactness is that the values $x_i(t)$ depend only on the previous values $b_j(t - \Delta t)$, $b_j(t - 2\Delta t)$, etc., and not on the current values $b_j(t)$.

To predict the values $x_i(T)$ at a moment T, we need $n = T/\Delta t$ iterations.

To update the values, we need to consider all possible combinations of $m+k+l$ variables $x_1(t), \ldots, x_m(t), a_1, \ldots, a_k, b_1(t), \ldots, b_l(t)$; so, to predict the values at moment $T = n \cdot \Delta t$ in the future for a given accuracy $\varepsilon > 0$, we need the running time $n \cdot C^{m+k+l} \sim n^{k+l+m+1}$. This is still polynomial in n.

Towards extension to p-boxes and classes of probability distributions. Often, in addition to the interval $\mathbf{x}_i$ of possible values of the inputs x_i, we also have partial information about the probabilities of different values $x_i \in \mathbf{x}_i$. An exact probability distribution can be described, e.g., by its cumulative distribution function (cdf) $F_i(z) = \text{Prob}(x_i \le z)$. In these terms, a partial information means that instead of a single cdf, we have a *class* $\mathcal{F}$ of possible cdfs.

A practically important particular case of this partial information is when, for each z, instead of the exact value $F(z)$, we know an interval $\mathbf{F}(z) = [\underline{F}(z), \overline{F}(z)]$ of possible values of $F(z)$. Such an "interval-valued" cdf is called a *probability box*, or a *p-box*, for short; see, e.g., [2].

Propagating p-box uncertainty via computations: a problem. Once we know the classes $\mathcal{F}_i$ of possible distributions for x_i, and data processing algorithms

$f(x_1, \ldots, x_n)$, we would like to know the class $\mathcal{F}$ of possible resulting distributions for $y = f(x_1, \ldots, x_n)$.

Idea. For problems like systems of ODEs, it is sufficient to keep and update, for all t, the set of possible joint distributions for the tuple $(x_1(t), \ldots, a_1, \ldots)$.

4 Conclusions

In many practical situations, for each quantity x_i, we only know the upper bound Δ_i on the measurement error $\Delta x_i \stackrel{\text{def}}{=} \widetilde{x}_i - x_i$; in this case, once we know the measurement result $\widetilde{x}_i$, the only information that we have about the actual (unknown) value x_i is that it belongs to the interval $\mathbf{x}_i = [\widetilde{x}_i - \Delta_i, \widetilde{x}_i + \Delta_i]$. For each quantity $y = f(x_1, \ldots, x_n)$, different values $x_i \in \mathbf{x}_i$ lead, in general, to different values y; it is therefore desirable to find the range $\mathbf{y}$ of all such values. In this paper, we show that for many problems, we can efficiently compute this range if we follow the original computation of y step-by-step with a rough set instead of a collection of exact values: we start with a box $\mathbf{x}_1 \times \ldots \times \mathbf{x}_n$, and then estimate rough sets corresponding to each intermediate result.

Acknowledgments. This work was supported in part by the National Science Foundation grants HRD-0734825 and DUE-0926721 and by Grant 1 T36 GM078000-01 from the National Institutes of Health. The author is thankful to Dominik Ślęzak and Sergey Kuznetsov for the invitation and for the helpful editing advise.

References

1. Ceberio, C., Ferson, S., Kreinovich, V., Chopra, S., Xiang, G., Murguia, A., Santillan, J.: How to take into account dependence between the inputs: from interval computations to constraint-related set computations. In: Proc. 2nd Int'l Workshop on Reliable Engineering Computing, Savannah, Georgia, February 22-24, pp. 127–154 (2006); final version: Journal of Uncertain Systems 1(1), 11–34 (2007)
2. Ferson, S.: RAMAS Risk Calc 4.0. CRC Press, Boca Raton (2002)
3. Ferson, S., Ginzburg, L., Kreinovich, V., Aviles, M.: Computing variance for interval data is NP-hard. ACM SIGACT News 33(2), 108–118 (2002)
4. Jaulin, L., Kieffer, M., Didrit, O., Walter, E.: Applied Interval Analysis. Springer, London (2001)
5. Klir, G., Yuan, B.: Fuzzy Sets and Fuzzy Logic. Prentice Hall, Upper Saddle River (1995)
6. Kreinovich, V., Lakeyev, A., Rohn, J., Kahl, P.: Computational Complexity and Feasibility of Data Processing and Interval Computations. Kluwer, Dordrecht (1997)
7. Pawlak, Z.: Rough Sets: Theoretical Aspects of Reasoning About Data. Kluwer, Dordrecht (1991)
8. Shary, S.P.: Solving tied interval linear systems. Siberian Journal of Numerical Mathematics 7(4), 363–376 (2004) (in russian)

Rough Set Based Uncertain Knowledge Expressing and Processing

Guoyin Wang

Institute of Computer Science and Technology
Chongqing University of Posts and Telecommunications,
Chongqing, 400065, China
wanggy@ieee.org

Abstract. Uncertainty exists almost everywhere. In the past decades, many studies about randomness and fuzziness were developed. Many theories and models for expressing and processing uncertain knowledge, such as probability & statistics, fuzzy set, rough set, interval analysis, cloud model, grey system, set pair analysis, extenics, etc., have been proposed. In this paper, these theories are discussed. Their key idea and basic notions are introduced and their difference and relationship are analyzed. Rough set theory, which expresses and processes uncertain knowledge with certain methods, is discussed in detail.

Keywords: uncertain knowledge expressing, uncertain knowledge processing, fuzzy set, rough set, cloud model.

1 Introduction of Uncertainty

The methods for uncertain knowledge expressing and processing have become one of the key problems of artificial intelligence. There are many kinds of uncertainties in knowledge, such as randomness, fuzziness, vagueness, incompleteness, inconsistency, etc. Randomness and fuzziness are the two most important and fundamental ones. Randomness implies a lack of predictability (causality). It is a concept of non-order or non-coherence in a sequence of symbols or steps, such that there is no intelligible pattern or combination. Fuzziness is the uncertainty caused by the boundary region, reflecting the loss of excluded middle law. There are many theories about randomness and fuzziness developed in the past decades. Many theories and models have been proposed, such as probability & statistics, fuzzy set [20], rough set [15], interval analysis [14], cloud model [13], grey system [6], set pair analysis [22], extenics [4], etc.

In this paper, we specifically discuss fuzzy set, rough set, type-2 fuzzy set, interval-valued fuzzy set, intuitionistic fuzzy set, cloud model, grey set, set pair analysis, interval analysis, and extenics. The key ideas and basic notions of these approaches are introduced and their differences and relationships are analyzed. Some further topics and problems related to expressing and processing uncertain knowledge based on rough set are discussed too.

S.O. Kuznetsov et al. (Eds.): RSFDGrC 2011, LNAI 6743, pp. 11–18, 2011.

2 Set Theory

A set is a collection of distinct objects. Set is one of the most fundamental concepts in mathematics. The basic operators of set theory are: intersection ($A \cap B$), union ($A \cup B$), subtraction ($A - B$), and complement (A^c).

3 Fuzzy Set Theory

Fuzzy set, which was proposed by Zadeh as an extension of the classical notion of set [20], whose elements have degrees of membership. In classical set theory, the membership of elements in a set is assessed in binary terms according to a bivalent condition — an element either belongs or does not belong to the set, i.e., the membership function of elements in the set is one or zero. By contrast, fuzzy set theory permits the gradual assessment of the membership of elements in a set. The membership function is valued in the real unit interval $[0, 1]$. The membership of an element x belonging to a fuzzy set A is defined as $\mu_A(x)$. Quite typically, fuzzy set operators of intersection, union, and complement are defined as $\mu_{A\cap B}(x) = \min\{\mu_A(x), \mu_B(x)\}$, $\mu_{A\cup B}(x) = \max\{\mu_A(x), \mu_B(x)\}$, and $\mu_{A^c}(x) = 1 - \mu_A(x)$, respectively.

3.1 Type-2 Fuzzy Set

In 1975, Zadeh proposed a type-2 fuzzy set [21]. In 1999, Mendel argued that "words mean different things to different people", and claimed that we need type-2 fuzzy set to handle "ambiguity" in natural language [11]. Type-2 fuzzy set is a fuzzy set whose membership grades themselves is fuzzy set.

Definition 1 [11]. A **type-2 fuzzy set**, denoted $\tilde{A}$, is characterized by a type-2 membership $\mu_{\tilde{A}}(x, u)$, where for each $x \in U$ and $u \in J_x \subseteq [0, 1]$ there is $0 \leq \mu_{\tilde{A}}(x, u) \leq 1$. $\tilde{A}$ takes a form of $\{((x, u), \mu_{\tilde{A}}(x, u))\}$ or $\int_{x\in X} \int_{u\in J_x} \mu_{\tilde{A}}(x, u)/(x, u)$, where $\int\int$ denotes the union over all admissible x and u.

Let $\tilde{A} = \int_{x\in X} \int_{u\in J_x} \mu_{\tilde{A}}(x, u)/(x, u), \tilde{B} = \int_{x\in X} \int_{w\in J_x} \mu_{\tilde{B}}(x, w)/(x, w)$ be two type-2 fuzzy sets on U, where $u, w \in J_x$ and $\mu_{\tilde{A}}(x, u), \mu_{\tilde{B}}(x, w) \in [0, 1]$. The operations of union, intersection, and complement are defined as $\mu_{\tilde{A}\cup\tilde{B}}(x) = \int_u \int_w \frac{\mu_{\tilde{A}}(x,u)*\mu_{\tilde{B}}(x,u)}{u\vee w}$, $\mu_{\tilde{A}\cap\tilde{B}}(x) = \int_u \int_w \frac{\mu_{\tilde{A}}(x,u)*\mu_{\tilde{B}}(x,u)}{u\wedge w}$, and $\mu_{\tilde{A}^c}(x) = \int_u \frac{\mu_{\tilde{A}}(x,u)}{1-u}$, respectively, where "$*$" denotes a $t-$norm.

3.2 Interval-Valued Fuzzy Set

The interval-valued fuzzy set, which was proposed by Zadeh, is defined by an interval-valued membership function.

Definition 2 [21]. Let U be a universe. Define a map $A : U \to Int([0, 1])$, where $Int([0, 1])$ is the set of closed intervals in $[0, 1]$. Then, A is called an **interval-valued fuzzy set** on U and the membership function of A can be denoted by $A(x) = [A^-(x), A^+(x)]$.

Operations take form of $A \cup B(x) = [\sup(A^-(x), B^-(x)), \sup(A^+(x), B^+(x))]$, $A \cap B(x) = [\inf(A^-(x), B^-(x)), \inf(A^+(x), B^+(x))]$, and $A^c = [1 - A^+(x), 1 - A^-(x)]$, where $A^- = \inf(A), A^+ = \sup(A)$ for any $A \subset [0,1]$. Interval-valued fuzzy set is sometimes called grey set proposed by Deng [6].

Definition 3 [18]. Let G be a **grey set** of U defined by two mappings of the upper membership function $\bar{\mu}_G(x)$ and the lower membership function $\underline{\mu}_G(x)$ as follows: $\bar{\mu}_G(x) : U \to [0,1]$; $\underline{\mu}_G(x) : U \to [0,1]$, where $\underline{\mu}_G(x) \leq \bar{\mu}_G(x)$, $x \in U$. When $\underline{\mu}_G(x) = \bar{\mu}_G(x)$, the grey set G becomes a fuzzy set.

3.3 Intuitionistic Fuzzy Set

In fuzzy set theory, the membership of an element to a fuzzy set is a single value between zero and one. But in real life, it may not always be certain that the degree of non-membership of an element to a fuzzy set is just equal to 1 minus the degree of membership, i.e., there may be some hesitation degree. So, as a generalization of fuzzy set, the concept of intuitionistic fuzzy set was introduced by Atanassov [1]. Bustince and Burillo [3] showed that vague set defined by Gau and Buehrer [8] is equivalent to intuitionistic fuzzy set.

Definition 4 [1]. $A = \{\langle x, \mu_A(x), \nu_A(x)\rangle | x \in U\}$ is called an **intuitionistic fuzzy set**, where $\mu_A : U \to [0,1]$ and $\nu_A : U \to [0,1]$ are such that $0 \leq \mu_A + \nu_A \leq 1$, and $\mu_A, \nu_A \in [0,1]$ denote degrees of membership and non-membership of $x \in A$, respectively. For each intuitionistic fuzzy set A in U, "hesitation margin"(or "intuitionistic fuzzy index") of $x \in A$ is given by $\pi_A(x) = 1 - (\mu_A(x) + \nu_A(x))$ which expresses a hesitation degree of whether x belongs to A or not.

Operations take form of $A \cup B = \{\langle x, \max(\mu_A(x), \mu_B(x)), \min(\nu_A(x), \nu_B(x))\rangle | x \in U\}$, $A \cap B = \{\langle x, \min(\mu_A(x), \mu_B(x)), \max(\nu_A(x), \nu_B(x))\rangle | x \in U\}$ and $A^c = \{\langle x, \nu_A(x), \mu_A(x)\rangle | x \in U\}$.

There are plenty of theories treating imprecision and uncertainty. Some of them are extensions of fuzzy set theory, such as type-2 fuzzy set, interval-valued fuzzy set, intuitionistic fuzzy set, etc., while the others try to handle imprecision and uncertainty in a different way. Kerre [12] gave a summary of the links between fuzzy sets and other mathematical models such as flou set, two-fold fuzzy set and $L-$fuzzy set [9]. Deschrijver [7] proved:

1. There exists an isomorphism between $L-$intuitionistic fuzzy set [2] and $L-$fuzzy set. If L is the interval $[0,1]$ provided with the usual ordering, an $L-$intuitionistic fuzzy set is an intuitionistic fuzzy set;
2. There exists an isomorphism between interval-valued intuitionistic fuzzy set and $L-$fuzzy set for some specific lattice;
3. Intuitionistic fuzzy set can be embedded in interval-valued intuitionistic fuzzy set, so interval-valued intuitionistic fuzzy set theory extends intuitionistic fuzzy set theory;
4. There exists an isomorphism between interval-valued fuzzy set and intuitionistic fuzzy set, so interval-valued fuzzy set theory is equivalent to intuitionistic fuzzy set theory.

4 Rough Set Theory

Although fuzzy set can express the phenomenon that the elements in the boundary region belong to the set partially, it can not solve the "vague" problems that there are some elements which can not be classified into either a subset or its complement. For example: no mathematical formula to calculate the number of vague elements; no formal method to calculate the membership of vague elements. Rough set, which was proposed by Pawlak in 1982 [15], uses two certain sets, that is the lower approximation set and the upper approximation set, to define the boundary region of an uncertain set based on an equivalence relation (indiscernibility relation). The "vagueness degree" and the number of the vague elements can be calculated by the boundary region of a rough set.

The information of most natural phenomenon has the following characteristics: incomplete, inaccurate, vague or fuzzy. Classical set theory and mathematical logic can not express and deal with uncertainty problems successfully. The rough set theory is designed for expressing and processing vague information. The main advantage of rough set theory in data analysis is that it does not need any preliminary or additional information about data.

Rough set theory deals with uncertain problems using precise boundary lines to express the uncertainty. For an indiscernibility relation R and a set X, it operates with $R-$lower approximation of X, $R-$upper approximation of X, and $R-$boundary region of X, which are defined as $\underline{R}X = \{x \in U | [x]_R \subseteq X\}$, $\overline{R}X = \{x \in U | [x]_R \cap X \neq \emptyset\}$, and $RN_R(X) = \overline{R}X - \underline{R}X$, respectively.

If the boundary region of a set is empty, it means that the set is crisp, otherwise the set is **rough** (inexact). Nonempty boundary region means that our knowledge about the set is not sufficient to define it precisely.

The lower approximation of X contains all objects of U that can be classified into the class of X according to knowledge R. The upper approximation of X is the set of objects that can be and may be classified into the class of X. The boundary region of X is the set of objects that can possibly, but not certainly, be classified into class of X. Basic properties of rough set are as follows [15]:

1. $\overline{R}(X \cup Y) = \overline{R}(X) \cup \overline{R}(Y)$, $\underline{R}(X \cup Y) \supseteq \underline{R}(X) \cup \underline{R}(Y)$;
2. $\overline{R}(X \cap Y) \subseteq \overline{R}(X) \cap \overline{R}(Y)$, $\underline{R}(X \cap Y) = \underline{R}(X) \cap \underline{R}(Y)$;
3. $\overline{R}(X - Y) \subseteq \overline{R}(X) - \underline{R}(X)$, $\underline{R}(X - Y) = \underline{R}(X) - \overline{R}(Y)$;
4. $\sim \overline{R}(X) = \underline{R}(\sim X)$, $\sim \underline{R}(X) = \overline{R}(\sim X)$.

5 The Relationships of Fuzzy Set and Rough Set

Both fuzzy set and rough set are generalizations of the classical set theory for modeling vagueness and uncertainty. A fundamental question concerning both theories is their connections and differences [16]. It is generally accepted that they are related but distinct and complementary theories [5]. The two theories model different types of uncertainty:

1. Rough set theory takes into consideration the indiscernibility between objects. The indiscernibility is typically characterized by an equivalence relation. Rough set is the result of approximating crisp sets using equivalence

classes. The fuzzy set theory deals with the ill-definition of the boundary of a class through a continuous generalization of set characteristic functions. The indiscernibility between objects is not used in fuzzy set theory.
2. Rough set deals with uncertain problems using a certain method, while fuzzy set uses an uncertain method.
3. Fuzzy membership function relies on experts' prior knowledge. Rough set theory doesn't. For uncertainty of boundary regions, fuzzy set theory uses membership to express it, while rough set theory uses precise boundary lines to express it. Hence, fuzzy set theory and rough set theory could complement each other's advantages in dealing with uncertainties.

6 Cloud Model

Languages and words are powerful tools for human thinking, and the use of them is the fundamental difference between human intelligence and the other creatures' intelligence. We have to establish the relationship between the human brains and machines, which is performed by formalization. To describe uncertain knowledge by concepts is more natural and more generalized than to do it by mathematics. Li proposed a cloud model based on the traditional fuzzy set theory and probability statistics, which can realize the uncertain transformation between qualitative concepts and quantitative values.

Definition 5 [13]. Let U be the universe of discourse, C be a qualitative concept related to U. The membership μ of x to C is a random number with a stable tendency: $\mu : U \rightarrow [0,1]$, $\forall x \in U, x \rightarrow \mu(x)$, then the distribution of x on U is defined as a **cloud**, and every x is defined as a **cloud drop**. Qualitative concept is identified by three digital characteristics: Ex (Expected value), En (Entropy) and He (Hyper entropy).

Ex is the expectation of cloud drops' distribution in the universe of discourse, which means the most typical sample in the quantitative space of the concept. En is the uncertainty measurement of qualitative concept, decided by the randomness and the fuzziness of the concept. En reflects the numerical range which can be accepted by this concept in the universe of discourse, and embodies the uncertain margin of the qualitative concept. He is a measurement of entropy's uncertainty. It reflects the stability of the drops. The special numerical characteristic of cloud lies in using three values to sketch the whole cloud constituted by thousands of cloud drops, and it integrates the fuzziness and randomness of language value represented by quality method.

In practice, the normal cloud model is the most important kind of cloud models. It is based on normal distribution, and was proved universally to represent linguistic terms in various branches of natural and social science.

7 Set Pair Analysis

The set pair analysis theory, proposed by Zhao [22], is a novel uncertainty theory that is different from traditional probability theory and fuzzy set theory. Set pair is a pair of two related sets and set pair analysis is a method to process many kinds of uncertainties. The two sets have three relations: identical, different and contrary, and the connecting degree is an integrated description of them.

Definition 6 [22]. Assuming $H = (A, B)$ is a set pair of two sets A and B. For some application, H has total N attributes and S of them are mutual attributes of A and B, and P of them are contrary attributes, residual $F = N - S - P$ attributes are neither mutual nor opposite, then the connection degree of H is defined as: $\mu = \frac{S}{N} + \frac{F}{N}i + \frac{P}{N}j$, where S/N is identical degree, F/N is different degree, and P/N is contrary degree. Usually, we use a, b and c denote them, respectively, and $a + b + c = 1$.

8 Interval Analysis

Moore proposed an interval analysis theory, the purpose of which is to process error analysis automatically [14]. Interval analysis implements the storing and computing of data using interval, and the computing results ensure including all the possible true values.

Definition 7 [14]. A continuous subset $X = [\underline{x}, \bar{x}]$ on a real number domain R is called a **real interval**, and the upper and lower endpoints of an interval are represented by $\sup(X)$ and $\inf(X)$, respectively.

Let $X = [\underline{x}, \bar{x}]$, $Y = [\underline{y}, \bar{y}]$ be real intervals. The set of operations $\{+, -, *, \div\}$ is provided as follows [14]: $X + Y = [\underline{x} + \underline{y}, \bar{x} + \bar{y}]$, $X - Y = [\underline{x} - \bar{y}, \bar{x} - \underline{y}]$, $X * Y = [\min\{\underline{x}\underline{y}, \underline{x}\bar{y}, \bar{x}\underline{y}, \bar{x}\bar{y}\}, \max\{\underline{x}\underline{y}, \underline{x}\bar{y}, \bar{x}\underline{y}, \bar{x}\bar{y}\}]$, and $X \div Y = X * \frac{1}{Y}$, where $\frac{1}{Y} = \{\frac{1}{y} | y \in Y\}$ if $0 \notin Y$.

Interval methods can effectively define the function scope and provide strict operation results in mathematical meaning, which enable it appropriate to solve the problems of certain nonlinear equations and global optimization [10]. In addition, uncertainty of data can be expressed by interval. It is suitable, e.g., for solving nonlinear problems of parameter uncertainty in auto control [23].

9 Extension Set

The classical set and the fuzzy set mainly describe the "static" things. For the description of transformation of object A with character A_1 to object B with character B_1, Wen proposed the extension set to solve the qualitative description for "yes (true)" and "no (false)" to quantitative description and also to the variation procedures of "from yes to no" and "from no to yes" in 1983 [4]. It provides a suitable mathematical tool for solving the contradiction problems.

Definition 8 [19]. Let U be a domain and k be a reflection from U to the real domain R. Denote by T_u, T_k, and T_U the transformation of element, transformation of correlation function, and transformation of domain, respectively. For $T \in \{T_U, T_k, T_u\}$, $\tilde{A}(T) = \{(u, y, y') | u \in U, y = k(u) \in R, y' = T_k k(T_u u)\}$ is called an **extension set** on U about T. $y = k(u)$ and $y' = T_k k(T_u u)$ are called the **correlation function** and **extension function** of $\tilde{A}(T)$, respectively.

Let $\tilde{A}_1(T_1), \tilde{A}_2(T_2)$ be extension sets for $T_i \in \{T_U^i, T_k^i, T_u^i\}$ $(i = 1, 2)$. Denote "or" and "and" as $T_1 \vee T_2$ and $T_1 \wedge T_2$, respectively. We can consider the following operations [17]:

1. $\tilde{A}_1(T_1) \cup \tilde{A}_2(T_2) = \{(u, y, y')|u \in U, y = k(u), y' = T_k k(T_u u)\}$, where $T = T_1 \vee T_2$ and $k(u) = k_1(u) \vee k_2(u)$;
2. $\tilde{A}_1(T_1) \cap \tilde{A}_2(T_2) = \{(u, y, y')|u \in U, y = k(u), y' = T_k k(T_u u)\}$, where $T = T_1 \wedge T_2$, $k(u) = k_1(u) \wedge k_2(u)$;
3. $\tilde{A}_1^c(T_1) = \{(u, y, y')|u \in U, y = -y_1, y' = -y_1'\}$.

10 Future Directions and Topics of Rough Set Based Uncertain Knowledge Expressing and Processing

Rough set itself and the integration of rough set and other methods, including vague set, neural network, SVM, swarm intelligence, GA, expert system, etc., can deal with difficult problems like fault diagnosis, intelligent decision-making, image processing, huge data processing, intelligent control, and so on. At the same time, there are also new research directions to be studied in the future:

1. The extension of equivalence relation: order relation, tolerance relation, similarity relation, etc.;
2. Granular computing based on rough set theory (Dynamic Granular Computing);
3. The interactions among attributes (features): interactions among redundant attributes might be meaningful for problem expressing and solving;
4. The generalization of rough set reduction: reduction leads to over fitting (over training) in the training samples space;
5. Domain explanation of knowledge generated from reduction: The knowledge generated from data does not correspond to the human's formal knowledge;
6. Rough set characterize the ambiguity of decision information systems, but the randomness is not studied. Extended rough set model through combing rough set and cloud model?
7. 3DM (Domain-oriented Data-driven Data Mining): Knowledge generated should be kept the same as existed in the data sets; Reduce the dependence of prior domain knowledge in data mining processes;
8. Granular computing based on cloud model: granules (concepts) could be extracted from data using the backward cloud generator automatically.

Acknowledgments. This paper is supported by National Natural Science Foundation of P. R. China under grant 61073146, Natural Science Foundation Project of CQ CSTC under grant 2008BA2041.

References

1. Atanassov, K.T.: Intuitionistic fuzzy sets. Fuzzy Sets and Systems 20, 87–96 (1986)
2. Atanassov, K.T.: Intuitionistic fuzzy sets. Physica-Verlag, Heidelberg (1999)
3. Bustince, H., Burillo, P.: Vague Sets are intuitionistic fuzzy sets. Fuzzy Sets and Systems 79, 403–405 (1996)
4. Cai, W.: The extension set and non-compatible problems. Journal of Science Explore (1), 83–97 (1983)
5. Chanas, S., Kuchta, D.: Further remarks on the relation between rough and fuzzy sets. Fuzzy Sets and Systems 47, 391–394 (1992)
6. Deng, J.L.: Grey systems. China Ocean Press, Beijing (1988)
7. Deschrijver, G., Kerre, E.E.: On the relationship between some extensions of fuzzy set theory. Fuzzy Sets and Systems 133, 227–235 (2003)
8. Gau, W.L., Buehrer, D.J.: Vague sets. IEEE Transaction on Systems Man Cybernetics 23(2), 610–614 (1993)
9. Goguen, J.A.: $L-$fuzzy sets. Journal of mathematical analysis and applications 18, 145–174 (1967)
10. Hansen, E.R.: Global optimization using interval analysis. Marcel Dekker, New York (1992)
11. Karnik, N.N., Mendel, J.M., Liang, Q.L.: Type-2 fuzzy logic systems. IEEE Transaction on Fuzzy Systems 7(6), 643–658 (1999)
12. Kerre, E.E.: A first view on the alternatives of fuzzy set theory. Computational Intelligence in Theory and Practice, 55–72 (2001)
13. Li, D.Y., Meng, H.J., Shi, X.M.: Membership clouds and cloud generators. Journal of Computer Research and Development 32, 32–41 (1995)
14. Moore, R.E.: Interval analysis, pp. 25–39. Prentice-Hall, Englewood Cliffs (1966)
15. Pawlak, Z.: Rough sets. International Journal of Computer and Information Sciences 5(11), 341–356 (1982)
16. Pawlak, Z.: Rough sets and fuzzy sets. Fuzzy Sets and Systems 17, 99–102 (1985)
17. Sun, H.: On operations of the extension set. Mathematics in Practice and Theory 37(11), 180–184 (2007)
18. Wu, Q., Liu, Z.T.: Real formal concept analysis based on grey-rough set theory. Knowledge-based Systems 22, 38–45 (2009)
19. Yang, C.Y., Cai, W.: New definition of extension set. ournal of Guangdong University of Technology 18(1), 59–60 (2001)
20. Zadeh, L.A.: Fuzzy sets. Information and Control 8, 338–353 (1965)
21. Zadeh, L.A.: The concept of a linguistic variable and its application to approximate reasoning-I. Information Sciences 8, 199–249 (1975)
22. Zhao, K.Q.: Set pair analysis and its primary application. Zhejiang Science and Technology Press, Hangzhou (2000)
23. Zettler, M., Garloff, J.: Robustness analysis of polynomials with polynomials parameter dependency using Bernstein expansion. IEEE Trans. on Automatic Control 43(3), 425–431 (1998)

What is a Fuzzy Concept Lattice? II*

Radim Belohlavek

Department of Computer Science, Palacky University, Olomouc
17. listopadu 12, CZ-771 46 Olomouc, Czech Republic
radim.belohlavek@acm.org

Abstract. This paper is a follow up to "Belohlavek, Vychodil: What is a fuzzy concept lattice?, Proc. CLA 2005, 34–45", in which we provided a then up-to-date overview of various approaches to fuzzy concept lattices and relationships among them. The main goal of the present paper is different, namely to provide an overview of conceptual issues in fuzzy concept lattices. Emphasized are the issues in which fuzzy concept lattices differ from ordinary concept lattices. In a sense, this paper is written for people familiar with ordinary concept lattices who would like to learn about fuzzy concept lattices. Due to the page limit, the paper is brief but we provide an extensive list of references with comments.

1 Why Fuzzy Concepts?

1.1 Concepts in Formal Concept Analysis

In formal concept analysis (FCA, [4,48,25]), the notion of concept is used in accordance with the Port-Royal logic [1], as an entity that consists of its extent (objects to which the concept applies) and its intent (attributes covered by the concept). In FCA, extents and intents are determined by a relation I between a set X of objects and a set Y of attributes; $\langle X, Y, I\rangle$ is called a *formal context.* $\langle X, Y, I\rangle$, which represents the input data table with binary attributes, induces two *concept-forming operators*, denoted here $^{\uparrow}$ and $^{\downarrow}$, and a *formal concept* of I is defined as a pair $\langle A, B\rangle$ of $A \subseteq X$ (*extent*) and $B \subseteq Y$ (*intent*) satisfying $A^{\uparrow} = B$ and $B^{\downarrow} = A$; here $A^{\uparrow} = \{y \in Y \mid \text{for each } x \in A : \langle x, y\rangle \in I\}$ and $B^{\downarrow} = \{x \in X \mid \text{for each } y \in B : \langle x, y\rangle \in I\}$. $\mathcal{B}(X, Y, I)$, the set of all formal concepts of I, ordered by inclusion $\subseteq$ of extents (or, by $\supseteq$ of intents) is a complete lattice, called the *concept lattice* of I.

1.2 Psychological Evidence

There exists a strong evidence, established in the 1970s in the psychology of concepts, see e.g. [33,46], that human concepts have a *graded structure* in that whether or not a concept applies to a given object is a matter of degree, rather than a yes-or-no question, and that people are capable of working with the degrees in a consistent way. This finding is intuitively quite appealing because people say "this product is more or less good" or "to a certain degree, he is a good athlete", implying the graded structure of concepts.

* Supported by Grant No. 202/10/0262 of the Czech Science Foundation.

S.O. Kuznetsov et al. (Eds.): RSFDGrC 2011, LNAI 6743, pp. 19–26, 2011.

1.3 Fuzzy Logic as a Natural Choice

In his classic paper [49], Zadeh called the concepts with a graded structure *fuzzy concepts* and argued that these concepts are a rule rather than an exception when it comes to how people communicate knowledge. Moreover, he argued that to model such concepts mathematically is important for the tasks of control, decision making, pattern recognition, and the like. Zadeh proposed the notion of a *fuzzy set* that gave birth to the field of *fuzzy logic*: A fuzzy set in a universe U is a mapping $A : U \to L$ where L is $[0, 1]$ or some other partially ordered set of truth degrees. $A(u) \in$ is interpreted as the degree to which u belongs to A (to which the fuzzy set A applies to u). Fuzzy sets and fuzzy logic are nowadays well established theoretically as well as in applications, see e.g. [31,32,35].

2 The Basic Approach

In its ordinary setting [25], FCA is designed to model "crisp" (term used in fuzzy logic; other terms: yes-or-no, bivalent) concepts, i.e. concepts that either apply or do not apply to any given object. To extend (generalize) FCA for graded concepts, fuzzy logic seems an obvious choice. The first paper in this line is [22] by Burusco and Fuentes-Gonzáles, followed by contributions by Pollandt (PhD thesis published as [45]) and Belohlavek (the first published note is [5]). The approach by Pollandt and Belohlavek is particularly important because it uses residuated structures of truth degrees and can be regarded as the basic, mainstream approach till now (even though various generalizations and variants exist). Further early contributions include [21,36]. Since then, many other papers appeared on FCA in a fuzzy setting. Some are listed in the references but we do not intend to provide a representative list in this paper. Rather, as mentioned above, we focus on differences from the ordinary case.

2.1 Basic Notions

We now present the basic approach. In fuzzy logic, one uses a set of truth degrees equipped with (truth functions of) logical connectives. The basic approach uses so-called complete residuated lattices, which are certain algebras $\mathbf{L} = \langle L, \wedge, \vee, \otimes, \to, 0, 1\rangle$ (introduced in [47] and brought in fuzzy logic by [30], for further information see [10,31,32,34]). Elements $a \in L$ are interpreted as *degrees of truth* [32] (0 stands for full falsity and 1 stands for full truth). $\otimes$ (multiplication) and $\to$ (residuum) serve as the truth functions of "fuzzy conjunction" and "fuzzy implication". A common choice of $\mathbf{L}$ is $L = [0, 1]$ or $L = \{0, \frac{1}{n}, \ldots, \frac{n-1}{n}, 1\}$ equipped with a $\bigvee$-preserving $\otimes$ and its residuum $\to$. Two examples are: Łukasiewicz ($a \otimes b = \max(0, a + b - 1)$, $a \to b = \min(1, 1 - a + b)$) and Gödel ($a \otimes b = \min(a, b)$, $a \to b = 1$ if $a \leq b$, $a \to b = b$ if $a > b$). Below, $\mathbf{L}$ refers to some complete residuated lattice, L^U denotes the set of all fuzzy sets in universe U, i.e. set of all mappings from U to L.

For a given $\mathbf{L}$, a *formal fuzzy context* (formal $\mathbf{L}$-context) is a triplet $\langle X, Y, I\rangle$ where I is a fuzzy relation between ordinary sets X and Y (of objects and attributes), i.e. $I : X \times Y \to L$ and $I(x,y) \in L$ is interpreted as the degree to which object $x \in X$ has attribute $y \in Y$. This is the basic difference from the ordinary case—one starts with a fuzzy (graded) relationship rather than a yes-or-no relationship, and the fuzziness then naturally enters all subsequent definitions. Typical examples of formal fuzzy contexts are data obtained from questionnaires (objects x are respondents, attributes y are products/services, $I(x,y)$ is the degree to which x considers y good) [20]. $\langle X,Y,I\rangle$ induces the concept forming operators $^{\uparrow} : L^X \to L^Y$ (assigns fuzzy sets of attributes to fuzzy sets of objects) and $^{\downarrow} : L^Y \to L^X$ (same, but in the other direction) by:

$$A^{\uparrow}(y) = \bigwedge_{x\in X}(A(x) \to I(x,y)) \text{ and } B^{\downarrow}(x) = \bigwedge_{y\in Y}(B(y) \to I(x,y)).$$

A *formal fuzzy concept* of I is a pair $\langle A,B\rangle$ consisting of fuzzy sets $A \in L^X$ and $B \in L^Y$ satisfying $A^{\uparrow} = B$ and $B^{\downarrow} = A$. Due to the basic rules of predicate fuzzy logic, $A^{\uparrow}(y)$ is the truth degree of "y is shared by all objects from A" and $B^{\downarrow}(x)$ is the truth degree of "x has all attributes from B". An important consequence is that the verbal description, i.e. the meaning, of the notion of a formal concept in a fuzzy setting is essentially the same as in the ordinary case. The second consequence is that for $L = \{0,1\}$ (the residuated lattice is then the two-element Boolean algebra of classical logic), formal fuzzy contexts and formal fuzzy concepts become the ordinary formal contexts and formal concepts (when identifying sets with their characteristic functions). Therefore, the approach under discussion *generalizes the notions of ordinary FCA*. Put

$$\mathcal{B}(X,Y,I) = \{\langle A,B\rangle \mid A^{\uparrow} = B,\ B^{\downarrow} = A\}$$

(set of all formal fuzzy concepts of I) and define on this set a binary relation $\le$ by

$$\langle A_1,B_1\rangle \le \langle A_2,B_2\rangle \text{ iff } A_1 \subseteq A_2 \text{ (iff } B_1 \supseteq B_2).$$

Here,

$$A_1 \subseteq A_2 \text{ means that } A_1(x) \le A_2(x) \text{ for all } x \in X; \qquad (*)$$

same for $B_1 \supseteq B_2$. The partial order $\le$ makes $\mathcal{B}(X,Y,I)$ a complete lattice, called the *fuzzy concept lattice* of I. There exists a basic theorem for fuzzy concept lattices (with two different proofs [8,10,45], one is discussed in Sec. 3.1), see also Sec. 3.3.

2.2 Related Approaches

Let us mention the following related approaches. Independently, [16,21,36] studied essentially the same notion, called crisply generated or one-sided fuzzy concepts, which are fuzzy concepts with crisp extent (alternatively, crisp intent); see [44] for a relationship to pattern structures. [16] shows that these are just particular fuzzy concepts and studies their structure within $\mathcal{B}(X,Y,I)$. Second, several approaches exist that generalize the basic approach in that they use different, more general residuated structures, see e.g. [12,17,29,37,38,39,42] (in some cases, the motivation is purely mathematical, in the others, it comes from some need, e.g. to reduce the number of formal concepts in a parameterized way [17]).

3 Mathematical Structures Behind

3.1 Closure Operators, Systems, and Galois Connections

For a fuzzy context $\langle X, Y, I\rangle$, one may consider the complete lattices $\langle L^X, \subseteq\rangle$ and $\langle L^Y, \subseteq\rangle$ where $\subseteq$ is the inclusion of fuzzy sets given by (*). As in the ordinary case, $\langle {}^{\uparrow}, {}^{\downarrow}\rangle$ forms a Galois connection between $\langle L^X, \subseteq\rangle$ and $\langle L^Y, \subseteq\rangle$. However, $\langle {}^{\uparrow}, {}^{\downarrow}\rangle$ satisfies more: It forms a *fuzzy Galois connection* [6] in that it is a Galois connection that is antitone w.r.t. *graded inclusion*. That is, it satisfies (i) $S(A_1, A_2) \leq S(A_2^{\uparrow}, A_1^{\uparrow})$ and (ii) $A \subseteq A^{\uparrow\downarrow}$, plus the dual conditions for ${}^{\downarrow}$. $S(A_1, A_2) = \bigwedge_{x\in X}(A_1(x) \rightarrow A_2(x))$ is the degree of inclusion of A_1 in A_2 (degree to which every element of A_1 is also an element of A_2). One has $S(A_1, A_2) = 1$ iff $A_1 \subseteq A_2$. S is therefore a graded generalization of the bivalent inclusion $\subseteq$ of fuzzy sets and (i) is stronger than saying that (i') $A_1 \subseteq A_2$ implies $A_2^{\uparrow} \subseteq A_1^{\uparrow}$. Now, with graded inclusion in the definition of a fuzzy Galois connection, things are as in the ordinary case [25]. For example, there is a one-to-one correspondence between fuzzy Galois connections and formal fuzzy contexts [6] (this is not true if one uses (i')).

Similar results hold true for closure operators involved in FCA: ${}^{\uparrow\downarrow}$ forms a closure operator in $\langle L^X, \subseteq\rangle$ that is even a *fuzzy closure operator* [9], i.e. satisfies (i) above; $S(A_1, A_2) \leq S(A_1^{\uparrow\downarrow}, A_2^{\uparrow\downarrow})$ (which is stronger than $A_1 \subseteq A_2$ implying $A_1^{\uparrow\downarrow} \subseteq A_2^{\uparrow\downarrow}$); and $A^{\uparrow\downarrow} = (A^{\uparrow\downarrow})^{\uparrow\downarrow}$. In the ordinary case, the sets of fixpoints of closure operators are just systems closed under arbitrary intersections, called closure systems. The systems of fixpoints of fuzzy closure operators, called *fuzzy closure systems*, are closed under intersection but also under so-called shifts. For $a \in L$, the a-shift of a fuzzy set $A \in L^X$ is a fuzzy set $a \rightarrow A$ defined by $(a \rightarrow A)(x) = a \rightarrow A(x)$. Closedness under intersections is weaker than closedness under intersections and shifts.

3.2 Reduction to the Ordinary Case

Two different ways of representing fuzzy Galois connections by ordinary Galois connections are known. First, a fuzzy Galois connection may be represented by a particular system of ordinary Galois connections indexed by truth values from L [6]. Another type of representation is presented in [8]: A fuzzy Galois connection induced by a fuzzy context $\langle X, Y, I\rangle$ may be represented by the Galois connection of the ordinary context $\langle X \times L, Y \times L, I^{\times}\rangle$ where

$$\langle\langle x, a\rangle, \langle y, b\rangle\rangle \in I^{\times} \quad \text{iff} \quad a \otimes b \leq I(x, y).$$

Importantly, the fuzzy concept lattice $\mathcal{B}(X, Y, I)$ is isomorphic to the ordinary concept lattice $\mathcal{B}(X \times L, Y \times L, I^{\times})$. This observation was utilized in [45] for proving indirectly the basic theorem for fuzzy concept lattices (for a direct proof, see e.g. [10]). Independently and within the context of Galois connections, these results appeared in [8]. $\langle X \times L, Y \times L, I^{\times}\rangle$ results by what may be regarded as a new type of scaling (double scaling), which works differently from the well-known ordinal scaling [25] (a fuzzy context may be ordinally scaled to an ordinary context, but the resulting ordinary concept lattice is then different from the fuzzy concept lattice; namely, it is isomorphic to the lattice of all crisply generated fuzzy concepts [16]).

3.3 Fuzzy Concept Lattice as a Lattice?

As was mentioned above, a fuzzy concept lattice is a complete lattice whose structure is described by a basic theorem for fuzzy concept lattices. Looking at things this way may be regarded not satisfactory from the mathematical viewpoint. For example, the well-known result saying that for a complete lattice $\langle V, \leq \rangle$, the ordinary concept lattice $\mathcal{B}(V, V, \leq)$ is isomorphic to $\langle V, \leq \rangle$ and more generally, that for a partially ordered set $\langle V, \leq \rangle$, $\mathcal{B}(V, V, \leq)$ is essentially the Dedekind-MacNeille completion, fails in a fuzzy setting if a fuzzy concept lattice is regarded as a lattice. In order for things to work as in the ordinary case, a many-valued (graded, fuzzy) partial order needs to be considered on the fuzzy concept lattice. This is studied in [10,11], [40] contains additional results; see also [50] (there exist further related papers).

4 Some Further Issues Different from Ordinary Case

4.1 Formal Concepts as Maximal Rectangles

As in the ordinary case, $\langle A, B \rangle$ is a formal fuzzy concept of I iff the Cartesian product of A and B (based on $\otimes$) is a maximal Cartesian subrelation of I, i.e. a "maximal rectangle of I" [6]. Different from the ordinary case is that the correspondence between concepts of I and maximal rectangles of I is no longer bijective: There may exist two (or more) different fuzzy concepts for which the corresponding rectangle is the same.

4.2 For Infinite Set of Truth Degrees, Fuzzy Concept Lattice over Finite Sets of Objects and Attributes May be Infinite

This is because in such a case the set $L^X \times L^Y$ of possible fixpoints is infinite and it may be indeed the case that the set of actual fixpoints is infinite (for instance for Łukasiewics operations, but not for Gödel). If only a part of the concept lattice is used, this may not be a problem. If the whole concept lattice is to be used, a pragmatic approach is to use a finite set L of truth degrees (using small L is reasonable also due to the well-known 7 ± 2 phenomenon [43]).

4.3 Reduction of a Fuzzy Context

In the ordinary case, the reduction of a finite context consists in clarification (so that there are no identical rows and columns in the input data table) and then removing objects and attributes (rows and columns) for which the object- and attribute-concepts are $\bigvee$-reducible and $\bigwedge$-reducible. That is, we delete objects whose rows are intersections of other rows; same for attributes. This is because we want to obtain the smallest set of rows that generates the same closure system as the original set of rows and because the generating operation for ordinary closure systems is the intersection. As was mentioned Sec. 3.1, in a fuzzy setting we work with fuzzy closure systems and in this case, there are two generating operations: intersection and a-shifts. Looking for the smallest generating set of the fuzzy closure system of the original rows may be regarded as computing a base in a certain space over $\mathbf{L}$ (analogous to computing a base of a linear subspace generated by a set of vectors) [13]. Note that [27], which studies reduction of many-valued contexts, deals with a different problem: in the construction of the concept lattice of [27], only intersection plays a role.

4.4 Antitone vs. Isotone Galois Connections Induced by *I*

In the ordinary case, an anotitone Galois connection $\langle {}^{\cap}, {}^{\cup} \rangle$ is induced by $\langle X, Y, I \rangle$ by $A^{\cap} = \{y \in Y \mid \text{ for some } x \in A : \langle x, y \rangle \in I\}$ and $B^{\cup} = \{x \in X \mid \text{ for each } y \in Y : \langle x, y \rangle \in I \text{ implies } y \in B\}$. It is well-known that due to the law of double negation, $\langle {}^{\cap}, {}^{\cup} \rangle$ and $\langle {}^{\uparrow}, {}^{\downarrow} \rangle$ are mutually reducible [26] (essentially, fixpoints of $\langle {}^{\uparrow}, {}^{\downarrow} \rangle$ induced by I may be identified with those of $\langle {}^{\cap}, {}^{\cup} \rangle$ induced by the complement of I). Such reduction fails in a fuzzy setting (because in fuzzy logic, the law of double negation does not hold). However, a unified approach leaving both $\langle {}^{\uparrow}, {}^{\downarrow} \rangle$ and $\langle {}^{\cap}, {}^{\cup} \rangle$ particular cases is still possible (see [12,29] for two different approaches).

5 Further Issues

We conclude by brief comments on three other issues.

Algorithms. Due to the reduction described in Sec. 3.2, a fuzzy concept lattice may be computed using existing algorithms for ordinary concept lattice. As is shown in [14], a direct approach is considerably more efficient. The investigation of algorithms for fuzzy concept lattices is, however, in its beginning.

Attribute Implications. This area is completely skipped in this paper (see [19] for an overview of some results). This is an interesting area with several differences from the ordinary case. Up to now, the results are presented in various proceedings of conferences on fuzzy logic.

Terminology. The terminology in the literature seems sometimes strange (this is subjective, of course). In our view, "fuzzy data", "fuzzy FCA", or "fuzzy formal concept" are not nice and perhaps make not much sense. Although we understand that the first two may be considered useful shorthands, the analysis is not fuzzy as suggested by "fuzzy FCA". More reasonable are "data with fuzzy attributes", "FCA of data with fuzzy attributes", an "formal fuzzy concept".

References

1. Arnauld, A., Nicole, P.: La logique ou l'art de penser. English: Logic or the Art of Thinking, vol. 1662. Cambridge University Press, Cambridge (1996)
2. Bandler, W., Kohout, L.J.: Mathematical relations, their products and generalized morphisms. Technical Report EES-MMS-REL 77-3, ManMachine Systems Laboratory, Dept. Electrical Engineering, University of Essex, Essex, Colchester (1977)
3. Bandler, W., Kohout, L.J.: Semantics of implication operators and fuzzy relational products. Int. J. Man-Machine Studies 12, 89–116 (1980)
4. Barbut, M., Monjardet, B.: L'ordre et la classification, algèbre et combinatoire, tome II, Paris, Hachette (1970)
5. Belohlavek, R.: Lattices generated by binary fuzzy relations (extended abstract). In: Abstracts of FSTA 1998, Liptovský Ján, Slovakia, p. 11 (1998)
6. Belohlavek, R.: Fuzzy Galois connections. Math. Log. Quart. 45(4), 497–504 (1999)
7. Belohlavek, R.: Similarity relations in concept lattices. J. Logic Computation 10(6), 823–845 (2000)
8. Belohlavek, R.: Reduction and a simple proof of characterization of fuzzy concept lattices. Fundamenta Informaticae 46(4), 277–285 (2001)
9. Belohlavek, R.: Fuzzy closure operators. J. Mathematical Analysis and Applications 262, 473–489 (2001)
10. Belohlavek, R.: Fuzzy Relational Systems: Foundations and Principles. Kluwer Academic/Plenum Publishers, New York (2002)
11. Belohlavek, R.: Concept lattices and order in fuzzy logic. Annals of Pure and Applied Logic 128, 277–298 (2004)
12. Belohlavek, R.: Sup-t-norm and inf-residuum are one type of relational product: unifying framework and consequences. Fuzzy Sets and Systems (to appear)
13. Belohlavek, R.: Reduction of formal contexts as computing base: the case of binary and fuzzy attributes (to be submitted)
14. Belohlavek, R., De Baets, B., Outrata, J., Vychodil, V.: Computing the lattice of all fixpoints of a fuzzy closure operator. IEEE Transactions on Fuzzy Systems 18(3), 546–557 (2010)
15. Belohlavek, R., Dvorak, J., Outrata, J.: Fast factorization by similarity in formal concept analysis of data with fuzzy attributes. J. Computer and System Sciences 73(6), 1012–1022 (2007)
16. Bělohlávek, R., Sklenář, V., Zacpal, J.: Crisply generated fuzzy concepts. In: Ganter, B., Godin, R. (eds.) ICFCA 2005. LNCS (LNAI), vol. 3403, pp. 269–284. Springer, Heidelberg (2005)
17. Belohlavek, R., Vychodil, V.: Reducing the size of fuzzy concept lattices by hedges. In: Proc. FUZZ-IEEE 2005, Reno, Nevada, pp. 663–668 (2005)
18. Belohlavek, R., Vychodil, V.: What is a fuzzy concept lattice? In: Proc. CLA 2005. CEUR WS, vol. 162, pp. 34–45 (2005)
19. Bělohlávek, R., Vychodil, V.: Attribute implications in a fuzzy setting. In: Missaoui, R., Schmidt, J. (eds.) Formal Concept Analysis. LNCS (LNAI), vol. 3874, pp. 45–60. Springer, Heidelberg (2006)
20. Belohlavek, R., Vychodil, V.: Factor Analysis of Incidence Data via Novel Decomposition of Matrices. In: Ferré, S., Rudolph, S. (eds.) ICFCA 2009. LNCS, vol. 5548, pp. 83–97. Springer, Heidelberg (2009)
21. Ben Yahia, S., Jaoua, A.: Discovering knowledge from fuzzy concept lattice. In: Kandel, A., Last, M., Bunke, H. (eds.) Data Mining and Computational Intelligence, pp. 167–190. Physica-Verlag, Heidelberg (2001)
22. Burusco, A., Fuentes-Gonzáles, R.: The study of the L-fuzzy concept lattice. Mathware & Soft Computing 3, 209–218 (1994)

23. Burusco, A., Fuentes-Gonzáles, R.: Concept lattice defined from implication operators. Fuzzy Sets and Systems 114(3), 431–436 (2000)
24. Ganter, B., Kuznetsov, S.O.: Pattern structures and their projections. In: Delugach, H.S., Stumme, G. (eds.) ICCS 2001. LNCS (LNAI), vol. 2120, pp. 129–142. Springer, Heidelberg (2001)
25. Ganter, B., Wille, R.: Formal Concept Analysis. Mathematical Foundations. Springer, Berlin (1999)
26. Gediga G., Düntsch I.: Modal-style operators in qualitative data analysis. In: Proc. IEEE ICDM 2002, p. 155 (Technical Report # CS-02-15, Brock University, 15 pp.) (2002)
27. Gély, A., Medina, R., Nourine, L.: Representing lattices using many-valued relations. Information Sciences 179(16), 2729–2739 (2009)
28. Georgescu, G., Popescu, A.: Concept lattices and similarity in non-commutative fuzzy logic. Fundamenta Informaticae 53(1), 23–54 (2002)
29. Georgescu, G., Popescu, A.: Non-dual fuzzy connections. Archive for Mathematical Logic 43, 1009–1039 (2004)
30. Goguen, J.A.: The logic of inexact concepts. Synthese 18, 325–373 (1968-1969)
31. Gottwald, S.: A Treatise on Many-Valued Logics. Research Studies Press, Baldock (2001)
32. Hájek, P.: Metamathematics of Fuzzy Logic. Kluwer, Dordrecht (1998)
33. Heider, E.R.: Universals in color naming and memory. J. of Experimental Psychology 93, 10–20 (1972)
34. Höhle, U.: On the fundamentals of fuzzy set theory. J. Mathematical Analysis and Applications 201, 786–826 (1996)
35. Klir, G.J., Yuan, B.: Fuzzy Sets and Fuzzy Logic. Theory and Applications. Prentice-Hall, Englewood Cliffs (1995)
36. Krajči, S.: Cluster based efficient generation of fuzzy concepts. Neural Network World 5, 521–530 (2003)
37. Krajči, S.: The basic theorem on generalized concept lattice. In: Bělohlávek, R., Snášel, V. (eds.) Proc. of 2nd Int. Workshop on CLA 2004, Ostrava, pp. 25–33 (2004)
38. Krajči, S.: A generalized concept lattice. Logic J. of IGPL 13, 543–550 (2005)
39. Krajči, S.: Every concept lattice with hedges is isomorphic to some generalized concept lattice. In: Proc. CLA 2005. CEUR WS, vol. 162, pp. 1–9 (2005)
40. Krupka, M.: Main theorem of fuzzy concept lattices revisited (submitted)
41. Lai, H., Zhang, D.: Concept lattices of fuzzy contexts: Formal concept analysis vs. rough set theory. Int. J. Approximate Reasoning 50(5), 695–707 (2009)
42. Medina, J., Ojeda-Aciego, M., Ruiz-Claviño, J.: Formal concept analysis via multiadjoint concept lattices. Fuzzy Sets and Systems 160, 130–144 (2009)
43. Miller, G.A.: The magical number seven, plus or minus two: Some limits on our capacity for processing information. Psychological Review 63(2), 343–355 (1956)
44. Pankratieva, V.V., Kuznetsov, S.O.: Relations between proto-fuzzy concepts, crisply generated fuzzy concepts, and interval pattern structures. In: Proc. CLA 2010. CEUR WS, vol. 672, pp. 50–59 (2010)
45. Pollandt, S.: Fuzzy Begriffe. Springer, Berlin (1997)
46. Rosch, E.: Natural categories. Cognitive Psychology 4, 328–350 (1973)
47. Ward, M., Dilworth, R.P.: Residuated lattices. Trans. AMS 45, 335–354 (1939)
48. Wille, R.: Restructuring lattice theory: an approach based on hierarchies of concepts. In: Rival, I. (ed.) Ordered Sets, pp. 445–470. Reidel, Dordrecht (1982)
49. Zadeh, L.A.: Fuzzy sets. Information and Control 8, 338–353 (1965)
50. Zhao, H., Zhang, D.: Many vaued lattice and their representations. Fuzzy Sets and Sytems 159, 81–94 (2008)

Rough Set Based Ensemble Classifier

C.A. Murthy, Suman Saha, and Sankar K. Pal

Machine Intelligence Unit, Indian Statistical Institute, Kolkata, India
murthy@isical.ac.in

Combining the results of a number of individually trained classification systems to obtain a more accurate classifier is a widely used technique in pattern recognition. In [1], we introduced a Rough Set Meta classifier (RSM) to classify web pages. It tries to solve the problems of representing less redundant ensemble of classifiers and making reasonable decision from the predictions of ensemble classifiers, using rough set attribute reduction and rule generation methods on a granular meta data generated by base classifiers from input data.

The proposed method consists of two parts. In the first part, the outputs of individually trained classifiers are considered for constructing a decision table, with each instance corresponding to a single row. Predictions made by individual classifiers are used as condition attribute values and actual class – as decision attribute value. In the second part, rough set attribute reduction and rule generation processes are used on that decision table to construct a meta classifier. The combination of classifiers corresponding to the features of minimal reduct is taken to form classifier ensemble for RSM classifier system. Going further, from the obtained minimal reduct we compute decision rules by finding mapping between decision attribute and condition attributes. Decision rules obtained by rough set techniques are then applied to perform classification task.

It is shown that (1) the performance of the meta classifier is better than the performance of every constituent classifier, and (2) the meta classifier is optimal with respect to a quality measure that we proposed. Some other theoretical results on RSM and comparison with Bayes decision rule are also described. There are several ensemble classifiers available in literature like Adaboost, Bagging, Stacking. Experimental studies show that RSM improves accuracy of classification uniformly over some benchmark corpora and beats other ensemble approaches in accuracy by a decisive margin, thus demonstrating the theoretical results. Apart from this, it reduces the CPU load compared to other ensemble techniques by removing redundant classifiers from the combination.

References

1. Saha, S., Murthy, C.A., Pal, S.K.: Rough set based ensemble classifier for web page classification. Fundamenta Informaticae 76(1-2), 171–187 (2007)

S.O. Kuznetsov et al. (Eds.): RSFDGrC 2011, LNAI 6743, p. 27, 2011.

The Use of Rough Set Methods in Knowledge Discovery in Databases
Tutorial Abstract

Marcin Szczuka*

Institute of Mathematics, The University of Warsaw
Banacha 2, 02-097 Warsaw, Poland
szczuka@mimuw.edu.pl

Knowledge Discovery in Databases (KDD) is a process involving many stages. One of them is usually Data Mining, i.e., the sequence of operations that leads to creation (discovery) of new, interesting and non-trivial patterns from data. Under closer examination one can identify several interconnected smaller steps that together make it possible to go from the original low-level data set(s) to high-level representation and visualisation of knowledge contained in it. That includes, among others, operations on data such as:

- Data preparation, in particular: feature selection, reduction, and construction.
- Data selection, in particular: data sampling, data reduction and decomposition of large data sets.
- Data filtering and cleaning, in particular: discretisation, quantisation, dealing with missing/distorted data points.
- Knowledge model construction and management, in particular: decision and/or association rule discovery, template discovery, rule set transformations.

While attempting to deal with some or all tasks listed above one may consider using various existing methods. In practice, one will resort to those paradigms and solutions, which are on one hand relevant for the given set of data and comprehensive but, on the other hand, have readily available and easy to use implementations. Quite frequently the choice of method for data analysis is determined mostly by the existence and ease-of-use of the software toolbox that has been prepared for the purpose. In this tutorial we would like to demonstrate that among various choices for methodology and tools one may want to consider those originating in the theory of Rough Sets.

Theory of Rough Sets (RS) has been around for nearly three decades (cf. [1,2,3]). During that time it has transformed from being purely the theory

* The author is supported by the grant N N516 077837 from the Ministry of Science and Higher Education of the Republic of Poland and by the National Centre for Research and Development (NCBiR) under Grant No. SP/I/1/77065/10 by the strategic scientific research and experimental development program: "Interdisciplinary System for Interactive Scientific and Scientific-Technical Information".

S.O. Kuznetsov et al. (Eds.): RSFDGrC 2011, LNAI 6743, pp. 28–30, 2011.

of reasoning about data [1] into comprehensive, multi-faceted field of research and practice (cf. [2,4]). Along the way it has absorbed and transformed several ideas from related fields (cf. [5,6]) and produced several methods and algorithms (cf. [7,8,9]). These algorithmic methods support various steps in KDD process and have proven to be novel, practical and useful on some types of data. More importantly, there exist several software libraries and toolboxes that make it possible to use rough set approach with minimal programming effort (see [10,11,12,13]).

In this short tutorial our goal will be to present a hands-on guide for using methods and algorithms that originated in the area of Rough Sets for the purposes of KDD. We will try to answer the common issue of choosing the right method for a given set of data and convince the audience that in some situations the algorithms originating in RS theory are best suited for the job. We will demonstrate how existing software tools may come handy at various steps of KDD process.

The tutorial is intended to be mainly a practical guide. Therefore, only few most fundamental and important notions from RS theory will be introduced in detail. We will concentrate on methods and algorithms, paying only marginal attention to (existing) theoretical results that justify their correctness and quality. Some simplification will be made in order to fit as much material as possible into the limited time frame. Hence, it is also assumed that the audience is somewhat familiar with general concepts in KDD, Data Mining and Machine Learning such as:

- tabular data representation, attribute-value space, sampling;
- learning from data, error rates, quality measures and evaluation models;
- typical tasks for Data Mining.

As a conclusion we will try to briefly point out possible new trends in both basic and applied research on using RS methods in KDD. We will also explain how the ideas originating in RS theory may influence areas other than KDD, for example data warehousing (cf. [14]).

References

1. Pawlak, Z.: Rough Sets - Theoretical Aspects of Reasoning about Data. Kluwer Academic Publishers, Dordrecht (1991)
2. Pawlak, Z., Skowron, A.: Rudiments of rough sets. Information Sciences 177, 3–27 (2007)
3. Wikipedia - the free Encyclopedia: Rough Set (2011), `http://en.wikipedia.org/wiki/Rough_set`
4. Grochowalski, P., Suraj, Z.: RSDS - the Rough Set Database System - a bibliographic database on wide aspects of rough sets. WWW Page (2009), `http://rsds.univ.rzeszow.pl/`
5. Bazan, J.G., Latkowski, R., Szczuka, M.S.: Missing template decomposition method and its implementation in rough set exploration system. In: Greco, S., Hata, Y., Hirano, S., Inuiguchi, M., Miyamoto, S., Nguyen, H.S., Słowiński, R. (eds.) RSCTC 2006. LNCS (LNAI), vol. 4259, pp. 254–263. Springer, Heidelberg (2006)

6. Nguyen, H.S.: Approximate boolean reasoning: Foundations and applications in data mining. In: Peters, J.F., Skowron, A. (eds.) Transactions on Rough Sets V. LNCS, vol. 4100, pp. 334–506. Springer, Heidelberg (2006)
7. Bazan, J.G., Nguyen, H.S., Nguyen, S.H., Synak, P., Wróblewski, J.: Rough set algorithms in classification problem. In: Rough Set Methods and Applications, pp. 49–88. Physica-Verlag, Heidelberg (2000)
8. Kotłowski, W., Dembczyński, K., Greco, S., Słowinski, R.: Stochastic dominance-based rough set model for ordinal classification. Information Sciences 178, 4019–4037 (2008)
9. Grzymała-Busse, J.W., Rząasa, W.: A local version of the MLEM2 algorithm for rule induction. Fundamenta Informaticae 100, 99–116 (2010)
10. Øhrn, A.: ROSETTA Development Team: The ROSETTA software toolkit. WWW Page (2009), http://www.lcb.uu.se/tools/rosetta/
11. Bazan, J., Szczuka, M.: The rough set exploration system - RSES. WWW Page (2006), http://logic.mimuw.edu.pl/~rses
12. Wojna, A.: The Rseslib 3.0 library. WWW Page (2011), http://rsproject.mimuw.edu.pl
13. Laboratory of Intelligent Decision Support Systems, Poznań Univ. of Technology: Software and other projects. WWW Page (2011), http://idss.cs.put.poznan.pl/site/software.html
14. Infobright, Inc.: Infobright Community Edition (ICE). WWW Page (2011), http://infobright.org

Fuzzy-Rough Data Mining

Richard Jensen

Dept. of Comp. Sci., Aberystwyth University,
Ceredigion, SY23 3DB, Wales, UK
rkj@aber.ac.uk

Abstract. It is estimated that every 20 months or so the amount of information in the world doubles. In the same way, tools that mine knowledge from data must develop to combat this growth. Fuzzy-rough set theory provides a framework for developing such applications in a way that combines the best properties of fuzzy sets and rough sets, in order to handle uncertainty. In this tutorial we will cover the mathematical groundwork required for an understanding of the data mining methods, before looking at some of the key developments in the area, including feature selection and classifier learning.

1 Introduction

Lately there has been great interest in developing methodologies which are capable of dealing with imprecision and uncertainty, and the resounding amount of research currently being done in the areas related to fuzzy [18] and rough sets [14] is representative of this. The success of rough set theory is due in part to three aspects of the theory. Firstly, only the facts hidden in data are analysed. Secondly, no additional information about the data is required for data analysis such as thresholds or expert knowledge on a particular domain. Thirdly, it finds a minimal knowledge representation for data. As rough set theory handles only one type of imperfection found in data, it is complementary to other concepts for the purpose, such as fuzzy set theory. The two fields may be considered analogous in the sense that both can tolerate inconsistency and uncertainty - the difference being the type of uncertainty and their approach to it; fuzzy sets are concerned with vagueness, rough sets are concerned with indiscernibility. Many relationships have been established [4,15] and more so, most of the recent studies have concluded at this complementary nature of the two methodologies, especially in the context of granular computing. Therefore, it is desirable to extend and hybridize the underlying concepts to deal with additional aspects of data imperfection. Such developments offer a high degree of flexibility and provide robust solutions and advanced tools for data analysis.

2 Fuzzy-Rough Feature Selection

Feature selection addresses the problem of selecting those input features that are most predictive of a given outcome; a problem encountered in many areas of computational intelligence. Unlike other dimensionality reduction methods,

S.O. Kuznetsov et al. (Eds.): RSFDGrC 2011, LNAI 6743, pp. 31–35, 2011.

feature selectors preserve the original meaning of the features after reduction. This has found application in tasks that involve datasets containing huge numbers of features (in the order of tens of thousands) which, for some learning algorithms, might be impossible to process further. Recent examples include text processing and web content classification [6].

There are often many features involved, and combinatorially large numbers of feature combinations, to select from. Note that the number of feature subset combinations with m features from a collection of N total features is $N!/[m!(N-m)!]$. It might be expected that the inclusion of an increasing number of features would increase the likelihood of including enough information to distinguish between classes. Unfortunately, this is not necessarily true if the size of the training dataset does not also increase rapidly with each additional feature included. A high-dimensional dataset increases the chances that a learning algorithm will find spurious patterns that are not valid in general. More features may introduce more measurement noise, and hence reduce performance (e.g. classification accuracy). Most techniques employ some degree of reduction in order to cope with large amounts of data, so an efficient and effective reduction method is required.

Fuzzy-rough feature selection (FRFS) [3,6,8,17] provides a means by which discrete or real-valued data (or a mixture of both) can be effectively reduced without the need for user-supplied information. Additionally, this technique can be applied to data with continuous or nominal decision attributes, and as such can be applied to regression as well as classification datasets. Noise is an important factor degrading the performance of reduction: a single misclassified object prevents rough set analysis from making any conclusive statements about all other objects it is related to. To reduce the impact of noise, the original rough set approach has been adapted by using VPRS approximations (see e.g. [19]), such that problematic elements are not taken into account as long as their relative proportion remains below a certain threshold. Recently [1], a vaguely quantified approach was proposed (VQRS) that goes one step further by relaxing this crisp threshold into a smoother region of tolerance towards classification errors. The approach has been integrated with FRFS approaches, providing a general model that is robust and effective [2].

3 Fuzzy-Rough Nearest Neighbour Classification

The K-nearest neighbour (KNN) algorithm [5] is a well-known classification technique that assigns a test object to the decision class most common among its K nearest neighbours, i.e., the K training objects that are closest to the test object. An extension of the KNN algorithm to fuzzy set theory (FNN) was introduced in [12]. It allows partial membership of an object to different classes, and also takes into account the relative importance (closeness) of each neighbour w.r.t. the test instance. However, as Sarkar correctly argued in [16], the FNN algorithm has problems dealing adequately with insufficient knowledge. In particular, when every training pattern is far removed from the test object, and hence there are no suitable neighbours, the algorithm is still forced to makeclear-cut predictions. This is because the predicted membership degrees to the various decision classes always need to sum up to 1.

To address this problem, Sarkar [16] introduced a so-called fuzzy-rough ownership function that, when plugged into the conventional FNN algorithm, produces class confidence values that do not necessarily sum up to 1. However, this method does not refer to the main ingredients of rough set theory, i.e., lower and upper approximation. Fuzzy-rough nearest neighbours (FRNN) [11] is an alternative approach, which uses a test object's nearest neighbours to construct the lower and upper approximation of each decision class, and then computes the membership of the test object to these approximations. The method is very flexible, as there are many options to define the fuzzy-rough approximations, including the traditional implicator/t-norm based model [15], as well as the vaguely quantified rough set model [1], which is more robust in the presence of noisy data.

4 Hybrid Fuzzy Rule Induction

Feature selection often precedes classification as a preprocessing step, simplifying a decision system by selecting those conditional attributes that are most pertinent to the decision, and eliminating those that are redundant and/or misleading.

A common strategy in rough set theory is to induce rules by overlaying decision reducts over the original (training) decision system and reading off the values. In other words, by partitioning the universe via the features present in a decision reduct, each resulting equivalence class forms a single rule. As the partitioning is produced by a reduct, it is guaranteed that each equivalence class is a subset of, or equal to, a decision concept, meaning that the attribute values that produced this equivalence class are good predictors of the decision concept. The use of a reduct also ensures that each object is covered by the set of rules. A disadvantage of this approach is that the generated rules are often too specific, as each rule antecedent always includes every feature appearing in the final reduct. For this reason, the rule induction step can be directly integrated into the feature selection process, generating rules on the fly [10]. In particular, the greedy hill-climbing algorithm used for subset search can be used such that, at each step, fuzzy rules that maximally cover the training objects, with a minimal number of attributes, are generated. For the purposes of combining rule induction and feature selection, rules are constructed from fuzzy tolerance classes (antecedents) and corresponding decision concepts (consequents).

5 Instance Selection

An additional hurdle faced by many of these techniques is the sheer volume of data that must be processed and analysed. This increases the chances that learning algorithms find spurious patterns that are not valid in general. Often, the problem encountered is the prohibitively high number of training instances present or conflicting information between them. In this case, instance selection is desired to make the volume of data manageable and to remove misleading training instances in an effort to improve learned models from this data. Fuzzy-rough instance selection (FRIS) approaches have been developed for this purpose [7]. The main idea behind these approaches is to remove instances that cause conflicts with other instances as determined by the fuzzy-rough positive region. By removing these instances, the quality of training data can be improved and classifier training time reduced.

6 Handling Missing Values

Central to traditional fuzzy-rough feature selection is the fuzzy tolerance relation. From this, the fuzzy-rough lower approximations are constructed which then form the fuzzy positive regions utilised in the degree of dependency measure. Thus, the starting point for the process, type-1 fuzzy tolerance, is critical for its success. It is recognised that type-1 approaches are unable to address particular types of uncertainty due to their requirement of totally crisp membership functions [13]. An interval-valued approach may therefore be able to better handle this uncertainty and at the same time model the uncertainty inherent in missing values. Currently, there is no way to handle such values in fuzzy-rough set theory. Thus, the starting point for this is the interval-valued tolerance relation. If an object contains a missing value for a particular feature, then the resulting degree of similarity with other objects is unknown. In an interval-valued context, this can be modeled by returning the unit interval when an attribute value is missing for one or both objects. On this basis, fuzzy-rough data mining algorithms can be constructed [9].

References

1. Cornelis, C., De Cock, M., Radzikowska, A.: Vaguely Quantified Rough Sets. In: An, A., Stefanowski, J., Ramanna, S., Butz, C.J., Pedrycz, W., Wang, G. (eds.) RSFDGrC 2007. LNCS (LNAI), vol. 4482, pp. 87–94. Springer, Heidelberg (2007)
2. Cornelis, C., Jensen, R.: A Noise-tolerant Approach to Fuzzy-Rough Feature Selection. In: Proceedings of the 17th International Conference on Fuzzy Systems (FUZZ-IEEE 2008), pp. 1598–1605 (2008)
3. Cornelis, C., Jensen, R., Hurtado Martín, G., Ślęzak, D.: Attribute Selection with Fuzzy Decision Reducts. Information Sciences 180(2), 209–224 (2010)
4. Dubois, D., Prade, H.: Rough fuzzy sets and fuzzy rough sets. International Journal of General Systems 17, 91–209 (1990)
5. Duda, R., Hart, P.: Pattern Classification and Scene Analysis. Wiley, New York (1973)
6. Jensen, R., Shen, Q.: Computational Intelligence and Feature Selection: Rough and Fuzzy Approaches. IEEE Press, Wiley & Sons (2008)
7. Jensen, R., Cornelis, C.: Fuzzy-rough instance selection. In: Proceedings of the 19th International Conference on Fuzzy Systems (FUZZ-IEEE 2010), pp. 1776–1782 (2010)
8. Jensen, R., Shen, Q.: New approaches to fuzzy-rough feature selection. IEEE Transactions on Fuzzy Systems 17(4), 824–838 (2009)

9. Jensen, R., Shen, Q.: Interval-valued Fuzzy-Rough Feature Selection in Datasets with Missing Values. In: Proceedings of the 18th International Conference on Fuzzy Systems (FUZZ-IEEE 2009), pp. 610–615 (2009)
10. Jensen, R., Cornelis, C., Shen, Q.: Hybrid Fuzzy-Rough Rule Induction and Feature Selection. In: Proceedings of the 18th International Conference on Fuzzy Systems (FUZZ-IEEE 2009), pp. 1151–1156 (2009)
11. Jensen, R., Cornelis, C.: Fuzzy-Rough Nearest Neighbour Classification. In: Peters, J.F., Skowron, A., Chan, C.-C., Grzymala-Busse, J.W., Ziarko, W.P. (eds.) Transactions on Rough Sets XIII. LNCS, vol. 6499, pp. 56–72. Springer, Heidelberg (2011)
12. Keller, J.M., Gray, M.R., Givens, J.A.: A fuzzy K-nearest neighbor algorithm. IEEE Trans. Systems Man Cybernet. 15(4), 580–585 (1985)
13. Mendel, J.M., John, R.I.: Type-2 Fuzzy Sets Made Simple. IEEE Transactions on Fuzzy Systems 10(2), 117–127 (2002)
14. Pawlak, Z.: Rough Sets: Theoretical Aspects of Reasoning About Data. Kluwer Academic Publishing, Dordrecht (1991)
15. Radzikowska, A.M., Kerre, E.E.: A comparative study of fuzzy rough sets. Fuzzy Sets and Systems 126, 137–156 (2002)
16. Sarkar, M.: Fuzzy-Rough nearest neighbors algorithm. Fuzzy Sets and Systems 158, 2123–2152 (2007)
17. Shen, Q., Jensen, R.: Selecting Informative Features with Fuzzy-Rough Sets and its Application for Complex Systems Monitoring. Pattern Recognition 37(7), 1351–1363 (2004)
18. Zadeh, L.: Fuzzy sets. Information and Control 8(3), 338–353 (1965)
19. Ziarko, W.: Decision Making with Probabilistic Decision Tables. In: Zhong, N., Skowron, A., Ohsuga, S. (eds.) RSFDGrC 1999. LNCS (LNAI), vol. 1711, pp. 463–471. Springer, Heidelberg (1999)

Dual Rough Approximations in Information Tables with Missing Values

Michinori Nakata[1] and Hiroshi Sakai[2]

[1] Faculty of Management and Information Science,
Josai International University
1 Gumyo, Togane, Chiba, 283-8555, Japan
nakatam@ieee.org

[2] Department of Mathematics and Computer Aided Sciences,
Faculty of Engineering, Kyushu Institute of Technology,
Tobata, Kitakyushu, 804-8550, Japan
sakai@mns.kyutech.ac.jp

Abstract. A method of possible equivalence classes has been developed under information tables with missing values. To deal with imprecision of rough approximations that comes from missing values, the concepts of certainty and possibility are used. When an information table contains missing values, two rough approximations, certain and possible ones, are obtained. The actual rough approximation lies between the certain and possible rough approximations. The method gives the same results as a method of possible worlds. This justifies the method of possible equivalence classes. Furthermore, the method is free from the restriction that missing values may occur to only some specified attributes. Hence, we can use the method of possible equivalence classes to obtain rough approximations between arbitrary sets of attributes having missing values.

Keywords: Rough sets, Incomplete information, Missing values, Possible equivalence classes, Lower and upper approximations.

1 Introduction

The framework of rough sets, proposed by Pawlak [20], is used in various fields [26,27]. The framework is characterized by rough approximations, which consist of lower and upper approximations, under using equivalence classes in information tables containing only complete information. Even if the information obtained from the real world does not contain any incomplete information, a derived approximation is not unique; namely, two approximations, lower and upper ones, are obtained. This comes from imprecision of knowledge that is obtained from information tables with complete information. Also, real tables usually contain incomplete information such as partial values, missing values, possibilistic values, and so on [19]. Lots of studies have been made for information tables with incomplete information [4,5,6,8,11,12,13,14,15,16,17,21,22,23,24,25].

The studies are broadly classified into two types. One is based on a method of possible worlds, where equivalence classes are used [17,21,22,23]. Every missing

S.O. Kuznetsov et al. (Eds.): RSFDGrC 2011, LNAI 6743, pp. 36–43, 2011.

value is replaced by some element of the domain and a set of possible tables is created. Procedures using equivalence classes, which are established in dealing with information tables containing only complete information, are applied to each possible table. Then, the results from each table are aggregated.

The other method is based on observation that when an information table contains incomplete information, we cannot obtain equivalence classes without creating possible tables. Then, some other types of classes are used to derive rough approximations, such as: tolerance classes [8], similarity classes [25], or maximal consistent blocks [11]. Using a class that is not an equivalence class is problematic, as rough approximations do not coincide with those from the method of possible worlds [16], definability cannot be expressed, and monotonicity of the accuracy of approximation may not hold [11].

Recently, an approach using possible equivalence classes without a need of creating possible tables was proposed [12,13]. The approach assumes that an object with missing values has only the possibility of being indiscernible with other objects. It is called a method of possible equivalence classes [16]. In this approach, rough approximations coincide with those from the method of possible worlds. Also, monotonicity of the accuracy of approximation holds.

In this paper, we extend the method of possible equivalence classes by introducing the concept of certainty in addition to possibility. The paper is organized as follows. In section 2, approaches based on rough sets are briefly addressed under complete information. In section 3, the method of possible worlds is described and a correctness criterion is shown for methods extended to deal with incomplete information. In section 4, we introduce the above-mentioned extension. In section 5, conclusions are addressed.

2 Rough Sets Under Complete Information

A data set is represented as a table, called an information table, where each row represents an object and each column does an attribute. The information table is pair (U, AT), where U is a non-empty finite set of objects called the universe and AT is a non-empty finit set of attributes such that $\forall a \in AT : U \rightarrow D(a)$ where set $D(a)$ is the domain of attribute a. Binary relation $IND(a)$ for indiscernibility of objects on attribute $a \in AT$ is $IND(a) = \{(o, o') \in U \times U \mid a(o) = a(o')\}$, where $a(o)$ is the value for attribute a of object o. From the relation, equivalence class $E_a(o)$ containing object o is obtained: $E_a(o) = \{o' \mid (o, o') \in IND(a)\}$. Finally, family $\mathcal{E}_a$ of equivalence classes on a is: $\mathcal{E}_a = \{E_a(o) \mid o \in O\}$. Equivalence class $E \in \mathcal{E}_a$ is characterized by value v that object $o \in E$ has on attribute a. This is expressed by $E_{a=v}$. Also, we can define the definability for set $\mathcal{O}$ of discernible objects by using equivalence class E_i on a such that $\mathcal{O}$ is a-definable if and only if $\exists_{E_1,\cdots,E_l} \cup_i E_i = \mathcal{O}$. Using the family of equivalence classes on a, lower approximation $\underline{Apr}_a(\mathcal{O})$ and upper approximation $\overline{Apr}_a(\mathcal{O})$ of set $\mathcal{O}$ of indiscernible objects are:

$$\underline{Apr}_a(\mathcal{O}) = \{E \mid E \in \mathcal{E}_a, E \subseteq \mathcal{O}\}, \quad \overline{Apr}_a(\mathcal{O}) = \{E \mid E \in \mathcal{E}_a, E \cap \mathcal{O} \neq \emptyset\}.$$

Generally, an object is specified by values for attributes. Let $\mathcal{E}_b(\mathcal{O})$[1] be the family of equivalence classes derived from $\mathcal{O}$ on attribute b. Lower approximation $\underline{Apr}_a(\mathcal{O}/b)$ and upper approximation $\overline{Apr}_a(\mathcal{O}/b)$ of set $\mathcal{O}$ of objects that are specified by values for b are obtained on a:

$$\underline{Apr}_a(\mathcal{O}/b) = \{E \mid E \in \underline{Apr}_a(\mathcal{O}'), \mathcal{O}' \in \mathcal{E}_b(\mathcal{O})\},$$
$$\overline{Apr}_a(\mathcal{O}/b) = \{E \mid E \in \overline{Apr}_a(\mathcal{O}'), \mathcal{O}' \in \mathcal{E}_b(\mathcal{O})\}.$$

Equivalence classes $E \in \underline{Apr}_a(\mathcal{O}/b)$ and $E \in \overline{Apr}_a(\mathcal{O}/b)$ are characterized by a pair of values u and v that object $o \in E$ has on attributes a and b, respectively. This is expressed by $E_{a=u \to b=v}$, which means that equivalence class $E_{a=u \to b=v}$ supports rule $a = u \to b = v$. Expressions in terms of a set of objects are:

$$\underline{apr}_a(\mathcal{O}/b) = \cup_{E \in \underline{Apr}_a(\mathcal{O}/b)} E, \quad \overline{apr}_a(\mathcal{O}/b) = \cup_{E \in \overline{Apr}_a(\mathcal{O}/b)} E.$$

apr is used for the expressions by a set of objects whereas *Apr* by a family of equivalence classes. Formulae on sets A and B of attributes are derived from formulae on single attributes a and b:

$$IND(A) = \cap_{a \in A} IND(a),\ \mathcal{E}_A = \{\cap_{a \in A} E_a \mid E_a \in \mathcal{E}_a\},$$
$$\underline{Apr}_A(\mathcal{O}/B) = \cap_{b \in B}\{\cap_{a \in A} E_a \mid E_a \in \underline{Apr}_a(\mathcal{O}/b)\},$$
$$\overline{Apr}_A(\mathcal{O}/B) = \cap_{b \in B}\{\cap_{a \in A} E_a \mid E_a \in \overline{Apr}_a(\mathcal{O}/b)\},$$
$$\underline{apr}_A(\mathcal{O}/B) = \cap_{a \in A, b \in B} \underline{apr}_a(\mathcal{O}/b),$$
$$\overline{apr}_A(\mathcal{O}/B) = \cap_{a \in A, b \in B} \overline{apr}_a(\mathcal{O}/b).$$

3 Methods of Possible Worlds

Set $rep_a(T)$ of possible tables on attribute a is obtained from every missing value on a being replaced by some element comprising the domain. Family $\mathcal{E}_a^{t_i}$ of equivalence classes on a is derived from each possible table t_i by using the method addressed in Section 2. We use two types of aggregation, union and intersection. Family $\bigcup \mathcal{E}_a$ of equivalence classes is the union of $\mathcal{E}_a^{t_i}$, which is based on values characterizing equivalence classes. The union is defined by:

$$\bigcup \mathcal{E}_a = \{\cup_i E_{a=v}^{t_i} \mid E_{a=v}^{t_i} \in \mathcal{E}_a^{t_i}, v \in D(a)\}.$$

When equivalence class E belongs to $\bigcup \mathcal{E}_a$, every object $o \in E$ has the same value on a in at least one possible table. On the other hand, Family $\bigcap \mathcal{E}_a$ of equivalence classes, the intersection of $\mathcal{E}_a^{t_i}$, is defined by:

$$\bigcap \mathcal{E}_a = \{\cap_i E_{a=v}^{t_i} \mid E_{a=v}^{t_i} \in \mathcal{E}_a^{t_i}, v \in D(a)\}.$$

When E belongs to $\bigcap \mathcal{E}_a$, every object $o \in E$ has the same value on a in all possible tables. Rough approximations $\bigcup \underline{Apr}_a(\mathcal{O}/b)$ and $\bigcup \overline{Apr}_a(\mathcal{O}/b)$ are:

$$\bigcup \underline{Apr}_a(\mathcal{O}/b) = \{\cup_i E_{a=u \to b=v}^{t_i} \mid E_{a=u \to b=v}^{t_i} \in \underline{Apr}_a(\mathcal{O}/b)^{t_i}, u \in D(a), v \in D(b)\},$$
$$\bigcup \overline{Apr}_a(\mathcal{O}/b) = \{\cup_i E_{a=u \to b=v}^{t_i} \mid E_{a=u \to b=v}^{t_i} \in \overline{Apr}_a(\mathcal{O}/b)^{t_i}, u \in D(a), v \in D(b)\}.$$

[1] $(\mathcal{O})$ is usually omitted when $\mathcal{O}$ is equal to U.

There exists an equivalence class that supports plural rules. In such a case, equivalence classes with different rules are described, although they are identical as a set. Rough approximations $\bigcap \underline{Apr}_a(\mathcal{O}/b)$ and $\bigcap \overline{Apr}_a(\mathcal{O}/b)$ are:

$$\bigcap \underline{\underline{Apr}}_a(\mathcal{O}/b) = \{\cap_i E^{t_i}_{a=u\rightarrow b=v} \mid E^{t_i}_{a=u\rightarrow b=v} \in \underline{\underline{Apr}}_a(\mathcal{O}/b)^{t_i}, u \in D(a), v \in D(b)\},$$
$$\bigcap \overline{Apr}_a(\mathcal{O}/b) = \{\cap_i E^{t_i}_{a=u\rightarrow b=v} \mid E^{t_i}_{a=u\rightarrow b=v} \in \overline{Apr}_a(\mathcal{O}/b)^{t_i}, u \in D(a), v \in D(b)\}.$$

We have two types of definability. We say that set $\mathcal{O} \subseteq U$ of discernible objects is certainly a-definable, if $\mathcal{O}$ is a-definable on a in every possible table, while $\mathcal{O}$ is possibly a-definable, if $\mathcal{O}$ is a-definable on a in some possible tables [23]. For expressions by a set of objects of rough approximations, we have:

$$\bigcup \underline{apr}_a(\mathcal{O}) = U - \bigcap \overline{apr}_a(U - \mathcal{O}), \quad \bigcap \underline{apr}_a(\mathcal{O}) = U - \bigcup \overline{apr}_a(U - \mathcal{O}),$$

where

$$\bigcup \underline{apr}_a(\mathcal{O}) = \cup_i \underline{apr}_a(\mathcal{O})^{t_i}, \bigcap \underline{apr}_a(\mathcal{O}) = \cap_i \underline{apr}_a(\mathcal{O})^{t_i},$$
$$\bigcap \overline{apr}_a(U - \mathcal{O}) = \cap_i \overline{apr}_a(U - \mathcal{O})^{t_i}, \bigcup \overline{apr}_a(U - \mathcal{O}) = \cup_i \overline{apr}_a(U - \mathcal{O})^{t_i}.$$

We adopt the results by the method of possible worlds as a correctness criterion for extended methods. This is usually used in the field of databases dealing with incomplete information [1,2,3,7,18,28]. This criterion is formally represented as $q_A(T) = \bigodot q'_A(rep_A(T))$ where q'_A is the classical method that is described in Section 2 and q_A is an extended method of q'_A, and $\bigodot$ is an aggregate operator.

4 Methods of Possible Equivalence Classes

An approach using possible equivalence classes was originally proposed in [12,13]. The approach is called a method of possible equivalence classes. In the field of databases with incomplete information, two types of sets, which mean certain and possible answers, are obtained in query processing to tables containing missing values [9,10]. The method of possible equivalence classes that was described in [16] deals with only possible rough approximations. Therefore, we extend it to handle certainty in addition to possibility.

Possible equivalence classes on attributes are made by adding objects with missing values to equivalence classes obtained from a set of objects with no missing values on those attributes. Let O^c_a and O^i_a be sets of objects that have no missing values and missing values on attribute a, respectively. For set O^c_a, we obtain family $\mathcal{E}(O^c_a)$[2] of equivalence classes on a from using the classical method addressed in Section 2. The family of the possible equivalence classes for object o belonging to $E \in \mathcal{E}(O^c_a)$ is $\{E \cup E' \mid E' \in \mathcal{P}(O^i_a)\}$, where $\mathcal{P}(O^i_a)$ is the power set of O^i_a. Clearly, the family of possible equivalence classes including an object that is an element of O^c_a is a lattice for $\subseteq$ on attribute a. For example, we suppose object o is included in equivalence class $E \in \mathcal{E}(O^c_a)$ and $O^i_a = \{o_1, o_2, o_3\}$. The lower bound is E and the upper bound is $E \cup O^i_a$ in the lattice. All equivalence classes contained in the family have common part $E \in \mathcal{E}(O^c_a)$.

[2] $\mathcal{E}(O^c_a)$ is formally $\mathcal{E}_a(O^c_a)$. The subscript a is omitted for simplicity.

In addition, there are families of equivalence classes with common part $\emptyset$. Element $E \in \mathcal{P}(O^i_a)$ is a possible equivalence class. $\mathcal{P}(O^i_a)$ is a lattice with the lower bound $\emptyset$ and the upper bound O^i_a. Finally, family $Pos(\mathcal{E}_a))$ of possible equivalence classes on attribute a is:

$$Pos(\mathcal{E}_a) = \{E \cup E' \mid E \in \mathcal{E}(O^c_a), E' \in \mathcal{P}(O^i_a)\} \cup \mathcal{P}(O^i_a) \backslash \{\emptyset\}.$$

In each family of possible equivalence classes with a common part, we call the lower bound and the upper bound the minimal possible equivalence class and the maximal possible equivalence class, respectively. Family $Pos(\mathcal{E}_a)_{max}$ of maximal possible equivalence classes on a is:

$$Pos(\mathcal{E}_a)_{max} = \{E \cup O^i_a \mid E \in \mathcal{E}(O^c_a)\} \cup O^i_a.$$

Family $Pos(\mathcal{E}_a)_{min}$ of minimal possible equivalence classes is:

$$Pos(\mathcal{E}_a)_{min} = \mathcal{E}(O^c_a).$$

Using maximal and minimal possible equivalence classes, possible rough approximations $Pos(\underline{Apr}_a(\mathcal{O}))$ and $Pos(\overline{Apr}_a(\mathcal{O}))$ of set $\mathcal{O}$ of indiscernible objects are:

$$Pos(\underline{Apr}_a(\mathcal{O})) = \{E \cap \mathcal{O} \mid E \in Pos(\mathcal{E}_a)_{max}, \exists_{E' \supseteq E} E' \in Pos(\mathcal{E}_a)_{min}, E' \subseteq \mathcal{O}\},$$
$$Pos(\overline{Apr}_a(\mathcal{O})) = \{E \mid E \in Pos(\mathcal{E}_a)_{max}, E \cap \mathcal{O} \neq \emptyset\}.$$

Lower approximation $Pos(\underline{Apr}_a(\mathcal{O}/b))$ and upper approximation $Pos(\overline{Apr}_a(\mathcal{O}/b))$ of set $\mathcal{O}$ of objects that are specified by values for attribute b are:

$$Pos(\underline{Apr}_a(\mathcal{O}/b)) = \{E \mid E \in Pos(\underline{Apr}_a(\mathcal{O}')), \mathcal{O}' \in Pos(\mathcal{E}_b(\mathcal{O}))_{max}\},$$
$$Pos(\overline{Apr}_a(\mathcal{O}/b)) = \{E \mid E \in Pos(\overline{Apr}_a(\mathcal{O}')), \mathcal{O}' \in Pos(\mathcal{E}_b(\mathcal{O}))_{max}\}.$$

Proposition 1. If $E \in Pos(\underline{Apr}_a(\mathcal{O}/b))$, then $\exists_{E' \supseteq E} E' \in Pos(\overline{Apr}_a(\mathcal{O}/b))$.

Proposition 2. If $C \subseteq A$ for sets A and C of attributes, then $Pos(\underline{apr}_C(\mathcal{O}/b)) \subseteq Pos(\underline{apr}_A(\mathcal{O}/b))$ and $Pos(\overline{apr}_A(\mathcal{O}/b)) \subseteq Pos(\overline{apr}_C(\mathcal{O}/b))$.

Namely, monotonicity of the accuracy of approximations holds for expressions based on objects.

Proposition 3. We have $Pos(\underline{Apr}_a(\mathcal{O}/b)) = \bigcup \underline{Apr}_a(\mathcal{O}/b)$ and $Pos(\overline{Apr}_a(\mathcal{O}/b)) = \bigcup \overline{Apr}_a(\mathcal{O}/b)$.

Certain rough approximations $Cer(\underline{Apr}_a(\mathcal{O}/b)$ and $Cer(\overline{Apr}_a(\mathcal{O}/b))$ of set $\mathcal{O}$ of indiscernible objects are:

$$Cer(\underline{Apr}_a(\mathcal{O})) = \{E \mid E \in Pos(\mathcal{E}_a)_{min}, \exists_{E' \supseteq E} E' \in Pos(\mathcal{E}_a)_{max}, E' \subseteq \mathcal{O}\},$$
$$Cer(\overline{Apr}_a(\mathcal{O})) = \{E \mid E \in Pos(\mathcal{E}_a)_{min}, E \cap \mathcal{O} \neq \emptyset\}.$$

Lower and upper approximations – $Cer(\underline{Apr}_a(\mathcal{O}/b))$ and $Cer(\overline{Apr}_a(\mathcal{O}/b))$ – of set $\mathcal{O}$ of objects that are specified by values for b are:

$$Cer(\underline{Apr}_a(\mathcal{O}/b)) = \{E \mid E \in Cer(\underline{Apr}_a(\mathcal{O}')), \mathcal{O}' \in Pos(\mathcal{E}_b(\mathcal{O}))_{min}\},$$
$$Cer(\overline{Apr}_a(\mathcal{O}/b)) = \{E \mid E \in Cer(\overline{Apr}_a(\mathcal{O}')), \mathcal{O}' \in Pos(\mathcal{E}_b(\mathcal{O}))_{min}\}.$$

Proposition 4. If $E \in Cer(\underline{Apr}_a(\mathcal{O}/b))$, then $\exists_{E' \supseteq E} E' \in Cer(\overline{Apr}_a(\mathcal{O}/b))$.

Proposition 5. If $C \subseteq A$ for sets A and C of attributes, then $Cer(\underline{apr}_C(\mathcal{O}/b)) \subseteq Cer(\underline{apr}_A(\mathcal{O}/b))$ and $Cer(\overline{apr}_A(\mathcal{O}/b)) \subseteq Cer(\overline{apr}_C(\mathcal{O}/b))$.

Namely, monotonicity of the accuracy of approximations holds for expressions based on objects.

Proposition 6. If $E \in Cer(\underline{Apr}_a(\mathcal{O}/b))$, then $\exists_{E' \supseteq E} E' \in Pos(\underline{Apr}_a(\mathcal{O}/b))$. If $E \in Cer(\overline{Apr}_a(\mathcal{O}/b))$, then $\exists_{E' \supseteq E} E' \in Pos(\overline{Apr}_a(\mathcal{O}/b))$.

Proposition 7. We have $Cer(\underline{Apr}_a(\mathcal{O}/b)) = \bigcap \underline{Apr}_a(\mathcal{O}/b)$ and $Cer(\overline{Apr}_a(\mathcal{O}/b)) = \bigcap \overline{Apr}_a(\mathcal{O}/b)$.

Definability in the method of possible equivalence classes is described as follows: set $\mathcal{O}$ of discernible objects is certainly a-definable, if $\cup_v E_{a=v} = \mathcal{O}$, where $v \in \cup_{o \in \mathcal{O}} a(o)$, missing value $*$ is replaced by $D(a)$, and $E_{a=v} \in Pos(\mathcal{E}_a)_{max}$; set $\mathcal{O}$ of discernible objects is possibly a-definable, if $\exists_{E_1,\cdots,E_l} \cup_i \ E_i = \mathcal{O}$, where $E_i \in Pos(\mathcal{E}_a)$. For expressions by a set of objects to set $\mathcal{O}$ of discernible objects, we have:

$$Pos(\underline{apr}_a(\mathcal{O})) = U - Cer(\overline{apr}_a(U - \mathcal{O})),$$
$$Cer(\underline{apr}_a(\mathcal{O})) = U - Pos(\overline{apr}_a(U - \mathcal{O})),$$

where

$$Cer(\underline{apr}_a(\mathcal{O})) = \{o \mid \forall_{v \in V_{a(o)}} \exists_{E_{a=v}} E_{a=v} \in Pos(\mathcal{E}_a)_{max}, E_{a=v} \subseteq \mathcal{O}\},$$
$$Cer(\overline{apr}_a(\mathcal{O})) = \{o \mid \forall_{v \in V_{a(o)}} \exists_{E_{a=v}} E_{a=v} \in Pos(\mathcal{E}_a)_{min}, E_{a=v} \cap \mathcal{O} \neq \emptyset\},$$
$$Pos(\underline{apr}_a(\mathcal{O})) = \{o \mid \exists_{v \in V_{a(o)}} \exists_{E_{a=v}} E_{a=v} \in Pos(\mathcal{E}_a), E_{a=v} \subseteq \mathcal{O}\},$$
$$Pos(\overline{apr}_a(\mathcal{O})) = \{o \mid \exists_{v \in V_{a(o)}} \exists_{E_{a=v}} E_{a=v} \in Pos(\mathcal{E}_a), E_{a=v} \cap \mathcal{O} \neq \emptyset\}.$$

where if $a(o) = *$ set $V_{a(o)}$ is equal to $D(a)$, otherwise $\{a(o)\}$.

Proposition 8. Set $\mathcal{O}$ of discernible objects is certainly a-definable in the method of possible equivalence classes, if and only if it is so in the method of possible worlds, whereas $\mathcal{O}$ is possibly a-definable in the method of possible equivalence classes, if and only if it is so in the method of possible worlds.

From Propositions 3 and 7, the method of possible equivalence classes gives the same rough approximations as the method of possible worlds.

It is also noticeable that any restriction such that only specified attributes may have missing values does not at all impose on the method. Thus, we can use the method of possible equivalence classes to obtain rough approximations between arbitrary sets of attributes having missing values in information tables.

5 Conclusions

We have extended the method of possible equivalence classes to deal with information tables containing missing values. The extension is based on the concept of certainty in addition to possibility.

Dual rough approximations, certain and possible rough approximations, are obtained, because we cannot derive unique rough approximations under incomplete information. The actual rough approximations lie between certain and possible rough approximations.

The method gives the same results as the method of possible worlds. Certain and possible rough approximations are equal to the intersection and the union of rough approximations in all possible tables, respectively. This justifies the method of possible equivalence classes.

Furthermore, the method is free from the restriction that missing values may occur to only some specified attributes. Hence, we can use the method of possible equivalence classes to obtain rough approximations between arbitrary sets of attributes having missing values.

Acknowledgments. This work has been partially supported by the Grant-in-Aid for Scientific Research (C), Japan Society for the Promotion of Science, No. 22500204.

References

1. Abiteboul, S., Hull, R., Vianu, V.: Foundations of Databases. Addison-Wesley Publishing Company, Reading (1995)
2. Bosc, P., Duval, L., Pivert, O.: An Initial Approach to the Evaluation of Possibilistic Queries Addressed to Possibilistic Databases. Fuzzy Sets and Systems 140, 151–166 (2003)
3. Grahne, G.: The Problem of Incomplete Information in Relational Databases. LNCS, vol. 554. Springer, Heidelberg (1991)
4. Greco, S., Matarazzo, B., Słowiński, R.: Handling missing values in rough set analysis of multi-attribute and multi-criteria decision problems. In: Zhong, N., Skowron, A., Ohsuga, S. (eds.) RSFDGrC 1999. LNCS (LNAI), vol. 1711, pp. 146–157. Springer, Heidelberg (1999)
5. Grzymala-Busse, J.W.: Data with Missing Attribute Values: Generalization of Indiscernibility Relation and Rule Induction. Transactions on Rough Sets I, 78–95 (2004)
6. Guan, Y.-Y., Wang, H.-K.: Set-valued Information Systems. Information Sciences 176, 2507–2525 (2006)
7. Imielinski, T., Lipski, W.: Incomplete Information in Relational Databases. Journal of the ACM 31, 761–791 (1984)
8. Kryszkiewicz, W.: Rules in Incomplete Information Systems. Information Sciences 113, 271–292 (1999)

9. Lipski, W.: On Semantics Issues Connected with Incomplete Information Databases. ACM Transactions on Database Systems 4, 262–296 (1979)
10. Lipski, W.: On Databases with Incomplete Information. Journal of the ACM 28, 41–70 (1981)
11. Leung, Y., Li, D.: Maximum Consistent Techniques for Rule Acquisition in Incomplete Information Systems. Information Sciences 153, 85–106 (2003)
12. Nakata, M., Sakai, H.: Checking Whether or Not Rough-Set-Based Methods to Incomplete Data Satisfy a Correctness Criterion. In: Torra, V., Narukawa, Y., Miyamoto, S. (eds.) MDAI 2005. LNCS (LNAI), vol. 3558, pp. 227–239. Springer, Heidelberg (2005)
13. Nakata, M., Sakai, H.: Rough Sets Handling Missing Values Probabilistically Interpreted. In: Ślęzak, D., Wang, G., Szczuka, M.S., Düntsch, I., Yao, Y. (eds.) RSFDGrC 2005. LNCS (LNAI), vol. 3641, pp. 325–334. Springer, Heidelberg (2005)
14. Nakata, M., Sakai, H.: Lower and Upper Approximations in Data Tables Containing Possibilistic Information. Transactions on Rough Sets VII, 170–189 (2007)
15. Nakata, M., Sakai, H.: Applying Rough Sets to Information Tables Containing Probabilistic Values. In: Torra, V., Narukawa, Y., Yoshida, Y. (eds.) MDAI 2007. LNCS (LNAI), vol. 4617, pp. 282–294. Springer, Heidelberg (2007)
16. Nakata, M., Sakai, H.: Rough Sets Approximations in Data Tables Containing Missing Values. In: Proceedings of FUZZ-IEEE 2008, pp. 673–680. IEEE Press, New York (2008)
17. Orlowska, E., Pawlak, Z.: Representation of Nondeterministic Information. Theoretical Computer Science 29, 313–324 (1984)
18. Paredaens, J., De Bra, P., Gyssens, M., Van Gucht, D.: The Structure of the Relational Database Model. Springer, Heidelberg (1989)
19. Parsons, S.: Current Approaches to Handling Imperfect Information in Data and Knowledge Bases. IEEE Transactions on Knowledge and Data Engineering 8, 353–372 (1996)
20. Pawlak, Z.: Rough Sets: Theoretical Aspects of Reasoning about Data. Kluwer Academic Publishers, Dordrecht (1991)
21. Sakai, H.: Effective Procedures for Handling Possible Equivalence Relation in Non-deterministic Information Systems. Fundamenta Informaticae 48, 343–362 (2001)
22. Sakai, H., Nakata, M.: An Application of Discernibility Functions to Generating Minimal Rules in Non-deterministic Information Systems. Journal of Advanced Computational Intelligence and Intelligent Informatics 10, 695–702 (2006)
23. Sakai, H., Okuma, A.: Basic Algorithms and Tools for Rough Non-deterministic Information Systems. Transactions on Rough Sets I, 209–231 (2004)
24. Slowiński, R., Stefanowski, J.: Rough Classification in Incomplete Information Systems. Mathematical and Computer Modelling 12, 1347–1357 (1989)
25. Stefanowski, J., Tsoukiàs, A.: Incomplete Information Tables and Rough Classification. Computational Intelligence 17, 545–566 (2001)
26. Szczuka, M., Kryszkiewicz, M., Ramanna, S., Jensen, R., Hu, Q. (eds.): RSCTC 2010. LNCS, vol. 6086. Springer, Heidelberg (2010)
27. Yu, J., Greco, S., Lingras, P., Wang, G., Skowron, A. (eds.): RSKT 2010. LNCS, vol. 6401. Springer, Heidelberg (2010)
28. Zimányi, E., Pirotte, A.: Imperfect Information in Relational Databases. In: Motro, A., Smets, P. (eds.) Uncertainty Management in Information Systems: From Needs to Solutions, pp. 35–87. Kluwer Academic Publishers, Boston (1997)

Rough Sets and General Basic Set Assignments

Tong-Jun Li and Wei-Zhi Wu

School of Mathematics, Physics and Information Science,
Zhejiang Ocean University, Zhoushan, Zhejiang 316004, P.R. China
{litj,wuwz}@zjou.edu.cn

Abstract. Rough sets based on binary relations are of generalized rough sets. Meanwhile, the rough sets based on serial relations can be expressed via basic set assignments. In this paper, the notion of general basic set assignment is proposed by omitting a condition satisfied by basic set assignment. By the new proposed notion, a generalized rough set model is given. The relationships between the new model and the binary relation based rough sets are examined in detail. The investigation shown that virous types of binary relations can be characterized by general basic set assignments clearly, and the new rough sets are of another form of binary relation based rough sets.

Keywords: Basic set assignments, Binary relations, Rough sets.

1 Introduction

Rough set theory proposed by Pawlak [6,8] is a useful tool for dealing with uncertainty in information systems. In this model, the connection among objects of universe is represented by an indiscernibility relations. Indiscernibility relation, lower and upper approximations are key notions of Pawlaks rough set model. However, equivalence relation seems to be a very stringent condition that may limit the applications of rough set theory. Therefore, some interesting and meaningful extensions of the Pawlak's rough set model have been proposed in the literature [2,4,5,16,18,19]. In the Meanwhile, important attention is paid on rough set model based on arbitrary binary relations [13,14,17].

As for binary relation based rough sets, Yao [11,12] peculiarly mention that interval structures and serial rough set algebra can be represented by basic set assignments. The notion of basic set assignment was also implicitly used by Fagin and Halpern [1], and by Harmanec et al. [3]. However, some type of binary relation based rough set algebra, for example, symmetric, transitive rough set algebra, can not be expressed by basic set assignments. In order to solve this problem, the generalization of rough sets via non-numeric functions is investigated in this paper.

The rest of the paper is organized as follows. In next section, Pawlak rough sets and its extensions based on binary relations are reviewed. Through generalizing the notion of basic set assignment, the concept of general basic set assignment is obtained in section 3. By the new defined notion we generalize Pawlak rough

S.O. Kuznetsov et al. (Eds.): RSFDGrC 2011, LNAI 6743, pp. 44–51, 2011.

sets. The relationship between binary relations and general basic set assignments is clarified. In section 4, various types of binary relations are characterized by general basic set assignments. Section 5 presents the conclusions of the paper.

2 Preliminaries

In this section, we review Pawlak rough sets and its generalization briefly.

2.1 Pawlak Rough Sets

In Pawlak rough set model, the knowledge on the universe is described by equivalence relations. The theoretic framework of Pawlak rough sets can be denoted as a ordered pair (U, R) referred to as *approximation space*, where U is a finite and nonempty set called universe of discourse, R an equivalence relation on U. For any $x \in U$, the equivalence class including x is $[x]_R = \{y \in U | (x, y) \in R\}$, by convention, simply denoted as $[x]$. The set of all equivalence classes on U, $\{[x] | x \in U\}$, is a partition of U, and denoted as U/R. In Pawlak rough set model, equivalence classes are basic knowledge of the universe. Pawlak call them *atoms* [7], or called *elementary sets*. Unions of atoms are called *definable sets* [7,8], all of definable sets forms a σ-algebra denoted as $\sigma(U/R)$, which represents the total basic knowledge of the universe.

Let (U, R) be Pawlak approximation space, by the definable sets of $\sigma(U/R)$, $X \subseteq U$ can be assigned with two definable sets, denoted as $\underline{R}(X)$ and $\overline{R}(X)$, called the *lower approximation* and the *upper approximation* of X, respectively:

$$\underline{R}(X) = \{x \in U | [x] \subseteq X\} = \bigcup \{[x] | [x] \subseteq X\},$$
$$\overline{R}(X) = \{x \in U | [x] \cap X \neq \emptyset\} = \bigcup \{[x] | [x] \cap X \neq \emptyset\}.$$

It follows that $\underline{R}(X) \subseteq X \subseteq \overline{R}(X)$. Hence $X \in \sigma(U/R)$ if and only if $\underline{R}(X) = \overline{R}(X)$, then $X \subseteq U$ is said to be *rough* if $\underline{R}(X) \neq \overline{R}(X)$, that is, $X \notin \sigma(U/R)$.

2.2 Extension of Pawlak Rough Sets

Considering binary relations between two universes, an extended rough set model can be obtained.

Let U and W be two finite and nonempty universes of discourse. $R \subseteq U \times W$ is a binary crisp relation (binary relation in short) from U to W, $(x, y) \in R$ is also denoted as xRy. For any $x \in U, y \in W$, we define xR as the set $\{y \in W | xRy\}$, Ry as $\{x \in U | xRy\}$, and xR and Ry are called the successor neighborhood of x and the predecessor neighborhood of y, respectively. The relation R is called *serial* if $xR \neq \emptyset$ for all $x \in U$; R is called *inverse serial* if $Ry \neq \emptyset$ for all $y \in W$. If $U = W$, R is called a binary relation on U. R is called *reflexive* if $x \in xR$ for all $x \in U$; R is called *symmetric* if for any $x, y \in U$, xRy implies yRx; R is called *transitive* if for any $x, y, z \in U$, xRy and yRz imply xRz.

Let U and W be two finite nonempty universes of discourse, and $R \subseteq U \times W$. The triple (U, W, R) is called a *generalized approximation space*. For $X \subseteq W$, *the lower and upper approximations* of X with respect to (w.r.t.) (U, W, R), denoted as $\underline{R}(X)$ and $\overline{R}(X)$, respectively, are defined by

$$\underline{R}(X) = \{x \in U | xR \subseteq X\}, \quad \overline{R}(X) = \{x \in U | xR \cap X \neq \emptyset\}).$$

Obviously, when $U = W$ and R is an equivalence relation on U, the above model is identical with Pawlak rough set model.

The properties for the approximation operators $\underline{R}$ and $\overline{R}$ are list as follows: $\forall X, Y \subseteq W$,

(L1) $\underline{R}(X) = \sim \overline{R}(\sim X)$, (U1) $\overline{R}(X) = \sim \underline{R}(\sim X)$;

(L2) $\underline{R}(W) = U$, (U2) $\overline{R}(\emptyset) = \emptyset$;

(L3) $\underline{R}(X \cap Y) = \underline{R}(X) \cap \underline{R}(Y)$, (U3) $\overline{R}(X \cup Y) = \overline{R}(X) \cup \overline{R}(Y)$;

(L4) $A \subseteq B \Rightarrow \underline{R}(X) \subseteq \underline{R}(Y)$, (U4) $A \subseteq B \Rightarrow \overline{R}(X) \subseteq \overline{R}(Y)$;

(L5) $\underline{R}(X \cup Y) \supseteq \underline{R}(X) \cup \underline{R}(Y)$, (U5) $\overline{R}(X \cap Y) \subseteq \overline{R}(X) \cap \overline{R}(Y)$.

3 Rough Sets Based on General Basic Set Assignments

In approximate reasoning, non-numeric functions usually be used to represent uncertainty. Formally, a non-numeric functions is a mapping from one power set to another. Different types of non-numeric functions satisfy corresponding axioms [9,10,15].

Basic set assignment is a special kind of non-numeric functions. A *basic set assignment* is a mapping j from power set $P(W)$ to another $P(U)$, and satisfies the following conditions:

(c1) $j(\emptyset) = \emptyset$,

(c2) $\bigcup_{A \in P(W)} j(A) = U$,

(c3) $A \neq B \Rightarrow j(A) \cap j(B) = \emptyset, A, B \in P(W)$.

If $W = U$, then j is called a *basic set assignment* on U. A subset $A \in P(W)$ with $j(A) \neq \emptyset$ is called a *focal set* [12,15]. By the above condition (c2) and (c3) we can see that the set of all j-images of all focal sets of j, $P_j = \{j(A) | A \subseteq W, j(A) \neq \emptyset\}$, forms a partition of U.

In order to characterize the relationship between the approximation operators w.r.t. generalized approximations space and basic set assignments more clearly, the notion of basic set assignment is generalized as follows:

Definition 1. *A non-numeric function j from $P(W)$ to $P(U)$ is called a general basic set assignment if j satisfies the conditions* (c2) *and* (c3).

In generalized approximation space (U, W, R), $\underline{R}$ and $\overline{R}$ can be expressed by a general basic set assignment j_R induced from R.

Let R be a binary relation between U and W. By R we construct a general basic set assignment $j_R : P(W) \to P(U)$,

$$j_R(A) = \{x \in U | xR = A\}, \quad A \in P(W). \tag{1}$$

It is easy to check that j_R satisfies the conditions (c2) and (c3). Hence j_R is a general basic set assignment. It should be noted that if R is a serial relation then j_R is a basic set assignment, otherwise j_R is a general basic set assignment, not a basic set assignment.

Proposition 1. *Let $R \subseteq U \times W$. Then*

(1) $x \in j_R(A) \Longleftrightarrow A = xR, x \in U, A \subseteq W$.

(2) $x \in j_R(xR), x \in U$.

(3) $Ry = \bigcup_{y \in A} j_R(A), y \in W$.

Making use of j_R, $\underline{R}$ and $\overline{R}$ can be rewriten as

$$\underline{R}(X) = \bigcup_{A \subseteq X} j_R(A), \quad \overline{R}(X) = \bigcup_{A \cap X \neq \emptyset} j_R(A). \tag{2}$$

Conversely, j_R can also be calculated by $\underline{R}$ as follows:

$$j_R(A) = \underline{R}(A) - \bigcup_{B \subset A} \underline{R}(B), \quad A \subseteq W.$$

In view of Eq. (2), based on a general basic set assignment j from $P(W)$ to $P(U)$, we can define the lower and upper approximations of $X \subseteq W$ as follows:

Definition 2. *Let j be a general basic set assignment from $P(W)$ to $P(U)$, and $X \subseteq W$. Then the lower and upper approximations of X, denoted as $\underline{j}(X)$ and $\overline{j}(X)$, respectively, are defined as*

$$\underline{j}(X) = \bigcup_{A \subseteq X} j(A), \quad \overline{j}(X) = \bigcup_{A \cap X \neq \emptyset} j(A).$$

In the above rough set model, we can regard elements of P_j as elementary sets. The empty set and the unions of one or more elementary sets are called definable sets. The family of all definable sets formed from P_j is denoted by $\sigma(P_j)$. Since W is finite, $\sigma(P_j)$ is the σ-algebra generated by P_j, P_j is its basis. Meanwhile, from Definition 2 we can see that the lower approximation $\underline{j}(X)$ and the upper approximation $\overline{j}(X)$ are definable sets.

In particular, if $W = U$ then the approximation operators defined in Def. 2 degenerated into the approximation operators on single universe U.

From Eqs. (2), the following conclusions is clear.

Theorem 1. *Let $R \subseteq U \times W$. Then*

$$\underline{R} = \underline{j_R}, \quad \overline{R} = \overline{j_R}.$$

Theorem 1 shows that binary relation based approximations are of general basic set assignment based approximations. Now a natural question is that whether the approximations based on general basic set assignment are also of binary relation based approximations, the answer is positive.

According to Eq. (1), given a binary relation R from U to W, we can construct a general basic set assignment j_R. Conversely, from a general basic set assignment j from $P(W)$ to $P(U)$, we can also define a relation R_j from U to W as follows:

$$xR_jy \Longleftrightarrow \exists A \subseteq W(y \in A, x \in j(A)), \quad x \in U, y \in W. \tag{3}$$

Hence we have the following proposition.

Proposition 2. *If j is a general basic set assignment from $P(W)$ to $P(U)$, then*

(1) $R_j = \bigcup\limits_{A \subseteq W} j(A) \times A$,

(2) $R_jy = \bigcup\limits_{y \in A} j(A), y \in W$.

Proposition 3. *If j is a general basic set assignment from $P(W)$ to $P(U)$, then*

$$xR_j = A \Longleftrightarrow x \in j(A), x \in U, A \subseteq W.$$

Proof. To consider two cases: $A = \emptyset$ and $A \neq \emptyset$.

Case 1. $A = \emptyset$.

($\Rightarrow$) From Eq. (3) we have

$$^{\neg}(xR_jy) \Longleftrightarrow \forall A \subseteq W(x \in j(A) \Rightarrow y \notin A), \; x \in U, y \in W.$$

If $xR_j = \emptyset$, the for any $y \in W$, $y \notin xR_j$, thus for any $A \subseteq W$, $x \in j(A)$ implies $y \notin A$. Since there is unique $j(B) \in P_j$ such that $x \in j(B)$, for any $y \in W$, $y \notin B$. Consequently, we know that $B = \emptyset$, this means that $x \in j(\emptyset)$.

($\Leftarrow$) Assume that $x \in j(\emptyset)$ and $xR_j \neq \emptyset$. Let $y \in xR_j$. Then there exists an $A \subseteq W$ such that $y \in A$ and $x \in j(A)$. As P_j is a partition on U, we realize that $A = \emptyset$. So we get a wrong result $y \in \emptyset$. Therefore we can conclude that if $x \in j(\emptyset)$ then $xR_j = \emptyset$.

Case 2. $A \neq \emptyset$.

($\Rightarrow$) Suppose that $xR_j = A$. By the definition of R_j, for any $y \in A$ there exists a $B \subseteq W$ such that $y \in B$ and $x \in j(B)$. Since P_j is a partition on U, we know B is unique. Thus $A \subseteq B$. On the other hand, for any $y \in B$, from $x \in j(B)$ and Eq. (3) we know xR_jy, that is, $y \in xR_j = A$. Hence $B \subseteq A$. Therefore $A = B$ holds, so $x \in j(A)$.

($\Leftarrow$) Assume $x \in j(A)$. By Eq. (3) we have $A \subseteq xR_j$. On the other hand, for any $y \in xR_j$, by the definition of R_j, there exists a $B \subseteq W$ such that $y \in B$ and $x \in j(B)$. $j(A) = j(B)$ follows from that P_j is a partition on U. According to property (c3) we can conclude that $A = B$, So $y \in A$. Hence $xR_j \subseteq A$. Combining $A \subseteq xR_j$ with $A \subseteq xR_j$ we obtain $A = xR_j$. □

By Eq. (1), Proposition 1 and 3, the following propositions can be given.

Proposition 4. $R_{j_R} = R, R \subseteq U \times W$.

Proposition 5. *If j is a general basic set assignment from $P(W)$ to $P(U)$, then $j_{R_j} = j$.*

The next conclusion directly follows from Proposition 4 and Theorem 1.

Theorem 2. *For any j, a general basic set assignment from $P(W)$ to $P(U)$, we have*

$$\underline{j} = \underline{R_j}, \quad \overline{j} = \overline{R_j}.$$

Theorem 2 illustrates that general basic set assignment based approximations are of binary relation based approximations.

Now that general basic set assignment based approximations are of binary relation based approximations, of course, general basic set assignment based approximation operators also satisfy the properties (L1-L5) and (U1-U5).

4 Binary Relations and General Basic Set Assignments

In this section, we investigate some characteristics of general basic set assignments corresponding to different types of binary relations.

Theorem 3. *If j is a general basic set assignment from $P(W)$ to $P(U)$, then R_j is serial if and only if $j(\emptyset) = \emptyset$.*

Proof. ($\Rightarrow$) If R_j is serial, then for any $x \in U$, $xR_j \neq \emptyset$. For any $x \in U$, since $xR_j = \emptyset$ is equivalent to $x \in j(\emptyset)$, it follows from $xR_j \neq \emptyset$ that $x \notin j(\emptyset)$. By the arbitrariness of x, we have $j(\emptyset) = \emptyset$.

($\Leftarrow$) Let $x \in U$. By Proposition 3 we have that $xR_j = \emptyset$ is equivalent to $x \in j(\emptyset)$. So if $j(\emptyset) = \emptyset$, then $xR_j \neq \emptyset$ follows from that $x \notin j(\emptyset)$. □

Theorem 3 illustrates that only when j is a basic set assignment, R_j just be a serial relation, that is to say, non-serial relations can not be generated by basic set assignments.

Theorem 4. *If j is a general basic set assignment from $P(W)$ to $P(U)$, then R_j is inverse serial if and only if there exists an $A \subseteq W$ such that $y \in A$ and $j(A) \neq \emptyset$, or equivalently $\bigcup_{y \notin A} j(A) \neq U$.*

Proof. It follows from Proposition (2) and the definition of inverse series. □

Theorem 5. *If j is a general basic set assignment on U, then R_j is reflexive if and only if $j(A) \subseteq A, A \subseteq U$.*

Proof. ($\Rightarrow$) Suppose R_j is reflexive. Then for any $x \in U$, $x \in xR_j$. For any $A \subseteq U$, if $x \in j(A)$, then $xR_j = A$ by Proposition 3. According to $x \in xR_j$ we have $x \in A$. Hence $j(A) \subseteq A$.

($\Leftarrow$) Assume that $j(A) \subseteq A$ for all $A \subseteq U$. For any $x \in U$, since $j(xR_j) \subseteq xR_j$, by Proposition 1 (2) and Proposition 5 we have $x \in j_{R_j}(xR_j) = j(xR_j)$. Hence, $x \in xR_j$ follows from $j(xR_j) \subseteq xR_j$. Therefore R_j is reflexive. □

Theorem 6. *If j is a general basic set assignment on U, then R_j is symmetric if and only if $j(A) \cap B \neq \emptyset$ implies $j(B) \subseteq A$.*

Proof. ($\Rightarrow$) Suppose R_j is symmetric. If $j(A) \cap B \neq \emptyset$, we can take $x \in j(A) \cap B$. Thus, $xR_j = A$ follows from $x \in j(A)$. For any $y \in j(B)$, $yR_j = B$ holds. Noting $x \in B$, we have $x \in yR_j$. Since R_j is symmetric, $y \in xR_j$ follows, that is, $y \in A$. By the arbitrariness of y, we can conclude $j(B) \subseteq A$.

($\Leftarrow$) For any $x, y \in U$, if $y \in xR_j$, then by denoting $A = yR_j$ and $B = xR_j$ we know that $y \in j(A)$ and $x \in j(B)$ hold. Since $y \in B$, we get $j(A) \cap B \neq \emptyset$. By the given condition we can gain $j(B) \subseteq A$. Since $x \in j(xR_j) = j(B)$, $x \in A$ holds, that is, $x \in yR_j$. This proves that R_j is symmetric. □

Theorem 7. *If j is a general basic set assignment on U, then R_j is transitive if and only if for any $j(A), j(B) \in P_j$, $j(A) \cap B \neq \emptyset$ implies $A \subseteq B$.*

Proof. ($\Rightarrow$) Suppose R_j is transitive. If for any $j(A), j(B) \in P_j$, $j(A) \cap B \neq \emptyset$, by taking $x \in j(A) \cap B$, we have $xR_j = A$. As $j(B) \neq \emptyset$, take $y \in j(B)$, so $yR_j = B$. For any $z \in A$, obviously $z \in xR_j$. It follows from $x \in B$ that $x \in yR_j$. Since R_j is transitive, we have $z \in yR_j$, that is, $z \in B$. From the arbitrariness of z we obtain $A \subseteq B$.

($\Leftarrow$) For any $x, y, z \in U$, if $y \in xR_j$ and $z \in yR_j$, denoting $A = yR_j$ and $B = xR_j$, we have $j(A) \neq \emptyset$ and $j(B) \neq \emptyset$. As $y \in j(yR_j) = j(A)$ and $y \in xR_j = B$, $j(A) \cap B \neq \emptyset$ holds. By the given condition we get $A \subseteq B$, that is, $yR_j \subseteq xR_j$. Thus, according to $z \in yR_j$ we can conclude $z \in xR_j$. Therefore it is proved that R_j is transitive. □

5 Conclusions

The theory of rough sets is typically studied based on the notion of an approximation space and the two induced non-numeric functions, lower and upper approximations of subsets of a universe. This paper focus on the extension of rough set model. By eliminating the normalization condition in the definition of basic set assignment, the notion of general basic set assignment is proposed, with the new proposed notion, new rough set model is established. Through making the one to one correspondence between binary relations and general basic set assignments clear, the equivalence between binary relation based rough sets and general basic set assignment based rough sets is understood. Finally, many types of binary relations are characterized by general basic set assignments. This paper's work may be helpful for other non-numeric analysis approaches, e.g. propositional logic, evidence theory, possibility theory, etc.

Acknowledgements. This work was supported by grants from the National Natural Science Foundation of China (Nos. 11071284 and 61075120) and the Key Project of Zhejiang Ocean University of China.

References

1. Fagin, R., Halpern, J.Y.: Uncertainty, belief, and probability. Comput. Intell. 7, 160–173 (1991)
2. Grzymala-Busse, J., Rzasa, W.: Definability and other properties of approximations for generalized indiscernibility relations. In: Peters, J.F., Skowron, A. (eds.) Transactions on Rough Sets XI. LNCS, vol. 5946, pp. 14–39. Springer, Heidelberg (2010)
3. Harmanec, D., Klir, J., Resconi, G.: On modal logic interpretation of Dempster-Shafer theory of evidence. Int. J. Intell. Syst. 9, 941–951 (1994)
4. Meng, Z., Shi, Z.: A fast approach to attribute reduction in incomplete decision systems with tolerance relation-based rough sets. Inform. Sci. 179(16), 2774–2793 (2009)
5. Slowinski, R., Vanderpooten, D.: A generalized definition of rough approximations based on similarity. IEEE Transactions on Knowledge and Data Engineering 12(2), 331–336 (2000)
6. Pawlak, Z.: Rough Sets. International Journal of Computer and Information Sciences 5, 341–356 (1982)
7. Pawlak, Z.: Rough sets. Theoretical Aspects of Reasoning About Data. Kluwer Academic Publishers, Dordrecht (1991)
8. Pawlak, Z.: Some Issues on Rough Sets. In: Peters, J.F., Skowron, A., Grzymała-Busse, J.W., Kostek, B.z., Świniarski, R.W., Szczuka, M.S. (eds.) Transactions on Rough Sets I. LNCS, vol. 3100, pp. 1–58. Springer, Heidelberg (2004)
9. Wong, S.K.M., Wang, L.S., Yao, Y.Y.: Non-numeric Belief structures. In: Proceedings of the Fourth International Conference on Computing and Information, pp. 274–277 (1992)
10. Wong, S.K.M., Wang, L.S., Yao, Y.Y.: On modelling uncertainty with interval structures. Comput. Intell. 11, 406–426 (1995)
11. Yao, Y.Y.: Two views of the theory of rough sets in finite universes. International Journal of Approximation Reasoning 15, 291–317 (1996)
12. Yao, Y.Y.: Interpretations of Belief Functions in the Theory of Rough Sets. Information Sciences 104, 81–106 (1998)
13. Yao, Y.Y.: Constructive and algebraic methods of the theory of rough sets. Information Sciences 109(1-4), 21–47 (1998)
14. Yao, Y.Y.: Relational Interpretations of Neighborhood Operators and Rough Set Approximation Operators. Information Sciences 111(1-4), 239–259 (1998)
15. Yao, Y.Y., Wong, S.K.M., Wang, L.S.: A non-numeric approach to uncertain reasoning. International Journal of General Systems 23, 343–359 (1995)
16. Zhu, P.: Covering rough sets based on neighborhoods: An approach without using neighborhoods. Int. J. Approx. Reason (2010), doi:10.1016/j.ijar.2010.10.005
17. Zhu, W.: Generalized rough sets based on relations. Information Sciences 177(22), 4997–5011 (2007)
18. Zhu, W.: Relationship among basic concepts in covering-based rough sets. Information Sciences 179(14), 2478–2486 (2009)
19. Ziarko, W.: Variable precision rough set model. Journal of Computer and System Sciences 46, 39–59 (1993)

General Tool-Based Approximation Framework Based on Partial Approximation of Sets

Zoltán Csajbók[1] and Tamás Mihálydeák[2]

[1] Department of Health Informatics, Faculty of Health, University of Debrecen, Sóstói út 2-4, H-4400 Nyíregyháza, Hungary
csajzo@de-efk.hu

[2] Department of Computer Science, Faculty of Informatics, University of Debrecen, Egyetem tér 1, H-4032 Debrecen, Hungary
mihalydeak.tamas@inf.unideb.hu

Abstract. Let us assume that we observe a class of objects and have some well-defined features with which an observed object possesses or not. In real life, two relevant groups of objects can be established determined by our current and necessarily constrained knowledge. In particular, a group whose elements really possess a feature in question, and another group whose elements substantially do not possess the same feature. In practice, as a rule, we can observe a feature of objects via only *tools* with which we are able to judge easily whether an object possesses a property or not. Of course, a property ascertained by tools does not coincide with a feature completely. To manage this problem, we propose a general tool-based approximation framework based on partial approximation of sets in which a positive feature and its negative one of any proportion of the observed objects can *simultaneously* be approximated.

Keywords: Approximations, approximation schemes, rough set theory, partial approximation of sets.

1 Introduction

At the very beginning, we assume that we observe a class of objects which is modelled as an abstract set, called the universe of discourse. In addition, let us assume that we have some well-defined, decidable features with which an observed object possesses or not. These features assign crisp subsets within the universe. In other words, we model an object of interest as the element of an abstract set, called the universe, and its property 'it possesses a feature' as 'it is the element of a crisp subset of this universe'.

In practice, two relevant groups of objects can be established determined by our currently available and necessarily constrained knowledge: a group whose elements really possess a feature in question, and another group whose elements do not substantially possess the same feature. Both groups correspond two crisp

S.O. Kuznetsov et al. (Eds.): RSFDGrC 2011, LNAI 6743, pp. 52–59, 2011.

subsets of the universe. They are disjoint, and, in general, the union of them does not add up to the whole universe. For obvious reasons, the former can be marked with the adjective *positive*, whereas the latter with *negative*.

In real life, we cannot normally observe the features of objects directly. We need *tools* at our disposal with which we are able to judge easily and unambiguously whether an object possesses a *property* ascertained by a tool or not. It is expected that these tools can be used simply and quickly. The objects classified by a tool can also be modelled as one or more crisp subsets of the universe. With a slight abuse of terminology, these subsets are also simply called tools.

As a rule, a property does not coincide with a feature completely. Different tools form usually different subsets, but they are not necessarily disjoint. Notice that the complement of a tool is not necessarily tool at the same time. For instance, let us take the tools being recursively enumerable. However, the complement of a recursively enumerable set is not necessarily recursively enumerable [10]. This significant fact confirms the partial nature of our approach [9].

Let us distinguish two types of tools: *positive* and *negative* ones. Positive (resp., negative) tools provide an opportunity to locate a positive (resp,. negative) subset. It is a natural assumption that the union of positive tools and the union of negative tools are disjoint.

In practice, a feature which defines a positive (negative) subset is complicated. In general, it is completely impossible to locate it with the only tool. Instead, we need tools of finite or infinite number. To manage this problem we need an *approximation framework*. It may be built on the *rough set theory* [8], because it provides a powerful foundation to reveal and discover important structures in data and classify complex objects.

The rough set theory was introduced by the Polish mathematician, Z. Pawlak in the early 1980s [11], [12]. It may be seen as a new mathematical approach to *vagueness* [6], [13], [15], [16]. According to Pawlak's idea, the vagueness of a subset within a finite universe U is defined by the difference of its upper and lower approximations with respect to a partition of U.

Using partitions, however, is a very strict requirement. Our starting point will be an *arbitrary* family of subsets of an *arbitrary* universe U. We will not assume that this family of sets covers the universe whether that the universe is finite. Our concepts of lower and upper approximations are straightforward *point-free* generalizations of Pawlak's ones [1]. We will apply it to build a tool-based approximation framework in which a positive feature and its negative one of any proportion of the observed objects can be approximated *simultaneously*.

The rest of the paper is organized as follows. In Section 2 we illustrate our approach with a simple running example. Section 3 sums up the basic principles of the partial approximation of sets. Only those facts will be considered which are definitely necessary in the following. The major contributions of the paper are covered in Section 4 in which we will propose a general tool-based approximation framework. Its main notions are illustrated in Section 5. Finally, in Section 6, we conclude the paper.

2 An Illustrative Example

To illustrate our point of view let us see a simple running example. We want to describe the safe behavior of a complex computer network system. All computer applications or the whole computer system have an anticipated *expected behavior* how they should or shouldn't work. Its range may extend from the informal presupposing activities of applications to more formal ones. The latter is posited in user manuals and other different artifacts.

We focus solely on externally observable execution traces sent out by the observed system, and model its expected behavior via these traces. Let U denote the set of all execution traces generated by the system.

A^+ is the set of *expected* execution traces describing expected behavior of the running system. $U \setminus A^+$ can be seen as the abnormal behavior of the system which deviate from the previously defined expected profile. Its elements are called anomalies [4]. According to our current available knowledge, however, only a subset $A^- \subseteq U \setminus A^+$ can really be modelled as *unexpected* behavior of the system. Its elements are usually called misuses. Of course, an unexpected behavior has its own right to be profiled (for more details, see [2], [3]).

To justify A^+ we have *positive tools* at our disposal whose elements are called *acceptable.* They can be seen as the *prescriptions* of the security policy which are modelled by sets of execution traces. To justify A^- we have *negative tools* at our disposal. Its elements are called *unacceptable.* They can be seen as the *proscriptions* of the security policy which are modelled by sets of execution traces too (see Fig. 1).

Notice that the subsets A^+, A^- and $T_1^+, \ldots, T_5^+, T_1^-, \ldots, T_4^-$ are all crisp, in addition $A^+ \cap A^- = \emptyset$, $\bigcup \mathfrak{T}^+ \cap \bigcup \mathfrak{T}^- = \emptyset$, $\bigcup \mathfrak{T}^+ \cup \bigcup \mathfrak{T}^- \subseteq U$.[1]

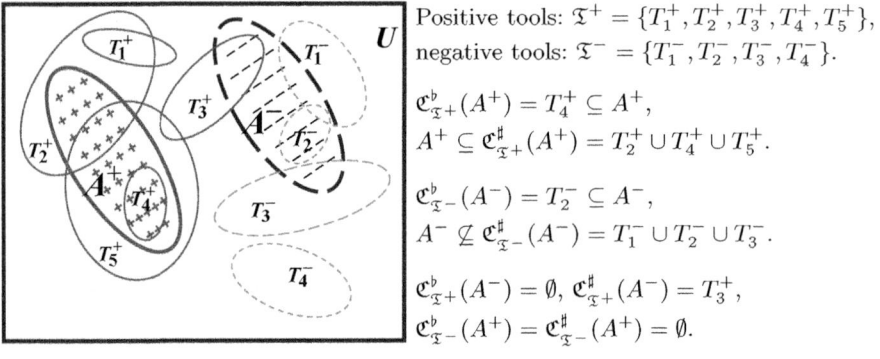

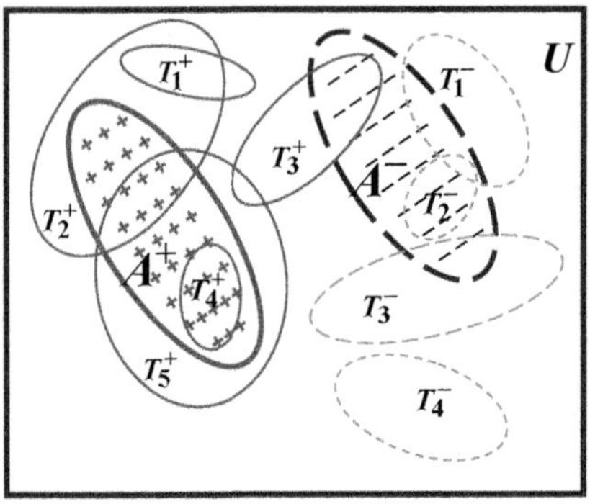

Positive tools: $\mathfrak{T}^+ = \{T_1^+, T_2^+, T_3^+, T_4^+, T_5^+\}$,
negative tools: $\mathfrak{T}^- = \{T_1^-, T_2^-, T_3^-, T_4^-\}$.

$\mathfrak{C}^{\flat}_{\mathfrak{T}^+}(A^+) = T_4^+ \subseteq A^+$,
$A^+ \subseteq \mathfrak{C}^{\sharp}_{\mathfrak{T}^+}(A^+) = T_2^+ \cup T_4^+ \cup T_5^+$.

$\mathfrak{C}^{\flat}_{\mathfrak{T}^-}(A^-) = T_2^- \subseteq A^-$,
$A^- \not\subseteq \mathfrak{C}^{\sharp}_{\mathfrak{T}^-}(A^-) = T_1^- \cup T_2^- \cup T_3^-$.

$\mathfrak{C}^{\flat}_{\mathfrak{T}^+}(A^-) = \emptyset$, $\mathfrak{C}^{\sharp}_{\mathfrak{T}^+}(A^-) = T_3^+$,
$\mathfrak{C}^{\flat}_{\mathfrak{T}^-}(A^+) = \mathfrak{C}^{\sharp}_{\mathfrak{T}^-}(A^+) = \emptyset$.

Fig. 1. Positive and negative tools (see Section 2). Some sample weak lower and upper approximations in $\mathfrak{T}^+$ and $\mathfrak{T}^-$-approximation spaces (see Section 3).

[1] If $\mathfrak{A} \subseteq 2^U$, we define $\bigcup \mathfrak{A} = \{x \mid \exists A \in \mathfrak{A}(x \in A)\}$, and $\bigcap \mathfrak{A} = \{x \mid \forall A \in \mathfrak{A}(x \in A)\}$. If $\mathfrak{A}$ is an empty family of sets, $\bigcup \emptyset = \emptyset$ and $\bigcap \emptyset = U$.

3 Fundamentals of Partial Approximation of Sets

First, let us see the basic concepts and properties of *rough set theory* [5], [12]. The pair (U, ε), where U is a finite universe of discourse and ε is an equivalence relation on U, is called *Pawlak's approximation space.* The elements of the partition generated by ε are called *ε-elementary sets.* $X \in 2^U$ is *ε-definable*, if it is a union of ε-elementary sets, otherwise X is *ε-undefinable.* By definition, the empty set is considered to be an ε-definable set.

The lower and upper ε-approximations of $X \in 2^U$ can be defined in two equivalent forms, in a *point-free* manner—based on the ε-elementary sets, and in a *point-wise* manner—based on the elements [17].

The *lower ε-approximation* of X is[2]

$$\underline{\varepsilon}(X) = \bigcup\{Y \mid Y \in U/\varepsilon, Y \subseteq X\} = \{x \in U \mid [x]_\varepsilon \subseteq X\},$$

and the *upper ε-approximation* of X is

$$\overline{\varepsilon}(X) = \bigcup\{Y \mid Y \in U/\varepsilon, Y \cap X \neq \emptyset\} = \{x \in U \mid [x]_\varepsilon \cap X \neq \emptyset\}.$$

The set $B_\varepsilon(X) = \overline{\varepsilon}(X) \setminus \underline{\varepsilon}(X)$ is the *ε-boundary* of X. X is *ε-crisp*, if $B_\varepsilon(X) = \emptyset$, otherwise X is *ε-rough.*

Let $\mathfrak{D}_{U/\varepsilon}$ denote the family of ε-definable subsets of U. Clearly, $\underline{\varepsilon}(X), \overline{\varepsilon}(X) \in \mathfrak{D}_{U/\varepsilon}$, the maps $\underline{\varepsilon}, \overline{\varepsilon} : 2^U \to \mathfrak{D}_{U/\varepsilon}$ are total, and many-to-one. It can easily be seen ([12], Proposition 2.2, points 1, 9, 10) that the map $\underline{\varepsilon}$ is contractive and $\overline{\varepsilon}$ is extensive, i.e., $\forall X \in 2^U(\underline{\varepsilon}(X) \subseteq X \subseteq \overline{\varepsilon}(X))$.

Turning to the partial approximation of sets [1], from now on let U be any nonempty set.

Definition 1. *Let $\mathfrak{B} \subseteq 2^U$ be a nonempty family of nonempty subsets of U called the* base system. *Its elements are the $\mathfrak{B}$*-sets.

A family of sets $\mathfrak{D} \subseteq 2^U$ is $\mathfrak{B}$-definable *if its elements are $\mathfrak{B}$-sets, otherwise $\mathfrak{D}$ is $\mathfrak{B}$*-undefinable. *A nonempty subset $X \in 2^U$ is $\mathfrak{B}$*-definable *if there exists a $\mathfrak{B}$-definable family of sets $\mathfrak{D}$ such that $X = \bigcup \mathfrak{D}$, otherwise X is $\mathfrak{B}$*-undefinable.

The empty set is considered to be a $\mathfrak{B}$-definable set.

Definition 2. *Let $\mathfrak{B} \subseteq 2^U$ be a base system and X be any subset of U.*

The weak lower $\mathfrak{B}$-approximation of X *is*

$$\mathfrak{C}^\flat_\mathfrak{B}(X) = \bigcup\{Y \mid Y \in \mathfrak{B}, Y \subseteq X\},$$

and the weak upper $\mathfrak{B}$-approximation of X *is*

$$\mathfrak{C}^\sharp_\mathfrak{B}(X) = \bigcup\{Y \mid Y \in \mathfrak{B}, Y \cap X \neq \emptyset\}.$$

---

[2] Let ϵ be an arbitrary binary relation on U. We define $[x]_\epsilon = \{y \in U \mid (x, y) \in \epsilon\}$, and U/ϵ denote the family of $[x]_\epsilon$.

Let $\mathfrak{D}_{\mathfrak{B}}$ denote the family of $\mathfrak{B}$-definable sets of U. The maps $\mathfrak{C}^{\flat}_{\mathfrak{B}} : 2^U \to \mathfrak{D}_{\mathfrak{B}}$, $\mathfrak{C}^{\sharp}_{\mathfrak{B}} : 2^U \to \mathfrak{D}_{\mathfrak{B}}$ are point-free generalizations of lower and upper ε-approximations. Obviously, $\mathfrak{C}^{\flat}_{\mathfrak{B}}(X), \mathfrak{C}^{\sharp}_{\mathfrak{B}}(X) \in \mathfrak{D}_{\mathfrak{B}}$, and the maps $\mathfrak{C}^{\flat}_{\mathfrak{B}}$, $\mathfrak{C}^{\sharp}_{\mathfrak{B}}$ are total, onto, and many-to-one. Both of them are monotone.

Clearly, $\forall X \in 2^U(\mathfrak{C}^{\flat}_{\mathfrak{B}}(X) \subseteq \mathfrak{C}^{\sharp}_{\mathfrak{B}}(X))$. It can easily be seen ([1], Theorem 17) that $\mathfrak{C}^{\flat}_{\mathfrak{B}}$ is contractive, but $\mathfrak{C}^{\sharp}_{\mathfrak{B}}$ is extensive if and only if $\mathfrak{B}$ covers the universe.

Definition 3. *Let the fixed base system $\mathfrak{B} \subseteq 2^U$ and maps $\mathfrak{C}^{\flat}_{\mathfrak{B}}$ and $\mathfrak{C}^{\sharp}_{\mathfrak{B}}$ be given. The quadruple $(U, \mathfrak{B}, \mathfrak{C}^{\flat}_{\mathfrak{B}}, \mathfrak{C}^{\sharp}_{\mathfrak{B}})$ is called a* weak $\mathfrak{B}$-approximation space.

4 A General Tool-Based Approximation Framework

Let U be any nonempty set. Let $A^+, A^- \in 2^U$ be nonempty subsets of U such that $A^+ \cap A^- = \emptyset$. A^+ and A^- are called the *positive reference set* and *negative reference set*, respectively. In general, $A^+ \cap A^- = \emptyset$ is the only requirement for A^+ and A^-. Of course, additional relations between them may be supposed.

Furthermore, let and $\mathfrak{T}^+, \mathfrak{T}^- \subseteq 2^U$ be nonempty families of subsets of U such that $\bigcup \mathfrak{T}^+ \cap \bigcup \mathfrak{T}^- = \emptyset$. $\mathfrak{T}^+$ is called *positive* or $\mathfrak{T}^+$*-tools*, $\mathfrak{T}^-$ is called *negative* or $\mathfrak{T}^-$*-tools*. For each subset $T^+ \in \mathfrak{T}^+$ (resp., $T^- \in \mathfrak{T}^-$) it is *easy* to decide whether an element of U belongs to T^+ (resp., T^-) or not.

The sets in $\mathfrak{T}^+$ are not necessarily pairwise disjoint, so they are not in $\mathfrak{T}^-$. Neither $\bigcup \mathfrak{T}^+$ nor $\bigcup \mathfrak{T}^-$ covers U.

Note that, the adjectives *positive* and *negative* claim nothing else but that the sets A^+ (resp., $\mathfrak{T}^+$) and A^- (resp., $\mathfrak{T}^-$) are well separated.

The quadruples $(U, \mathfrak{T}^+, \mathfrak{C}^{\flat}_{\mathfrak{T}^+}, \mathfrak{C}^{\sharp}_{\mathfrak{T}^+})$ and $(U, \mathfrak{T}^-, \mathfrak{C}^{\flat}_{\mathfrak{T}^-}, \mathfrak{C}^{\sharp}_{\mathfrak{T}^-})$ form a weak $\mathfrak{T}^+$-approximation space and a weak $\mathfrak{T}^-$-approximation space, respectively.

Borrowing the terminology from the inductive logic programming [7], the mutual relationships between A^+ and A^- can be characterized by available $\mathfrak{T}^+$ and $\mathfrak{T}^-$-tools as follows. It is said that A^+ is

- $\mathfrak{T}^+$*-complete* if $A^+ \subseteq \mathfrak{C}^{\sharp}_{\mathfrak{T}^+}(A^+)$, $\mathfrak{T}^+$*-incomplete* otherwise;
- $\mathfrak{T}^-$*-consistent* if $\mathfrak{C}^{\sharp}_{\mathfrak{T}^-}(A^+) = \emptyset$, $\mathfrak{T}^-$*-inconsistent* otherwise.

According to previous name conventions, a positive reference set A^+ may be

- $\mathfrak{T}^+$*-complete* and $\mathfrak{T}^-$*-consistent*, $\mathfrak{T}^+$*-complete* and $\mathfrak{T}^-$*-inconsistent*;
- $\mathfrak{T}^+$*-incomplete* and $\mathfrak{T}^-$*-consistent*, $\mathfrak{T}^+$*-incomplete* and $\mathfrak{T}^-$*-inconsistent*.

Similar specifications can be defined to the negative reference set A^-. From a pure combinatorial point of view, there may be in sum $4 \cdot 4 = 16$ different situations. However, by the constraints $A^+ \cap A^- = \emptyset$ and $\bigcup \mathfrak{T}^+ \cap \bigcup \mathfrak{T}^- = \emptyset$, some of them are impossible. Owing to limitations of length of this paper, all possible different cases are not taken into account completely here.

The general framework can be used in three consecutive steps:

1. *Justifying reference sets* to reveal (in)consistencies and (in)complete regions in terms of partial approximation of sets based on $\mathfrak{T}^+$ and $\mathfrak{T}^-$.

2. *Rebuilding positive and negative tools* to resolve inconsistencies and eliminate incomplete regions, if necessary.
   - In the case of consistency, there is nothing to be done.
   - In the case of inconsistency, we have to *decide* within the context of the system if A^+ (the concerned negative tool) is reasonable or not, and/or whether A^- (the concerned positive tool) is reasonable or not.
   - In the case of completeness, we remove the covered positive and/or negative reference sets from the framework.
   - In the case of incompleteness, we may *decide* within the context of the system either to remove the uncovered subset from A^+ (resp., A^-) or to augment the positive (resp., negative) tools with new subsets whose elements are patterned upon one or more elements of the uncovered subset of A^+ (resp., A^-). These new subsets may contain any element of the universe, provided that they can easily be determined. For the new $\mathfrak{T}^+$-tools and/or $\mathfrak{T}^-$-tools, $\bigcup \mathfrak{T}^+ \cap \bigcup \mathfrak{T}^- = \emptyset$ should also be fulfilled.

   By the end of steps 1 and 2, we obtain rebuilt tools $\mathfrak{T}_r^+$ and $\mathfrak{T}_r^-$.
3. *Applying rebuilt tools* to justify any subset of the universe in terms of partial approximation of sets based on rebuilt positive and negative tools.

5 Applying General Framework to the Running Example

Examples depend on the actual observed real applications. However, there is no room here to elaborate such examples, so we show these in an abstract way.

Step 1. Justifying reference sets in terms of partial approximation of sets based on $\mathfrak{T}^+$ and $\mathfrak{T}^-$ (see Fig. 1):

- A^+ is $\mathfrak{T}^+$-complete ($A^+ \subseteq \mathfrak{C}^\sharp_{\mathfrak{T}^+}(A^+)$), $\mathfrak{T}^-$-consistent ($\mathfrak{C}^\sharp_{\mathfrak{T}^-}(A^+) = \emptyset$).
- A^- is $\mathfrak{T}^-$-incomplete ($A^- \not\subseteq \mathfrak{C}^\sharp_{\mathfrak{T}^-}(A^-)$), $\mathfrak{T}^+$-inconsistent ($\mathfrak{C}^\sharp_{\mathfrak{T}^+}(A^-) \neq \emptyset$).

Step 2. Rebuilding positive and negative tools (see Fig. 2):

- Since A^+ was $\mathfrak{T}^+$-complete, A^+ was removed from the framework.
- Since A^+ was $\mathfrak{T}^-$-consistent, there was nothing to be done.
- A^- was $\mathfrak{T}^-$-incomplete, we *decided* that we augmented negative tools with T_5^+, T_6^+ patterned upon elements of the uncovered subset of A^+.
- A^- was $\mathfrak{T}^+$-inconsistent, we *decided* that the positive tool T_3^+ was reasonable.

By the end of Steps 1 and 2, we obtained the rebuilt positive tools $\mathfrak{T}_r^+ = \mathfrak{T}^+$, and the rebuilt negative tools $\mathfrak{T}_r^- = \mathfrak{T}^- \cup \{T_5^-, T_6^-\}$.

Step 3. We apply the rebuilt tools to justify snapshots of the system as follows (sample snapshots S_1, S_2, S_3 are depicted in Fig. 2):

- $\mathfrak{C}^\flat_{\mathfrak{T}_r^+}(S_2)$ is the set of all acceptable execution traces which *certainly* belong to S_2 with respect to the prescriptions of the security policy.

As $\mathfrak{C}^{\flat}_{\mathfrak{T}_r^+}(S_2) = T_4^+$, T_4^+ is the only prescription which in full belongs to S_2.

- $\mathfrak{C}^{\sharp}_{\mathfrak{T}_r^+}(S_2)$ is the set of all acceptable execution traces which *possibly* belong to S_2 with respect to the prescriptions of the security policy.
  As $\mathfrak{C}^{\sharp}_{\mathfrak{T}_r^+}(S_2) = T_4^+ \cup T_5^+$, only the prescriptions T_4^+, T_5^+ of the security policy can be associated with S_2.
- $\bigcup \mathfrak{T}_r^+ \setminus \mathfrak{C}^{\sharp}_{\mathfrak{T}_r^+}(S_2)$ is the set of all acceptable execution traces which *certainly does not* belong to S_2 with respect to the prescriptions of the security policy.
  As $T_1^+ \subsetneqq \bigcup \mathfrak{T}_r^+ \setminus \mathfrak{C}^{\sharp}_{\mathfrak{T}_r^+}(S_2)$, T_1^+ is the only prescription which does not belong to S_2.
- Acceptable execution traces in $\mathfrak{C}^{\sharp}_{\mathfrak{T}_r^+}(S_2) \setminus \mathfrak{C}^{\flat}_{\mathfrak{T}_r^+}(S_2) = T_5^+ \setminus T_4^+$ are *abstained* because they cannot be uniquely classified either as belonging to S_2 or as not belonging to S_2 with respect to the prescriptions of the security policy. Notice that the set $\mathfrak{C}^{\sharp}_{\mathfrak{T}_r^+}(S_2) \setminus \mathfrak{C}^{\flat}_{\mathfrak{T}_r^+}(S_2)$ can also be approximated. For instance, $\mathfrak{C}^{\sharp}_{\mathfrak{T}_r^+}(\mathfrak{C}^{\sharp}_{\mathfrak{T}_r^+}(S_2) \setminus \mathfrak{C}^{\flat}_{\mathfrak{T}_r^+}(S_2)) = T_2^+ \cup T_3^+$.
- As $S_2 \not\subseteq \mathfrak{C}^{\sharp}_{\mathfrak{T}_r^+}(S_2)$, the execution traces in $S_2 \setminus \mathfrak{C}^{\sharp}_{\mathfrak{T}_r^+}(S_2)$ are anomalous. Since $\mathfrak{C}^{\sharp}_{\mathfrak{T}_r^-}(S_2) = T_3^- \cup T_6^-$, a subset of the anomalous traces are unexpected with respect to the proscriptions T_3^- and T_6^- of the security policy.

Very similar statements can be made to the snapshot S_1. Note that the snapshot S_3 cannot be justified with all the available tools at all.

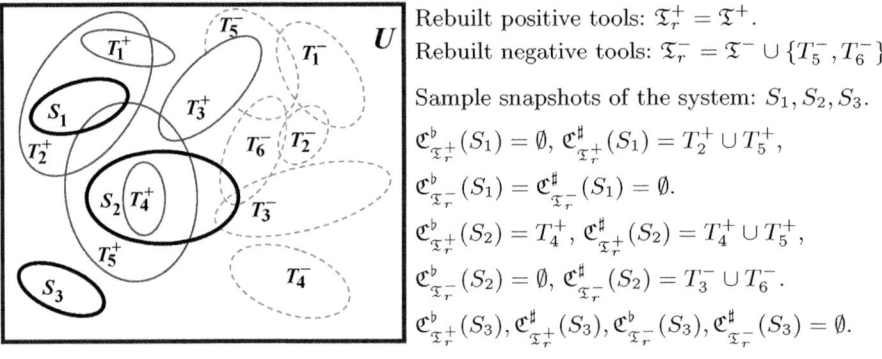

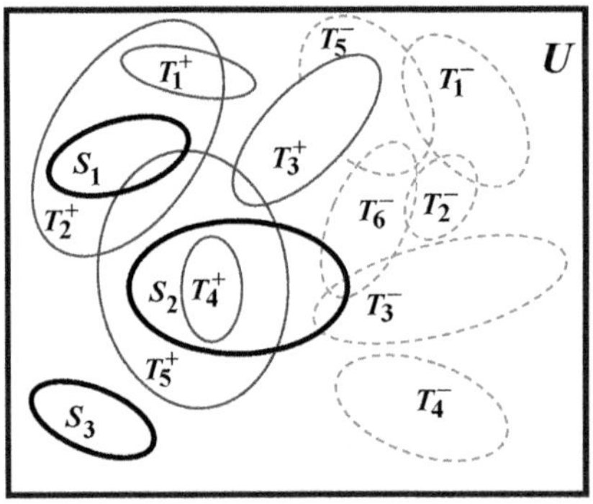

Rebuilt positive tools: $\mathfrak{T}_r^+ = \mathfrak{T}^+$.
Rebuilt negative tools: $\mathfrak{T}_r^- = \mathfrak{T}^- \cup \{T_5^-, T_6^-\}$.

Sample snapshots of the system: S_1, S_2, S_3.
$\mathfrak{C}^{\flat}_{\mathfrak{T}_r^+}(S_1) = \emptyset$, $\mathfrak{C}^{\sharp}_{\mathfrak{T}_r^+}(S_1) = T_2^+ \cup T_5^+$,
$\mathfrak{C}^{\flat}_{\mathfrak{T}_r^-}(S_1) = \mathfrak{C}^{\sharp}_{\mathfrak{T}_r^-}(S_1) = \emptyset$.
$\mathfrak{C}^{\flat}_{\mathfrak{T}_r^+}(S_2) = T_4^+$, $\mathfrak{C}^{\sharp}_{\mathfrak{T}_r^+}(S_2) = T_4^+ \cup T_5^+$,
$\mathfrak{C}^{\flat}_{\mathfrak{T}_r^-}(S_2) = \emptyset$, $\mathfrak{C}^{\sharp}_{\mathfrak{T}_r^-}(S_2) = T_3^- \cup T_6^-$.
$\mathfrak{C}^{\flat}_{\mathfrak{T}_r^+}(S_3), \mathfrak{C}^{\sharp}_{\mathfrak{T}_r^+}(S_3), \mathfrak{C}^{\flat}_{\mathfrak{T}_r^-}(S_3), \mathfrak{C}^{\sharp}_{\mathfrak{T}_r^-}(S_3) = \emptyset$.

Fig. 2. Rebuilt positive and negative tools. Sample snapshots of the system.

6 Conclusion

In our paper, we have presented a framework in which many questions concerning approximation problems without exact knowledge can be represented *uniformly*. In addition to this, positive features and their substantially negative features of observed objects can *simultaneously* be approximated.

References

1. Csajbók, Z.: Partial approximative set theory: A generalization of the rough set theory. In: Martin, T., Muda, A.K., Abraham, A., Prade, H., Laurent, A., Laurent, D., Sans, V. (eds.) Proceedings of SoCPaR 2010, Cergy Pontoise / Paris, France, December 7-10, pp. 51–56. IEEE, Los Alamitos (2010)
2. Csajbók, Z.: A security model for personal information security management based on partial approximative set theory. In: Ganzha, M., Paprzycki, M. (eds.) Proceedings of IMCSIT 2010, Wisła, Poland, October 18-20, vol. 5, pp. 839–845. PTI – IEEE Computer Society Press, Katowice, Poland – Los Alamitos, USA (2010)
3. Csajbók, Z.: Simultaneous anomaly and misuse intrusion detections based on partial approximative set theory. In: Cotronis, Y., Danelutto, M., Papadopoulos, G.A. (eds.) Proceedings of PDP 2011, Ayia Napa, Cyprus, February 9-11, pp. 651–655. IEEE Computer Society Press, Los Alamitos (2011)
4. Denning, D.E.: An intrusion-detection model. IEEE Transactions on Software Engineering SE 13(2), 222–232 (1987)
5. Järvinen, J.: Lattice theory for rough sets. In: Peters, J.F., Skowron, A., Düntsch, I., Grzymała-Busse, J.W., Orłowska, E., Polkowski, L. (eds.) Transactions on Rough Sets VI. LNCS, vol. 4374, pp. 400–498. Springer, Heidelberg (2007)
6. Keefe, R.: Theories of Vagueness. Cambridge Studies in Philosophy. Cambridge University Press, Cambridge (2000)
7. Lavrač, N., Džeroski, S.: Inductive Logic Programming: Techniques and Applications. Ellis Horwood, New York (1994)
8. Marek, V.W., Truszczyński, M.: Approximation schemes in logic and artificial intelligence. In: Peters, J.F., Skowron, A., Rybiński, H. (eds.) Transactions on Rough Sets IX. LNCS, vol. 5390, pp. 135–144. Springer, Heidelberg (2008)
9. Mihálydeák, T.: On tarskian models of general type-theoretical languages. In: Drossos, C., Peppas, P., Tsinakis, C. (eds.) Proceedings of the 7th Panhellenic Logic Symposium, pp. 127–131. Patras University Press, Patras (2009)
10. Odifreddi, P.: Classical Recursion Theory. In: The Theory of Functions and Sets of Natural Numbers. Studies in Logic and the Foundations of Mathematics, vol. 125. Elsevier, Amsterdam (1989)
11. Pawlak, Z.: Rough sets. International Journal of Information and Computer Science 11(5), 341–356 (1982)
12. Pawlak, Z.: Rough Sets: Theoretical Aspects of Reasoning about Data. Kluwer Academic Publishers, Dordrecht (1991)
13. Pawlak, Z., Polkowski, L., Skowron, A.: Rough sets: An approach to vagueness. In: Rivero, L.C., Doorn, J., Ferraggine, V. (eds.) Encyclopedia of Database Technologies and Applications, pp. 575–580. Idea Group Inc., Hershey (2005)
14. Pawlak, Z., Skowron, A.: Rudiments of rough sets. Information Sciences 177(1), 3–27 (2007)
15. Russell, B.: Vagueness. Australasian Journal of Philosophy and Psychology 1, 84–92 (1923)
16. Skowron, A.: Vague concepts: A rough-set approach. In: Beats, B.D., Caluwe, R.D., Tré, G.D., Fodor, J., Kacprzyk, J., Zadrożny, S. (eds.) Proceedings of EUROFUSE 2004, Warszawa, Poland, September 22-25, pp. 480–493. Akademicka Oficyna Wydawnicza EXIT, Warszawa (2004)
17. Yao, Y.Y.: On generalizing rough set theory. In: Wang, G., Liu, Q., Yao, Y., Skowron, A. (eds.) RSFDGrC 2003. LNCS (LNAI), vol. 2639, pp. 44–51. Springer, Heidelberg (2003)

An Improved Variable Precision Model of Dominance-Based Rough Set Approach

Weibin Deng[1,2], Guoyin Wang[2], and Feng Hu[1,2]

[1] School of Information Science and Technology
Southwest Jiaotong University, Chengdu, 610031, China
d_w_b@163.com, hufeng@cqupt.edu.cn
[2] Institute of Computer Science and Technology
Chongqing University of Posts and Telecommunications, Chongqing, 400065, China
wanggy@ieee.org

Abstract. The classification, ranking and sorting performance of Dominance-based rough set approach (DRSA) will be affected by the inconsistencies of the decision tables. Two relaxation models (VC-DRSA and VP-DRSA) have been proposed by Greco and Inuiguchi respectively to relax the strict dominance principle. But these relaxation methods are not always suitable for treating inconsistencies. Especially, some objects which should be included in lower-approximations are excluded. After analyzing the inadequacies of the two models, an improved variable precision model, which is called ISVP-DRSA, based on inclusion degree and supported degree is proposed in this paper. The basic concepts are defined and the properties are discussed. Furthermore, the lower approximations of ISVP-DRSA are the union of those of VC-DRSA and VP-DRSA, and the upper approximations are the intersection of those of the two models. Then more objects will be included in lower approximations and the quality of approximation classification is not poor than the above two models. Finally, the efficiency of ISVP-DRSA is illustrated by an example.

Keywords: Rough set, dominance based rough set approach, variable precision, inclusion degree, supported degree.

1 Introduction

In order to deal with information systems with preference-ordered attributes, Greco, Matarazzo and Słowiński [1,2] presented a generalization of the rough set approach, Dominance-based Rough Set Approach (DRSA). Similarly to the original rough sets, DRSA is based on approximations of partitions of the objects into pre-defined categories. However differently to the original model, the categories are ordered from the best to the worst and the approximations are constructed using a dominance relation instead of an indiscernibility relation. DRSA can be not only used to classify, but also to choice, rank, and sort [2]. The variable-consistency dominance-based rough set approach (VC-DRSA) has

S.O. Kuznetsov et al. (Eds.): RSFDGrC 2011, LNAI 6743, pp. 60–67, 2011.

been proposed in [3] to overcome the shortcomings of DRSA for treating inconsistent information systems. But as analyzing in [4], it does not always work well when the decision table including outliers. Inuiguchi and Yoshioka introduced a variable precision dominance-based rough set approach (VP-DRSA) [4] which is based on supported degree to treat errors and missing condition attributes. But it has been found that the marginal objects of classes will not be included in the lower approximations when the decision table including outliers. Therefore, few objects will support the weak decision rules, i.e. supported by few objects from approximations. In order to overcome the shortcomings of VC-DRSA and VP-DRSA, an improved variable precision model of DRSA is proposed, which is based on inclusion degree and supported degree. The lower and upper approximations are defined respectively and the properties are discussed. The lower approximations are not less than VC-DRSA and VP-DRSA, while the upper approximations are not larger than those of the two models. Therefore, the quality of approximation classification is not poor than the above two models.

The paper is organized as follows. In Section 2, some basic concepts of VC-DRSA and VP-DRSA are reviewed. In Section 3, the improved variable precision model of dominance-based rough set approach is introduced and an example is given to illustrate how the improved model overcomes the shortcomings of VC-DRSA and VP-DRSA. The paper is summarized in Section 4.

2 VC-DRSA and VP-DRSA

A decision table is defined as 4-tuple $S = (U, R, V, f)$, where U is a finite set of objects and $R = C \cup d$ is a finite set of attributes, C is the condition attribute set and d is the decision attribute set. With every attribute $a \in R$, set of its values V_a is associated. Each attribute a determines a function $f_a : U \to V_a$. In multicriteria classification, condition attributes are criteria. The notion of criterion involves a preference order in its domain. E.g. in Table 1, we have $U = \{S_1, S_2, \ldots, S_{17}\}, C = \{\text{Mathematics } (Math), \text{Literature } (Lit)\}$, d=Passing Status (PS) and V={Utterly Bad (UB), Very Bad (VB), Bad (B), Medium (M), Good (G), Very Good (VG), Excellent (E), Yes (Y), No (N)}.

See [1,2] for the basic concepts of DRSA.

2.1 VC-DRSA

Greco et al. proposed the variable consistency model of dominance-based rough set approach (DRSA) [3] to treat the inconsistency problem. The main concepts and properties are as follows.

For any $P \subseteq C$, we say that $x \in U$ belongs to $Cl_t^{\geq}$ at consistency level $l \in (0, 1]$, if $x \in Cl_t^{\geq}$ and at least $l \times 100\%$ of all objects y$\in U$ dominating x with respect to P also belong to $Cl_t^{\geq}$, shown more formally as follows:

$$\beta = \frac{|D_P^+(x) \cap Cl_t^{\geq}|}{|D_P^+(x)|} \geq l. \tag{1}$$

Table 1. A decision table of student evaluation

Student	Mathematics	Literature	Passing Status
S_1	Excellent	Very Good	Yes
S_2	Excellent	Medium	Yes
S_3	Very Good	Very Good	No
S_4	Very Good	Good	Yes
S_5	Very Good	Bad	Yes
S_6	Very Good	Utterly Bad	No
S_7	Good	Excellent	Yes
S_8	Medium	Excellent	Yes
S_9	Medium	Bad	Yes
S_{10}	Bad	Medium	No
S_{11}	Bad	Very Bad	No
S_{12}	Very Bad	Medium	No
S_{13}	Very Bad	Very Bad	No
S_{14}	Very Bad	Utterly Bad	No
S_{15}	Utterly Bad	Bad	No
S_{16}	Utterly Bad	Very Bad	Yes
S_{17}	Utterly Bad	Utterly Bad	No

The level l controls the degree of consistency between objects qualified as belonging to $Cl_t^{\geq}$. In other words, if l <1, then (1-l)×100% of all objects $y \in U$ dominating x with respect to P may not belong to $Cl_t^{\geq}$ and thus contradict the inclusion of x in $Cl_t^{\geq}$. Analogously, for any $P \subseteq C$ we say that $x \in U$ belongs to $Cl_t^{\leq}$ at consistency level $l \in (0, 1]$, if $x \in Cl_t^{\leq}$ and at least l×100% of all the objects $y \in U$dominated by x with respect to P also belong to $Cl_t^{\leq}$.

The definition of P-lower approximations of the unions of classes $Cl_t^{\geq}$ and $Cl_t^{\leq}$ at some consistency level l, respectively,

$$\begin{aligned} \underline{P}^l_{VC}(Cl_t^{\geq}) &= \{x \in Cl_t^{\geq} : \tfrac{|D_P^+(x) \cap Cl_t^{\geq}|}{|D_P^+(x)|} \geq l\}, \\ \underline{P}^l_{VC}(Cl_t^{\leq}) &= \{x \in Cl_t^{\leq} : \tfrac{|D_P^-(x) \cap Cl_t^{\leq}|}{|D_P^-(x)|} \geq l\}. \end{aligned} \quad (2)$$

The P-upper approximations of $Cl_t^{\geq}$ and $Cl_t^{\leq}$ can be defined by

$$\overline{P}^l_{VC}(Cl_t^{\geq}) = U - \underline{P}^l_{VC}(Cl_{t-1}^{\leq}), \overline{P}^l_{VC}(Cl_t^{\leq}) = U - \underline{P}^l_{VC}(Cl_{t+1}^{\geq}). \quad (3)$$

For more details on VC-DRSA see [3]. Above definitions hold also for DRSA, however, they can be simplified since in that case consistency level l equals 1.

2.2 VP-DRSA

Consider a decision table given in Table 1. The data of S_3 and S_{16} can be regarded as two outliers. These inconsistencies may be caused by some errors in recording or observation.

From Table 1, we have $Cl_Y^{\geq} = \{S_1, S_2, S_4, S_5, S_7, S_8, S_9, S_{16}\}$ and $Cl_N^{\leq} = \{S_3, S_6, S_{10}, S_{11}, S_{12}, S_{13}, S_{14}, S_{15}, S_{17}\}$. Let $P = C$, l=0.8. We have $\underline{P}_{VC}^{0.8}(Cl_Y^{\geq})= \{\mathbf{S_1}, \mathbf{S_2}, S_5, \mathbf{S_7}, \mathbf{S_8}, S_9\}$, $\beta(S_5)$=0.8, and $\beta(S_9)$=0.875, as well as $\underline{P}_{VC}^{0.8}(Cl_N^{\leq}) = \{\mathbf{S_6}, S_{10}, S_{11}, S_{12}, \mathbf{S_{14}}, \mathbf{S_{17}}\}$. The objects present in the lower approximations obtained for consistency degree β=1 are in bold. Note $S_4 \in Cl_Y^{\geq}$ ($\beta(S_4)$=0.667) which takes better values in both $Math$ and Lit than S_5 and S_9 is not included in $\underline{P}_{VC}^{0.8}(Cl_Y^{\geq})$, but S_5 and S_9 are included in $\underline{P}_{VC}^{0.8}(Cl_Y^{\geq})$. Analogously, S_{13} and $S_{15} \in Cl_Y^{\geq}$ which takes worse values in both $Math$ and Lit than S_{10}, S_{11} and S_{12} are not included in $\underline{P}_{VC}^{0.8}(Cl_N^{\leq})$, but S_{10}, S_{11} and S_{12} are.

In order to treat the inconsistencies caused by errors in recording, measurement, observation, and so on, a variable-precision dominance-based rough set approach (VP-DRSA) [4] was proposed by Inuiguchi, et al. They defined the precision of $x \in Cl_t^{\geq}$ denoted by

$$\beta = \frac{|D_P^-(x) \cap Cl_t^{\geq}|}{|D_P^-(x) \cap Cl_t^{\geq}| + |D_P^+(x) \cap Cl_{t-1}^{\leq}|}. \tag{4}$$

Then, given a precision level $l \in (0, 1]$, corresponding to the P-lower approximation of $Cl_t^{\geq}$, a P-lower approximation of $Cl_t^{\geq}$ with respect to $P \subseteq C$ is defined as a set of objects $x \in Cl_t^{\geq}$ whose degrees of precision are not less than l. In order to correspond to the definitions of VC-DRSA, the lower approximations of VP-DRSA are minor adjusted by us, shown as follows:

$$\underline{P}_{VP}^{l}(Cl_t^{\geq}) = \{x \in Cl_t^{\geq} : \frac{|D_P^-(x) \cap Cl_t^{\geq}|}{|D_P^-(x) \cap Cl_t^{\geq}| + |D_P^+(x) \cap Cl_{t-1}^{\leq}|} \geq l\}, \tag{5}$$

$$\underline{P}_{VP}^{l}(Cl_t^{\leq}) = \{x \in Cl_t^{\leq} : \frac{|D_P^+(x) \cap Cl_t^{\leq}|}{|D_P^+(x) \cap Cl_t^{\leq}| + |D_P^-(x) \cap Cl_{t+1}^{\geq}|} \geq l\}. \tag{6}$$

By using the duality, corresponding to P-lower approximations, P-upper approximations of $Cl_t^{\geq}$ and $Cl_t^{\leq}$ with respect to $P \subseteq C$ can be defined by

$$\overline{P}_{VP}^{l}(Cl_t^{\geq}) = \{x \in U : \frac{|D_P^-(x) \cap Cl_t^{\geq}|}{|D_P^-(x) \cap Cl_t^{\geq}| + |D_P^+(x) \cap Cl_{t-1}^{\leq}|} \geq 1 - l\}, \tag{7}$$

$$\overline{P}_{VP}^{l}(Cl_t^{\leq}) = \{x \in U : \frac{|D_P^+(x) \cap Cl_t^{\leq}|}{|D_P^+(x) \cap Cl_t^{\leq}| + |D_P^-(x) \cap Cl_{t+1}^{\geq}|} \geq 1 - l\}. \tag{8}$$

2.3 Inadequacies of VP-DRSA

From Table 1, for $P = C$, l=0.8, we obtain $\underline{P}_{VP}^{0.8}(Cl_Y^{\geq}) = \{\mathbf{S_1}, \mathbf{S_2}, S_4, \mathbf{S_7}, \mathbf{S_8}\}$, $\beta(S_4)$=0.8, $\underline{P}_{VP}^{0.8}(Cl_N^{\leq}) = \{\mathbf{S_6}, S_{13}, \mathbf{S_{14}}, S_{15}, \mathbf{S_{17}}\}$, $\beta(S_{13})$=0.83, $\beta(S_{15})$=0.8.

The following decision rules with consistency degree can be induced:

- if $(f(x, Math) \succeq E)$ then $x \in Cl_Y^{\geq}$ $[\alpha = 1]$,

- if $(f(x, Lit) \succeq E)$ then $x \in Cl_Y^{\geq}$ $[\alpha = 1]$,
- if $(f(x, Math) \succeq M)$ and $(f(x, Lit) \succeq M)$ then $x \in Cl_Y^{\geq}$ $[\alpha = 0.83]$,
- if $(f(x, Lit) \preceq UB)$ then $x \in Cl_N^{\leq}$ $[\alpha = 1]$,
- if $(f(x, Math) \preceq VB)$ and $(f(x, Lit) \preceq B)$ then $x \in Cl_N^{\leq}$ $[\alpha = 0.8]$.

Note S_5 and S_9 should be included in $\underline{P}_{VP}^{0.8}(Cl_Y^{\geq})$. But S_5 and S_9 are marginal objects of upward union of class $Cl_Y^{\geq}$, i.e. $|D_P^-(S_5) \cap Cl_Y^{\geq}|$ and $|D_P^-(S_9) \cap Cl_Y^{\geq}|$ are very small. Owing to the existing of S_3, the supported degrees of S_5 and S_9 endorses included in $Cl_Y^{\geq}$ are smaller than the precision level l. Then they are excluded from the P-lower approximation of $Cl_Y^{\geq}$. Analogously, S_{10} and S_{11} are not included in $\underline{P}_{VP}^{0.8}(Cl_N^{\leq})$ owing to the outlier of S_{16}. Therefore, S_5, S_9, S_{10} and S_{11} can not be classified by the decision rules which are induced from the lower approximations. The classification accuracy is not satisfied.

3 Improved VP-DRSA Based on Inclusion Degree and Supported Degree (ISVP-DRSA)

3.1 Definitions and Properties of ISVP-DRSA

In order to overcome the shortcomings of VC-DRSA and VP-DRSA, an improved VP-DRSA which is based on inclusion degree and supported degree (simply, ISVP-DRSA) is proposed.

The consistency degree of an object $x \in U$ belongs to $Cl_t^{\geq}$ with respect to $P \subseteq C$ is defined by

$$\beta = \max\left(\frac{|D_P^+(x) \cap Cl_t^{\geq}|}{|D_P^+(x)|}, \frac{|D_P^-(x) \cap Cl_t^{\geq}|}{|D_P^-(x) \cap Cl_t^{\geq}| + |D_P^+(x) \cap Cl_{t-1}^{\leq}|}\right). \tag{9}$$

The degree of inclusion is a particular case of inclusion in a degree (rough inclusion) in rough mereology. Inclusion degree in rough set data analysis was proposed in [5]. $|D_P^+(x) \cap Cl_t^{\geq}|/|D_P^+(x)|$ denotes the inclusion degree of the P-dominating set of x belongs to $Cl_t^{\geq}$. But this parameter cannot treat the inconsistency caused by outliers, as shown in Section 2.2. The concept of classification supported degree in DRSA was proposed in [4]. But only according to the supported degree to determine whether an object should be included in the lower approximations, the marginal objects will be excluded from the lower approximations. Hence, so as to enlarge lower approximations and to treat the inconsistencies caused by hesitation and errors, an object should be included in the lower approximations if its inclusion degree or the supported degree is not less than l.

Then, given a consistency level $l \in (0, 1]$, a P-lower approximation of $Cl_t^{\geq}$ with respect to $P \subseteq C$ is defined as a set of objects $x \in Cl_t^{\geq}$ whose inclusion degrees or supported degrees are not less than l, i.e.

$$\underline{P}^l_{ISVP}(Cl^{\geq}_t) = \{x \in Cl^{\geq}_t : \max(\frac{|D^+_P(x)\cap Cl^{\geq}_t|}{|D^+_P(x)|}, \frac{|D^-_P(x)\cap Cl^{\geq}_t|}{|D^-_P(x)\cap Cl^{\geq}_t|+|D^+_P(x)\cap Cl^{\leq}_{t-1}|}) \geq l\}. \quad (10)$$

A P-lower approximation of $Cl^{\leq}_t$ with respect to $P \subseteq C$ is defined by

$$\underline{P}^l_{ISVP}(Cl^{\leq}_t) = \{x \in Cl^{\leq}_t : \max(\frac{|D^-_P(x)\cap Cl^{\leq}_t|}{|D^-_P(x)|}, \frac{|D^+_P(x)\cap Cl^{\leq}_t|}{|D^+_P(x)\cap Cl^{\leq}_t|+|D^-_P(x)\cap Cl^{\geq}_{t+1}|}) \geq l\}. \quad (11)$$

By using the duality, P-upper approximations of $Cl^{\geq}_t$ and $Cl^{\leq}_t$ with respect to $P \subseteq C$ can be defined by

$$\overline{P}^l_{ISVP}(Cl^{\geq}_t) = U - \underline{P}^l_{ISVP}(Cl^{\leq}_{t-1}),\ \overline{P}^l_{ISVP}(Cl^{\leq}_t) = U - \underline{P}^l_{ISVP}(Cl^{\geq}_{t+1}). \quad (12)$$

$\overline{P}^l_{ISVP}(Cl^{\geq}_t)$ can be interpreted as the set of all the objects belong to $Cl^{\geq}_t$, possibly ambiguous at consistency level l. $\overline{P}^l_{ISVP}(Cl^{\leq}_t)$ can be interpreted as the set of all the objects belonging to $Cl^{\leq}_t$, possibly ambiguous at consistency level l. The P-boundaries (P-doubtful regions) of $Cl^{\geq}_t$ and $Cl^{\leq}_t$ are defined by

$$\begin{aligned} Bn^l_{P_{ISVP}}(Cl^{\geq}_t) &= \overline{P}^l_{ISVP}(Cl^{\geq}_t) - \underline{P}^l_{ISVP}(Cl^{\geq}_t), \\ Bn^l_{P_{ISVP}}(Cl^{\leq}_t) &= \overline{P}^l_{ISVP}(Cl^{\leq}_t) - \underline{P}^l_{ISVP}(Cl^{\leq}_t). \end{aligned} \quad (13)$$

According to definitions of ISVP-DRSA, the following properties hold:

$$\overline{P}^l_{ISVP}(Cl^{\geq}_t) = Cl^{\geq}_t \cup \{x \in Cl^{\leq}_{t-1} : \min(\frac{|D^-_P(x)\cap Cl^{\geq}_t|}{|D^-_P(x)|}, \frac{|D^-_P(x)\cap Cl^{\geq}_t|}{|D^-_P(x)\cap Cl^{\geq}_t|+|D^+_P(x)\cap Cl^{\leq}_{t-1}|}) \geq 1-l\}. \quad (14)$$

$$\overline{P}^l_{ISVP}(Cl^{\leq}_t) = Cl^{\leq}_t \cup \{x \in Cl^{\geq}_{t+1} : \min(\frac{|D^+_P(x)\cap Cl^{\leq}_t|}{|D^+_P(x)|}, \frac{|D^+_P(x)\cap Cl^{\leq}_t|}{|D^+_P(x)\cap Cl^{\leq}_t|+|D^-_P(x)\cap Cl^{\geq}_{t+1}|}) \geq 1-l\}. \quad (15)$$

$$\begin{aligned} \underline{P}^l_{ISVP}(Cl^{\geq}_t) \subseteq Cl^{\geq}_t \subseteq \overline{P}^l_{ISVP}(Cl^{\geq}_t), \\ \underline{P}^l_{ISVP}(Cl^{\leq}_t) \subseteq Cl^{\leq}_t \subseteq \overline{P}^l_{ISVP}(Cl^{\leq}_t). \end{aligned} \quad (16)$$

$$Bn^l_{P_{ISVP}}(Cl^{\geq}_t) = Bn^l_{P_{ISVP}}(Cl^{\leq}_{t-1}),\ for\ t = 2, \cdots, n. \quad (17)$$

$$Bn^l_{P_{ISVP}}(Cl^{\leq}_t) = Bn^l_{P_{ISVP}}(Cl^{\geq}_{t+1}),\ for\ t = 1, \cdots, n-1. \quad (18)$$

Furthermore, according to the definitions of VC-DRSA, VP-DRSA and ISVP-DRSA, the following theorems can be proved:

$$\begin{aligned} \underline{P}^l_{ISVP}(Cl^{\geq}_t) &= \underline{P}^l_{VC}(Cl^{\geq}_t) \cup \underline{P}^l_{VP}(Cl^{\geq}_t), \\ \underline{P}^l_{ISVP}(Cl^{\leq}_t) &= \underline{P}^l_{VC}(Cl^{\leq}_t) \cup \underline{P}^l_{VP}(Cl^{\leq}_t). \end{aligned} \quad (19)$$

$$\overline{P}^l_{ISVP}(Cl_t^{\geq}) = \overline{P}^l_{VC}(Cl_t^{\geq}) \cap \overline{P}^l_{VP}(Cl_t^{\geq}),$$
$$\overline{P}^l_{ISVP}(Cl_t^{\leq}) = \overline{P}^l_{VC}(Cl_t^{\leq}) \cap \overline{P}^l_{VP}(Cl_t^{\leq}). \quad (20)$$

$$\gamma^l_{P_{ISVP}}(Cl) \geq \max(\gamma^l_{P_{VC}}(Cl), \gamma^l_{P_{VP}}(Cl)). \quad (21)$$

3.2 Illustrative Example

An illustrative example is presented in this section. The data is shown in Table 1. As discussed in Section 2.2 and Section 2.3, it is inappropriate to use VC-DRDA and VP-DRSA for decision tables including outliers. It will be seen how Table 1 is analyzed appropriately by the proposed ISVP-DRSA.

Let $P = C$, l=0.8, then we obtain (the objects present in the lower approximations obtained for consistency degree β=1 are in bold): $\underline{P}^{0.8}_{ISVP}(Cl_Y^{\geq}) = \{\mathbf{S_1}, \mathbf{S_2}, S_4, S_5, \mathbf{S_7}, \mathbf{S_8}, S_9\}$, $\beta(S_4)$=0.8, $\beta(S_5)$=0.8, $\beta(S_9)$= 0.875. $\underline{P}^{0.8}_{ISVP}(Cl_N^{\leq})$= $\{\mathbf{S_6}, S_{10}, S_{11}, S_{12}, S_{13}, \mathbf{S_{14}}, S_{15}, \mathbf{S_{17}}\}$, $\beta(S_{10})$=0.875, $\beta(S_{11})$ =0.8, $\beta(S_{12})$=0.83, $\beta(S_{13})$=0.83, $\beta(S_{15})$=0.8.

The following decision rules with precision degrees can be induced:

- if $(f(x, Math) \succeq E)$ then $x \in Cl_Y^{\geq}$ $[\alpha = 1]$,
- if $(f(x, Lit) \succeq E)$ then $x \in Cl_Y^{\geq}$ $[\alpha = 1]$,
- if $(f(x, Math) \succeq M)$ and $(f(x, Lit) \succeq B)$ then $x \in Cl_Y^{\geq}$ $[\alpha = 0.875]$,
- if $(f(x, Lit) \preceq UB)$ then $x \in Cl_N^{\leq}$ $[\alpha = 1]$,
- if $(f(x, Math) \preceq B)$ then $x \in Cl_N^{\leq}$ $[\alpha = 0.875]$.

From the $P-$lower approximations, it has been found that except the outliers S_3 and S_{16}, all the objects are included in the lower approximations of ISVP-DRSA. Then the classification accuracy is improved.

Besides, P-upper approximations, $P-$boundaries and the quality of approximation classification can be computed as follows: $\overline{P}^{0.8}_{ISVP}(Cl_Y^{\geq}) = \{S_1, S_2, S_3, S_4, S_5, S_7, S_8, S_9, S_{16}\}$, $\overline{P}^{0.8}_{ISVP}(Cl_N^{\leq}) = \{S_3, S_6, S_{10}, S_{11}, S_{12}, S_{13}, S_{14}, S_{15}, S_{16}, S_{17}\}$. $Bn^{0.8}_{P_{ISVP}}(Cl_Y^{\geq}) = Bn^{0.8}_{P_{ISVP}}(Cl_N^{\leq}) = \{S_3, S_{16}\}$. Also, $\gamma^{0.8}_{P_{ISVP}}(Cl) = 0.882$, as well as $\gamma^{0.8}_{P_{VC}}(Cl) = 0.705$ and $\gamma^{0.8}_{P_{VP}}(Cl)$=0.588.

The P-lower approximations, P-boundaries and the approximation classification qualities of the three models are shown as Table 2.

The above results illustrate the properties (14)-(18) and theorems (19)-(21).

Table 2. The results of the three models

	VC-DRSA	VP-DRSA	ISVP-DRSA
$\underline{P}^{0.8}(Cl_Y^{\geq})$	$S_1, S_2, S_5, S_7, S_8, S_9$	S_1, S_2, S_4, S_7, S_8	$S_1, S_2, S_4, S_5, S_7, S_8, S_9$
$\underline{P}^{0.8}(Cl_N^{\leq})$	$S_6, S_{10}, S_{11}, S_{12}, S_{14}, S_{17}$	$S_6, S_{13}, S_{14}, S_{15}, S_{17}$	$S_6, S_{10}, S_{11}, S_{12}, S_{13}, S_{14}, S_{15}, S_{17}$
Boundaries	$S_3, S_4, S_{13}, S_{15}, S_{16}$	$S_3, S_5, S_9, S_{10}, S_{11}, S_{12}, S_{16}$	S_3, S_{16}
$\gamma_P^{0.8}(Cl)$	0.706	0.588	0.882

4 Conclusions

After analyzing the inadequacies of VC-DRSA and VP-DRSA proposed by Greco and Inuiguchi respectively in this paper, an improved variable precision DRSA (ISVP-DRSA) which is based on inclusion degree and supported degree is proposed. The properties of lower approximations and upper approximations are investigated. And it is found that more objects will be included in the lower approximations in ISVP-DRSA. Then more useful decision rules can be induced and higher quality of approximation classification can be obtained. In the future, we will concentrate on developing useful classification algorithms based on ISVP-DRSA.

Acknowledgments. This work has been partially supported by National Natural Science Foundation of P. R. China under grant 61073146, Inter-governmental Science and Technology Cooperation of P. R. China and Poland under Grant No. 34-5 and Chongqing Key Lab of Computer Network and Communication Technology under grant CY-CNCL-2010-05.

References

1. Greco, S., Matarazzo, B., Słowiński, R.: Rough approximation of a preference relation by dominance relations. Eur. J. of Operational Research 117(1), 63–68 (1999)
2. Greco, S., Matarazzo, B., Słowiński, R.: Rough sets theory for multicriteria decision analysis. Eur. J. of Operational Research 129(1), 1–47 (2001)
3. Greco, S., Matarazzo, B., Słowiński, R., Stefanowski, J.: Variable Consistency Model of Dominance-Based Rough Sets Approach. In: Ziarko, W., Yao, Y.Y. (eds.) RSCTC 2000. LNCS (LNAI), vol. 2005, pp. 170–181. Springer, Heidelberg (2001)
4. Inuiguchi, M., Yoshioka, Y.: Variable-Precision Dominance-Based Rough Set approach. In: Greco, S., Hata, Y., Hirano, S., Inuiguchi, M., Miyamoto, S., Nguyen, H.S., Słowiński, R. (eds.) RSCTC 2006. LNCS (LNAI), vol. 4259, pp. 203–212. Springer, Heidelberg (2006)
5. Xu, Z.B., Liang, J.Y., Dang, C.Y., Chin, K.S.: Inclusion degree: a perspective on measures for rough set data analysis. Information Sciences 141(3), 229–238 (2002)
6. Ziarko, W.: Variable Precision Rough Set Model. J. of Computer and System Science 46(1), 39–59 (1993)
7. Pawlak, Z.: Rough Sets. Int. J. of Information & Computer Science 11(5), 341–356 (1982)
8. Wang, G.Y., Wang, Y.: 3DM: Domain-oriented Data-driven Data Mining. Fundamenta Informaticae 90(4), 395–426 (2009)

Rough Numbers and Rough Regression

Marcin Michalak

Silesian University of Technology, ul. Akademicka 16, 44-100 Gliwice, Poland
Marcin.Michalak@polsl.pl

Abstract. In this article a new model of regression is defined. On the basis of the rough sets theory a notion of rough number is defined. Typical real numbers calculations do not keep the additional information like the uncertainty or the error of input data. Rough numbers remove this limitation. It causes that rough numbers seem to be interested as the basis of the new way of regression: rough regression.

Keywords: rough sets, rough numbers, machine learning, nonparametric regression, rough regression.

For over forty years people have been trying to specify the inexactitude with notions like the fuzzyness [7] or roughness [3]. This human need of problems simplification by generalization is associated with the price of loosing the accuracy. In this article a new approach of inexact data representation is described that extends the limited Pawlak definition of a rough number [4]. With the wider definition of the approximation space this definition makes it possible to define basic arithmetical operation on rough numbers based on the whole $\mathbb{R}$ set. It finds the application in the nonparametric regression.

1 Classical and Extended Models of Rough Numbers

Rough numbers [4] are based on the several following notions. For the set $\mathbb{R}^+$ a sequence of nonnegative reals $S=(x_i)_{i=1}^n$ such that $x_i<x_j, i<j$ is called the categorization. The *approximation space* is the ordered pair $A=(\mathbb{R}^+, S)$. Every categorization S of $\mathbb{R}^+$ induces partition $\pi(s)$ on $\mathbb{R}^+$ defined as $\pi(S) = \{0, (0, x_1), x_1, \ldots, x_i, (x_i, x_{i+1}), \ldots\}$ where (x_i, x_{i+1}) denotes the open interval. $S(x)$ denotes the block of partition $\pi(S) : x \in \pi(S)$. Let $x \in (x_i, x_{i+i})$. The closed interval $\overline{S}(x) = \langle x_i, x_{i+1}\rangle$ is called the closure $S(x)$. $Q(x)$ is the closed interval $\langle 0, x\rangle$. For a given approximation space $A(\mathbb{R}^+, S)$ for every $Q(x)$ its lower and upper approximation may be defined, denoted as $S_*(Q(x))$ and $S^*(Q(x))$ respectively. $S_*(Q(x))=\{y \in \mathbb{R}^+ \colon S(y) \subseteq Q(x)\}$; $S^*(Q(x)) = \{y \in \mathbb{R}^+ \colon S(y) \cap Q(x) \neq \emptyset\}$ As the *S-lower* and *S-upper* approximation of the nonnegative real number x the following values are considered: $S_*(x) = Sup\{y \in S : y \leq x\}$; $S^*(x) = Inf\{y \in S : y \geq x\}$. All notions above lead to the definition of an approximation of a real number x on the basis of a set of real numbers S: $S(x) = (S_*(x), S^*(x))$. The number is exact in $A = (\mathbb{R}^+, S)$ iff $S_*(x) = S^*(x)$, otherwise it is inexact (rough) in A. Every inexact number may be represented as a pair of exact numbers

S.O. Kuznetsov et al. (Eds.): RSFDGrC 2011, LNAI 6743, pp. 68–71, 2011.

or as the interval. Now, let $\mathcal{S}$ is the categorization of $\mathbb{R}$: $\mathcal{S} \subseteq \mathbb{R}$. An approximation space $A = (\mathbb{R}, \mathcal{S})$ will be denoted as $\mathcal{A}_{\mathcal{S}}$. For a given $\mathcal{S} = (x_i)_{i=1}^n$ the generalized projection Π is defined: $\Pi(S) = \{(-\infty, x_1), x_1, (x_1, x_2), x_2, \ldots\}$. Notions $\mathcal{S}(x)$ and $\overline{\mathcal{S}}(x)$ remain as above. As $\mathcal{A}_{\mathcal{S}}$ is based on $\mathbb{R}$, $Q(x)$ should be defined as $Q(x) = \langle -\infty, x\rangle$ what causes that definitions of its $\mathcal{S}$-*lower* and $\mathcal{S}$-*upper* approximations remain unchanged. Also definitions of real number x remain the same as in [4]. This extension of the Pawlak definitions leads us to the notion of rough real number. Every $\mathcal{A}_{\mathcal{S}}$ gives $\overline{\overline{\mathcal{S}}}$ exact numbers and $\overline{\overline{\mathcal{S}}} + 1$ rough numbers. As the exact number x is also the rough number $\langle x, x\rangle$ we say that every $\mathcal{A}_{\mathcal{S}}$ introduces $2\overline{\overline{\mathcal{S}}} + 1$ rough real numbers. Two of them has infinite limits: $(-\infty, x_1\rangle$, $\langle x_{max}, \infty)$ and will be called nonfinite rough numbers. All other will be called finite rough numbers. Exact rough number will be denoted as x : $\langle x, x\rangle = \underline{x}$; inexact finite rough number x : $\langle \mathcal{S}_*(x), \mathcal{S}^*(x)\rangle$; inexact nonfinite rough number x : $\langle -\infty, x_1\rangle$ and $\langle x_{max}, \infty\rangle$; set of finite rough numbers: $\mathcal{R}$; set of rough numbers (finite and nonfinite): $\mathcal{R}_\infty$.

2 Rough Calculations and Rough Regression

Let ρ_1, ρ_2 are finite rough numbers: $\rho_1 = \langle a_1, a_2\rangle$, $\rho_2 = \langle b_1, b_2\rangle$. Consider: **addition**: $\rho_1 + \rho_2 = \langle a_1 + b_1, a_2 + b_2\rangle$; **multiplication**: $\rho_1 \cdot \rho_2 = \langle a_1 \cdot b_1, a_2 \cdot b_2\rangle$; **mirroring**: ρ_1 and ρ_2 will be called mirrored iff $a_1 = b_2$ and $a_2 = b_1$. The function μ is called a mirror function: $\mu : \mathcal{R} \to \mathcal{R}; \mu(\langle a, b\rangle) = \langle b, a\rangle$. It is easy to prove that $\rho + \mu(\rho) = \underline{0}$. Addition and multiplication are commutative, have neutral and opposite elements, and satisfy distributivity of the multiplication over the addition. For every kernel function $K : \mathbb{R} \to \mathbb{R}$ a rough kernel function $\mathcal{K} : \mathcal{R} \to \mathbb{R}$ may be defined as follows: $\mathcal{K}(\langle \rho_1, \rho_2\rangle) = (\rho_2 - \rho_1)^{-1} \int_{\rho_1}^{\rho_2} K(x)dx$. Rough kernel function should take real values for real arguments. It is easy to prove that $\mathcal{K}(\langle \rho, \rho\rangle) = K(\rho)$ and that $\mathcal{K}(\rho) = \mathcal{K}(\mu(\rho))$. One of the most popular kernel estimators of the regression function is Nadraya–Watson estimator [2][6]: $\widetilde{f}(x) = [\sum_{i=1}^n y_i K((x - x_i)/h)]/[\sum_{i=1}^n K((x - x_i)/h)]$ where (x_i, y_i) are some known training points $(x_i, y_i \in \mathbb{R})$, K is a kernel function and h is the smoothing parameter. In this section the extension of NW estimator for rough numbers is presented. The process of roughing the real data is as follows: for the set of real valued observations $U = \{(x_i, y_i)\}_{i=1}^n$ the set of $2n - 1$ rough valued observations is defined in the following way: $\mathcal{R}(U) = \{(\langle x_1, x_1\rangle, \langle y_1, y_1\rangle), (\langle x_1, x_2\rangle, \langle y_1, y_2\rangle), (\langle x_2, x_2\rangle, \langle y_2, y_2\rangle), \ldots, (\langle x_n, x_n\rangle, \langle y_n, y_n\rangle)\}$. As the set X is a categorization of $\mathbb{R}$ we may define the $\mathcal{X}-lower$ and $\mathcal{X}-upper$ function regression. If $\widetilde{f}(\rho)$ is the rough kernel regressor then the rough function $\mathcal{F}_X$ may be defined as follows $\mathcal{F}_X = \{(u, \widetilde{f}(u)) : u \in \mathcal{R}(X)\}$ or: $\mathcal{F}_X = \{(\langle u_{i1}, u_{i2}\rangle, \langle \widetilde{f}(u_i)_1, \widetilde{f}(u_i)_2\rangle) : u_i = \langle u_{i1}, u_{i2}\rangle \in \mathcal{R}(X), \widetilde{f}(u_i) = \langle \widetilde{f}(u_i)_1, \widetilde{f}(u_i)_2\rangle, i = 1, \ldots, 2\overline{\overline{X}} - 1\}$. From the definition above we may define the $\mathcal{X}$-lower and $\mathcal{X}$-upper rough estimators of the function:

$$\underline{\mathcal{F}}_X = \{(\langle u_{i1}, u_{i2}\rangle, \langle \widetilde{f}(u_i)_1, \widetilde{f}(u_i)_1\rangle) : (\langle u_{i1}, u_{i2}\rangle, \langle \widetilde{f}(u_i)_1, \widetilde{f}(u_i)_2\rangle) \in \mathcal{F}_X\}$$
$$\overline{\mathcal{F}}_X = \{(\langle u_{i1}, u_{i2}\rangle, \langle \widetilde{f}(u_i)_2, \widetilde{f}(u_i)_2\rangle) : (\langle u_{i1}, u_{i2}\rangle, \langle \widetilde{f}(u_i)_1, \widetilde{f}(u_i)_2\rangle) \in \mathcal{F}_X\}$$

3 Experiments and Results

As the kernel function the Epanechnikov kernel was used: $K(x) = 0.75(1 - x^2), x \in \langle -1, 1\rangle$. The smoothing parameter was calculated with the following formula [5]: $h = 1,06 \min(\widetilde{\sigma}, \widetilde{R}/1.34)n^{-1/5}$. For the illustration of rough regression four noised data sets were used [1]: $D_1 : y = x + 2\exp(-16x^2)$; $D_2 : y = \sin 2x + 2\exp(-16x^2)$; $D_3 : y = 0.3\exp(-4(x+1)^2) + 0.7\exp(-16(x-1)^2)$; $D_4 : y = 0.4x + 1$. The distribution of the noise value was normal with the standard deviation as follows: $0.4, 0.3, 0.1, 0.15$. The domain of every data set was the closed interval $[-2, 2]$. Each data set contained 101 pairs of objects ($x_1 = -2, x_2 = -1.96, \ldots, x_{101} = 2$) what caused that for every set the same value of smoothing parameter was calculated ($h_{FG1} = h_{FG2} = h_{FG3} = h_{FG4} = 1.055$). After the roughing step all data set contained 201 pairs of rough numbers.

For the analysis of the regression error a popular $RMSE$ measure was used: $RMSE = (n^{-1}\sum_{i=1}^{n}(y - \widetilde{y})^2)^{0.5}$. It is easy to prove that this definition may be extended for rough regression error. On the left side of the Fig. 1 the partial result of rough regression of the first data set is shown (rough regression for the interval $[-0.4, 0.18]$). The solid black line represents the original analytical dependence. Blue points represent data with noise and black ones mean the value of the standard Nadraya–Watson kernel estimator. Finally, two red characteristics illustrate the $\mathcal{X}$-lower and the $\mathcal{X}$-upper regression. Real and rough regression error for all data sets are presented in table 1.

Comparison of real and rough regression, as the comparison of real and rough $RMSE$, is shown on the right side of the Fig. 1. Let us assume that $err = y - \widetilde{y}$ is the real error and $rerr = y - \widetilde{y}_{\mathcal{R}}$ is the analogical rough error. Rough regression

Table 1. Regression errors for D_1 and D_2 datasets

Data set	$RMSE$	rough $RMSE$	Data set	$RMSE$	rough $RMSE$
D_1	0.4370	$\langle 0.4384, 0.4374\rangle$	D_3	0.1179	$\langle 0.1182, 0.1183\rangle$
D_2	0.3790	$\langle 0.3809, 0.3801\rangle$	D_4	0.1426	$\langle 0.1430, 0.1427\rangle$

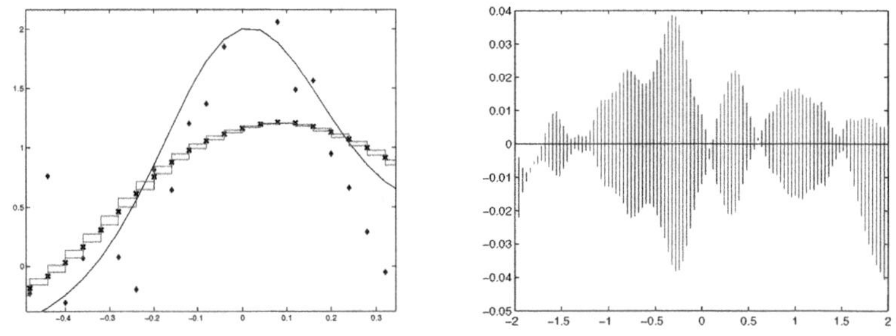

Fig. 1. Left: $\mathcal{X}$-upper and $\mathcal{X}$-lower regression for the slice of the FG1 set (red lines), noised data (blue diamonds), real kernel regression (black x-es). Right: rough $RMSE$ error with regards to real $RMSE$.

will give better results than real regression if $err - rerr > 0$. On the Fig. 1 we may see that on the ends of the interval rough regression gives worse results.

4 Conclusions and Further Works

This short paper tries to redefine existing and define not existing notions which become the basis of the rough regression. These notions lead to the definition of the rough number that is the ordered pair of real values. In this case the nonparametric Nadaraya–Watson kernel regressor was shown as its application.

From the interpretative point of view the rough estimator gives more information that real estimator: for the rough argument it gives the rough output that is thinner if the local data dependence is more stable (there is no local trend, it's rather oscillating) or wider otherwise. For a rough argument of the estimator the output rough value describes the range of possible values taken by the dependence for the pointed values. For both arguments (real and rough numbers) the result may be interpreted that the estimator of the value is the number between the $\mathcal{X}$-lower and the $\mathcal{X}$-upper function regression.

Acknowledgements. This work was supported by the European Community from the European Social Fund. Special thanks to Beata Sikora for some mathematical advice.

References

1. Fan, J., Gijbels, I.: Variable bandwith and local linear regression smoothers. The Annals of Statistics 20(4), 2008–2036 (1992)
2. Nadaraya, E.A.: On estimating regression. Theory of Probability and Its Applications 9(1), 141–142 (1964)
3. Pawlak, Z.: Rough Sets: Theoretical Aspects of Reasoning About Data. Kluwer Academic Publishers, Dordrecht (1991)
4. Pawlak, Z.: Rough sets, rough relations and rough functions. Fundamenta Informaticae 27(2-3), 103–108 (1996)
5. Silverman, B.: Density Estimation for Statistics and Data Analysis. Chapman and Hall, Boca Raton (1986)
6. Watson, G.S.: Smooth regression analysis. Sankhya - The Indian Journal of Statistics 26(4), 359–372 (1964)
7. Zadeh, L.A.: Fuzzy sets. Information and Control 8(3), 338–353 (1965)

Covering Numbers in Covering-Based Rough Sets

Shiping Wang[1], Fan Min[2], and William Zhu[2,*]

[1] School of Mathematical Sciences,
University of Electronic Science and Technology of China, Chengdu, 611731, China
[2] Lab of Granular Computing,
Zhangzhou Normal University, Zhangzhou, 363000, China

Abstract. Rough set theory provides a systematic way for rule extraction, attribute reduction and knowledge classification in information systems. Some measurements are important in rough sets. For example, information entropy, knowledge dependency are useful in attribute reduction algorithms. This paper proposes the concepts of the lower and upper covering numbers to establish measurements in covering-based rough sets which are generalizations of rough sets. With covering numbers, we establish a distance structure, two semilattices and a lattice for covering-based rough sets. The new concepts are helpful in studying covering-based rough sets from topological and algebraical viewpoints.

Keywords: Rough set, Covering, Measurement, Quantitative, Distance, Covering number, Lattice.

1 Introduction

As a technique for data mining, rough set theory [1] has been used for rule extraction, knowledge classification and so on. In order to characterize rough sets, a number of measurements have been proposed, such as information entropy [2,3] and knowledge dependency [4,5]. Based on these measurements, rough sets can be studied quantitatively, and efficient algorithms for attribute reduction [6] can be designed.

Covering-based rough set theory [7,8] is a generalization of rough set theory. However, those measurements in rough sets can not be used in covering-based rough sets. Therefore, there is much need to construct some measurements in covering-based rough sets. Furthermore, the structure of covering-based rough sets [9,10,11] have been a hotspot of study. It includes topological structures [12,13] and algebraical structures [14,15,16,17].

This paper proposes a measurement to study covering-based rough sets quantitatively. The concepts of the upper covering number and the lower covering number are defined to measure a set in covering-based rough sets. With covering numbers, we establish a distance, a lattice and two semilattice structures in covering-based rough sets. These structures are helpful for characterizing covering-based rough sets quantitatively. Specifically, the distance is used to

* Corresponding author. williamfengzhu@gmail.com

S.O. Kuznetsov et al. (Eds.): RSFDGrC 2011, LNAI 6743, pp. 72–78, 2011.

reveal the relationship between two subsets of a domain from a quantitative point of view.

The rest of this paper is arranged as follows. Section 2 reviews some fundamental concepts. Section 3 proposes two concepts of the upper and lower covering numbers, and studies their properties. With covering numbers, a distance structure, a lattice and two semilattices are established. Section 4 concludes this paper.

2 Basic Definitions

This section presents some fundamental concepts to be used in this paper. The classical rough set theory is based on equivalence relations. An equivalence relation corresponds to a partition, while a covering is an extension of a partition.

Definition 1. *(Covering [7]) Let U be a domain of discourse, $\mathbf{C}$ a family of subsets of U. If none of subsets in $\mathbf{C}$ is empty and $\bigcup \mathbf{C} = U$, $\mathbf{C}$ is called a covering of U.*

Distance is a fundamental concept for computer science and mathematics. The most commonly used distance is the Euclid distance. Posets and lattices are also basic concepts in computer science.

Definition 2. *(Distance [18]) Let X be a nonempty set, $d : X \times X \longrightarrow R$ a real function. d is called a distance on X, if $\forall x, y, z \in X$, the following three conditions hold:*
(1) $d(x, y) \geq 0$, and $d(x, y) = 0$ if and only if $x = y$;
(2) $d(x, y) = d(y, x)$;
(3) $d(x, y) \leq d(x, z) + d(z, y)$.

Definition 3. *(Poset [18]) A relation R on a set P is called a partial order if R is reflective, antisymmetric, and transitive. If R is a partial order on P, then (P, R) is called a poset.*

Definition 4. *(Semilattice [18]) An upper-semilattice is a poset (P, R) in which every subset $\{a, b\}$ has a least upper bound $a \vee b$. A lower-semilattice is a poset (P, R) in which every subset $\{a, b\}$ has a greatest lower bound $a \vee b$. The upper-semilattice and the lower-semilattice are also called semilattices.*

Definition 5. *(Lattice [18]) A lattice is a poset (P, R) which is an upper-lattice and a lower-semilattice.*

Here $\vee$, $\wedge$ are binary operations, and $(P, \vee, \wedge)$ is an algebraic system induced by the lattice $(P, \leq)$. Sometimes, we also call $(P, \vee, \wedge)$ a lattice.

3 Covering Numbers

Various measurements have been proposed to characterize rough sets quantitatively [5,19]. Similarly, we establish some measurements to describe covering-based rough sets quantitatively.

3.1 Definitions and Properties of Covering Numbers

The upper covering number of a subset of a domain is the minimal number of some elements in a covering which can cover the subset. The lower covering number of a subset is the maximal number of some elements in a covering which can be included in the subset.

Definition 6. *(Covering numbers) Let* $\mathbf{C}$ *be a covering of* U*. For all* $X \subseteq U$*, we define*
$N^*(X|\mathbf{C}) = min\{|\mathbf{B}||(X \subseteq \bigcup \mathbf{B}) \wedge (\mathbf{B} \subseteq \mathbf{C})\}$,
$N_*(X|\mathbf{C}) = |\{K \in \mathbf{C}|K \subseteq X\}|$.
$N^*(X|\mathbf{C})$ *and* $N_*(X|\mathbf{C})$ *are called the upper and lower covering numbers of* X *with respect to* $\mathbf{C}$*. When there is no confusion,* $N^*(X|\mathbf{C})$ *is denoted simply by* $N^*(X)$*, and* $N_*(X|\mathbf{C})$ *by* $N_*(X)$*.*

Example 1. Let $U = \{a, b, c, d\}$, $K_1 = \{a, b\}$, $K_2 = \{a, c\}$, $K_3 = \{b, c\}$, $K_4 = \{d\}$, $\mathbf{C} = \{K_1, K_2, K_3, K_4\}$, $X = \{a, d\}$, $Y = \{a, b, c\}$. Then $\mathbf{B}_1 = \{K_1, K_4\}$, $\mathbf{B}_2 = \{K_2, K_4\}$, $\mathbf{B}_3 = \{K_1, K_2, K_4\}$, $\mathbf{B}_4 = \{K_1, K_3, K_4\}$, $\mathbf{B}_5 = \{K_2, K_3, K_4\}$, and $\mathbf{B}_6 = \{K_1, K_2, K_3, K_4\}$ are also coverings of X; in other words, $X \subseteq \bigcup \mathbf{B}_i$ for $i \in \{1, \cdots, 6\}$. So $N^*(X) = min\{|\mathbf{B}_i||1 \leq i \leq 6\} = 2$. $N_*(X) = |\{K \in \mathbf{C}|K \subseteq X\}| = |\{K_4\}| = 1$. Similarly, $N^*(Y) = 2$ and $N_*(Y) = 3$.

In particular, according to Definition 6, we know that $N^*(\emptyset) = 0$ since $\emptyset \subseteq \bigcup\{\emptyset\}$ and $\{\emptyset\} \subseteq \mathbf{C}$. The result makes the concept of the covering numbers more reasonable.

In the following, we study the properties of covering numbers in detail, such as monotonicity.

Proposition 1. *Let* $\mathbf{C}$ *be a covering of* U*.* $\forall X, Y \subseteq U$*, the following properties about the upper covering number hold:*
(1) If $X \subseteq Y$*, then* $N^*(X) \leq N^*(Y)$*;*
(2) $N^*(X \bigcup Y) \leq N^*(X) + N^*(Y)$*;*
(3) $N^*(X) = 0$ *if and only if* $X = \emptyset$*;*
(4) If $K \in \mathbf{C}$*, then* $N^*(K_1) = 1$*, for all* $K_1 \subseteq K$*, and* $K_1 \neq \emptyset$*.*

Proof. (1) Suppose $N^*(Y) = |\mathbf{B}_Y|$. If $\mathbf{B}_Y$ is not the only one, then we choose any one satisfying the condition. So $X \subseteq Y \subseteq \bigcup \mathbf{B}_Y$ and $\mathbf{B}_Y \subseteq \mathbf{C}$. Hence $N^*(X) = min\{|\mathbf{B}||(X \subseteq \bigcup \mathbf{B}) \wedge (\mathbf{B} \subseteq \mathbf{C})\} \leq |\mathbf{B}_Y| = N^*(Y)$.
(2) If $X = \emptyset$ or $Y = \emptyset$, the conclusion holds. More generally, suppose $X \neq \emptyset$ and $Y \neq \emptyset$. We assume $N^*(X) = |\mathbf{B}_X|$ and $N^*(Y) = |\mathbf{B}_Y|$, which imply $X \subseteq \bigcup \mathbf{B}_X$ and $Y \subseteq \bigcup \mathbf{B}_Y$. Hence $X \bigcup Y \subseteq (\bigcup \mathbf{B}_X) \bigcup (\bigcup \mathbf{B}_Y) = \bigcup (\mathbf{B}_X \bigcup \mathbf{B}_Y)$. On the other hand, $\mathbf{B}_X \bigcup \mathbf{B}_Y \subseteq \mathbf{C}$ since $\mathbf{B}_X \subseteq \mathbf{C}$ and $\mathbf{B}_Y \subseteq \mathbf{C}$. Therefore, $N^*(X \bigcup Y) \leq |\mathbf{B}_X \bigcup \mathbf{B}_Y| \leq |\mathbf{B}_X| + |\mathbf{B}_Y| = N^*(X) + N^*(Y)$.
(3) and (4) are straightforward.

As an illustration, we consider the upper covering numbers of an element in a covering and a set having only one element, respectively.

Corollary 1. *Let* $\mathbf{C}$ *be a covering of* U*. For all* $K \in \mathbf{C}$*,* $N^*(K) = 1$*.*

Corollary 2. *Let* $\mathbf{C}$ *be a covering of* U*. For all* $x \in U$, $N^*(\{x\}) = 1$.

Similarly, we list some fundamental properties about the lower covering number.

Proposition 2. *Let* $\mathbf{C}$ *be a covering of* U *and* $|\mathbf{C}| = n$*. The following properties about the lower covering number hold:* $\forall X, Y \subseteq U$,
(1) $0 \leq N_*(X) \leq n$;
(2) $N_*(U) = n$;
(3) If $X \subseteq Y$*, then* $N_*(X) \leq N_*(Y)$;
(4) $N_*(X) + N_*(Y) \leq N_*(X \bigcup Y) + N_*(X \bigcap Y)$;
(5) $N_*(X) + N_*(X^c) \leq n$*, where* X^c *is the complement of* X.

Based on the properties of the upper and lower covering numbers, the relationship between covering numbers is explored as follows.

Proposition 3. *Let* $\mathbf{C}$ *be a covering of* U *and* $|\mathbf{C}| = n$*. For all* $X \subseteq U$, $N^*(X) + N_*(X^c) \leq n$*, where* X^c *is the complement of* X.

Proof. $N^*(X) + N_*(X^c) \leq |\{K \in \mathbf{C} | K \bigcap X \neq \emptyset\}| + N_*(X^c) = n$.

The relationships between covering numbers and partitions should also be studied. A sufficient and necessary condition is obtained; that is, a covering of a domain is degenerated to a partition if and only if the upper covering number of any subset of the domain and the lower covering number of its complement equals to the number of elements in the covering.

Proposition 4. *Let* $\mathbf{C}$ *be a covering of* U *and* $|\mathbf{C}| = n$. $\mathbf{C}$ *is a partition of* U *if and only if* $N^*(X) + N_*(X^c) = n$ *for all* $X \subseteq U$.

Proof. ($\Longrightarrow$): For all $X \subseteq U$, if $\mathbf{C}$ is a partition of U, then $N^*(X) = |\{K \in \mathbf{C} | K \bigcap X \neq \emptyset\}|$. Hence $N^*(X) + N_*(X^c) = |\{K \in \mathbf{C} | K \bigcap X \neq \emptyset\}| + N_*(X^c) = n$.
($\Longleftarrow$): If $N^*(X) + N_*(X^c) = n$ for all $X \subseteq U$, then according to Corollary 2, $N^*(\{x\}) + N_*(\{x\}^c) = n$, which implies $N^*(\{x\}) = |\{K \in \mathbf{C} | x \in K\}| = 1$ for all $x \in U$. Hence $\mathbf{C}$ is a partition of U.

When a covering is degenerated to a partition, some new characteristics about covering numbers are presented. Particularly, the upper covering number of a subset of a domain is greater than the lower covering number.

Proposition 5. *If* $\mathbf{C}$ *is a partition of* U*, then* $N_*(X) \leq N^*(X)$ *for all* $X \subseteq U$.

3.2 Distance Established with Covering Numbers

The distance is a numerical description of how far apart objects are, and it is used in many fields as a basic topological structure. In this subsection, we establishes a distance structure in covering-based rough sets with covering numbers.

Definition 7. *Let* $\mathbf{C}$ *be a covering of* U*. For all* $X, Y \subseteq U$*, we define a binary operation* d,
$d : 2^U \times 2^U \longrightarrow N$,
$d(X, Y) = N^*(X - Y) + N^*(Y - X)$.

Proposition 6. *d is a distance on 2^U.*

Proof. Firstly, $\forall X, Y \subseteq U$, we know that $d(X,Y) \geq 0$. $d(X,Y) = 0$ if and only if $N^*(X-Y) = 0$ and $N^*(Y-X) = 0$. According to Proposition 1, we can get that $d(X,Y) = 0$ if and only if $X = Y$. Secondly, $d(Y,X) = N^*(Y-X) + N^*(X-Y) = N^*(X-Y) + N^*(Y-X) = d(X,Y)$. Thirdly, because $X - Y = X \bigcap Y^C = X \bigcap (Z \bigcup Z^C) \bigcap Y^C = (X \bigcap Z \bigcap Y^C) \bigcup (X \bigcap Z^C \bigcap Y^C) \subseteq (Z \bigcap Y^C) \bigcup (X \bigcap Z^C) = (Z-Y) \bigcup (X-Z)$ for all $X, Y, Z \subseteq U$ where Y^C, Z^C are the complements of Y, Z with respect to U, respectively. $N^*(X-Y) \leq N^*[(X-Z) \bigcup (Z-Y)] \leq N^*(X-Z) + N^*(Z-Y)$. This completes the proof.

3.3 Lattice Established with Covering Numbers

Lattices are important algebraical structures, and have a variety of applications in the real world. This subsection establishes a lattice structure and two semi-lattices in covering-based rough sets with covering numbers.

Definition 8. *Let $\mathbf{C}$ be a covering of U. For all $X, Y \subseteq U$, if $X \subseteq Y$ and $N^*(X) = N^*(Y)$, we call Y an upper-set of X, and X a lower-set of Y.*

The family of all upper-sets and the family of all lower-sets are semilattices.

Proposition 7. *Let $\mathbf{C}$ be a covering of U. For all $X \subseteq U$, we call τ_X, $\tau_X^{'}$ the family of all upper-sets, lower-sets of X, respectively, i.e.,*

$$\tau_X = \{Y \subseteq U | (X \subseteq Y) \wedge (N^*(X) = N^*(Y))\},$$
$$\tau_X^{'} = \{Y \subseteq U | (Y \subseteq X) \wedge (N^*(X) = N^*(Y))\}.$$

Then $(\tau_X, \bigcap)$, and $(\tau_X^{'}, \bigcup)$ are semilattices.

Proof. In fact, we only need to prove $Y_1 \bigcap Y_2 \in \tau$ for all $Y_1, Y_2 \in \tau$, and $Y_1 \bigcup Y_2 \in \tau^{'}$ for all $Y_1, Y_2 \in \tau^{'}$. For all $Y_1, Y_2 \in \tau$, $N^*(Y_1) = N^*(X)$, $X \subseteq Y_1$ and $N^*(Y_2) = N^*(X)$, $X \subseteq Y_2$. So $X \subseteq Y_1 \bigcap Y_2 \subseteq Y_1$. Thus $N^*(X) \leq N^*(Y_1 \bigcap Y_2) \leq N^*(Y_1) = N^*(X)$, that is, $N^*(Y_1 \bigcap Y_2) = N^*(X)$. Therefore, $Y_1 \bigcap Y_2 \in \tau$. Similarly, we can prove $Y_1 \bigcup Y_2 \in \tau^{'}$ for all $Y_1, Y_2 \in \tau^{'}$.

There is widespread interest in breaking large objects into smaller, more easily understood, pieces. A detached-set of a domain divides the domain and the covering into two disjoint parts.

Definition 9. *Let $\mathbf{C}$ be a covering of U and $|\mathbf{C}| = n$. For $X \subseteq U$, if $N_*(X) + N_*(X^c) = n$, we call X a detached-set of U with respect to $\mathbf{C}$.*

With the detached-set, a covering is divided into two smaller coverings of two smaller domains. Moreover, the concept of the detached-set leads to a lattice structure.

Proposition 8. *Let $\mathbf{C}$ be a covering of U and $|\mathbf{C}| = n$. ψ is denoted as the family of all detached-sets of U, i.e.,*

$$\psi = \{X \subseteq U | N_*(X) + N_*(X^c) = n\}.$$

Then $(\psi, \bigcup, \bigcap)$ is a lattice.

Proof. For all $X, Y \in \psi$, we have $N_*(X)+N_*(X^c) = n$ and $N_*(Y)+N_*(Y^c) = n$. So $2n = (N_*(X) + N_*(X^c)) + (N_*(Y) + N_*(Y^c)) \leq (N_*(X \bigcup Y) + N_*(X \bigcap Y)) + (N_*(X^c \bigcup Y^c)) + N_*(X^c \bigcap Y^c) = [N_*(X \bigcup Y) + N_*((X \bigcup Y)^c)] + [N_*(X \bigcap Y) + N_*((X \bigcap Y)^c)]$. It means that $N_*(X \bigcup Y) + N_*((X \bigcup Y)^c) = n$ and $N_*(X \bigcap Y) + N_*((X \bigcap Y)^c) = n$. Thus $X \bigcup Y \in \psi$ and $X \bigcap Y \in \psi$.

4 Conclusions

This paper proposes the concepts of covering numbers to characterize covering-based rough sets quantitatively. Covering-based rough sets can be finely depicted by covering numbers. Specifically, a sufficient and necessary condition on a partition is obtained with covering numbers. Moreover, with covering numbers, a distance, two semilattices and a lattice structures are established in covering-based rough sets. Therefore, this work is helpful in describing covering-based rough sets from topological and algebraical viewpoints.

Acknowledgements. This work is in part supported by National Nature Science Foundation of China under grant No.60873077/F020107.

References

1. Pawlak, Z.: Rough sets. International Journal of Computer and Information Sciences 11, 341–356 (1982)
2. Wang, G., Yu, H., Yang, D.: Decision table reduction based on conditional information entropy. Chinese Journal of Computers 2, 759–766 (2002)
3. Qian, Y., Liang, J., Pedrycz, W., Dang, C.: Positive approximation: An accelerator for attribute reduction in rough set theory. Artificial Intelligence 174, 597–618 (2010)
4. Banerjee, M., Mitra, S., Pal, S.: Rough fuzzy mlp: knowledge encoding and classification. IEEE Transactions on Neural Networks 9, 1203–1216 (1998)
5. Qian, Y., Liang, J., Yao, Y., Dang, C.: Mgrs: A multi-granulation rough set. Information Sciences 180, 949–970 (2010)
6. Min, F., Liu, Q.: A hierarchical model for test-cost-sensitive decision systems. Information Sciences 179, 2442–2452 (2009)
7. Zhu, W., Wang, F.: Reduction and axiomization of covering generalized rough sets. Information Sciences 152, 217–230 (2003)
8. Zhu, W.: Relationship between generalized rough sets based on binary relation and covering. Information Sciences 179, 210–225 (2009)
9. Zhu, W.: Generalized rough sets based on relations. Information Sciences 177, 4997–5011 (2007)
10. Zhu, W.: Relationship among basic concepts in covering-based rough sets. Information Sciences 17, 2478–2486 (2009)
11. Wang, S., Zhu, P., Zhu, W.: Structure of covering-based rough sets. International Journal of Mathematical and Computer Sciences 6, 147–150 (2010)
12. Zhu, W.: Topological approaches to covering rough sets. Information Sciences 177, 1499–1508 (2007)
13. Wang, S., Zhu, W., Zhu, P.: Poset approaches to covering-based rough sets. Rough Sets and Knowledge Technology, 25–29 (2010)

14. Liu, G., Zhu, W.: The algebraic structures of generalized rough set theory. Information Sciences 178, 4105–4113 (2008)
15. Yao, Y.: Constructive and algebraic methods of theory of rough sets. Information Sciences 109, 21–47 (1998)
16. Zhu, W., Wang, F.-Y.: Binary relation based rough sets. In: Wang, L., Jiao, L., Shi, G., Li, X., Liu, J. (eds.) FSKD 2006. LNCS (LNAI), vol. 4223, pp. 276–285. Springer, Heidelberg (2006)
17. Wang, S., Zhu, W., Zhu, P.: Abstract interdependency in rough sets. Journal of Nanjing University 46, 507–510 (2010) (in chinese)
18. Rajagopal, P., Masone, J.: Discrete mathematics for computer science. Saunders College, Canada (1992)
19. Min, F., Liu, Q., Fang, C.: Rough sets approach to symbolic value partition. International Jounal on Approximate Reasoning 49, 689–700 (2008)

On Coverings of Rough Transformation Semigroups

S.P. Tiwari and Shambhu Sharan

Department of Applied Mathematics
Indian School of Mines
Dhanbad-826004, India
{sptiwarimaths,shambhupuremaths}@gmail.com
http://www.ismdhanbad.ac.in

Abstract. The purpose of this paper is towards the algebraic study of rough finite state machines, i.e., to introduce the concept of homomorphisms between two rough finite state machines, to associate a rough transformation semigroup with a rough finite state machine and to introduce the concept of coverings of rough finite state machines as well as rough transformation semigroups.

Keywords: Rough finite state machine, Rough transformation semigroup, Homomorphism, Covering.

1 Introduction and Preliminaries

The concepts of homomorphism, transformation semigroup and covering play prominent role in the study of finite state machines [2]. Much later, Malik, Mordeson and Sen [3] introduced these ideas for fuzzy finite state machines and explored their algebraic properties (cf., [5] for more details). Inspired from [6], Basu [1] has recently introduced the concept of rough finite state automaton (a concept resembles to rough finite state machine) and tried to design a recognizer that accepts imprecise statements, while Tiwari, Srivastava and Sharan [7] studied the separated and connectedness properties of rough finite state automata. The purpose of this paper is to introduce the concepts of homomorphism, transformation semigroup and covering for rough finite state machines. In particular, we establish a congruence relation to associate a semigroup with given rough finite state machine, which lead us to associate a rough transformation semigroup to each rough finite state machine. Lastly, we introduce the notion of coverings for both rough finite state machines and rough transformation semigroups.

Now, we collect some concepts associated with rough set theory, which are useful in the next sections. We start from the following concept of approximation space.

Definition 1. [6] *Let X be a nonempty set and R be an equivalence relation on X. Then the pair (X, R) is called an* ***approximation space****.*

S.O. Kuznetsov et al. (Eds.): RSFDGrC 2011, LNAI 6743, pp. 79–86, 2011.

Definition 2. [6] *Let (X, R) be an approximation space and $[x]_R$ be the equivalence class of x under R. Then* ***lower approximation*** *and* ***upper approximation*** *of $A \subseteq X$ are, respectively, defined to be the sets*

$$\underline{A} = \{x \in X \mid [x]_R \subseteq A\}, and$$
$$\overline{A} = \{x \in X \mid [x]_R \cap A \neq \phi\}.$$

For an approximation space (X, R), $A \subseteq X$ is called a **definable set** if it is an union of equivalence classes under R and a pair (L, U) of definable sets is called a **rough set** in (X, R) if $L \subseteq U$, and if any equivalence class of x is a singleton set $\{x\}$ such that $\{x\} \in U$, then $\{x\} \in L$.

2 Rough Finite State Machines

In this section, we recall some concepts related to a rough finite state machine and introduce the concept of homomorphism between two rough finite state machines.

Definition 3. [1] *A* ***rough finite state machine*** *(or RFSM) is a 4-tuple $M = (Q, R, X, \delta)$, where Q is a nonempty finite set (the* **set of states** *of M), R is a given equivalence relation on Q, X is a nonempty finite set (the* **set of inputs***) and δ is a map $\delta : Q \times X \to D \times D$. D being the collection of all definable sets in the approximation space (Q, R), if $\delta(q, a) = (D_1, D_2)$, where $q \in Q$ and $a \in X$, then (D_1, D_2) is a rough set with $D_1 = \underline{A}$ and $D_2 = \overline{A}$, for some $A \subseteq Q$.*

If $\delta(q, a) = (D_1, D_2)$ with $D_1 = \underline{A}$ and $D_2 = \overline{A}$, then by the abuse of notation, we identify A as $\delta(q, a)$; thus $D_1 = \underline{\delta(q, a)}$ and $D_2 = \overline{\delta(q, a)}$ i.e., $\delta(q, a) = (\underline{\delta(q, a)}, \overline{\delta(q, a)})$.

Definition 4. [1] *For any equivalence class (block) B of R and $a \in X$, $Ba = (\underline{Ba}, \overline{Ba})$ is called* ***block transition****, where $\underline{Ba} = \bigcup\{\underline{qa} \mid q \in B\}$, $\overline{Ba} = \bigcup\{\overline{qa} \mid q \in B\}$ and for a definable set D, $\underline{Da} = \bigcup\{\underline{Ba} \mid B$ is a block of R and $B \subseteq D\}$, $\overline{Da} = \bigcup\{\overline{Ba} \mid B$ is a block of R and $B \subseteq D\}$.*

Let X^* be the set of all *words* on X (i.e., finite strings of elements of X, which form a monoid under concatenation of strings) including the empty word (which we shall denote by ϵ). Throughout, $\forall x \in X^*$, $|x|$ denotes the length of string x.

Definition 5. [7] *For any equivalence class (block) B of R and $x \in X^*$, $Bx = (\underline{Bx}, \overline{Bx})$, where $\underline{Bx} = \bigcup\{\underline{qx} \mid q \in B\}$, $\overline{Bx} = \bigcup\{\overline{qx} \mid q \in B\}$ and for a definable set D, $\underline{Dx} = \bigcup\{\underline{Bx} \mid B$ is a block of R and $B \subseteq D\}$, $\overline{Dx} = \bigcup\{\overline{Bx} \mid B$ is a block of R and $B \subseteq D\}$.*

Definition 6. [1] *Let $M = (Q, R, X, \delta)$ be a RFSM. Define $\delta^* : Q \times X^* \to D \times D$ as follows:*

1. *$\forall q \in Q$, $\delta^*(q, \epsilon_X) = ([q]_R, [q]_R)$, where $[q]_R$ is the equivalence class of q under R, and $\forall x \in X^* \backslash \{\epsilon_X\}, \delta^*(q, x) = \delta^*([q]_R, x)$, and*

2. *$\forall q \in Q$, $\forall x \in X^*$ and $\forall a \in X$, $\delta^*(q, xa) = (\underline{\delta^*(q, xa)}, \overline{\delta^*(q, xa)})$, where $\underline{\delta^*(q, xa)} = \underline{\delta(\underline{\delta^*(q, x)}, a)} = \bigcup\{\underline{Ba} \mid B$ is a block under R and $B \subseteq \underline{\delta^*(q, x)}\}$ and $\overline{\delta^*(q, xa)} = \overline{\delta(\overline{\delta^*(q, x)}, a)} = \bigcup\{\overline{Ba} \mid B$ is a block under R and $B \subseteq \overline{\delta^*(q, x)}\}$.*

In [1], it is shown that the transition function δ of a RFSM (Q, R, X, δ) can be recursively extended to a function $\delta^* : Q \times X^* \to D \times D$ as follows:

1. $\forall q \in Q$, $\delta^*(q, \epsilon_X) = ([q]_R, [q]_R)$, where $[q]_R$ is the equivalence class of q under R, and $\forall x \in X^* \backslash \{\epsilon_X\}, \delta^*(q, x) = \delta^*([q]_R, x)$, and
2. $\forall q \in Q$ and $\forall x, y \in X^*$, $\delta^*(q, xy) = (\underline{\delta^*(q, xy)}, \overline{\delta^*(q, xy)})$, where $\underline{\delta^*(q, xy)} = \underline{\delta^*(\underline{\delta^*(q, x)}, y)} = \bigcup\{\underline{By} \mid B$ is a block under R and $B \subseteq \underline{\delta^*(q, x)}\}$ and $\overline{\delta^*(q, xy)} = \overline{\delta^*(\overline{\delta^*(q, x)}, y)} = \bigcup\{\overline{By} \mid B$ is a block under R and $B \subseteq \overline{\delta^*(q, x)}\}$.

Inspired from [2], we now introduce the following definition of homomorphism between two rough finite state machines.

Definition 7. *A **homomorphism** from a RFSM $M_1 = (Q_1, R_1, X_1, \delta_1)$ to RFSM $M_2 = (Q_2, R_2, X_2, \delta_2)$ is a pair of maps (f, g), where $f : Q_1 \to Q_2$ and $g : X_1 \to X_2$ are functions such that*

1. *$(p, q) \in R_1 \Rightarrow (f(p), f(q)) \in R_2$, $\forall p, q \in Q_1$,*
2. *$(f(\underline{\delta_1(q, a)}), f(\overline{\delta_1(q, a)})) \subseteq (\underline{\delta_2(f(q), g(a))}, \overline{\delta_2(f(q), g(a))})$, $\forall q \in Q_1$ and $\forall a \in X_1$.*

Let $M_1 = (Q_1, R_1, X_1, \delta_1)$, $M_2 = (Q_2, R_2, X_2, \delta_2)$ be two rough finite state machines and $(f, g) : M_1 \to M_2$ be a homomorphism. Let $g^* : X_1^* \to X_2^*$ be a map such that $g^*(\epsilon_{X_1}) = \epsilon_{X_2}$ and $g^*(ua) = g^*(u)g(a)$, $\forall u \in X_1^*$ and $a \in X_1$. Then we have the following lemma.

Lemma 1. *Let $M_1 = (Q_1, R_1, X_1, \delta_1)$, $M_2 = (Q_2, R_2, X_2, \delta_2)$ be two rough finite state machines and $(f, g) : M_1 \to M_2$ be a homomorphism. Then $g^*(xy) = g^*(x)g^*(y), \forall x, y \in X_1^*$.*

Proof. Let $x, y \in X_1^*$. We prove the result by induction on $|y| = n$. If $n = 0$, then $y = \epsilon_{X_1}$ and so $xy = x\epsilon_{X_1} = x$. Thus $g^*(xy) = g^*(x) = g^*(x)\epsilon_{X_2} = g^*(x)g^*(\epsilon_{X_1}) = g^*(x)g^*(y)$, whereby, the result is true for $n = 0$. Also, let the result be true $\forall z \in X_1^*$ such that $|z| = n - 1, n > 0$ and $y = za$, where $a \in X_1$. Then $g^*(xy) = g^*(xza) = g^*(xz)g(a) = g^*(x)g^*(z)g(a) = g^*(x)g^*(za) = g^*(x)g^*(y)$. Hence the result is true for $|y| = n$. □

Proposition 1. *Let $M_1 = (Q_1, R_1, X_1, \delta_1)$, $M_2 = (Q_2, R_2, X_2, \delta_2)$ be two rough finite state machines and $(f, g) : M_1 \to M_2$ be a homomorphism. Then $(f(\underline{\delta_1^*(q, x)}), f(\overline{\delta_1^*(q, x)})) \subseteq (\underline{\delta_2^*(f(q), g^*(x))}, \overline{\delta_2^*(f(q), g^*(x))})$, $\forall q \in Q_1$ and $\forall x \in X_1^*$.*

Proof. Let $q_1 \in Q_1$ and $x \in X_1^*$. We prove the result by induction on $|x| = n$. If $n = 0$, then $x = \epsilon_{X_1}$ and $g^*(x) = g^*(\epsilon_{X_1}) = \epsilon_{X_2}$. Now $\forall q \in Q_1$, $(f(\underline{\delta_1^*(q, \epsilon_{X_1})}), f(\overline{\delta_1^*(q, \epsilon_{X_1})})) = (f([q]_{R_1}), f([q]_{R_1})) \subseteq ([f(q)]_{R_2}, [f(q)]_{R_2}) = (\underline{\delta_2^*(f(q), \epsilon_{X_2})}, \overline{\delta_2^*(f(q), \epsilon_{X_2})})$ (as $(q, q) \in R_1 \Rightarrow (f(q), f(q)) \in R_2$, $\forall q \in Q_1$). Let the result be true for all $y \in X_1^*$ such that $|y| = n - 1, n > 0$ and $x = ya$, where $a \in X_1$. Then $(f(\underline{\delta_1^*(q, x)}), f(\overline{\delta_1^*(q, x)})) = (f(\underline{\delta_1^*(q, ya)}), f(\overline{\delta_1^*(q, ya)})) = (f(\underline{\delta_1(\underline{\delta_1^*(q, y)}, a)}), f(\overline{\delta_1(\overline{\delta_1^*(q, y)}, a)})) \subseteq \underline{\delta_2(f(\underline{\delta_1^*(q, y)}), g(a))}, \overline{\delta_2(f(\overline{\delta_1^*(q, y)}), g(a))}) \subseteq (\underline{\delta_2(\underline{\delta_2^*(f(q), g^*(y))}, g(a))}, \overline{\delta_2(\overline{\delta_2^*(f(q), g^*(y))}, g(a))}) = (\underline{\delta_2(\delta_2^*(f(q), g^*(ya)))}, \overline{\delta_2(\delta_2^*(f(q), g^*(ya)))}) = (\underline{\delta_2^*(f(q), g^*(ya))}, \overline{\delta_2^*(f(q), g^*(ya))}) = (\underline{\delta_2^*(f(q), g^*(x))}, \overline{\delta_2^*(f(q), g^*(x))})$. □

3 Rough Transformation Semigroups

The notion of transformation semigroups has been introduced and studied in both finite state machines and fuzzy finite state machines (cf., [2], [5]). In this section, we introduce analogous notion for rough finite state machines and discuss their properties.

Recall from [5] that an equivalence relation $\sim$ on a semigroup $(X, *)$ is called a **congruence relation** on X if, $\forall a, b, c \in X$, $a \sim b \Rightarrow a*c \sim b*c$ and $c*a \sim c*b$.

Let (Q, R, X, δ) be a RFSM. Define a relation $\simeq$ on X^* by $x \simeq y \Leftrightarrow (\underline{\delta^*(q, x)}, \overline{\delta^*(q, x)}) = (\underline{\delta^*(q, y)}, \overline{\delta^*(q, y)})$, $\forall q \in Q$, $\forall x, y \in X^*$. Then we have the following.

Proposition 2. *Let (Q, R, X, δ) be a RFSM. Then the relation $\simeq$ is a congruence relation on X^*.*

Proof. It is obvious that the relation $\simeq$ is an equivalence relation on X^*. Let $x, y \in X^*$ such that $x \simeq y$ and $z \in X^*$. Then $\forall q \in Q$, $\delta^*(q, xz) = (\underline{\delta^*(q, xz)}, \overline{\delta^*(q, xz)}) = (\underline{\delta^*(\underline{\delta^*(q, x)}, z)}, \overline{\delta^*(\overline{\delta^*(q, x)}, z)}) = (\underline{\delta^*(\underline{\delta^*(q, y)}, z)}, \overline{\delta^*(\overline{\delta^*(q, y)}, z)}) = (\underline{\delta^*(q, yz)}, \overline{\delta^*(q, yz)}) = \delta^*(q, yz)$. Thus $xz \simeq yz$. Similarly, $zx \simeq zy$. Hence $\simeq$ is a congruence relation on X^*. □

For given RFSM $M = (Q, R, X, \delta)$, let $[x] = \{y \in X^* : x \simeq y\}$ and $E(M) = \{[x] : x \in X^*\}$. Define a binary operation $*$ on $E(M)$ by $[x] * [y] = [xy]$, $\forall$ $[x], [y] \in E(M)$. Then we have the following.

Proposition 3. *For given RFSM $M = (Q, R, X, \delta)$, $(E(M), *)$ is a finite semigroup with identity.*

Proof. Associativity of the $*$ is trivial. For $[x] \in E(M)$, we have $[x] * [\epsilon_X] = [x\epsilon_X] = [x] = [\epsilon_X x] = [\epsilon_X] * [x]$, whereby $[\epsilon_X]$ is the identity of $(E(M), *)$. Thus $(E(M), *)$ is a semigroup with identity. The finiteness of $E(M)$ follows from the fact that Q is finite and by the definition of relation $\simeq$. Hence $(E(M), *)$ is a finite semigroup with identity. □

Definition 8. *A* ***rough transformation semigroup*** *(or RTS) is a 4-tuple* $A = (Q, R, S, \lambda)$, *where* Q *is a nonempty finite set (the* **set of states** *of* A*),* R *is an equivalence relation on* Q*,* S *is a nonempty finite semigroup and* $\lambda : Q \times S \to D \times D$*, where* D *is the collection of all definable sets in the approximation space* (Q, R)*, such that* $\forall q \in Q$*,*

1. *if* S *contains the identity* e*, then* $\lambda(q, e) = ([q]_R, [q]_R)$*, where* $[q]_R$ *is the equivalence class of* q *under* R*,* and
2. $\lambda(q, uv) = (\underline{\lambda(q, uv)}, \overline{\lambda(q, uv)})$*, where* $\underline{\lambda(q, uv)} = \underline{\lambda(\underline{\lambda(q, u)}, v)} = \bigcup\{\underline{Bv} \mid B \text{ is a block under } R \text{ and } B \subseteq \underline{\lambda(q, u)}\}$ *and* $\overline{\lambda(q, uv)} = \overline{\lambda(\overline{\lambda(q, u)}, v)} = \bigcup\{\overline{Bv} \mid B \text{ is a block under } R \text{ and } B \subseteq \overline{\lambda(q, u)}\}$*,* $\forall u, v \in S$*.*

If, in addition, $\forall q \in Q$ and $u, v \in S$, $(\underline{\lambda(q, u)}, \overline{\lambda(q, u)}) = (\underline{\lambda(q, v)}, \overline{\lambda(q, v)}) \Rightarrow u = v$, holds. Then (Q, R, S, λ) is called **faithful** RTS.

Let $A = (Q, R, S, \lambda)$ be a RTS which is not faithful. Define a relation $\sim$ on S by $u \sim v \Leftrightarrow (\underline{\lambda(q, u)}, \overline{\lambda(q, u)}) = (\underline{\lambda(q, v)}, \overline{\lambda(q, v)})$, $\forall u, v \in S$ and $\forall q \in Q$. Then, it can be easily seen that $\sim$ is an equivalence relation on S. Also, let $u, v, w \in S$ and $u \sim v$. Then $\lambda(q, uw) = (\underline{\lambda(q, uw)}, \overline{\lambda(q, uw)}) = (\underline{\lambda(\underline{\lambda(q, u)}, w)}, \overline{\lambda(\overline{\lambda(q, u)}, w)}) = (\underline{\lambda(\underline{\lambda(q, v)}, w)}, \overline{\lambda(\overline{\lambda(q, v)}, w)}) = (\underline{\lambda(q, vw)}, \overline{\lambda(q, vw)}) = \lambda(q, vw)$. Thus $uw \sim vw$. Similarly, $wu \sim wv$. Hence $\sim$ is a congruence relation on S.

Let $[u]$ be the equivalence class of u induced by the relation $\sim$ and $S/\sim = \{[u] : u \in S\}$. Define $\mu : Q \times S/\sim \to D \times D$ by $\mu(q, [x]) = \lambda(q, x)$, i.e., $(\underline{\mu(q, [x])}, \overline{\mu(q, [x])}) = (\underline{\lambda(q, x)}, \overline{\lambda(q, x)})$, $\forall q \in Q$ and $\forall [x] \in S/\sim$. Now $\mu(q, [e]) = ([q]_R, [q]_R)$. Also, $(\underline{\mu(q, [x][y])}, \overline{\mu(q, [x][y])}) = (\underline{\mu(q, [xy])}, \overline{\mu(q, [xy])}) = (\underline{\lambda(q, xy)}, \overline{\lambda(q, xy)}) = (\underline{\lambda(\underline{\lambda(q, x)}, y)}, \overline{\lambda(\overline{\lambda(q, x)}, y)}) = (\underline{\mu(\underline{\mu(q, [x])}, [y])}, \overline{\mu(\overline{\mu(q, [x])}, [y])})$, $\forall [x], [y] \in S/\sim$. Again, let $(\underline{\mu(q, [x])}, \overline{\mu(q, [x])}) = (\underline{\mu(q, [y])}, \overline{\mu(q, [y])})$, $\forall q \in Q$. Then $(\underline{\lambda(q, x)}, \overline{\lambda(q, x)}) = (\underline{\lambda(q, y)}, \overline{\lambda(q, y)})$, $\forall q \in Q$. Thus $x \sim y$, whereby $[x] = [y]$, showing that $(Q, R, S/\sim, \mu)$ is a faithful RTS.

Proposition 4. *Let* $M = (Q, R, X, \delta)$ *be a RFSM. Then* $(Q, R, E(M), \lambda)$ *is a faithful RTS, where* $(\underline{\lambda(q, [x])}, \overline{\lambda(q, [x])}) = (\underline{\delta^*(q, x)}, \overline{\delta^*(q, x)})$*,* $\forall q \in Q$ *and* $\forall x \in X^*$*.*

Proof. In view of Proposition 3.2, $E(M)$ is a finite semigroup with identity $[\epsilon_X]$. Obviously, $\lambda(q, [\epsilon_X]) = ([q]_R, [q]_R)$. Let $q \in Q$ and $[x], [y] \in E(M)$. Then $\lambda(q, [x] * [y]) = \lambda(q, [xy]) = \delta^*(q, xy) = (\underline{\delta^*(q, xy)}, \overline{\delta^*(q, xy)}) = (\underline{\delta^*(\underline{\delta^*(q, x)}, y)}, \overline{\delta^*(\overline{\delta^*(q, x)}, y)}) = (\underline{\lambda(\underline{\lambda(q, [x])}, [y])}, \overline{\lambda(\overline{\lambda(q, [x])}, [y])})$. Also, let $\lambda(q, [x]) = \lambda(q, [y])$, i.e., $(\underline{\lambda(q, [x])}, \overline{\lambda(q, [x])}) = (\underline{\lambda(q, [y])}, \overline{\lambda(q, [y])})$, $\forall q \in Q$. Then $\delta^*(q, x) = \delta^*(q, y)$, i.e., $(\underline{\delta^*(q, x)}, \overline{\delta^*(q, x)}) = (\underline{\delta^*(q, y)}, \overline{\delta^*(q, y)})$, $\forall q \in Q$. Thus $x \simeq y$ or $[x] = [y]$. Hence $(Q, R, E(M), \lambda)$ is a faithful RTS. $\square$

For RFSM $M = (Q, R, X, \delta)$, We shall denote by RTS(M), the RTS $(Q, R, E(M), \lambda)$, and call it the RTS associated with M.

Definition 9. *A* ***homomorphism*** *from a RTS $A_1 = (Q_1, R_1, S_1, \lambda_1)$ to RTS $A_2 = (Q_2, R_2, S_2, \lambda_2)$ is a pair of maps (α, β), where $\alpha : Q_1 \to Q_2$ and $\beta : S_1 \to S_2$ are functions such that*

1. $(p, q) \in R_1 \Rightarrow (\alpha(p), \alpha(q)) \in R_2$, $\forall p, q \in Q_1$,
2. $\beta(uv) = \beta(u)\beta(v)$, $\forall u, v \in S_1$,
3. *if S_1 and S_2 contain the identity e_1 and e_2 respectively, then $\beta(e_1) = e_2$, and*
4. $(\alpha(\underline{\lambda_1(q, u)}), \alpha(\overline{\lambda_1(q, u)})) \subseteq (\underline{\lambda_2(\alpha(q), \beta(u))}, \overline{\lambda_2(\alpha(q), \beta(u))})$, $\forall q \in Q_1$ *and* $\forall u \in S_1$.

A homomorphism $(\alpha, \beta) : A_1 \to A_2$ is called an **isomorphism** if α and β are both one-one and onto.

Let S be a semigroup with identity e and (Q, R, S, λ) be a faithful RTS. Define RFSM $M = (Q, R, S, \delta)$ by taking $\delta = \lambda$. Consider RTS $(M) = (Q, R, E(M), \rho)$, where $E(M) = S^* / \sim$ and $(\underline{\rho(q, [u])}, \overline{\rho(q, [u])}) = (\underline{\delta^*(q, u)}, \overline{\delta^*(q, u)})$. Now, for all $q \in Q$, $(\underline{\rho(q, [e])}, \overline{\rho(q, [e])}) = (\underline{\delta^*(q, e)}, \overline{\delta^*(q, e)}) = (\underline{\lambda(q, e)}, \overline{\lambda(q, e)}) = ([q]_R, [q]_R)$. Hence $(\underline{\rho(q, [e])}, \overline{\rho(q, [e])}) = (\underline{\rho(q, [\Lambda])}, \overline{\rho(q, [\Lambda])})$, where Λ is the empty word in S^*. Thus $[e] = [\Lambda]$.

Proposition 5. *Let $M = (Q, R, X, \delta)$ be a RFSM and S be a semigroup with identity e. Then RTS(M) is isomorphic to faithful RTS $A = (Q, R, S, \lambda)$.*

Proof. Let $\alpha : Q \to Q$ and $\beta : S \to E(M)$ be maps such that $\alpha(q) = q$ and $\beta(u) = [u]$, $\forall q \in Q$ and $\forall u \in S$. Let $p, q \in Q$. Then $(p, q) \in R \Rightarrow (\alpha(p), \alpha(q)) \in R$ holds from the definition of α. Let $\bullet$ be the binary operation of S and for $a, b \in S$, $a \bullet b \in S$, $ab \in S^*$. Then $\delta^*(q, a \bullet b) = (\underline{\delta^*(q, a \bullet b)}, \overline{\delta^*(q, a \bullet b)}) = (\underline{\delta(q, a \bullet b)}, \overline{\delta(q, a \bullet b)}) = (\underline{\lambda(q, a \bullet b)}, \overline{\lambda(q, a \bullet b)}) = (\underline{\lambda(\underline{\lambda(q, a)}, b)}, \overline{\lambda(\overline{\lambda(q, a)}, b)}) = (\underline{\delta(\underline{\delta(q, a)}, b)}, \overline{\delta(\overline{\delta(q, a)}, b)}) = (\underline{\delta(q, ab)}, \overline{\delta(q, ab)}) = \delta^*(q, ab)$, $\forall q \in Q$. Thus $[a \bullet b] = [ab]$, showing that $\beta(a \bullet b) = [a \bullet b] = [ab] = [a][b] = \beta(a)\beta(b)$. Also, $(\underline{\rho(\alpha(q), \beta(u))}, \overline{\rho(\alpha(q), \beta(u))}) = (\underline{\rho(q, [u])}, \overline{\rho(q, [u])}) = (\underline{\delta^*(q, u)}, \overline{\delta^*(q, u)}) = (\underline{\delta(q, u)}, \overline{\delta(q, u)}) = (\underline{\lambda(q, u)}, \overline{\lambda(q, u)})$. Now it remains to show that α is one-one and onto. Let $u, v \in S$ be such that $\beta(u) = \beta(v)$. Then $[u] = [v]$. Thus $(\underline{\delta^*(q, u)}, \overline{\delta^*(q, u)}) = (\underline{\delta^*(q, v)}, \overline{\delta^*(q, v)})$, or that $(\underline{\delta(q, u)}, \overline{\delta(q, u)}) = (\underline{\delta(q, v)}, \overline{\delta(q, v)})$, implying that $(\underline{\lambda(q, u)}, \overline{\lambda(q, u)}) = (\underline{\lambda(q, v)}, \overline{\lambda(q, v)})$, or that $u = v$, as A is faithful. Thus β is one-one. Also, it can be easily see that if $c_i \in S$, $i \in [1, n]$, then $[c_1 \bullet c_2 \bullet \ldots\ldots \bullet c_n] = [c_1 c_2 \ldots\ldots c_n]$ by the induction. Lastly, let $[x] \in E(M)$. If $x = \Lambda$, then $[\Lambda] = [e]$ and $\beta(e) = [\Lambda]$. Let $x = a_1 a_2 \ldots\ldots a_n$, $a_i \in S, i \in [1, n]$. Then $\beta(a_1 \bullet a_2 \bullet \ldots\ldots \bullet a_n) = [a_1 \bullet a_2 \bullet \ldots\ldots \bullet a_n] = [a_1 a_2 \ldots\ldots a_n] = [x]$. Thus β is onto. □

4 Coverings

The concept of coverings for both finite state machines and fuzzy finite state machines have introduced in (cf., [2], [5]). In this section, we introduce analogous notion for RFSM.

Definition 10. *Let $M_1 = (Q_1, R_1, X_1, \delta_1)$ and $M_2 = (Q_2, R_2, X_2, \delta_2)$ be two rough finite state machines. Let $\eta : Q_2 \to Q_1$ be an onto map and $\xi : X_1 \to X_2$ be a map. Then the pair (η, ξ) is called a* **covering** *of M_1 by M_2, if*

1. *$(p, q) \in R_2 \Rightarrow (\eta(p), \eta(q)) \in R_1, \forall p, q \in Q_2$, and*
2. *$(\underline{\delta_1^*(\eta(q_2), x)}, \overline{\delta_1^*(\eta(q_2), x)}) \subseteq (\underline{\eta(\delta_2^*(q_2, \xi^*(x)))}, \overline{\eta(\delta_2^*(q_2, \xi^*(x)))}), \forall q_2 \in Q_2$ and $\forall x \in X_1^*$, where $\xi^* : X_1^* \to X_2^*$ is a map such that $\xi^*(\epsilon_X) = \epsilon_X$ and $\xi^*(x) = \xi(x_1)\xi(x_2)...\xi(x_n), \forall x = x_1 x_2 ... x_n \in X_1^*$.*

We shall denote by $M_1 \preceq M_2$, the covering of M_1 by M_2 .

Definition 11. *Let $A_1 = (Q_1, R_1, S_1, \lambda_1)$ and $A_2 = (Q_2, R_2, S_2, \lambda_2)$ be two rough transformation semigroups. An onto map $\eta : Q_2 \to Q_1$ is called* **covering** *of A_1 by A_2, if*

1. *$(p, q) \in R_2 \Rightarrow (\eta(p), \eta(q)) \in R_1$, $\forall p, q \in Q_2$, and*
2. *$\forall s \in S_1$, $\exists\ t_s \in S_2$ such that $(\underline{\lambda_1(\eta(q_2), s)}, \overline{\lambda_1(\eta(q_2), s)}) \subseteq (\underline{\eta(\lambda_2(q_2, t_s))}, \overline{\eta(\lambda_2(q_2, t_s))})$, $\forall q_2 \in Q_2$.*

We shall denote by $A_1 \preceq A_2$, the covering of A_1 by A_2 .

Proposition 6. *(i) Let M_1, M_2 and M_3 be rough finite state machines. Then $M_1 \preceq M_2$, $M_2 \preceq M_3 \Rightarrow M_1 \preceq M_3$.*
(ii) Let A_1, A_2 and A_3 be rough transformation semigroups. Then $A_1 \preceq A_2$, $A_2 \preceq A_3 \Rightarrow A_1 \preceq A_3$.

Proof. The proof is straightforward. □

Proposition 7. *Let M_1 and M_2 be two rough finite state machines such that $M_1 \preceq M_2$, then $RTS(M_1) \preceq RTS(M_2)$.*

Proof. Let $M_1 = (Q_1, R_1, X_1, \delta_1)$ and $M_2 = (Q_2, R_2, X_2, \delta_2)$ be two rough finite state machines such that $M_1 \preceq M_2$. Then there exists an onto map $\eta : Q_2 \to Q_1$ and a map $\xi^* : X_1^* \to X_2^*$ such that $(\underline{\delta_1^*(\eta(q_2), x)}, \overline{\delta_1^*(\eta(q_2), x)}) \subseteq (\underline{\eta(\delta_2^*(q_2, \xi^*(x)))}, \overline{\eta(\delta_2^*(q_2, \xi^*(x)))})$, $\forall q_2 \in Q_2$ and $\forall x \in X_1^*$. Let $RTS(M_1) = (Q_1, R_1, E(M_1), \lambda_1)$ and $RTS(M_2) = (Q_2, R_2, E(M_2), \lambda_2)$ be rough transformation semigroups associated with rough finite state machines M_1 and M_2 respectively. Also, let $s \in E(M_1)$. Then $\exists\ x \in X_1^*$ such that $s \in [x]$. Again, let $t_s = [\xi^*(x)] \in E(M_2)$ and $q_2 \in Q_2$. Then $(\underline{\lambda_1(\eta(q_2), s)}, \overline{\lambda_1(\eta(q_2), s)}) = (\underline{\lambda_1(\eta(q_2), [x])}, \overline{\lambda_1(\eta(q_2), [x])}) = (\underline{\delta_1^*(\eta(q_2), x)}, \overline{\delta_1^*(\eta(q_2), x)}) \subseteq (\underline{\eta(\delta_2^*(q_2), \xi^*(x))}, \overline{\eta(\delta_2^*(q_2), \xi^*(x))}) = (\underline{\eta(\lambda_2(q_2), [\xi^*(x)])}, \overline{\eta(\lambda_2(q_2), [\xi^*(x)])}) = (\underline{\eta(\lambda_2(q_2), t_s)}, \overline{\eta(\lambda_2(q_2), t_s)})$, showing that $RTS(M_1) \preceq RTS(M_2)$. □

5 Conclusion

Chiefly inspired from [3] and [5], we tried to present here the concept of homomorphism between two rough finite state machines, rough transformation semigroup associated with a rough finite state machine and coverings of rough finite state machines as well as rough transformation semigroups for the study of algebraic rough machines theory. Even, much more can be done in this direction by introducing the concept of different products of rough finite state machines, as in [4].

Acknowledgments. The authors gratefully thank the reviewers for their observations.

References

1. Basu, S.: Rough finite state automata. Cybernetics and Systems 36, 107–124 (2005)
2. Holcombe, W.M.L.: Algebraic automata theory. Cambridge University Press, Cambridge (1982)
3. Malik, D.S., Mordeson, J.N., Sen, M.K.: Semigroups of fuzzy finite state machines. In: Wang, P.P. (ed.) Advances in Fuzzy Theory and Technology, vol. II, pp. 87–98 (1994)
4. Malik, D.S., Mordeson, J.N., Sen, M.K.: Products of fuzzy finite state machines. Fuzzy Sets and Systems 92, 95–102 (1997)
5. Mordeson, J.N., Malik, D.S.: Fuzzy Automata and Languages: Theory and Applications. Chapman and Hall/CRC, Boca Raton (2002)
6. Pawlak, Z.: Rough sets. Int. J. Inf. Comp. Sci. 11, 341–356 (1982)
7. Tiwari, S.P., Srivastava, A.K., Sharan, S.: Characterizations of rough finite state automata. Communicated

Covering Rough Set Model Based on Multi-granulations

Caihui Liu[1,2] and Duoqian Miao[1]

[1] Department of Computer Science and Technology, Tongji University Shanghai, P.R.China, 201804
[2] Department of Mathematics and Computer Science, Gannan Normal University Ganzhou, Jiangxi Province, P.R. China, 341000
{liu_caihui,miaoduoqian}@163.com

Abstract. The paper extends the covering rough set model based on single granulation to the covering rough set model based on multi- granulations, which is named CMGRS. The lower and upper approximations of a set in a covering approximation space are defined based on multi-granulations, and some basic properties are investigated.

Keywords: Rough sets, Multi-granulations, Covering.

1 Introduction

Rough set theory [3] is a useful mechanism for uncertainty processing. Several extensions have been proposed, such as variable precision rough set model [10], graded rough set model [5,6], or rough set covering models [1,9].

In the view of granular computing [7], an equivalence relation on a universe can be regarded as a granulation, and a partition on the universe – as a granulation space [4]. Qian and Liang [2] introduced the rough set model based on multi-granulations, where multi-equivalence relations must be used because of user requirements or problem specification.

This paper proposes the rough set model based on multi-covering relations. The lower and upper approximations of a set are defined by multi-covering relations on the universe, and some basic properties are introduced. One can see that the rough set model based on multi-covering relations is an extension of the rough set model based on multi-granulations.

2 Preliminaries

In this section, we present some necessary concepts and preliminaries required in the sequel of our work. The detailed descriptions of the related theories can be found in the source papers [2,3,8,9]. Qian and Liang [2] proposed the rough set model based on multi-granulations called MGRS.

Definition 1 [2]. Let $K = (U, \mathbf{R})$ be a knowledge base and $\mathbf{R}$ be a family of equivalence relations on U. For any $X \subseteq U$ and $P, Q \in \mathbf{R}$, the lower and upper

S.O. Kuznetsov et al. (Eds.): RSFDGrC 2011, LNAI 6743, pp. 87–90, 2011.

approximations of X can be defined as $\underline{P+Q}X = \{x \in U | [x]_P \subseteq X \vee [x]_Q \subseteq X\}$ and $\overline{P+Q}X = \sim \underline{P+Q}(\sim X)$, where $\sim X$ is the complement of X in U.

Definition 2 [9]. Let U be a universe of discourse and C be a covering of U. The pair $< U, C >$ is called a covering approximation space. For $\forall x \in U$, the set $md(x)$ is called the minimal description of x, where $md(x) = \{K \in C | x \in K \wedge (\forall S \in C \wedge x \in S \wedge S \subseteq K \Rightarrow K = S)\}$.

Definition 3 [9]. Let $< U, C >$ be a covering approximation space. For $X \subseteq U$, the covering lower and upper approximations of X are defined as $\underline{C}(X) = \{x \in U | \cap md(x) \subseteq X\}$ and $\overline{C}(X) = \{x \in U | (\cap md(x)) \cap X \neq \emptyset\}$.

Proposition 1 [8]. The covering lower and upper approximations have the following properties: (1) $\underline{C}X \subseteq X \subseteq \overline{C}X$; (2) $\underline{C}\emptyset = \overline{C}\emptyset = \emptyset$ and $\underline{C}U = \overline{C}U = U$; (3) $\underline{C}(X \cap Y) = \underline{C}X \cap \underline{C}Y$ and $\overline{C}(X \cup Y) = \overline{C}X \cup \overline{C}Y$; (4) $\underline{C}(\underline{C}X) = \underline{C}X$ and $\overline{C}(\overline{C}X) = \overline{C}X$; (5) If $X \subseteq Y$, then $\underline{C}X \subseteq \underline{C}Y$ and $\overline{C}X \subseteq \overline{C}Y$; (6) $\overline{C}(X) = \sim \underline{C}(\sim X)$.

Theorem 1 [8]. Let C be a covering of U. C and reduct(C) generate the same lower and upper approximation operators, where reduct(C) is the reduction of C on U.

3 Approximation Operators of CMGRS

In this section, we extend MGRS to covering rough set model based on multi-granulations, called CMGRS. Some basic properties of the lower and upper approximations are discussed.

Definition 4. Let $< U, \mathbf{C} >$ be a knowledge base, $\mathbf{C}$ be a family of coverings on U and $P, Q \in \mathbf{C}$. For any $X \subseteq U$, its lower and upper approximations with respect to P, Q are defined as $\underline{P+Q}X = \{x \in U | \cap md_P(x) \subseteq X \vee \cap md_Q(x) \subseteq X\}$ and $\overline{P+Q}X = \sim \underline{P+Q}(\sim X)$. If $\underline{P+Q}(X) = \overline{P+Q}X$, then X is definable. Otherwise X is called a covering rough set with respect to P, Q. The pair $(\underline{P+Q}X, \overline{P+Q}X)$ is called a covering rough set based on P, Q.

Definitions 1 and 4 are the same when $\mathbf{C}$ is a family of partitions. We show the difference between the two models using example 1.

Example 1. Let $< U, \mathbf{C} >$ be a knowledge base, where $U = \{a, b, c, d\}$, and for $P, Q \in \mathbf{C}$ there is $P = \{\{a, b\}, \{b, c, d\}, \{c, d\}\}$ and $Q = \{\{a, c\}, \{b, d\}, \{a, b, d\}\}$. If $X = \{a, d\}$, then $P \cap Q = \{\{a\}, \{b\}, \{c\}, \{d\}, \{a, b\}, \{b, d\}\}$. Surely, $P \cap Q$ is also a covering of U. There is $\underline{P \cap Q}X = \{a, d\}$, $\overline{P \cap Q}X = \{a, d\}$, $\underline{P+Q}X = \{a\}$, and $\overline{P+Q}X = \{a, c, d\}$. So $\underline{P \cap Q}X \neq \underline{P+Q}X$ and $\overline{P \cap Q}X \neq \overline{P+Q}X$.

Lemma 1. Let $< U, \mathbf{C} >$ be a knowledge base, $P, Q \in \mathbf{C}$. Then for each $X \subseteq U$ we have $\underline{P+Q}X = \underline{P}X \cup \underline{Q}X$ and $\overline{P+Q}X = \overline{P}X \cap \overline{Q}X$.

Proposition 2. Let $< U, \mathbf{C} >$ be a knowledge base, $P, Q \in \mathbf{C}$. For each $X \subseteq U$, the following properties hold: (1) $\underline{P+Q}X \subseteq X \subseteq \overline{P+Q}X$; (2) $\underline{P+Q}\emptyset = \overline{P+Q}\emptyset = \emptyset$ and $\underline{P+Q}U = \overline{P+Q}U = U$; (3) $\underline{P+Q}(\sim X) = \sim \overline{P+Q}X$ and $\overline{P+Q}(\sim X) = \sim \underline{P+Q}X$; (4) $\underline{P+Q}(\underline{P+Q}X) = \underline{P+Q}X \subseteq \overline{P+Q}(\underline{P+Q}X)$; (5) $\underline{P+Q}(\overline{P+Q}X) \subseteq \overline{P+Q}X = \overline{P+Q}(\overline{P+Q}X)$; (6) $\underline{P+Q}X = \underline{Q+P}X$ and $\overline{P+Q}X = \overline{Q+P}X$.

Proposition 3. Let $< U, \mathbf{C} >$ be a knowledge base, $P, Q \in \mathbf{C}$. For each $X, Y \subseteq U$, the following properties hold: (1) $\underline{P+Q}(X \cap Y) = (\underline{P}X \cap \underline{P}Y) \cup (\underline{Q}X \cap \underline{Q}Y)$; (2) $\overline{P+Q}(X \cup Y) = (\overline{P}X \cup \overline{P}Y) \cap (\overline{Q}X \cup \overline{Q}Y)$; (3) $\underline{P+Q}(X \cap Y) \subseteq \underline{P+Q}X \cap \underline{P+Q}Y$; (4) $\overline{P+Q}(X \cup Y) \supseteq \overline{P+Q}X \cup \overline{P+Q}Y$; (5) If $X \subseteq Y$, then $\underline{P+Q}X \subseteq \underline{P+Q}Y$ and $\overline{P+Q}X \subseteq \overline{P+Q}Y$; (6) $\underline{P+Q}(X \cup Y) \supseteq \underline{P+Q}X \cup \underline{P+Q}Y$ and $\overline{P+Q}(X \cap Y) \subseteq \overline{P+Q}X \cap \overline{P+Q}Y$.

Example 2. (Continued from Example 1) If $Y = \{a, b\}$, then $X \cap Y = \{a\}$ and $X \cup Y = \{a, b, d\}$. There is $\underline{P}X = \emptyset$, $\overline{P}X = \{a, c, d\}$, $\underline{Q}X = \{a\}$, $\overline{Q}X = \{a, b, c, d\}$, $\underline{P}Y = \{a, b\}$, $\overline{P}Y = \{a, b\}$, and $\underline{Q}Y = \{a\}$, $\overline{Q}Y = \{a, b, c, d\}$. Then $\underline{P+Q}(X \cap Y) = \{a\} = (\underline{P}X \cap \underline{P}Y) \cup (\underline{Q}X \cap \underline{Q}Y)$, $\overline{P+Q}(X \cup Y) = \{a, b, c, d\} = (\overline{P}X \cup \overline{P}Y) \cap (\overline{Q}X \cup \overline{Q}Y)$, $\underline{P+Q}(X \cap Y) = \{a\} \subseteq \underline{P+Q}X \cap \underline{P+Q}Y = \{a\}$, $\overline{P+Q}(X \cup Y) = \{a, b, c, d\} \supseteq \overline{P+Q}X \cup \overline{P+Q}Y = \{a, c, d\}$, $Y = \{a, b\} \subset X \cup Y = \{a, b, d\}$, $\underline{P+Q}Y = \{a, b\} \subset \underline{P+Q}(X \cup Y) = \{a, b, d\}$, $\overline{P+Q}Y = \{a, c, d\} \subset \overline{P+Q}(X \cup Y) = \{a, b, c, d\}$, $\underline{P+Q}(X \cup Y) = \{a, b, d\} \supseteq \underline{P+Q}X \cup \underline{P+Q}Y = \{a, b\}$, and $\overline{P+Q}(X \cap Y) = \{a\} \subseteq \overline{P+Q}X \cap \overline{P+Q}Y = \{a, c, d\}$.

Definition 5. Let $< U, \mathbf{C} >$ be a knowledge base, $P, Q \in \mathbf{C}$, $reduct(P) = \{P_1, \ldots, P_p\}$, and $reduct(Q) = \{Q_1, \ldots, Q_q\}$. For each $x \in U$, if $x \in P_i (1 \leq i \leq p)$ and $x \in Q_j (1 \leq j \leq q)$, then there is $P_i \subseteq Q_j$, we say *reduct*(P) is finer than *reduct*(Q), denoted by $P \lhd Q$.

Theorem 2. Let $< U, \mathbf{C} >$ be a knowledge base, $P, Q \in \mathbf{C}$. If $P \lhd Q$, then, for each $X \subseteq U$, we have: (1) $\underline{P+Q}X = \underline{P}X$ and (2) $\overline{P+Q}X = \overline{P}X$.

Definition 6. Let $< U, \mathbf{C} >$ be a knowledge base, $\mathbf{C}$ be a family of coverings on U and $P_i \in \mathbf{C}$, $i = 1, \ldots, n$. For any $X \subseteq U$, the lower and upper approximations of X in U with respect to $P_1, \ldots, P_n$ are defined as $\underline{\sum_i P_i}X = \cup\{x \in U | \cap md_{P_i}(x) \subseteq X, i \leq n\}$ and $\overline{\sum_i P_i}X = \sim \underline{\sum_i P_i}(\sim X)$.

Definition 6 implies the following properties of the lower and upper approximation operators.

Proposition 4. Let $< U, \mathbf{C} >$ be a knowledge base, $\mathbf{C}$ be a family of coverings on U, $P_i \in \mathbf{C}$, $i = 1, \ldots, n$. For any $X, Y \subseteq U$, the following properties are satisfied: (1) $\underline{\sum_i P_i}X = \bigcup_i \underline{P_i}X$ and $\overline{\sum_i P_i}X = \bigcap_i \overline{P_i}X$; (2) $\underline{\sum_i P_i}(\sim X) = \sim \overline{\sum_i P_i}X$ and $\overline{\sum_i P_i}(\sim X) = \sim \underline{\sum_i P_i}X$; (3) If $X \subseteq Y$, then $\underline{\sum_i P_i}X \subseteq \underline{\sum_i P_i}Y$ and $\overline{\sum_i P_i}X \subseteq \overline{\sum_i P_i}Y$.

Proposition 5. Let $< U, \mathbf{C} >$ be a knowledge base, $\mathbf{C}$ be a family of coverings on U, $P_i \in \mathbf{C}$, $i = 1, \ldots, n$. For any $X_j \subseteq U, j = 1, \ldots, m$, the following properties are satisfied: (1) $\underline{\sum_i P_i}(\bigcap_j X_j) = \bigcup_i (\bigcap_j \underline{P_i} X_j)$; (2) $\overline{\sum_i P_i}(\bigcup_j X_j) = \bigcap_i (\bigcup_j \overline{P_i} X_j)$; (3) $\underline{\sum_i P_i}(\bigcap_j X_j) \subseteq \bigcap_j (\underline{\sum_i P_i} X_j)$; (4) $\overline{\sum_i P_i}(\bigcup_j X_j) \supseteq \bigcup_j (\overline{\sum_i P_i} X_j)$; (5) $\underline{\sum_i P_i}(\bigcup_j X_j) \supseteq \bigcup_j (\underline{\sum_i P_i} X_j)$; (6) $\underline{\sum_i P_i}(\bigcap_j X_j) \subseteq \bigcap_j (\overline{\sum_i P_i} X_j)$.

Theorem 3. Let $< U, \mathbf{C} >$ be a knowledge base, $\mathbf{C}$ be a family of coverings on U, $P_i \in \mathbf{C}$, $i = 1, \ldots, n$. For any $X \subseteq U$, the following properties are satisfied: (1) If $P_1 \lhd P_i, i = 1, \ldots, n$, then $\underline{\sum_i P_i} X = \underline{P_1} X$; (2) If $P_1 \lhd P_i, i = 1, \ldots, n$, then $\overline{\sum_i P_i} X = \overline{P_1} X$.

4 Conclusion

The extension of Pawlak rough set model is an important direction of research. In this paper, along the line of Qian and Liang [2], we proposed the covering rough set model based on multi-granulations and we discussed some of its interesting properties that may be important for future real life applications.

Acknowledgements. The research is supported by the National Natural Science Foundation of China under grant No: 60970061, 61075056.

References

1. Bonikowski, Z., Bryniarski, E., Skardowska, U.: Extension and intentions in the rough set theory. Information Sciences 107, 149–167 (1998)
2. Qian, Y.H., Liang, J.Y.: Rough set method based on multi-granulations. In: Proc. of 5th IEEE International Conference on Cognitive Informatics, pp. 297–304 (2006)
3. Pawlak, Z.: Rough sets. International Journal of Computer and Information Sciences 11, 341–356 (1982)
4. Yao, Y.Y.: Perspectives of Granular Computing. In: Proc. of the 2005 IEEE International Conference on Granular Computing, vol. 1, pp. 85–90 (2005)
5. Yao, Y.Y., Lin, T.Y.: Generalization of rough sets using modal logic. Intelligent Automation and Soft Computing 2(2), 103–120 (1996)
6. Yao, Y.Y., Lin, T.Y.: Graded rough set approximations based on nested neighborhood systems. In: Proc. of 5th European Congress on Intelligent Techniques and Soft Computing, vol. 1, pp. 196–200 (1997)
7. Zadeh, L.A.: Toward a theory of fuzzy information granulation and its centrality in human reasoning and fuzzy logic. Fuzzy sets and Systems 90, 111–127 (1997)
8. Zhu, W.: Relationship among basic concepts in covering-based rough sets. Information Sciences 179, 2478–2486 (2009)
9. Zhu, W., Wang, F.Y.: Reduction and axiomatization of covering generalized rough sets. Information Sciences 152, 217–230 (2003)
10. Ziarko, W.: Variable precision rough set model. Journal of Computer and System Science 46(1), 39–59 (1993)

A Descriptive Language Based on Granular Computing – Granular Logic

Qing Liu and Lan Liu

Department of Computer Science
Nanchang University, Nanchang, Jiangxi 330031, China

Abstract. In this article, a granular logic defined in granular space is studied. Value of any individual variable in the logic is token as a granule or granulation, so the logic is viewed as second logic. The logic will be applied to describe the theorems in granular mathematics and its proof and to describe clinic experience and its reasoning of medicinal experts.

Keywords: Rough sets, Fuzzy sets, Granular computing, Operations on granulations, Granular logic.

1 Introduction

According to the research and development of rough sets and rough logic [1 – 4], the concepts of information granulation and granular computing are proposed. Hobbs published an article "granularity" [5] in 1985, and applied it to the problem solving in AI. It was defined by the predicate in classical logic, but the operation rules on granularities were not defined. Skowron described information granules in 2001, formulating the concept of granular language for the first time [6], but its syntax and semantics were not defined. Polkowski studied another type of logic in 2004 [7] and reported the granular logic in granular space in 2006 [8]. However, that logic was not defined completely.

In this paper, real granulation and its relative operations in real granular space R^* are introduced. Secondly, syntax and semantics of the granular logic are defined in R^*.

2 Granulation in Real Sets and Related Operations

Based on Robinson's nonstandard analysis [9] and Zadeh's granular mathematics [10], we may find out that a super real number in nonstandard analysis or a real number in granular mathematics is essentially a real granulation in granular real space R^*. The operations on granulations in R^* may be discussed as follows.

2.1 Real Granular Space

Let us discuss the granulations in one-dimensional real granular space R^*. The granulations in two or more dimensions may be considered analogously.

S.O. Kuznetsov et al. (Eds.): RSFDGrC 2011, LNAI 6743, pp. 91–94, 2011.

Let R be a real space, where each real number a could be constructed as a real granulation. Hence, we have a real granular space R^* based on R. Obviously, $R \subseteq R^*$. Any real number a and positive infinitely small number ϵ constructs an interval $(a-\epsilon, a+\epsilon)$. It is called a granulation, denoted by a^*.

2.2 Indiscernibility Relation in R^*

Let $a^*, b^* \in R^*$. a^* and b^* are indiscernible, denoted by $a^* \infty b^*$, iff

$$a^* \ominus b^* = \epsilon^* \in R^*$$

where $\ominus$ is subtraction symbol in R^* and ϵ^* is infinitely small granulation in R^*. It means that the finite parts of a^* are same as finite parts of b^*, as well as infinite parts of a^* are different from infinite parts of b^* but they are infinete approached each other, that is, a_{n+1} is infinite approached to with $b_{n+1}, \cdots, a_{n+m}$ is infinite approached to with b_{n+m}.

The operation laws of granulations in R^* are defined as follows:

$$a^* \bigoplus b^* = (a+b)^* \qquad a^* \ominus b^* = (a-b)^*$$
$$a^* \bigotimes b^* = (a \times b)^* \qquad a^* \odot b^* = (a \div b)^*$$

We may prove that the operation laws are right according to the properties of interval operations in fuzzy mathematics [11].

3 Granular Logic

Granular logic is supposed to be a good tool for studying information reasoning [7, 8, 11, 12]. Here we discuss only the granular logic defined in real granular space. The logic is abbreviated as L_U.

3.1 Syntax

Let R be real space and R^* be real granular space. If U is a sub-space of R, then U^* is a sub-space of R^*.

(1) Symbol Sets

Constants: All elements in U;
Variables: $x_1, \ldots, x_n, \ldots$;
Predicates: $\infty, \in$;
Conjunctives: $\wedge, \neg$;
Quantifier: $\exists$;
Square brackets: $[,]$.

(2) Well-Formed Formulae

Atomic formulae:

$$u_1 \infty u_2, u_1 \infty x_i, x_i \infty u_2, x_i \infty x_j; \qquad u_1 \in u_2, u_1 \in x_i, x_i \in u_2, x_i \in x_j.$$

The formula with no conjunctive or quantifier is called atomic. All atomic formulae are the formulae in L_U. Let α, β be the formulae in L_U. Then

$\neg[\alpha], \exists x_i[\alpha], [\alpha] \wedge [\beta], \neg[\neg[\alpha]], \neg[\exists x_i[\alpha]], \exists x_i[\neg[\alpha]], \exists x_i[\exists x_j[\alpha]]$ are in L_U.

(3) Sentences
Formulae that do not include any free variables are called sentences.

(4) Axioms and Inferences
The above logic's axioms and inferences should be analogous with other non-standard logics [1 – 4].

3.2 Semantics

(1) The truth values of sentences or formulae
Let $K \subseteq L_U$ be the set of all sentences. The map t is defined in $K, t : K \rightarrow \{true, false\}$. If α is a sentence, then it must be one of the following cases:

- There is no quantifier: $u_1 \in u_2$, $t(\alpha) = true$, or $u_1 \notin u_2$, $t(\alpha) = false$. $u_1 \infty u_2$, $t(\alpha) = true$, or $u_1 \neg\infty u_2$, $t(\alpha) = false$;
- α is $\neg[\beta]$, or $[\beta] \wedge [\gamma]$, or $\exists x_i[\delta]$, where β, γ are sentences. If δ is the sentence including only free variable x_i, then $t(\alpha) = true$ or $t(\alpha) = false$;
- $t(\alpha) = true$, written in $\models_\infty \alpha, t(\alpha) = false$, denoted by $\models_\infty \neg[\alpha]$.

For example, let $u \in U$ be granular individual variable. Let formula α be $[u \neg\infty \emptyset] \wedge [\neg\exists x_i[x_i \in u]]$. Then $t(\alpha) = true$.

(2) $[\alpha] \rightarrow [\beta]$ iff $\neg[[\alpha] \wedge [\neg[\beta]]]$;

(3) $[\alpha] \vee [\beta]$ iff $\neg[[\neg[\alpha]] \wedge [\neg[\beta]]]$;

(4) $\forall x_i[\alpha]$ iff $\neg[\exists x_i[\neg[\alpha]]]$;

(5) $[\alpha] \leftrightarrow [\beta]$ iff $[[\alpha] \rightarrow [\beta]] \wedge [[\beta] \rightarrow [\alpha]]$ iff $[\neg[[\alpha] \wedge [\neg[\beta]]]] \wedge [\neg[[\beta] \wedge [\neg[\alpha]]]]$.

4 Granular Mathematical Theorems in $\boldsymbol{R^*}$

Let f be continuous at the granulation $a^* \in A^* \subseteq R^*$, i.e., for each $\epsilon^* > 0^*$, there is $\delta^* > 0^*$, such that for all $x^* \in A^*$ such that $\mid x^* \ominus a^* \mid\in \delta^*$, we have

$$\mid f^*(x^*) \ominus f^*(a^*) \mid\in \epsilon^*.$$

The granular mathematical theorem is described with the formula in L_U as

$$\vdash [\forall\epsilon^*[\exists\delta^*\forall x^*[>(\delta^*, 0^*) \wedge x^* \in A^* \wedge \mid x^* \ominus a^* \mid\in \delta^*]] \\ \rightarrow [>(\epsilon^*, 0^*) \wedge \mid f^*(x^*) \ominus f^*(a^*) \mid\in \epsilon^*]]$$

For example, the problems of modeling the experience and reasoning of medical experts may be described using the above principles [11, 12].

5 Perspective of Studying Granular Logic

The presented approach can be used in the future to provide a new methodology for the research on granular computing. Also, it may offer some new ideas for practical applications of both classical and nonstandard logics.

Acknowledgements. This study is supported by the Natural Science Foundation of China (NSFC #61070139).

References

1. Pawlak, Z.: Rough logic. Bull. of Polish Acad. of Sci. 35(5-6), 253–259 (1987)
2. Orłowska, E.: A logic of indicernibility relation. In: Skowron, A. (ed.) SCT 1984. LNCS, vol. 208, pp. 177–186. Springer, Heidelberg (1985)
3. Chakraborty, M.K., Banerjee, M.: Rough logic with rough quantifiers. Warsaw University of Technology, ICS Research Report 49/93 (1993)
4. Nakamura, A.: Graded modalities in rough logic. In: Polkowski, L., Skowron, A. (eds.) Rough Sets, Knowledge Discovery, vol. 1, pp. 192–208. Physica-Verlag, Heidelberg (1998)
5. Hobbs, J.R.: Granularity. In: Proc. of IJCAI 1985, Los Angeles, pp. 432–435 (1985)
6. Skowron, A.: Toward intelligent systems: calculi of information granules. In: Proceedings of International Workshop on Rough Set Theory and Granular Computing (RSTGC- 2001), vol. 5(1/2), pp. 9–30. Bulleting of International Rough Set Society, Japan (2001)
7. Polkowski, L.: Toward rough set foundations. Mereological approach. In: Tsumoto, S., Słowiński, R., Komorowski, J., Grzymała-Busse, J.W. (eds.) RSCTC 2004. LNCS (LNAI), vol. 3066, pp. 8–25. Springer, Heidelberg (2004)
8. Polkowski, L.: A Calculus on Granules from Rough Inclusions in Information Systems. In: Polkowski, L. (ed.) The Proceedings of International Forum on Theory of GrC from Rough Set Perspective, IFTGrCRSP 2006, Nanchang, China, pp. 22–27 (June 2006)
9. Robinson, A.: Nonstandard Analysis. Princeton University Press, Princeton (1965)
10. Zadeh, L.A.: Granular Computing with Uncertain, Imprecise and Partially True Data. In: Proceedings of 3th International Conference, GrCC 2007, USA (November 2007)
11. Liu, Q., Sun, H., Wang, H.: The Present Studying State of Granular Computing and Studying of Granular Computing Based on the Semantics of Rough Logic. Chinese Journal of Computer 31(4), 543–555 (2008) (in Chinese)
12. Liu, Q., Jiang, F., Deng, D.Y.: Design and Implement for the Diagnosis Software of Blood Viscosity Syndrome Based on Hemorheology on GrC. In: Wang, G., et al. (eds.) RSFDGrC 2003. LNCS (LNAI), vol. 2639, Springer, Heidelberg (2003)

Optimization and Adaptation of Dynamic Models of Fuzzy Relational Cognitive Maps

Grzegorz Słoń[1] and Alexander Yastrebov[2]

Kielce University of Technology, al. Tysiąclecia P. P. 7, 25-314 Kielce, Poland
{enegs,jastri}@tu.kielce.pl

Abstract. The article is devoted to the analysis of dynamic models of fuzzy relational cognitive maps. Therefore, the selection process was analyzed, in a way of optimal (in some sense) parameters of these models, in particular such quantities as linguistic variables and fuzzy relations. Two-stage approach was applied to the optimization issue, composed of conditional optimization and adaptation. Certain results of simulation research concerning the designed method were also adduced.

1 Introduction

Available information about elements of the analyzed object, as well as connections between them plays a crucial role, when elaborating the dynamic systems (technical, economical, sociological and others). The availability level of information, which is necessary for the elaborating process, has a significant impact on the calculation difficulty degree, when it comes to computer implementation of elaborated models.

When creating dynamic models with incomplete information, dynamic models of relational cognitive maps [1,2,3,4,6,7,8,9,10,11] are often used in recent time. Models of this type determine connections between significant concepts of that system and are based on the relation between temporary increases in the value of the concepts [4,8,10,11]. In practical (computer) realization of such models, computational difficulties associated with the arithmetic of fuzzy numbers and uncertainty as to the nature of the relationship between concepts play a significant role. That problem brings to the attempt to choose the optimal parameters of these models.

The article carries a certain approach to the problem of optimal choice of parameters and relations in dynamic models of fuzzy relational cognitive maps. Other approaches to solving similar problems were presented e.g. in [1,4,7].

2 Dynamic Models of Fuzzy Relational Cognitive Maps

The term "cognitive maps" is usually used for certain mathematical and IT models, intended to formalize the examined problem of complex system, as a set of concepts imitating variables system (features) and cause - effect relations

S.O. Kuznetsov et al. (Eds.): RSFDGrC 2011, LNAI 6743, pp. 95–102, 2011.

between them, taking into account interaction (static or dynamic) and relation alterations [1,2,3,4,8,10].

In accordance to the above mentioned definition, cognitive maps may be presented in the form of following sets:

$$< \mathrm{X}, \mathrm{R} > \tag{1}$$

where: $\mathrm{X} = [X_1, ..., X_n]^T$ – value set of map concepts (state vector); $\mathrm{R} = \{R_{i,j}\}$ – relation matrix between variables X_i and X_j; n – number of concepts; $i, j = 1, ..., n$.

Matrix R can take different forms and one of them is a form of fuzzy relations $R_{i,j}$with the corresponding member functions and the set universum scope.

The dynamic model of fuzzy relational cognitive map is described by following equation:

$$X_k(t+1) = X_k(t) \oplus \bigoplus_{i=1}^{n} \left[\Big(X_i(t) \ominus X_i(t-1) \Big) \circ R_{i,k} \right] \tag{2}$$

where: k – number of considered output concept ($k = 1, ..., n$); t – discrete time; n – the number of concepts; $\oplus$ – fuzzy summation operation; $\ominus$ – fuzzy subtraction operation; $R_{i,k}$ – single fuzzy relation between fuzzy concepts with i and k numbers; $\circ$ – maxmin fuzzy composition operation.

The fuzzyfication of concepts and relations results in a problem of choice of in some sense optimal (in terms of calculations) parameters.

3 Fuzzyfication Parameters Optimization Problem

Parametric optimization of the system described with a cognitive map (and so, often impossible for analytical description) encounters additional problems arising from the operation nature of such objects. The primary one is fundamental inability for mathematical determination of the optimal operation point of the model, resulting from, e.g. mathematical determined minimum of the optimization criterion, may prove to be impractical in the real employ system (mainly due to the time needed by the numerical system to do calculations in too complex model). Therefore, an approximate approach is proposed in this article, to solve this problem, composed of two stages:

- selection of conditionally approximate optimal parameters of the fuzzyfication (number of linguistic variables and elementary fuzziness coefficient of linguistic variables) - the decision as to achieve the "proper" point of the operation model is made by an expert,
- the parameters adaptation of individual fuzzy relations between concepts in dynamic model - the decision is also made by an expert, but it may be automated by adopting appropriate, limiting criterion value.

This chapter is devoted to the description of the first stage solution for optimization task.

In the analyzed problem of applying fuzzy cognitive maps, Gaussian-type allocation function is being introduced [5]:

$$\mu_{X_i}(x) = e^{-\left(\frac{x-\overline{X}_i}{\sigma_i}\right)^2} \tag{3}$$

where: $\overline{X}_i$ – centre of the i-th linguistic variable member function; $i = 1, ..., K$; K – number of linguistic variables; σ_i – fuzziness coefficient of the i-th linguistic variable (it's possible when $\sigma_1 = \sigma_2 = \ = \sigma_K$)

Model (1) can be represented graphically as shown Fig. 1.

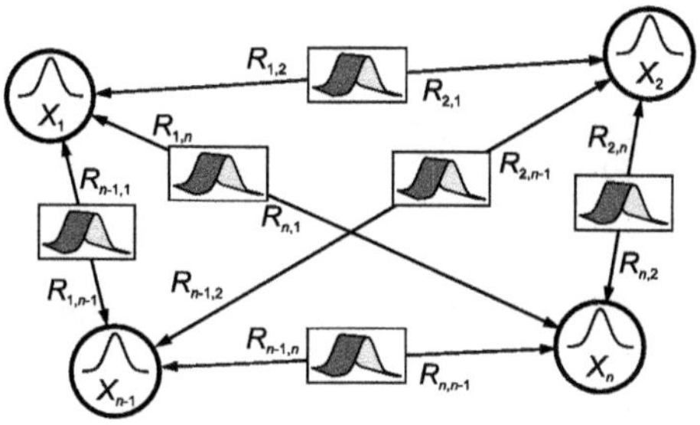

Fig. 1. Visualization of fuzzy cognitive map (1), where: X_i – fuzzy value of the i-th concept, $R_{i,j}$ - fuzzy relation between i and j concepts ($i, j = 1, ..., n$)

Creating the individual fuzzy relations $R_{i,j}$, mentioned in (1), (2) and shown in Fig. 1, is a separate issue, which was discussed, among others, in [10]. Their construction is based on two-argument Gaussian-type member functions $\mu_{R_{i,j}}$ (although triangular, trapezoid and other functions also can be used [10]) of following type:

$$\mu_{R_{i,j}}(x_1, x_2) = e^{-\left(\frac{x_2 - r_{i,j}(x_1)}{\sigma_{i,j}}\right)^2} \tag{4}$$

where: $\mu_{R_{i,j}}$ – member function of the fuzzy relation between concepts i-th and j-t; x_1, x_2 - axes of fuzzy relation universum; $\sigma_{i,j}$ - fuzziness coefficient of fuzzy relation between concepts i and j; $r_{i,j}$ - fuzzy relation power coefficient ($r_{i,j}(1)$ corresponds with relation power in crisp model).

In the article, for the choice of conditionally optimal parameters of the fuzzyfication (first stage of the optimization), the compliance criterion of fuzzyfication and defuzzyfication in following symbolic form [10] was applied:

$$J(Q) = \|X^w - X^o\|^2 \Rightarrow \min_Q \tag{5}$$

$$Q_{l+1} = Q_l - \alpha_l \Delta J(Q_l) \tag{6}$$

where: X^w – value of fuzzyfied concept (after defuzzyfication); X^o – the value of reference concept (crisp); $Q = [\overline{X}_1, ..., \overline{X}_K, \sigma_1, ..., \sigma_K, K]^T$ – vector of the fuzzyfication parameters; $\|\|$ – selected norm, e.g.: $J(Q) = \frac{1}{2}\left(X_j^w(Q) - X_j^o(Q)\right)^2$; $\alpha_l > 0$ – algorithm step; j – number of the selected concept; $\Delta J(Q_l)$ – direction rate of $J(Q)$ increment; $l = 0, ..., L$.

To solve problem (5) one can, in principle, use different optimization algorithms based on e.g. gradient, genetic and other methods [1,4,7].

4 Adaptation Algorithm of the Relations Parameters

When implementing the algorithm of fuzzy relations $R_{i,j}$ parameters adaptation different types of learning algorithms can be applied what makes possible to change different features of the relations.

General algorithm of the adaptation of fuzzy relations functions is presented below as:
– dynamic connections between $X_i(t)$ and $X_i(t+1)$ in the object and the adaptation model of a type (2);
– adaptation criterion:

$$J_1(P) = \|X(t) - X^M(t)\|^2 \Rightarrow \min_P \tag{7}$$

where: $P = \{R_{j,i}\}$ – adaptation parameters matrix; $\|\|$ – selected norm; $X(t) = [X_1(t), ..., X_n(t)]^T$; $X^M(t) = [X_1^M(t), ..., X_n^M(t)]^T$ – values of concepts in the object and the model; $X(0) = X^M(0)$; $t = 0, ..., T-1$; T – discrete time range; $i, j = 1, ..., n$;

– formal algorithm of relation $R_{j,i}(t)$ adaptation, type:

$$R_{j,i}(t+1) = R_{j,i}(t) \oplus \Delta R_{j,i}(t) \tag{8}$$

where: $R_{j,i}(t)$ – fuzzy relation between j and i concepts of the system's model in t adaptation step; $\Delta R_{j,i}(t)$ – "increase" of fuzzy relation; $R(0)$ – initially set relation; $t = 0, ..., T-1$; $i, j = 1, ..., n$.

Algorithm (8) can be attended as well as unattended. Adaptation process (8) formally consists of numeriac and functional adaptations. The specific example of algorithm (8) will be presented in the simulation research results description (chapter 5).

5 Selected Simulation Results

This chapter presents some results of the simulation for the second stage (adaptation of individual relations' parameters) of the model optimization (described in chapters 3 and 4) for the example of relational cognitive map from Fig. 2.

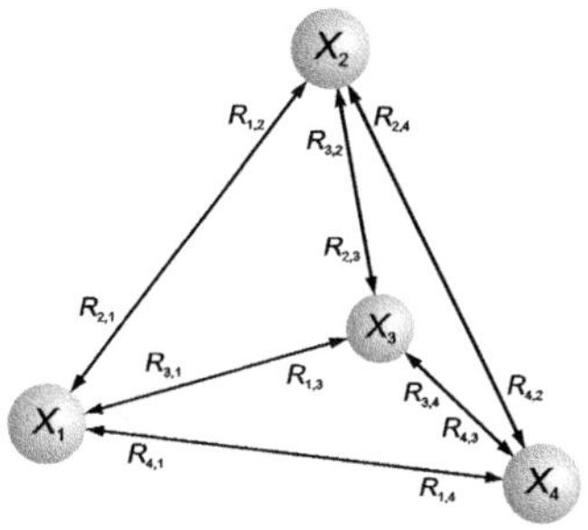

Fig. 2. An exemplary fuzzy relational cognitive map. X_1, ..., X_4 – fuzzy values of concepts; $R_{i,j}$ – fuzzy relation between i and j concepts of the object map

5.1 Model Building Assumptions

The following assumptions were introduced into the model:

Powers of relations – presented in Table 1 in a scalar form.

Table 1. A table of fuzzy relations' powers (crisp relations matrix)

r	X_1	X_2	X_3	X_4
X_1	0	0	0.5	0.4
X_2	0.3	0	0.3	0
X_3	0	0	0	0.4
X_4	0	-0.6	0	0

The initial values of concepts
The model's operation consisted of a single stimulation of selected concepts using external signals, of which values are presented in Table 2.

Table 2. Values of external stimulating signals

Concept	X_1	X_2	X_3	X_4
Stimulation value	0.6	0.3	0	0

Fuzzyfication universum
For the purpose of analysis, normalized values from the range [-1, 1] were used. Concepts' and relations' values were fuzzyfied on a universum in the range of [-2, 2].
Type of member functions for fuzzy concepts and relations
Gaussian-type member functions were used in accordance to (3) for fuzzy values of concepts and in accordance to (4) for fuzzy relations.
Dynamic model of fuzzy cognitive map has the form of (2) for $n = 4$ and uses adaptation criterion in the form (7).

5.2 Parameters Adaptation of Individual Relations

After selecting the optimal set of general parameters combination one can go to the modification of individual fuzzy relations' parameters.

The adaptation algorithm type (8) for the t cycle of signals circulation inside the dynamic model (t step of discrete time) consists in recurrent executing of the following consecutive steps:

1. Execute t cycles of signals circulation in crisp reference model – obtaining reference values of examined concepts (they can be got in a different way, e.g. from expert knowledge or measurement data).
2. Select initial values of parameters change coefficients Δr and $\Delta\sigma$.
3. Execute t cycles of signals circulation in a fuzzy model.
4. Calculate defuzzyfied values of concepts aberrations from crisp reference values (determined in step 1).
5. For each tested concept (e.g. X_p) and each relation, by which takes the signals from other concepts (e.g. $R_{i,p}$) investigate whether it's possible to reach the concept value closer to the reference value after increasing or decreasing the relation power $r_{i,p}$ with Δr. If so, a new $r_{i,p}$ value has to be accepted.
6. For each examined concept (e.g. X_p) and each relation, by which takes the signals from other concepts (e.g. $R_{i,p}$) investigate whether it's possible to reach the concept value closer to the reference value after increasing or decreasing the fuziness coefficient $\sigma_{i,p}$ with $\Delta\sigma$. If so, a new $\sigma_{i,p}$ value has to be accepted.
7. After modification of all fuzzy relations investigate whether the assumed model accuracy was obtained. If not, go back to step 3. If it's deemed necessary, you can change (reduce) parameters Δr and $\Delta\sigma$ value.
8. Repeat above steps until assumed criterion of the algorithm end is achieved.

The described above algorithm was used to designate the time waveforms of powers and fuzziness coefficients of fuzzy relations for the model of Fig. 2 with parameters as in Table 1 with initial values as in Table 2. Adaptation of fuzzy relations have been made for subsequent steps (within the range from the 5th till the 30th step) discrete time in fuzzy model with universum range of [-2, 2]. For the model purpose, $K = 17$ and $\sigma = 0.6$ (for all fuzzyfied concepts and relations) were used. As a result sets of values of relations' powers and fuzziness coefficients were obtained for all considered steps of discrete time. Fig. 3 presents selected courses (for fuzzy relations that influenced X_2) of these parameters changes depending on discrete time step chosen as reference moment.

Initial (before the adaptation process) values of $r_{1,2}$, $r_{3,2}$ and $r_{4,2}$ parameters are presented in Table 1.

It is worth to notice, that adjusting of the model parameters in adaptation process led to "creation" of new relations (initial zero values of powers of fuzzy relations $R_{1,2}$ and $R_{3,2}$ are replaced with the others – nonzero).

To illustrate the correctness of the method, three simulations of the course of concept X_2 value were carried out. There were respected conditions as in Table 2: for crisp model, fuzzy model without adaptation (with parameters from Table 1)

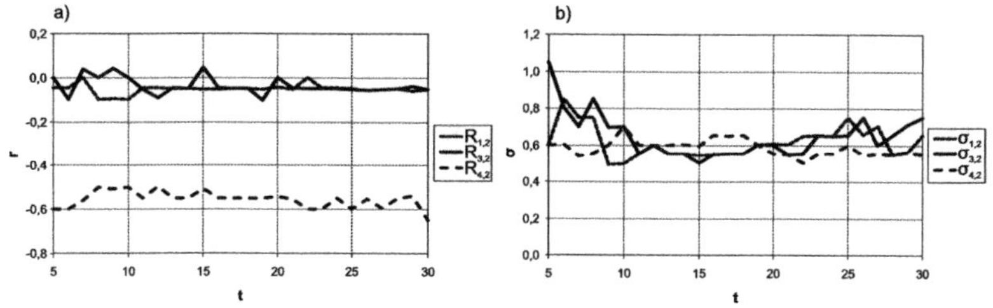

Fig. 3. Courses of changes of selected fuzzy relation powers: a) and fuzziness coefficients: b) obtained as a result of the relation adaptation in relational fuzzy cognitive map from Fig. 2. t - discrete time; r - relation power; σ - relation fuzziness coefficient

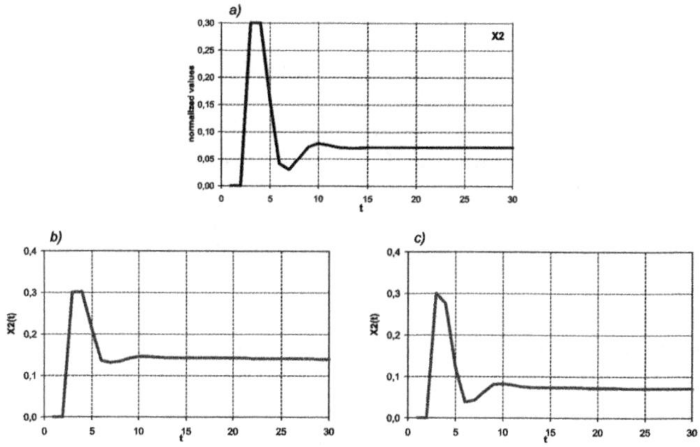

Fig. 4. Parameters adaptation result in model from Fig. 2, made for a single step of discrete time (t = 30). Time courses of X_2 value in a crisp model a) and (after defuzzyfication) in fuzzy model: b) before adaptation c) after adaptation

and fuzzy model after adaptation (with parameters obtained for the selected – 30th step of discrete time). The result of reference crisp model operation is shown in Fig. 4a). Comparison results of courses in fuzzy models (after defuzzyfication) are presented in Fig. 4b) and c).

Comparison of courses from Figs.: 4a), 4b) and 4c) proves significant convergence of concepts values (in the 30th step of discrete time) of crisp and fuzzy models after adaptation of relations parameters.

6 Conclusions

Optimization and adaptation of fuzzy model parameters is crucial for its accuracy and, hence, its usefulness for modelling the activity of real systems. The approach to the optimization of the model parameters, presented in the article, allows creating fuzzy cognitive maps that meet two main conditions of the model efficiency: possibly high accuracy and possibly low number of linguistic variables.

Selected results of the research, adduced in the article, indicate advantages of the described approach. The modification algorithm (in continual mode) of selected fuzzy relations' parameters is planned to be worked out in the next stage, to make them optimal not only in the one selected, but in each step of discrete time.

References

1. Borisov, V.V., Kruglow, V.V., Fedulov, A.C.: Fuzzy Models and Networks. Publishing house Telekom, Moscow, Russia (2004) (in Russian)
2. Kandasamy, W.B.V., Smarandache, F.: Fuzzy Cognitive Maps and Neutrosophic Cognitive Maps, Xiquan, Phoenix, AZ, USA (2003)
3. Kosko, B.: Fuzzy cognitive maps. Int. Journal of man-Machine Studies 24, 65–75 (1986)
4. Papageorgiou, E.I., Stylios, C.D.: Fuzzy Cognitive Maps. In: Pedrycz, W., Skowron, A., Kreinovich, V. (eds.) Handbook of Granular Computing. John Wiley & Son Ltd, Publication Atrium, Chichester, England (2008)
5. Piegat, A.: Fuzzy Modelling and Control. Physica-Verlag, Springer-Verlag Company, Heidelberg (2001)
6. Słoń, G.: Analysis of Selected Adaptation Algorithms for Relations in Fuzzy Cognitive Maps. Measurement Automation and Monitoring 56(12), 1445–1448 (2010) (in Polish)
7. Stach, W., Kurgan, L., Pedrycz, W., Reformat, M.: Genetic Learning of Fuzzy Cognitive Maps. Fuzzy Sets and Systems 153, 371–401 (2005)
8. Yastrebov, A., Słoń, G.: Fuzzy Cognitive Maps in Relational Modelling Monitoring Systems. In: Kowalczuk Z. (red.) Systems for Faults Detection, Analysis and Toleration. PWNT, Gdańsk, Poland 2009, pp. 217-224 (in Polish)
9. Yastrebov, A., Słoń, G.: Granular Calculations in Modeling Imprecise Objects Using Relational Fuzzy Cognitive Maps. Measurement Automation and Monitoring 56(12), 1449–1452 (2010) (in Polish)
10. Yastrebov, A., Słoń, G.: Synthesis and Computational Analysis of Intelligent Models Based on Cognitive Maps – Part I. Synthesis. In: Yasrebov, A. (ed.) Computer Science at XXI Century Time. Information Technologies and Their Applications, pp. 77–86. Science Publishing House of Exploitation Technology Institute – State Research Institute, Radom (2010) (in Polish)
11. Yastrebov, A., Słoń, G.: Synthesis and Computational Analysis of Intelligent Models Based on Cognitive Maps – Part II. Analysis. In: Yastrebov, A. (ed.) Computer Science at XXI Century Time. Information Technologies and Their Applications, pp. 77–86. Science Publishing House of Exploitation Technology Institute – State Research Institute, Radom (2010) (in Polish)

Sensitivity Analysis for Fuzzy Linear Programming Problems

Amit Kumar and Neha Bhatia

School of Mathematics and Computer Applications
Thapar University, Patiala, 147004, India
amit_rs_iitr@yahoo.com, neha26bhatia@gmail.com

Abstract. Kheirfam and Hasani [5] proposed a method for the sensitivity analysis for fuzzy linear programming problem with fuzzy variables. In this paper, a new method is proposed for solving same type of problems. The main advantage of proposed method over the existing method is that if the same problems are solved by both the existing as well as proposed method then the obtained results will be same, while, it is easy and less time consuming to apply the proposed method as compared to existing method. To illustrate the proposed method and to show the advantage of proposed method the numerical examples, solved by Kheirfam and Hasani, are solved by the proposed method and the obtained results are compared.

Keywords: Fuzzy linear programming problems, Ranking function, Sensitivity analysis, Trapezoidal fuzzy numbers.

1 Introduction

The fuzzy set theory [13] is being applied massively in many fields these days. One of these is linear programming problems. Sensitivity analysis is well-explored area in classical linear programming. Sensitivity analysis is a basic tool for studying perturbations in optimization problems. There is considerable research on sensitivity analysis for some operations research and management science models such as linear programming and investment analysis.

Fuzzy linear programming (FLP) provides the flexibility in values. But even after formulating the problem as FLP problem, one cannot stick to all the values for a long time or it is quite possible that the wrong values got entered. With time the factors like cost, required time or availability of product etc. changes widely. Sensitivity analysis for FLP problem needs to be applied in that case. Sensitivity analysis is one of the interesting researches in FLP problem.

Zimmermann [14] attempted to fuzzify a linear program for the first time, fuzzy numbers being the source of flexibility. Zimmermann also presented a fuzzy approach to multi-objective linear programming problem and its sensitivity analysis. Sensitivity analysis in FLP problem with crisp parameters and soft constraints was first considered by Hamacher et al. [4].

Tanaka and Asai [9] proposed a method of allocating the given investigation cost to each fuzzy coefficients by using sensitivity analysis. Tanaka et al. [10] formulated

S.O. Kuznetsov et al. (Eds.): RSFDGrC 2011, LNAI 6743, pp. 103–110, 2011.

a FLP problem with fuzzy coefficients and the value of information was discussed via sensitivity analysis. Sakawa and Yano [8] presented a fuzzy approach for solving multi- objective linear fractional programming problem via sensitivity analysis.

Fuller [2] proposed that the solution to FLP problems with symmetrical triangular fuzzy numbers is stable with respect to small changes of centers of fuzzy numbers. Perturbations occur due to calculation errors or just to answer managerial questions "What if …". Such questions propose after the simplex method and the related research area refers to as basis invariance sensitivity analysis.

Dutta et al. and several other authors [1], [11] studied sensitivity analysis for fuzzy linear fractional programming problem. Gupta and Bhatia [3] studied the measurement of sensitivity for changes of violations in the aspiration level for the fuzzy multi-objective linear fractional programming problem. Precup and Preitl [7] performed the sensitivity analysis for some fuzzy control systems. Lotfi et al. [6] developed a sensitivity analysis approach for the additive model. Kheirfam and Hasani [5] studied the basis invariance sensitivity analysis for FLP problems.

In this paper, the shortcomings of the existing methods [5] are pointed out. To overcome these shortcomings, a new method is proposed for the sensitivity analysis for fuzzy variable linear programming problem. To illustrate the proposed method, numerical examples are solved and the obtained results are discussed.

This paper is organized as follows: A new method for sensitivity analysis for fuzzy variable linear programming problems is proposed in Section 2. In Section 3, proposed method is explained with the help of numerical examples. Conclusions are discussed in Section 4.

2 Proposed Method

Kheirfam and Hasani [5] proposed a method for the sensitivity analysis of the following type of FLP problems with fuzzy variables:

Maximize (or Minimize) $\tilde{z} =_{\Re} C^T \tilde{X}$,

Subject to

$$A\tilde{X} \leq_{\Re} \text{ or } =_{\Re} \text{ or } \geq_{\Re} \tilde{b}, \tag{1}$$

$\tilde{X}$ is non-negative trapezoidal fuzzy vector.

where, $\tilde{b} = [\tilde{b}_j]_{m\times 1}$, $\tilde{X} = [\tilde{x}_j]_{n\times 1}$, $A = [a_{ij}]_{m\times n}$, $C^T = [c_j]_{1\times n}$, $\Re$ is a linear ranking function, $\tilde{x}_j = (a_j, b_j, c_j, d_j)$ and $\tilde{b}_j = (p_j, q_j, r_j, s_j)$ are the trapezoidal fuzzy numbers.

In this section, a new method is proposed for dealing with same type of problems. The steps of proposed method are as follows:

Step 1. Convert the FLP problem (1), into the following crisp linear programming (CLP) problem (2).

Maximize (or Minimize) $\Re(C^T \tilde{X})$,

Subject to

$$\Re(A\tilde{X}) \leq or = or \geq \Re(\tilde{b}),\ \Re(\tilde{X}) \geq \tilde{0}. \tag{2}$$

Step 2. Solve the CLP problem, obtained in Step 1, to find the optimal solution a_j, b_j, c_j and d_j.

Step 3. Find the fuzzy optimal solution by substituting the values of a_j, b_j, c_j and d_j, obtained from Step 2, in $\tilde{x}_j = (a_j, b_j, c_j, d_j)$.

Step 4. Check that which of the following case is to be considered:

1. Change in requirement vector $\tilde{b}$.
2. Addition of a new fuzzy variable.
3. Addition of new fuzzy constraint.
4. Deletion of a fuzzy variable.
5. Deletion of a fuzzy constraint.

Case 1: Change in Requirement Vector $\tilde{b}$

If the change in right hand side (RHS) or requirement vector is made i.e., $\tilde{b}$ is changed to $\tilde{b}'$ in (1) then, replace $\Re(\tilde{b})$ by $\Re(\tilde{b}')$ in CLP problem (2) to obtain (3).

Maximize (or Minimize) $\Re(C^T \tilde{X})$,

Subject to

$$\Re(A\tilde{X}) = \Re(\tilde{b}'),\ \Re(\tilde{X}) \geq 0 \tag{3}$$

where, $\tilde{b}' = [\tilde{b}_j]_{m\times 1}, \tilde{X} = [\tilde{x}_j]_{n\times 1}, A = [a_{ij}]_{m\times n}, \tilde{C}^T = [\tilde{c}_j]_{1\times n}$ and $\Re$ is a linear ranking function, $\tilde{x}_j = (a_j, b_j, c_j, d_j)$ and $\tilde{b}_j = (p_j, q_j, r_j, s_j)$ are the trapezoidal fuzzy numbers. Now apply the existing sensitivity analysis technique to find the optimal solution of (3) with the help of optimal solution of (2) and use Step 3 of the proposed method to find the fuzzy optimal solution of the resulting FLP problem.

Case 2: Addition of a New Fuzzy Variable

Suppose a new fuzzy variable, say $\tilde{x}_{n+1}$ having $\Re(\tilde{x}_{n+1}) \geq 0$ be added in (1). Assume that c_{n+1} is the cost and A_{n+1} is the column associated with $\tilde{x}_{n+1}$ then replace $\Re(A\tilde{X})$ by $\Re(A\tilde{X} \oplus A_{n+1}\tilde{x}_{n+1})$ and $\Re(C^T \tilde{X})$ by $\Re(C^T \tilde{X} \oplus c_{n+1}\tilde{x}_{n+1})$ in CLP problem (2) to obtain the new CLP problem (4).

Maximize (or Minimize) $\Re(C^T \tilde{X} \oplus c_{n+1}\tilde{x}_{n+1})$,

Subject to

$$\Re(A\tilde{X} \oplus A_{n+1}\tilde{x}_{n+1}) = \Re(\tilde{b}),\ \Re(\tilde{X}) \geq 0, \Re(\tilde{x}_{n+1}) \geq 0. \tag{4}$$

where, $\tilde{b} = [\tilde{b}_j]_{m\times 1}, \tilde{X} = [\tilde{x}_j]_{n\times 1}, A = [a_{ij}]_{m\times n}, C^T = [c_j]_{1\times n}$ and $\Re$ is a linear ranking function. Now apply the existing sensitivity analysis technique to find the optimal solution of (4) with the help of optimal solution of (2) and use Step 3 of the proposed method to find the fuzzy optimal solution of the resulting FLP problem.

Case 3: Addition of a New Fuzzy Constraint

Suppose a new fuzzy constraint is added in the original FLP problem (1) then, replace $\Re(A\tilde{X}) = \Re(\tilde{b})$ by $\Re(A'\tilde{X}) = \Re(\tilde{b}')$ in CLP problem (2) to obtain new CLP problem (5).

Maximize (or Minimize) $\Re(C^T \tilde{X})$,

Subject to

$$\Re(A'\tilde{X}) = \Re(\tilde{b}'),\ \Re(\tilde{X}) \geq 0. \qquad (5)$$

where, $\tilde{b}' = [\tilde{b}_j]_{(m+1)\times 1}, \tilde{X} = [\tilde{x}_j]_{n\times 1}, A' = [a_{ij}]_{(m+1)\times n}, C^T = [c_j]_{1\times n}$ and $\Re$ is a linear ranking function. Now apply the existing sensitivity analysis technique to find the optimal solution of (5) with the help of optimal solution of (2) and use Step 3 of the proposed method to find the fuzzy optimal solution of the resulting FLP problem.

Case 4: Deletion of a Fuzzy Variable

Suppose a fuzzy variable $\tilde{x}_n$ having $\Re(\tilde{x}_n) \geq 0$ is deleted from the original FLP problem (1) then, replace $\Re(A\tilde{X})$ by $\Re(A'\tilde{X}')$ and $\Re(C^T\tilde{X})$ by $\Re(C^T\tilde{X}')$ in CLP problem (2) to obtain new CLP problem (6).

Maximize (or Minimize) $\Re(C^T \tilde{X}')$,

Subject to

$$\Re(A'\tilde{X}') = \Re(\tilde{b}),\quad \Re(\tilde{X}') \geq 0. \qquad (6)$$

where, $\tilde{b} = [\tilde{b}_j]_{m\times 1}, \tilde{X}' = [\tilde{x}_j]_{(n-1)\times 1}, A = [a_{ij}]_{m\times(n-1)}, C^T = [c_j]_{1\times(n-1)}$ and $\Re$ is a linear ranking function. Now apply the existing sensitivity analysis technique to find the optimal solution of (6) with the help of optimal solution of (2) and use Step 3 of the proposed method to find the fuzzy optimal solution of the resulting FLP problem.

Case 5: Deletion of a Fuzzy Constraint

Suppose a fuzzy constraint is to be deleted from original FLP problem (1) then, replace $\Re(A\tilde{X}) = \Re(\tilde{b})$ by $\Re(A'\tilde{X}) = \Re(\tilde{b}')$ in CLP problem (2) to obtain new CLP problem (7).

Maximize (or Minimize) $\Re(C^T \tilde{X})$,

Subject to

$$\Re(A'\tilde{X}) = \Re(\tilde{b}'),\ \Re(\tilde{X}) \geq 0. \qquad (7)$$

where, $\tilde{b} = [\tilde{b}_j]_{(m-1)\times 1}, \tilde{X} = [\tilde{x}_j]_{n\times 1}, A = [a_{ij}]_{(m-1)\times n}, C^T = [c_j]_{1\times n}$ and $\Re$ is a linear ranking function. Now apply the existing sensitivity analysis technique to find the optimal solution of (7) with the help of optimal solution of (2) and use Step 3 of the proposed method to find the fuzzy optimal solution of the resulting FLP problem.

Remark: The definitions of trapezoidal fuzzy numbers, non-negative trapezoidal fuzzy numbers, equality of trapezoidal fuzzy numbers, arithmetic operations between trapezoidal fuzzy numbers and ranking index can be seen in Kheirfam and Hasani [5].

3 Numerical Examples

In this section, proposed method is illustrated with the help of a numerical example:

Example 3.1 Consider the FLP problem,

Minimize $\tilde{z} =_{\Re} \Theta\ \tilde{x}_1\ \Theta\ \tilde{x}_2 \Theta\ 2\tilde{x}_3,$

Subject to

$$\tilde{x}_1 \oplus \tilde{x}_2 \oplus 2\tilde{x}_3 \leq_{\Re} (3,5,8,13),\ \tilde{x}_2 \leq_{\Re} (4,6,10,16),\ \tilde{x}_1 \Theta \tilde{x}_1 \oplus \tilde{x}_2 \oplus \tilde{x}_3 \leq_{\Re} (-6,1,6,14),$$
$$\tilde{x}_1, \tilde{x}_2, \tilde{x}_3 \geq_{\Re} \tilde{0}. \tag{8}$$

(a) Discuss the effect of changing the requirement vector from $(3,5,8,13),(4,6,10,16),(-6,1,6,14)$ to $(8,10,12,15),(1,3,5,6),(2,4,8,11)$ on the fuzzy optimal solution of resulting FLP problem.

(b) Find the effect of addition of a new fuzzy variable $\tilde{x}_4$ having $\Re(\tilde{x}_4)\geq 0$ with cost 8 and column vectors $(0,5,7)^T$ respectively on the current fuzzy optimal solution.

(c) Find the effect of addition of a new fuzzy constraint $2\tilde{x}_1 \oplus \tilde{x}_2 \oplus \tilde{x}_3 \leq_{\Re} (1,2,3,4)$ on the current fuzzy optimal solution.

(d) If a fuzzy variable $\tilde{x}_3$ having $\Re(\tilde{x}_3)\geq 0$ is deleted from the given FLP problem (8) find the fuzzy optimal solution of resulting FLP problem.

(e) Discuss the effect of deletion of a fuzzy constraint $\tilde{x}_1 \oplus \tilde{x}_2 \oplus \tilde{x}_3 \leq_{\Re} (-6,1,6,14)$ from given FLP problem (8) on the fuzzy optimal solution of resulting FLP problem.

Solution:

Assuming $\tilde{x}_1 = (a_1, b_1, c_1, d_1)$, $\tilde{x}_2 = (a_2, b_2, c_2, d_2)$ and $\tilde{x}_3 = (a_3, b_3, c_3, d_3)$ and using Step 1 of the proposed method the FLP problem (8) is converted into the following CLP problem:

Minimize $\frac{1}{4}(-a_1 - a_2 - 2a_3 - b_1 - b_2 - 2b_3 - c_1 - c_2 - 2c_3 - d_1 - d_2 - 2d_3),$

Subject to

$$\begin{aligned}
&a_1 + a_2 + 2a_3 + b_1 + b_2 + 2b_3 + c_1 + c_2 + 2c_3 + d_1 + d_2 + 2d_3 \leq 29,\\
&a_1 - a_2 + b_1 - b_2 + c_1 - c_2 + d_1 - d_2 \leq 36,\\
&a_1 + a_2 + a_3 + b_1 + b_2 + b_3 + c_1 + c_2 + c_3 + d_1 + d_2 + d_3 \leq 15,\\
&a_1 + b_1 + c_1 + d_1 \geq 0,\ a_2 + b_2 + c_2 + d_2 \geq 0,\ a_3 + b_3 + c_3 + d_3 \geq 0,\\
&b_1 - a_1 \geq 0,\ c_1 - b_1 \geq 0, d_1 - c_1 \geq 0,\ b_2 - a_2 \geq 0,\ c_2 - b_2 \geq 0, d_2 - c_2 \geq 0,\\
&b_3 - a_3 \geq 0,\ c_3 - b_3 \geq 0, d_3 - c_3 \geq 0.
\end{aligned} \tag{9}$$

The optimal solution of the CLP problem (9) is:

$a_1 = 0, a_2 = 0, a_3 = 0, b_1 = 0, b_2 = 0, b_3 = 0, c_1 = 0, c_2 = 0, c_3 = 0, d_1 = 0, d_2 = 0, d_3 = 14.5$ and the optimal value is -7.25. Using Step 3 of the proposed method the fuzzy optimal solution is given by

$$\tilde{x}_1 = (0,0,0,0), \tilde{x}_2 = (0,0,0,0), \tilde{x}_3 = (0,0,0,14.5).$$

Since, $\Re(\tilde{x}_1)=0, \Re(\tilde{x}_2)=0, \Re(\tilde{x}_3)=3.625$ so all the trapezoidal fuzzy numbers $\tilde{x}_1, \tilde{x}_2, \tilde{x}_3$ whose rank will be 0, 0 and 3.625 respectively will be the fuzzy optimal solution of the resulting FLP problem and putting all such values of $\tilde{x}_1, \tilde{x}_2, \tilde{x}_3$ in objective function, a fuzzy number representing the fuzzy optimal value of resulting FLP problem, may be obtained. In all cases, rank of obtained trapezoidal fuzzy number will be -7.25 i.e. there will be an infinite numbers of fuzzy optimal solutions of the resulting problem. Some of the fuzzy optimal solutions of the resulting FLP problem, different from solution obtained by Kheirfam and Hasani [5] are shown below:

(i) $\tilde{x}_1 = (-2,-1,1,2), \tilde{x}_2 = (-2,-1,1,1), \tilde{x}_3 = (1,2.5,4,7)$ and rank of fuzzy optimal value is -7.25.

(ii) $\tilde{x}_1 = (-3,0,1,2), \tilde{x}_2 = (-3,0,1,2), \tilde{x}_3 = (2.5,3,4,5)$ and rank of fuzzy optimal value is -7.25 etc.

(a) Since the requirement vector is changed from $(3,5,8,13), (4,6,10,16)$ and $(-6,1,6,14)$ to $(8,10,12,15), (1,3,5,6)$ and $(2,4,8,11)$ in the original FLP problem (8) so replacing $\Re(3,5,8,13), \Re(4,6,10,16)$ and by $\Re(8,10,12,15), \Re(1,3,5,6)$ and $\Re(2,4,8,11)$ i.e., 29, 36 and 15 by 45, 15 and 25 respectively in (9). Applying existing sensitivity analysis technique, the optimal solution of resulting CLP problem with the help of optimal solution of CLP problem (9) is

$a_1 = 0, a_2 = 0, a_3 = 0, b_1 = 0, b_2 = 0, b_3 = 0, c_1 = 0, c_2 = 0, c_3 = 0, d_1 = 0, d_2 = 0, d_3 = 22.5$ and the optimal value is -11.25. Using Step 3, the fuzzy optimal solution is given by

$$\tilde{x}_1 = (0,0,0,0), \tilde{x}_2 = (0,0,0,0), \tilde{x}_3 = (0,0,0,22.50).$$

(b) Suppose a new fuzzy variable $\tilde{x}_4$ having $\Re(\tilde{x}_4) \geq 0$ with cost 8 and column $(0,5,7)^T$ respectively is added in the original FLP problem (8) then, first convert the CLP problem (9) into a new CLP problem as discussed in Case 2. Applying existing sensitivity analysis technique, the optimal solution of the resulting CLP problem with the help of optimal solution of CLP problem (9) is:

$a_1 = 0, a_2 = 0, a_3 = 0, b_1 = 0, b_2 = 0, b_3 = 0, c_1 = 0, c_2 = 0, c_3 = 0, d_1 = 0, d_2 = 0, d_3 = 14.5,$
$a_4 = 0, b_4 = 0, c_4 = 0, d_4 = 0$

and the optimal value is -7.25. Using Step 3, the fuzzy optimal solution is given by

$$\tilde{x}_1 = (0,0,0,0), \tilde{x}_2 = (0,0,0,0), \tilde{x}_3 = (0,0,0,14.5), \tilde{x}_4 = (0,0,0,0).$$

(c) Suppose a new fuzzy constraint $2\tilde{x}_1 \oplus \tilde{x}_2 \oplus \tilde{x}_3 \leq_{\Re} (1,2,3,4)$ is added to the original FLP problem (8) then, add the constraint:

$2a_1+a_2+a_3+2b_1+b_2+b_3+2c_1+c_2+c_3+2d_1+d_2+d_3\leq 10$ to CLP problem (9). Applying existing sensitivity analysis technique, the optimal solution of resulting CLP problem with the help of optimal solution of CLP problem (9) is:

$a_1=0, a_2=0, a_3=0, b_1=0, b_2=0, b_3=0, c_1=0, c_2=0, c_3=0, d_1=0, d_2=0, d_3=10$ and the optimal value is -5. Using Step 3, the fuzzy optimal solution is given by

$$\tilde{x}_1=(0,0,0,0), \tilde{x}_2=(0,0,0,0), \tilde{x}_3=(0,0,0,10)$$

(d) Suppose a fuzzy variable $\tilde{x}_3$ having $\Re(\tilde{x}_3)\geq 0$ is deleted from the given FLP problem (8) then, first convert the CLP problem (9) into a new CLP problem as discussed in Case 4. Applying existing sensitivity analysis technique, the optimal solution of resulting CLP problem with the help of optimal solution of CLP problem (9) is:

$a_1=0, a_2=0, b_1=0, b_2=0,\ c_1=0, c_2=0,\ d_1=0, d_2=15$ and the optimal value is $z=-3.75$. Using Step 3, the fuzzy optimal solution is given by

$$\tilde{x}_1=(0,0,0,0), \tilde{x}_2=(0,0,0,15)$$

(e) Suppose a fuzzy constraint $\tilde{x}_1 \oplus \tilde{x}_2 \oplus \tilde{x}_3 \leq_{\Re} (-6,1,6,14)$ is deleted from the original FLP problem (8) then, delete the constraint:

$a_1+a_2+a_3+b_1+b_2+b_3+c_1+c_2+c_3+d_1+d_2+d_3\leq 15$ from CLP problem (9). Applying existing sensitivity analysis technique, the optimal solution of resulting CLP problem with the help of optimal solution of CLP problem (9) is:

$a_1=0, a_2=0, a_3=0, b_1=0, b_2=0, b_3=7.25, c_1=0, c_2=0, c_3=7.25, d_1=0, d_2=0, d_3=0$

and the optimal value is -7.25. Using Step 3, the fuzzy optimal solution is given by

$$\tilde{x}_1=(0,0,0,0), \tilde{x}_2=(0,0,0,0), \tilde{x}_3=(0,7.25,7.25,0).$$

4 Conclusions

In this paper, a new method is proposed for sensitivity analysis for fuzzy variable linear programming problems. By using proposed method the fuzzy optimal solution of FLP problems can be easily obtained and also it is less time consuming as compared to the existing method [6]. To illustrate the proposed method numerical examples are solved.

Acknowledgements

The authors would like to thank to the anonymous reviewers for their suggestions. Special thanks go to Ms. Mehar Kaur.

References

[1] Dutta, D., Rao, J.R., Tiwari, R.N.: Sensitivity analysis in fuzzy linear fractional programming problem. Fuzzy Sets and Systems 48, 211–216 (1992)
[2] Fuller, R.: On stability in fuzzy linear programming problems. Fuzzy Sets and Systems 30, 339–344 (1989)
[3] Gupta, P., Bhatia, D.: Sensitivity analysis in fuzzy multi-objective linear fractional programming problem. Fuzzy Sets and Systems 122, 229–236 (2001)
[4] Hamacher, H., Leberling, H., Zimmermann, H.J.: Sensitivity analysis in fuzzy linear programming. Fuzzy Sets and Systems 1, 269–281 (1978)
[5] Kheirfam, B., Hasani, F.: Sensitivity analysis for fuzzy linear programming problems with fuzzy variables. Advanced Modeling and Optimization 12, 257–272 (2010)
[6] Lotfi, F.H., Jondabeh, M.A., Faizrahnemoon, M.: Senstivity analysis in fuzzy environment. Applied Mathematical Sciences 4, 1635–1646 (2010)
[7] Precup, R.E., Preitl, S.: Stability and sensitivity analysis of fuzzy control systems. Mechatronics Applications 3, 61–76 (2006)
[8] Sakawa, M., Yano, H.: An interactive fuzzy satisfying method for multi-objective linear fractional programming problems. Fuzzy Sets and Systems 28, 129–144 (1988)
[9] Tanaka, H., Asai, K.: Fuzzy linear programming with fuzzy number. Fuzzy Sets and Systems 13, 1–10 (1984)
[10] Tanaka, H., Ichihashi, H., Asai, K.: A value of information in fuzzy linear programming problems via sensitivity analysis. Fuzzy Sets and Systems 18, 119–129 (1986)
[11] Verdegay, J.L., Aguado, C.G.: On the sensitivity of membership function for fuzzy linear programming problem. Fuzzy Sets and Systems 56, 47–49 (1993)
[12] Yager, R.R.: A procedure for ordering fuzzy subsets of the unit interval. Information Sciences 24, 143–161 (1981)
[13] Zadeh, L.A.: Fuzzy sets. Information and Control 8, 338–353 (1965)
[14] Zimmerman, H.J.: Description and optimization of fuzzy systems. International Journal of General Systems 2, 209–215 (1976)

Estimation of Parameters of the Empirically Reconstructed Fuzzy Model of Measurements

Tatiana Kopit and Alexey Chulichkov

Department of Computer Methods of Physics, Faculty of Physics, Moscow State University, Moscow, 119991 Russia
kopit_tanya@mail.ru, achulichkov@gmail.com

Abstract. In this paper we introduce a method for the empirical reconstruction of a fuzzy model of measurements on the basis of testing measurements using a possibility-theoretical approach. The method of measurement reduction is developed for solving a problem of an estimation of parameters of a fuzzy system. It is shown that such problems are reduced to minimax problems. If the model is unknown it can be restored from testing experiments and can be applied for handling the problems of the type of forecasting the behavior of a system.

Keywords: mathematical modeling, fuzzy sets, decision making, analysis and interpretation of data, measurement and computing systems.

1 Introduction

The probability methods actually proved to be ineffective in the modeling of complex physical, technical, social, and economic objects, subjective opinions, etc. This explains an increased interest from the middle of the last century to the improbability models of randomness, fuzziness, and uncertainty [1,2,3,4,5,6].

In the present paper we used the possibility-theoretical approach [7] for an empirical reconstruction of the fuzzy model of measurements. Methods are developed for solving a problem of an estimation of parameters of fuzzy systems. These problems come from the analysis and interpretation of the data received in measuring experiments. The class of similar problems in the mathematical statistics has received the name of problems of regress.

In the monograph [7] were developed methods for the reduction of fuzzy measurements where the model of the measuring device is given accurately by known linear operator $A \in (\mathcal{R}_N \to \mathcal{R}_n)$. Also, there were studied the optimal properties for the estimates of a maximum possibility.

In this paper a model of a measuring device A is unknown and information about it is extracted from the results of test experiments by measuring accurately the known input test signals [8].

Consider the possibilistic model of the measuring experiment, where the input of the measuring device A receives a signal f from the measured object. The measurement of its output Af is accompanied by an additive error of z, and the measurement result is a vector x. It is considered that signals x, f and z are

S.O. Kuznetsov et al. (Eds.): RSFDGrC 2011, LNAI 6743, pp. 111–118, 2011.

implementations of the fuzzy vectors $\xi \in \mathcal{R}_n$, $\varphi \in \mathcal{R}_N$, $\nu \in \mathcal{R}_n$, where $\mathcal{R}_N$ and $\mathcal{R}_n$ are linear spaces. The model of the measuring device is a fuzzy element Λ in the space $(\mathcal{R}_N \to \mathcal{R}_n)$ of linear operators. Its output signal is a fuzzy vector $\Lambda\varphi$. Thus a scheme of the measuring experiment is written as

$$\xi = \Lambda\varphi + \nu. \tag{1}$$

The problems of the interpretation of measurements (1) are considered in the theory of measurement and computing systems [9] which consists in the most accurate estimation of the parameters η of a studied object

$$\eta = U\varphi. \tag{2}$$

It is assumed that a linear operator U whose output signal $U\varphi$ should be synthesized by a measuring computing system and should give the most accurate estimation of the parameters η.

The possibility-theoretical model is defined by the joint distribution of the possibilities of the following fuzzy elements: the output signal $\xi \in \mathcal{R}_n$ of the measurement component of the measuring-computing system, the fuzzy operator $\Lambda \in (\mathcal{R}_N \to \mathcal{R}_n)$ which is a model of measuring device, the fuzzy vector $\varphi \in \mathcal{R}_N$ which simulates the input signal, the noise ν and the fuzzy vector $\eta \in \mathcal{R}_M$ of parameters of the object under investigation.

$$\pi^{\xi,\Lambda,\varphi,\eta}(x, A, f, u), \quad (x, A, f, u) \in \mathcal{R}_n \times (\mathcal{R}_N \to \mathcal{R}_n) \times \mathcal{R}_N \times \mathcal{R}_M. \tag{3}$$

The value $\pi^{\xi,\Lambda,\varphi,\eta}(x, A, f, u)$ is equal to the possibility of the equalities $\xi = x$, $\Lambda = A$, $\varphi = f$, $\eta = u$.

The marginal distribution

$$\pi^{\xi,\eta}(x, u) = \sup_{f \in \mathcal{R}_N, A \in (\mathcal{R}_N \to \mathcal{R}_n)} \pi^{\xi,\Lambda,\varphi,\eta}(x, A, f, u), \quad (x, u) \in \mathcal{R}_n \times \mathcal{R}_M, \tag{4}$$

defines a model for interpreting the measurement. In particular, this model allows us to obtain an estimate for the parameters η based on the measurement result $\xi = x$ as an estimate of the maximum of possibility

$$\widehat{u}(x) = \sup_{u \in \mathcal{R}_M} \pi^{\xi,\eta}(x, u), \quad x \in \mathcal{R}_n. \tag{5}$$

The consistency of the model of experiment can be estimated based on the prior distribution of the signal ξ

$$\pi^{\xi}(x) = \sup_{u \in \mathcal{U}} \pi^{\xi,\eta}(x, u), \quad x \in \mathcal{R}_n. \tag{6}$$

If for example $\xi = x$ is the measurement result and $\pi^{\xi}(x) = 0$ then the model (3) has to be recognized as an inadequate model of experiment.

The problem of reduction when the model of measurements exactly known was solved in the paper [7]. There is considered that $\Lambda = A$ and the link between fuzzy elements ξ and φ for finding distribution (3) is defined by the distribution of transition possibilities

$$\pi^{\xi|\varphi}(x \mid f), x \in \mathcal{R}_n, \quad f \in \mathcal{R}_N, \tag{7}$$

determining the dependence of the distribution of the fuzzy element ξ on values $f \in \mathcal{R}_N$ of fuzzy element φ, and the distribution

$$\pi^{\varphi}(f), \quad f \in \mathcal{R}_N, \tag{8}$$

represents a priori information about the possible values the signal φ.

The equality

$$\pi^{\xi,\varphi}(x, f) = \min(\pi^{\xi|\varphi}(x \mid f), \pi^{\varphi}(f)), \quad (x, f) \in \mathcal{R}_n \times \mathcal{R}_N,$$

determines the joint distribution of ξ, φ.

The estimate of the maximum possibility of fuzzy element η is obtained by using the distribution of transition possibility

$$\pi^{\eta|\varphi}(u \mid f), \quad u \in \mathcal{R}_M, f \in \mathcal{R}_N, \tag{9}$$

and with the a priori distribution (8).

At the same time

$$\pi^{\eta,\varphi}(u, f) = \min(\pi^{\eta|\varphi}(u \mid f), \pi^{\varphi}(f)), \quad (f, u) \in \mathcal{R}_N \times \mathcal{R}_M.$$

Since the distribution of transition possibility

$$\pi^{\xi|\varphi,\eta}(x \mid f, u) = \pi^{\xi|\varphi}(x \mid f), \quad x \in \mathcal{R}_n, \quad u \in \mathcal{R}_M, f \in \mathcal{R}_N, \tag{10}$$

doesn't depend on u, the distribution (3) is determined by the following relations:

$$\begin{aligned} \pi^{\xi,\varphi,\eta}(x, f, u) &= \min(\pi^{\xi|\varphi}(x \mid f), \pi^{\varphi,\eta}(f, u)) = \\ &= \min(\pi^{\xi|\varphi}(x \mid f), \pi^{\eta|\varphi}(u \mid f), \pi^{\varphi}(f)), \quad x \in \mathcal{R}_n, \quad u \in \mathcal{R}_M, \quad f \in \mathcal{R}_N. \end{aligned} \tag{11}$$

In this case, the consistency of the model experiment is defined by the possibility $\pi^{\xi}(x) = \sup_{f \in \mathcal{R}_N} \pi^{\xi,\varphi}(x, f)$ of the result of measurement $\xi = x \in \mathcal{R}_n$.

2 The Problem of Reduction and Reconstruction of the Model

Now suppose that the model of measuring device is unknown and information about it can be extracted from the measurements of known test signals $f_1, \ldots, f_m$. A scheme of the testing measuring experiment is written as

$$\xi_j = A f_j + \nu_j, \quad j = 1, \ldots, m; \tag{12}$$

where $\nu_j \in R_n$ is the fuzzy element that characterizes the measurement error.

There are a set of test signals $f_1, \ldots, f_m$, a set of results of there registration $\xi_1, \ldots, \xi_m$ and a set of measurement errors $\nu_1, \ldots, \nu_m$. Let us introduce the linear operator $F \in (\mathcal{R}_m \to \mathcal{R}_N)$ and the fuzzy linear operators $\Xi \in (\mathcal{R}_m \to \mathcal{R}_n)$, $N \in (\mathcal{R}_m \to \mathcal{R}_n)$, that are defined for any vector $t = (t_1, \dots, t_m) \in \mathcal{R}_m$ by the equations

$$Ft = \sum_{j=1}^{m} \tilde{f}_j t_j, \;\; \Xi t = \sum_{j=1}^{m} x_j t_j, \;\; Nt = \sum_{j=1}^{m} \nu_j t_j. \tag{13}$$

Using the notation (13), we write the scheme of test measurements as (12) in the form

$$\Xi = \Lambda F + N. \tag{14}$$

The expression (14) enables us to write a transition possibility $\pi^{\Xi|\Lambda}(\cdot \mid \cdot)$ in the form

$$\pi^{\Xi|\Lambda}(X \mid A) = \pi^N(X - AF).$$

Now consider the measurement scheme of the fuzzy vector $\varphi \in \mathcal{R}_n$

$$\xi = \Lambda\varphi + \nu, \tag{15}$$

where the result we have to reduce to a form that is peculiar to the measurement f by the scheme (2). Suppose that a distribution $\pi^{\varphi}(\cdot)$ and $\pi^{\nu}(\cdot)$ of the fuzzy vectors $\varphi \in \mathcal{R}_N$ and $\nu \in \mathcal{R}_n$ are given. Then, considering that the transition possibilities takes the form of $\pi^{\xi|\varphi,\Lambda}(x \mid f, A) = \pi^{\nu}(x - Af)$, where $x \in \mathcal{R}_n$ is the measurement result of the fuzzy element ξ in (15), we obtained the following relations for the joint distribution $\pi^{\xi,\varphi,\Lambda,\Xi}(x, f, A, X)$:

$$\begin{aligned}
&\pi^{\xi,\varphi,\Lambda,\Xi}(x, f, A, X) = \min(\pi^{\xi|\varphi,\Lambda,\Xi}(x \mid f, A, X), \pi^{\varphi,\Lambda,\Xi}(f, A, X)) = \\
&= \min(\pi^{\nu}(x - Af), \pi^{\Xi|\varphi,\Lambda}(X \mid f, A), \pi^{\varphi,\Lambda}(f, A) = \\
&= \min(\pi^{\nu}(x - Af), \pi^N(X - AF), \pi^{\varphi}(f), \pi^{\Lambda}(A)).
\end{aligned} \tag{16}$$

It is considered that the distribution of ξ for the fixed Λ and φ doesn't depend on Ξ, the distribution of Ξ at a fixed Λ doesn't depend on φ, and the fuzzy elements φ and Λ are independent. So if we know the a priori distribution of $\pi^{\varphi}(f)$ and $\pi^{\Lambda}(A)$, then $\pi^{\varphi,\Lambda}(f, A) = \min(\pi^{\varphi}(f), \pi^{\Lambda}(A))$.

Further, we note that the distribution of $\pi^{\eta|\xi,\varphi,\Lambda,\Xi}(u \mid x, f, A, X)$ for fixed φ doesn't depend on ξ, Λ, Ξ, so

$$\pi^{\xi,\varphi,\Lambda,\Xi,\eta}(x, f, A, X, u) = \min(\pi^{\eta|\varphi}(u \mid f), \pi^{\xi,\varphi,\Lambda,X}(x, f, A, X)),$$

and taking into account (16) we obtained:

$$\pi^{\xi,\varphi,\Lambda,\Xi,\eta}(x, f, A, X, u) = \begin{cases} \min(\pi^{\nu}(x - Af), \pi^N(X - AF), \pi^{\varphi}(f), \pi^{\Lambda}(A)), & \text{if } u = Uf, \\ 0, & \text{if } u \neq Uf. \end{cases}$$

The problem of reduction and empirical reconstruction of the model of measurements now is reduced to the estimation of the maximum of possibility of the fuzzy elements η and Λ. That is, essentially, the calculation of the estimates $\widehat{A}$ and $\widehat{f}$ of the fuzzy elements Λ and φ from the maximin problem:

$$(\widehat{A}, \widehat{f}) = \arg\max_{A,f} \min(\pi^{\nu}(x - Af), \pi^N(X - AF), \pi^{\varphi}(f), \pi^{\Lambda}(A)) \tag{17}$$

and the subsequent calculation of the estimate $\widehat{u} = U\widehat{f}$.

The reduction of measurements with the a priori fuzzy constraints to coordinates of the signals and the matrix elements Λ. Let the vectors of the spaces $\mathcal{R}_N$, $\mathcal{R}_n$ and $\mathcal{R}_M$ are given by their coordinates, the operators from $(\mathcal{R}_N \to \mathcal{R}_n)$ are given by its matrixes and the a priori distribution of possibilities of the fuzzy vectors, and the operator Λ are defined as the fuzzy constraints on the coordinates and matrix elements by the following relations: for the vectors $\nu = (\nu_1, \ldots, \nu_n) \in \mathcal{R}_n$ and $\varphi = (\varphi_1, \ldots, \varphi_N) \in \mathcal{R}_N$ from (15)

$$\pi^{\nu}(x_1, \ldots, x_n) = \mu_0\left(\min_{i=1,\ldots,n} \frac{|x_i|}{\sigma_i}\right) \tag{18}$$

and

$$\pi^{\varphi}(f_1, \ldots, f_N) = \mu_0\left(\min_{i=1,\ldots,N}\left(\frac{|f_i - f_{0,i}|}{\sigma_i^{(\varphi)}}\right)\right) \tag{19}$$

respectively, for the for vectors $\nu_j = (\nu_{j_1}, \ldots, \nu_{j_n}) \in \mathcal{R}_n$, $j = 1, \ldots, m$ from (12), for the matrix $N_{ij} = (n_{j_i})$,

$$\pi^{N}(x_{11}, \ldots, x_{mn}) = \mu_0\left(\min_{i=1,\ldots,n;\, j=1,\ldots,m}\left(\frac{|x_{j_i}|}{\sigma_{ij}}\right)\right), \tag{20}$$

for the matrix (Λ_{ij}), $i = 1, \ldots, n$, $j = 1, \ldots, N$, of linear operator Λ

$$\pi^{\Lambda}(A_{11}, \ldots, A_{nN}) = \mu_0\left(\min_{i=1,\ldots,n;\, j=1,\ldots,N}\left(\frac{|A_{ij} - A_{0,ij}|}{\sigma_{ij}^{(A)}}\right)\right). \tag{21}$$

Where $\mu_0(\cdot) : [0, \infty) \to [0, 1]$ is a strictly decreasing function, $\mu_0(0) = 1$, $\lim_{z\to\infty} \mu_0(z) = 0$. The constants in the denominators of the formulas (18)–(21) are given by values that determine the value of "fuzziness" of relevant variables. And constants $f_{0,i}$ and $A_{0,ij}$, $i = 1, \ldots, n$, $j = 1, \ldots, N$ define the most possible value of the vector $\varphi \in \mathcal{R}_N$ and the matrix of the operator Λ respectively.

Then the problem (17) reduces to

$$(\widehat{A}, \widehat{f}) = \arg\inf_{A,f}\left(\max\left(\max_{i=1,\ldots,n}\left(\frac{|x_i - \sum_{k=1}^{N} A_{ik} f_k|}{\sigma_i}\right), \max_{s=1,\ldots,n;t=1,\ldots,m}\left(\frac{|\sum_{k=1}^{N} A_{sk} F_{kt} - X_{st}|}{\sigma_{st}}\right),\right.\right.$$
$$\left.\left.\max_{q=1,\ldots,N}\left(\frac{|f_q - f_{0,q}|}{\sigma_q^{(\varphi)}}\right), \max_{p=1,\ldots,n;\, l=1,\ldots,N}\left(\frac{|A_{pl} - A_{0,pl}|}{\sigma_{pl}^{(A)}}\right)\right)\right). \tag{22}$$

For a fixed vector $f \in \mathcal{R}_N$ the minimax problem determines the matrix elements of $\widehat{A}_{ij}$ the matrix of $\widehat{A}$ (22) reduces to a linear programming problem [10]. The minimization with respect to $f \in \mathcal{R}_N$ is carried out numerically.

If the operator Λ and the input signal f of the reducible measurement a priori arbitrary, then $\pi^{\Lambda}(A) = 1$ for every $A \in (\mathcal{R}_N \to \mathcal{R}_n)$ and $\pi^{\varphi}(f) = 1$ for every $f \in \mathcal{R}_N$, and the problem (17) takes the simpler form

$$(\widehat{A}, \widehat{f}) = \\ = \arg\inf_{A,f} \left(\max \left(\max_{i=1,\dots,n} \left(\frac{|x_i - \sum_{k=1}^{N} A_{ik} f_k|}{\sigma_i} \right), \max_{s=1,\dots,n;t=1,\dots,m} \left(\frac{|\sum_{k=1}^{N} A_{sk} F_{kt} - X_{st}|}{\sigma_{st}} \right) \right) \right). \tag{23}$$

The solution of minimax problem is carried out in two stages. The minimization of $A \in (\mathcal{R}_N \to \mathcal{R}_n)$ is reduced to the solving linear programming problem. Then the numerical minimization of $f \in \mathcal{R}_N$ is realized.

Let us to summarize the results.

Theorem 1. *If $(\widehat{A}, \widehat{f})$ is the solution of the problem* (22) (*or* (23) *in the absence of a priori constraints on the possible values of the operator Λ and the vector f*), *then the reduction of the measurement ξ* (15) *to the form* (2) *is equal to $\widehat{u} = U\widehat{f}$. The possibility that the model of measurements* (14) *and* (15) *consents with measurement results $\xi = x$ and $\Xi = X$ is equal to $\mu_0(z)$, where z is the minimax value that is obtained by solving the problem* (22) (*or* (23)).

The method for calculating the estimate of the maximum a posteriori features. The method for solving the problem (22) (or (23) in the absence of a priori knowledge about the possible values of the operator Λ) consists of two phase. At the first phase for each fixed vector $f \in \mathcal{R}_N$ the value of the function $q(f)$ is defined as the value that obtained by solving the minimax problem (22) (or (23)) in relation to the matrix elements $\widehat{A}_{ij}$ of the matrix $\widehat{A}$. This minimax problem for fixed f reduces to a linear programming problem. At the second phase we numerically minimize the function $q(\cdot)$ by $f \in \mathcal{R}_N$. If a minimum of $q(\cdot)$ for $f \in \mathcal{R}_N$ is attained at $f = \widehat{f}$ then the desired estimate is $\widehat{u} = U\widehat{f}$.

Example. The fuzzy value of $\varphi \in \mathcal{R}_1$ is measured in the experiment scheme

$$\xi = kx + b + \nu, \tag{24}$$

and information about the linear dependence of $y = kx + b$, $k, b \in \mathcal{R}_1$, contained in the test measurements

$$\xi_j = kx_j + b + \nu_j, \quad j = 1, \dots, m,$$

the fuzzy measurement errors ν and ν_j are characterized by distribution of possibilities

$$\pi^{\nu}(z) = \mu_0(z), \quad \pi^{\nu}(z_j) = \mu_0(z_j), \quad j = 1, \dots, m. \tag{25}$$

It is required to estimate the value of the argument x by the result of $\xi = y$ measurement (24) and the results of $\xi_j = y_j$ test measurements (25). The constants k, b of the linear dependence (25) and value of x in (24) are a priori arbitrary.

We insert the linear operator $A(k\ b) \in (R_2 \to \mathcal{R}1)$ and vector $f = \begin{pmatrix} x \\ 1 \end{pmatrix} \in \mathcal{R}^2$ for writing the measurement scheme (24) and (25) in the following form

$$\xi = Af + \nu, \quad \xi_j = Af_j + \nu_j, \quad j = 1, \dots, m.$$

As consistent with Theorem 1 for reduction of measurement (24) we should solve the minimax problem:

$$\inf_{k,b,x} \max \left(|y - kx - b|, \max_{j=1,\dots,m} (|y_j - kx_j - b|) \right). \tag{26}$$

The minimum of k, b for the fixed x is the computed solution of the linear programming problem in which you have to find the minimum of the linear function $q = (l, g)$, $l = (1, 0, 0)$, $g = (q, k, b)$, for subset of R_2, that is dedicated by linear inequalities:

$$(b, g) \geq y, \quad (d, g) \geq -y, \quad (b_j, g) \geq y_j, \quad (d_j, g) \geq -y_j, \quad j = 1, \dots, m, \tag{27}$$

where

$$b = (1, x, 1), \ d = (1, -x, -1), \ b_j = (1, x_j, 1), \ d_j = (1, -x_j, -1), \quad j = 1, \dots, m. \tag{28}$$

If $g(x) = (q(x), k(x), b(x))$ is its solution, the solution (26) is obtained from the minimization of $q(x)$ by $x \in R_1 : q(\widehat{x}) = \min_x q(x)$.

The value of $\widehat{x}$ is an estimate of the maximum of possibility of required to define argument of the linear function (24). And its concomitant reconstructed model of the linear dependence is given by the coefficients $k_* = k(\widetilde{x})$, $b_* = b(\widetilde{x})$. With regard to adequacy, if $\mu_0(q(\widehat{x})) = 0$ then the model experiment have to be recognized as out of keeping to the results of the measurements and these estimates have to be considered as inadequate.

Computer Experiment. Consider the numerical experiment that implements the example described above. It is based on the model of the photosynthetic system. It is measured the synthesized ATP under the influence of light. According to the numerical experiment we can predict by the exit of ATP the time when it has received. In this case, the measurements don't have a stochastic component and the system is a complex, multidimensional and evolving over time. We can't apply a probabilistic approach for interpretation of such measurements. Therefore the fuzzy method was applied to estimate the model parameters.

The measurement model corresponds to the linear equation (24). The information about the linear dependence of $y = kx + b$, where $k, b \in \mathcal{R}_1$, is contained in the test measurements. Where y is the number of synthesized ATP, k is the rate of ATP synthesis complex ATP synthase, x is the time in milliseconds. It is required to restore the fuzzy model of measurements and then estimate the value of the argument x the time t, when a certain amount of ATP had been obtained by result $\xi = y$ of measurement (24).

When solving linear programming problems was obtained the estimate $\hat{x} = 0.46$ of the maximum of possibility value of the argument of linear function (24). The reconstructed model is given by the set of coefficients $k_* = 3.1738$ $b_* = 0.0031$. The estimate of time of the maximum of possibility for the ATP exit that was equal to 99 ATP was 31.19 ms and at the experiment we obtained the ATP exit that was equal to 99 ATP at 31 and 32 ms.

3 Conclusions

In this paper we consider the empirical reconstruction of the fuzzy model of measurements on the basis of testing measurements using the possibility-theoretical approach. It is shown that if the fuzzy model is unknown, it can be restored from testing experiments and can be applied to the decision of problems of type of forecasting the behavior of a system. The decision of these problems is received numerically on the basis of effective methods and algorithms. This work was supported by the Russian Foundation for Basic Research (project no. 11-07-00338-a).

References

1. Savage, L.J.: The Foundations of Statistics. Dover, New York (1972)
2. Dempster, A.P.: Upper and Lower Probabilities Induced by a Multivalued Mapping. Ann. Math. Statist. 38, 25–39 (1967)
3. Shafer, G.: A Mathematical Theory of Evidence. Princeton Univ. Press, Prinston (1976)
4. Zadeh, L.A.: Fuzzy Sets as a Basis for a Theory of Possibility. Fuzzy Sets Syst. (1), 3–28 (1978)
5. de Cooman, G.: Possibility Theory. I, II, III. Int. J. General Syst. 25, 291–371 (1997)
6. Wolkenhauer, O.: Possibility Theory with Applications to Data Analysis. Wiley, New York (1998)
7. Pytev, Y.P.: Possibility as an Alternative of Probability. Fizmatlit, Moscow (2007) (in russian)
8. Cheremukhin, E.A., Chulichkov, A.I.: Vestn. Mosk. Univ. Fiz. Astron. vol. 3, pp. 15–18 (2004)
9. Pytev, Y.P.: Methods of Mathematical Simulation of Measuring–Computing Systems, Fizmatlit, Moscow (2002) (in russian)
10. Zhuchko, O.V., Pytev, Y.P.: Restoration of the Functional Dependence by Possibility–Theoretic Methods. Zh. Vychisl. Mat. Mat. Fiz. 43(5), 765–781 (2003)

Dominance-Based Rough Set Approach for Possibilistic Information Systems *

Tuan-Fang Fan[1], Churn-Jung Liau[2], and Duen-Ren Liu[3]

[1] Department of Computer Science and Information Engineering
National Penghu University of Science and Technology
Penghu 880, Taiwan.
dffan@npu.edu.tw

[2] Institute of Information Science
Academia Sinica, Taipei 115, Taiwan
liaucj@iis.sinica.edu.tw

[3] Institute of Information Management
National Chiao-Tung University, Hsinchu 300, Taiwan
dliu@iim.nctu.edu.tw

Abstract. In this paper, we propose a dominance-based fuzzy rough set approach for the decision analysis of a preference-ordered possibilistic information systems, which is comprised of a finite set of objects described by a finite set of criteria. The domains of the criteria may have ordinal properties that express preference scales. In the proposed approach, we first compute the degree of dominance between any two objects based on their possibilistic evaluations with respect to each criterion. This results in a fuzzy dominance relation on the universe. Then, we define the degree of adherence to the dominance principle by every pair of objects and the degree of consistency of each object. The consistency degrees of all objects are aggregated to derive the quality of the classification, which we use to define the reducts of an information system. In addition, the upward and downward unions of decision classes are fuzzy subsets of the universe. The lower and upper approximations of the decision classes based on the fuzzy dominance relation are thus fuzzy rough sets. By using the lower approximations of the decision classes, we can derive two types of decision rules that can be applied to new decision cases.

1 Introduction

The rough set theory proposed in [13] provides an effective tool for extracting knowledge from information systems. When rough set theory is applied to *multi-criteria decision analysis* (MCDA), it is crucial to deal with preference-ordered attribute domains and decision classes [3,4,5,17]. The original rough set theory cannot handle inconsistencies arising from violations of the dominance principle due to its use of the indiscernibility relation. Therefore, the indiscernibility

* This work was partially supported by NSC (Taiwan) Grants: 99-2410-H-346-001 (T.F. Fan) and 98-2221-E-001-013-MY3(C.J. Liau).

S.O. Kuznetsov et al. (Eds.): RSFDGrC 2011, LNAI 6743, pp. 119–126, 2011.

relation is replaced by a dominance relation to solve the multi-criteria sorting problem; and the information system is replaced by a pairwise comparison table to solve multi-criteria choice and ranking problems. The approach is called the *dominance-based rough set approach* (DRSA). For MCDA problems, DRSA can induce a set of decision rules from sample decisions provided by decision-makers. The induced rules form a comprehensive preference model and can provide recommendations about a new decision-making environment.

A strong assumption about information systems is that each object takes exactly one value with respect to an attribute. However, in practice, we may only have incomplete information about the values of an object's attributes. Thus, more general information systems are needed to represent incomplete information. For example, set-valued and interval-valued information systems have been introduced to represent incomplete information [9,10,11,12,18]. DRSA has also been extended to deal with missing or uncertain values in MCDA problems [3,17,2,8]. Since an information system with missing or uncertain values is a special case of possibilistic information systems, we propose further extending DRSA to the decision analysis of possibilistic information systems. In this paper, we investigate such an extension based on the fuzzy dominance principle.

In the proposed approach, we first compute the degree of dominance between any two objects based on their possibilistic evaluations with respect to each criterion. This results in a fuzzy dominance relation on the universe. Then, we define the degree of adherence to the dominance principle by every pair of objects and the degree of consistency of each object. The consistency degrees of all objects are aggregated to derive the quality of the classification, which we use to define the reducts of possibilistic information systems. In addition, the upward and downward unions of decision classes are fuzzy subsets of the universe. The lower and upper approximations of the decision classes based on the fuzzy dominance relation are thus fuzzy rough sets. By using the lower approximations of the decision classes, we can derive two types of decision rules that can be applied in new decision-making environments.

The remainder of the paper is organized as follows. In Section 2, we review the dominance-based rough set approach. In Section 3, we present the extension of DRSA for decision analysis of possibilistic information systems. Section 4 contains some concluding remarks.

2 Review of Rough Set Theory and DRSA

The basic construct of rough set theory is an *approximation space*, which is defined as a pair (U, R), where U is a finite universe and $R \subseteq U \times U$ is an equivalence relation on U. We write an equivalence class of R as $[x]_R$ if it contains the element x. For any subset X of the universe, the lower approximation and upper approximation of X are defined as $\underline{R}X = \{x \in U \mid [x]_R \subseteq X\}$ and $\overline{R}X = \{x \in U \mid [x]_R \cap X \neq \emptyset\}$ respectively.

Although an approximation space is an abstract framework used to represent classification knowledge, it can easily be derived from a concrete information system. In [14], an information system[1] is defined as a tuple $T = (U, A, \{V_i \mid i \in A\}, \{f_i \mid i \in A\})$, where U is a nonempty finite set, called the universe; A is a nonempty finite set of primitive attributes; for each $i \in A$, V_i is the domain of values of i; and for each $i \in A$, $f_i : U \to V_i$ is a total function. An attribute in A is usually denoted by the lower-case letters i or a. In decision analysis (and throughout this paper), we assume the set of attributes is partitioned into $\{d\} \cup (A - \{d\})$, where d is called the *decision attribute*, and the remaining attributes in $C = A - \{d\}$ are called *condition attributes*. Given a subset of attributes B, the *indiscernibility relation* with respect to B is defined as $ind(B) = \{(x, y) \mid x, y \in U, f_i(x) = f_i(y) \forall i \in B\}$. Obviously, for each $B \subseteq A$, $(U, ind(B))$ is an approximation space.

For MCDA problems, each object in an information system can be seen as a sample decision, and each condition attribute is a criterion for that decision. Since a criterion's domain of values is usually ordered according to the decision-maker's preferences, we define a preference-ordered information system (POIS) as a tuple $T = (U, A, \{(V_i, \succeq_i) \mid i \in A\}, \{f_i \mid i \in A\})$, where $(U, A, \{V_i \mid i \in A\}, \{f_i \mid i \in A\})$ is a classical information system; and for each $i \in A$, $\succeq_i \subseteq V_i \times V_i$ is a binary relation over V_i. The relation $\succeq_i$ is called a *weak preference relation* or *outranking* on V_i, and represents a preference over the set of objects with respect to the criterion i [17]. The weak preference relation $\succeq_i$ is supposed to be a complete preorder, i.e., a complete, reflexive, and transitive relation. In addition, we assume that the domain of the decision attribute is a finite set $V_d = \{1, 2, \cdots, n\}$ such that r is strictly preferred to s if $r > s$ for any $r, s \in V_d$.

To deal with inconsistencies arising from violations of the dominance principle, the indiscernibility relation is replaced by a dominance relation in DRSA. Let P be a subset of criteria. Then, we can define the *P-dominance relation* $D_P \subseteq U \times U$ as follows:

$$(x, y) \in D_P \Leftrightarrow f_i(x) \succeq_i f_i(y) \forall i \in P. \tag{1}$$

When $(x, y) \in D_P$, we say that x P-dominates y, and that y is P-dominated by x. We usually use the infix notation xD_Py to denote $(x, y) \in D_P$. Given the dominance relation D_P, the P-dominating set and P-dominated set of x are defined as $D_P^+(x) = \{y \in U \mid yD_Px\}$ and $D_P^-(x) = \{y \in U \mid xD_Py\}$ respectively. In addition, for each $t \in V_d$, we define the decision class Cl_t as $\{x \in U \mid f_d(x) = t\}$. Then, the *upward and downward unions of classes* are defined as $Cl_t^{\geq} = \bigcup_{s \geq t} Cl_s$ and $Cl_t^{\leq} = \bigcup_{s \leq t} Cl_s$ respectively. We can then define the P-lower and P-upper approximations of $Cl_t^{\geq}$ and $Cl_t^{\leq}$ by using the P-dominating sets and P-dominated sets instead of the equivalence classes.

[1] Also called knowledge representation systems, data tables, or attribute-value systems.

3 DRSA for Possibilistic Information Systems

3.1 Preference-Ordered Possibilistic Information Systems

A general approach to specify the uncertainty of information is to use possibility distributions. A possibility distribution on a domain V is simply a function $\pi : V \to [0,1]$. Intuitively, π specifies the degree of possibility of each element in the domain V. $\pi(v) = 1$ and $\pi(v) = 0$ mean that the element v is fully possible and totally impossible respectively, while the intermediate values in $(0,1)$ mean partial possibilities of v. We usually assume that a possibility distribution is *normalized*, i.e., $\sup_{v \in V} \pi(v) = 1$. Let π_1 and π_2 be two possibility distributions on V. Then, we say that π_1 is *at least as specific as* π_2, denoted by $\pi_1 \leq \pi_2$, if $\pi_1(v) \leq \pi_2(v)$ for each $v \in V$. Let us denote the set of all normalized possibility distributions on V by $(V \to [0,1])^+$. Then, a preference-ordered possibilistic information system (POPIS) is a tuple $T = (U, A, \{(V_i, \succeq_i) \mid i \in A\}, \{f_i \mid i \in A\})$, where $U, A, \{(V_i, \succeq_i) \mid i \in A\}$ are defined as above, and for each $i \in A$, $f_i : U \to (V_a \to [0,1])^+$.

3.2 Fuzzy Dominance Relation

In a POPIS, the objects may have imprecise evaluations with respect to the condition criteria and imprecise assignments to decision classes. Thus, the dominance relation between objects can not be determined with certainty. Instead, since possibility information for each value is available in a POPIS, we can use the extension principle in fuzzy set theory to compute the degree of dominance[19]. The extension principle extends an operation or a relation on a base domain to the class of all fuzzy sets or possibility distributions on the domain. In our context, we use the extension principle to extend the preference relation $\succeq_i$ on V_i to a fuzzy preference relation between two possibility distributions on V_i. Consequently, the dominance relation between two objects with respect to the criterion i is determined by their respective possibility distributions on the domain of the criterion. Let $\otimes$, $\oplus$ and $\to$ denote, respectively, a t-norm operation, an s-norm operation and an implication operation[2] on $[0,1]$. Then, the dominance relation with respect to the criterion i is a fuzzy relation $D_i : U \times U \to [0,1]$ such that

$$D_i(x,y) = \sup_{v_1, v_2 \in V_i} \{f_i(x)(v_1) \otimes f_i(y)(v_2) \mid v_1 \succeq_i v_2\}. \tag{2}$$

After deriving the fuzzy dominance relation for each criterion, we can aggregate all the relations into P-dominance relations for any subset of criteria P. Thus, the fuzzy P dominance relation $D_P : U \times U \to [0,1]$ is defined as

$$D_P(x,y) = \bigotimes_{i \in P} D_i(x,y). \tag{3}$$

---

[2] For the properties of these operations, see a standard reference on fuzzy logic, such as [7].

Since the dominance relation is a fuzzy relation, the satisfaction of the dominance principle is a matter of degree. Thus, the *degree of adherence* of (x, y) to the dominance principle with respect to a subset of condition criteria P is defined as

$$\delta_P(x, y) = D_P(x, y) \rightarrow D_d(x, y), \tag{4}$$

and the degree of P-consistency of x is defined as

$$\delta_P(x) = \bigotimes_{y \in U} (\delta_P(x, y) \otimes \delta_P(y, x)). \tag{5}$$

Let T be a POPIS. Then, the *quality of the classification* of T based on the set of criteria P is defined as

$$\gamma_P(T) = \frac{\sum_{x \in U} \delta_P(x)}{|U|}. \tag{6}$$

Note that $\gamma_P(T)$ is monotonic with respect to P, i.e., $\gamma_Q(T) \leq \gamma_P(T)$ if $Q \subseteq P$. Thus, we can define every minimal subset $P \subseteq C$ such that $\gamma_P(T) = \gamma_C(T)$ as a *reduct* of C, where $C = A - \{d\}$ is the set of all condition criteria. In addition, the degree of P-consistency is monotonic with respect to P, so a reduct is also a minimal subset $P \subseteq C$ such that $\delta_P(x) = \delta_C(x)$ for all $x \in U$. However, because $\delta_P(x)$ is less sensitive to individual changes in $\delta_P(x, y)$, we can not guarantee that a reduct will preserve the degree of adherence to the dominance principle for each pair of objects. To overcome such difficulty, we can adopt the following alternative definition of the quality of the classification:

$$\eta_P(T) = \frac{\sum_{x,y \in U} \delta_P(x, y)}{|U|^2}. \tag{7}$$

The reducts can also be defined in terms of this kind of definition.

3.3 Dominance-Based Fuzzy Rough Approximations

In a POPIS, the assignment of a decision label to an object may be imprecise, so the decision classes may be fuzzy subsets of the universe. Their membership functions are derived from the possibility distributions associated with the assignments of the objects. Specifically, for each $t \in V_d$, the decision class $Cl_t : U \rightarrow [0, 1]$ is defined by

$$Cl_t(x) = f_d(x)(t). \tag{8}$$

Then, the upward and downward unions of classes are defined by

$$Cl_t^{\geq}(x) = \sup_{v \geq t} f_d(x)(v) = \Pi_x(\{v \geq t\}) \tag{9}$$

and

$$Cl_t^{\leq}(x) = \sup_{v \leq t} f_d(x)(v) = \Pi_x(\{v \leq t\}) \tag{10}$$

respectively, where Π_x is the possibility measure corresponding to the possibility distribution $f_d(x)$. Finally, since our dominance relation is a fuzzy relation and the decision classes are fuzzy sets, the lower and upper approximations of the classes are defined in the same way as those for fuzzy rough sets[1,15]. More specifically, the P-lower and P-upper approximations of $Cl_t^{\geq}$ and $Cl_t^{\leq}$ for each $t \in V_d$ are defined as fuzzy subsets of U with the following membership functions:

$$\underline{P}(Cl_t^{\geq})(x) = \bigotimes_{y \in U}(D_P(y,x) \to Cl_t^{\geq}(y)), \tag{11}$$

$$\overline{P}(Cl_t^{\geq})(x) = \bigoplus_{y \in U}(D_P(x,y) \otimes Cl_t^{\geq}(y)), \tag{12}$$

$$\underline{P}(Cl_t^{\leq})(x) = \bigotimes_{y \in U}(D_P(x,y) \to Cl_t^{\leq}(y)), \tag{13}$$

$$\overline{P}(Cl_t^{\leq})(x) = \bigoplus_{y \in U}(D_P(y,x) \otimes Cl_t^{\leq}(y)). \tag{14}$$

3.4 Decision Rules

To represent knowledge discovered from a POPIS, we consider a preference-ordered possibilistic decision logic (POPDL). The well-formed formulas (wff) of POPDL are Boolean combinations of atomic formulas of the form $(\geq_i, \pi_i)$ or $(\leq_i, \pi_i)$, where $i \in A$ and $\pi_i \in (V_i \to [0,1])^+$. When π_i is a singleton possibility distribution such that $\pi(x) = 1$ if $x = v$ and $\pi(x) = 0$ if $x \neq v$, we abbreviate $(\geq_i, \pi_i)$ (resp. $(\leq_i, \pi_i)$) as $(\geq_i, v)$ (resp. $(\leq_i, v)$).

Let P denote a reduct of a POPIS and let $t \in V_d$. Then, for each object x, where $\underline{P}(Cl_t^{\geq})(x) > 0$ (or above some pre-determined threshold), we can derive the first type of fuzzy rule:

$$\underline{P}(Cl_t^{\geq})(x) : \bigwedge_{i \in P}(\geq_i, f_i(x)) \longrightarrow (\geq_d, t); \tag{15}$$

and for each object x, where $\underline{P}(Cl_t^{\leq})(x) > 0$ (or above some pre-determined threshold), we can derive the second type of fuzzy rule:

$$\underline{P}(Cl_t^{\leq})(x) : \bigwedge_{i \in P}(\leq_i, f_i(x)) \longrightarrow (\leq_d, t), \tag{16}$$

where $\underline{P}(Cl_t^{\geq})(x)$ and $\underline{P}(Cl_t^{\leq})(x)$ are the respective degrees of truth of the rules.

Now, for a new decision case with evaluations based on the condition criteria P, we can apply these two types of rules to derive the case's decision label assignment. Specifically, let x be a new object such that, for each criterion $i \in P$, $f_i(x) \in (V_i \to [0,1])^+$ is given; and let α be a rule $c : \bigwedge_{i \in P}(\geq_i, \pi_i) \longrightarrow (\geq_d, t)$ discovered by the proposed approach. Then, according to the rule α, we can derive that the degree of satisfaction of $f_d(x) \succeq_d t$ is $\varepsilon(\alpha, f_d(x) \succeq_d t) = c \otimes$

$\bigotimes_{i\in P}\bigoplus_{v_1\succeq_i v_2}(f_i(x)(v_1)\otimes\pi_i(v_2))$. Let $\mathcal{R}_t^{\geq}$ denote the set of all rules with a consequent $(\geq_d, t')$ such that $t'\geq t$. Then, the final degree of $f_d(x)\succeq_d t$ is $\bigoplus_{\alpha\in\mathcal{R}_t^{\geq}}\varepsilon(\alpha, f_d(x)\succeq_d t')$. We can derive the degree of $f_d(x)\preceq_d t$ from the second type of rule in a similar manner.

Mathematically, the evaluations and assignments in a POPIS are possibility distributions, so the atomic formulas of POPDL may also include any possibility distributions on the domain. However, in general, the set of all (normalized) possibility distributions is infinite, even though the domain is finite. This may result in a very large set of rules. Moreover, most of the possibility distributions may lack semantically meaningful interpretation for human users; hence, the induced rules may be hard to use. To resolve the difficulty, the standard practice in fuzzy logic is to use a set of meaningful linguistic labels whose interpretations are simply possibility distributions on the domain. Thus, the evaluations and assignments given in a POPIS are restricted to the (usually finite) set of linguistic labels, so the set of atomic formulas in our POPDL only contains $(\geq_i, \pi_i)$ or $(\leq_i, \pi_i)$, where π_i is a linguistic label. For example, if the evaluated criterion is "score" and its domain is [0,100], then the set of linguistic labels may be {poor, fair, good, excellent}, and their corresponding interpretations are possibility distributions on the domain.

4 Conclusion

The work reported in this paper extends DRSA to a dominance-based fuzzy rough set approach (DFRSA), which can be applied to the reduction of criteria and the induction of rules for decision analysis in a POPIS.

In contrast to other approaches that deal with imprecise evaluations and assignments, DFRSA induces fuzzy rules instead of qualitative rules. Thus, it would be worthwhile to compare DFRSA with other extensions of DRSA for handling uncertain information systems, e.g., those proposed in [6,16].

Since DFRSA is a general framework, we do not specify the t-norm operations used in the aggregation of consistency degrees or the implication operations used in the definition of adherence to the dominance principle. Hence, we do not present detailed algorithms for the computation of reducts. The computational aspects of DFRSA for specialized t-norm and implication operations will also be addressed in a future work.

References

1. Dubois, D., Prade, H.: Rough fuzzy sets and fuzzy rough sets. International Journal of General Systems 17, 191–209 (1990)
2. Fan, T.F., Liau, C.J., Liu, D.R.: Dominance-based rough set analysis of uncertain data table. In: Proc. of the International Fuzzy Systems Association (IFSA) World Congress and the European Society for Fuzzy Logic and Technology (EUSFLAT) Conference, pp. 294–299 (2009)
3. Greco, S., Matarazzo, B., Słowiński, R.: Rough set theory for multicriteria decision analysis. European Journal of Operational Research 129(1), 1–47 (2001)

4. Greco, S., Matarazzo, B., Słowiński, R.: Rough sets methodology for sorting problems in presence of multiple attributes and criteria. European Journal of Operational Research 138(2), 247–259 (2002)
5. Greco, S., Matarazzo, B., Słowiński, R.: Axiomatic characterization of a general utility function and its particular cases in terms of conjoint measurement and rough-set decision rules. European Journal of Operational Research 158(2), 271–292 (2004)
6. Greco, S., Matarazzo, B., Słowiński, R.: Dominance-based rough set approach as a proper way of handling graduality in rough set theory. Transactions on Rough sets VII, 36–52 (2007)
7. Hájek, P.: Metamathematics of Fuzzy Logic. Kluwer Academic Publishers, Dordrecht (1998)
8. Inuiguchi, M.: Rough set approach to rule induction from imprecise decision tables. In: Di Gesù, V., Pal, S.K., Petrosino, A. (eds.) WILF 2009. LNCS, vol. 5571, pp. 68–76. Springer, Heidelberg (2009)
9. Kryszkiewicz, M.: Properties of incomplete information systems in the framework of rough sets. In: Polkowski, L., Skowron, A. (eds.) Rough Sets in Knowledge Discovery, pp. 422–450. Physica-Verlag, Heidelberg (1998)
10. Kryszkiewicz, M., Rybiński, H.: Reducing information systems with uncertain attributes. In: Raś, Z.W., Michalewicz, M. (eds.) ISMIS 1996. LNCS, vol. 1079, pp. 285–294. Springer, Heidelberg (1996)
11. Kryszkiewicz, M., Rybiński, H.: Reducing information systems with uncertain real value attributes. In: Proc. of the 6th IPMU, pp. 1165–1169 (1996)
12. Lipski, W.: On databases with incomplete information. Journal of the ACM 28(1), 41–70 (1981)
13. Pawlak, Z.: Rough sets. International Journal of Computer and Information Sciences 11(15), 341–356 (1982)
14. Pawlak, Z.: Rough Sets–Theoretical Aspects of Reasoning about Data. Kluwer Academic Publishers, Dordrecht (1991)
15. Radzikowska, A.M., Kerre, E.E.: A comparative study of fuzzy rough sets. Fuzzy Sets and Systems 126(2), 137–155 (2002)
16. Sakai, H., Ishibashi, R., Nakata, M.: Lower and upper approximations of rules in non-deterministic information systems. In: Chan, C.C., Grzymala-Busse, J.W., Ziarko, W. (eds.) RSCTC 2008. LNCS (LNAI), vol. 5306, pp. 299–309. Springer, Heidelberg (2008)
17. Słowiński, R., Greco, S., Matarazzo, B.: Rough set analysis of preference-ordered data. In: Alpigini, J.J., Peters, J.F., Skowron, A., Zhong, N. (eds.) RSCTC 2002. LNCS (LNAI), vol. 2475, pp. 44–59. Springer, Heidelberg (2002)
18. Yao, Y.Y., Liu, Q.: A generalized decision logic in interval-set-valued information tables. In: Zhong, N., Skowron, A., Ohsuga, S. (eds.) RSFDGrC 1999. LNCS (LNAI), vol. 1711, pp. 285–293. Springer, Heidelberg (1999)
19. Zadeh, L.A.: The concept of a linguistic variable and its applications in approximate reasoning. Information Sciences 8, 199–251 (1975)

Creating Fuzzy Concepts: The One-Sided Threshold, Fuzzy Closure and Factor Analysis Methods

Valerie Cross and Meenakshi Kandasamy

Computer Science and Software Engineering
Miami University
Oxford, OH 45056 USA
{crossv,kandasm}@muohio.edu

Abstract. The two main approaches to fuzzy concept lattice creation are the one-sided threshold approach and the fuzzy closure approach. These two methods are applied to gene annotation data files that are converted into fuzzy formal contexts by translating evidence codes into a degree of evidence strength in [0, 1]. Fuzzy factor analysis is also applied to this same test data. These three methods are briefly described and then compared based on their results using the gene annotation data files.

Keywords: Fuzzy concept lattices, threshold, fuzzy closure operator, factor analysis.

1 Introduction

Numerous researchers have proposed approximate reasoning techniques to handle these imprecision in the Semantic Web [1]. In the past few years research has been supporting the integration of fuzzy logic for knowledge representation on the Semantic Web [2]. A result of this research is fuzzy ontologies and a variety of methods to create them, one of which is fuzzy formal concept analysis (FFCA). A fuzzy formal context is specified by a matrix with a set of objects X for its rows and a set of attributes or properties Y for its column. The relation R between the objects and the attributes is a fuzzy binary relation. The degree is taken from the interval [0, 1] and indicates that the object x_i possesses the attribute y_j to a certain degree.

Two main approaches can be found for creating fuzzy concept lattices. The one-sided or α-cut threshold method is used in a variety of applications to create fuzzy ontologies [3]-[5]. The fuzzy formal context is viewed as a set of fuzzy sets over the attributes if the objects are given preference. Each fuzzy set describes one of the objects in the fuzzy formal context. Alternatively, the attributes can be given preference so that the fuzzy formal context is viewed as a set of fuzzy sets over the objects. When used to construct fuzzy ontologies, the objects have typically been given preference. The second approach recently presented in [6], the fuzzy closure operator approach, views a fuzzy formal context as a whole, i.e., it does not give preference to either the objects or the attributes of the fuzzy formal context and does not use thresholds. Experiments with this approach have been performed using

S.O. Kuznetsov et al. (Eds.): RSFDGrC 2011, LNAI 6743, pp. 127–134, 2011.

generated fuzzy formal contexts that have certain characteristics such as various distributions of membership degrees within the fuzzy formal context in order to evaluate its practical performance.

The fuzzy closure method generates a huge number of fuzzy concepts, even for small fuzzy formal contexts and sparse distributions. In [7] an approximation algorithm decomposes a fuzzy formal context into a product of two graded matrices with the number of factors as small as possible. These factors correspond to fuzzy concepts that summarize the information and permit a simpler interpretation than that of the complete fuzzy concept lattice. This paper summarizes the results of comparing the two FFCA approaches and extends the work in [8] and [9]. Gene annotation data files which play an important role in bioinformatics research are used to create real world fuzzy formal contexts. The fuzzy factor analysis algorithm is also used on this same data to gain insights into the relationships between these three algorithms.

2 Fuzzy Concept Lattice Creation and Fuzzy Factor Analysis

Each approach is briefly described below. More details for the threshold approach are in [3], for the fuzzy closure approach in [6], and for factor analysis in [7].

2.1 One-Sided Threshold Method

If objects are given preference, an object $x_i \in X$, the set of all objects, in a fuzzy formal context K with attribute set Y is represented as a fuzzy set $\Phi(x_i)$ as $\{ R(x_i,y_1)/y_1, R(x_i,y_2)/y_2, \ldots, R(x_i,y_n)/y_n \}$ where $R(x_i,y_j)$ is the degree to which object x_i possesses attribute y_j. Then a fuzzy formal concept is defined by specifying an α value, also referred to as a confidence threshold. For $A \subseteq X$, then the crisp definition simply is modified to require $R(x, y) \geq \alpha$ instead of $R(x, y) = 1$ as given:

$$A' = \{ y \in Y \mid \forall x \in A, R(x, y) \geq \alpha \} \quad (1)$$

and similarly, $B \subseteq Y$

$$B' = \{ x \in X \mid \forall y \in B, R(x, y) \geq \alpha \} \quad (2)$$

The threshold α eliminates from the original fuzzy formal context all (x, y) pairs with degree less than the threshold resulting in a crisp formal context for all remaining pairs. FCA is then used to create the concepts and the concept lattice structure, but no fuzziness exists at that point. Each crisp formal concept is changed into a fuzzy formal concept (A_f, B) where $A'=B$ and $B' = A$ and each $x \in A$ has a membership degree defined as

$$A_f(x) = \min_{y \in B} R(x, y). \quad (3)$$

If B is the empty set, then $A_f(x) = 1$ for every x. A is the extent and B is the intent of the fuzzy formal concept. The term fuzzy concept is used in the remainder of the paper as an abbreviation for fuzzy formal concept. When the object is given preference, the extent of the fuzzy concept is a fuzzy set. Each object's membership degree specifies the minimum degree to which it possesses all the attributes in the intent. The partial ordering of fuzzy concepts $\leq$ is defined such that $(A1_f, B1) \leq (A2_f, B2)$ if and only if $A1_f \subseteq A2_f$ and equivalently $B2 \subseteq B1$. This approach creates a set of fuzzy concepts and a fuzzy concept lattice based on the $\leq$ ordering. The one-sided (object preference) threshold method produces fuzzy set extents and crisp intents, i.e. all attributes in the intent have membership degrees of 1.0.

2.2 Fuzzy Closure Method

The fuzzy closure method produces fuzzy concepts with both fuzzy extents and fuzzy intents. It requires $R(x, y)$ values be taken from a complete residuated lattice L in the real unit interval [0, 1]. The lattice structure requires the fuzzy conjunction $\otimes$ and a residuum $\rightarrow$, i.e., a fuzzy implication operator. In [6], they use Łukasiewicz operators

$$a \otimes b = \max(0, a + b - 1)$$
$$a \rightarrow b = \min(1 - a + b, 1) \tag{4}$$

The set of all fuzzy sets using L in a universe X is denoted by L^X where a fuzzy set A means a mapping $A: X \rightarrow L$ assigning to any $x \in X$ a truth degree $A(x) \in L$ to which x belongs to A. Given the fuzzy formal context $<X, Y, R>$, for fuzzy sets $A \in L^X$ and $B \in L^Y$, , fuzzy sets $A\uparrow \in L^Y$ and $B\downarrow \in L^X$ are defined as

$$A\uparrow(y) = min_{x \in X} (A(x) \rightarrow R(x,y))$$
$$B\downarrow(x) = min_{y \in Y} (B(y) \rightarrow R(x,y)) \tag{5}$$

$A\uparrow(y)$ specifies the truth degree that the attribute y is possessed by all objects in the fuzzy set A. Similarly, $B\downarrow(x)$ specifies the truth degree that the object x possesses all attributes in the fuzzy set B. $<A, B>$ is a fuzzy formal concept if $A\uparrow = B$ and $B\downarrow = A$. Denote $\boldsymbol{B} <X, Y, R> = \{<A, B> / A\uparrow = B, B\downarrow = A\}$ as the set of all fuzzy concepts for the fuzzy formal context $<X, Y, R>$. The conceptual hierarchy is modeled by the relation $\leq$ defined on $\boldsymbol{B} <X, Y, R>$ by $<A_1, B_1> \leq <A_2, B_2>$ if and only if $A_1 \subseteq A_2$ (or, equivalently $B_1 \supseteq B_2$). The set $\boldsymbol{B}<X, Y, R>$ under the relation $\leq$ is a fuzzy concept lattice. Since fuzzy concepts are simply the fixpoints of a particular fuzzy closure operator [6], computing all fuzzy concepts reduces to computing all fixpoints of a fuzzy closure operator. The set of all fixed points of C is denoted by $fix(C)$, such that $fix(C) = \{A \in L^X \mid A=C(A)\}$. Given a formal context $<X, Y, R>$, the compound mapping $\uparrow\downarrow: L^X \rightarrow L^X$ is a fuzzy closure operator in X and $\downarrow\uparrow: L^Y \rightarrow L^Y$ is a fuzzy closure operator in Y. The fixpoints of $\uparrow\downarrow$ and the fixpoints of $\downarrow\uparrow$ are just the extents and intents respectively of the fuzzy formal concepts of $<X, Y, R>$. The algorithm in [6] was implemented and used for comparison purposes in this paper.

2.3 Fuzzy Factor Analysis Method

In [9] factor analysis using tradition matrix decomposition is extended for the fuzzy formal context. Matrix decomposition produces two matrices A and B from a single n x m matrix $\boldsymbol{R}$ representing the relationship between the n objects and the m attributes such that $R = A \circ B$.The n x k object-factor matrix $\boldsymbol{A}$ explains how the hidden k factors are related to the objects, and the k x m factor-attribute matrix B explains how the hidden k factors are related to the attributes. A goal is to keep k small. The grades of memberships used in R, A, and B must be from a bounded scale. That is, L must be a finite set of membership degrees. Entries A_{il} and B_{lj} are the degree to which factor l applies to object i and the degree to which attribute j is a manifestation of factor l, respectively. The object-attribute relation is as following: object i has attribute j if there is a factor l which applies to i and for which j is one of its manifestations. To produce the matrices A and B, the matrix composition operation $\circ$ used in this algorithm is not the usual matrix product. Instead, this algorithm uses a t-norm $\otimes$ and is defined as:

$$(A \circ B)_{ij} = \vee_{l=1}^{k} A_{il} \otimes B_{lj}. \tag{6}$$

The factors are fuzzy concepts of R. Assume (C, D) is a fuzzy formal concept of R with C the fuzzy extent and D the fuzzy intent. Then $C(i)$ is the degree to which object i possesses all attributes in D. $D(j)$ is the degree to which attribute j is possessed by all objects in C. Now given a set of fuzzy formal concepts for $F = \{(C_1, D_1), \ldots (C_k, D_k)\}$, the A_F matrix is created as an n x k matrix where l^{th} column specifies degrees assigned to objects in C_l, i.e., $(A_F)_{il} = C_l(i)$. Similarly, the B_F matrix is created as a k x m matrix where l^{th} row specifies the degrees assigned to attributes in D_l, i.e., $(B_F)_{lj} = D_l(j)$. For the lth column in $C_l(i)$, each degree is the degree to which factor l applies to object i. For the lth row, $D_l(j)$ is the degree to which factor l applies to attribute j. If $R = A_F \circ B_F$, then F represents the set of factors which can completely explain the data.

3 Experiments and Analysis of Results

In [5] the threshold approach is used to create a fuzzy concept lattice to develop a fuzzy ontology. In [6], experiments using the fuzzy closure method look at its practical performance based on characteristics of the formal context such as the distribution of the truth degrees in randomly generated contexts. The experiments described here use real-world data to compare the results of these methods.

3.1 Experimental Datasets

When a gene is annotated with terms from the Gene Ontology (GO) [10], an evidence code is provided. It can be interpreted as the degree of association strength between a GO term and a gene product. These codes are translated into membership degrees as done in [11]. Traceable Author Statement (TAS) is 1.0. Inferred from Sequence Similarity (ISS) is 0.8. Inferred from Electronic Annotation (IEA) is 0.6. Non-traceable Author Statement (NAS) is 0.4. Not Documented (ND) and Not Recorded (NR) were modified from 0.1 to 0.2 since the fuzzy closure operator requires that the set of membership degrees in L must form a closed lattice, i.e., they must be closed on the implication operator so that $L = \{0, 0.2, 0.4. 0.6, 0.8, 1.0\}$.

A fuzzy formal context is created with the genes as the objects, the GO terms as the attributes, and the numeric evidence codes as the degree of association of the GO term with the gene. The gene annotation data file GPD 194 data set [11] contains 194 human gene products (proteins). Here genes are classified into three families:

collagen, myotubularin, and receptor precursor. The fuzzy formal context created using the collagen family gene annotation data is shown in Figure 1. Due to space limitations, the other family annotation data is not provided. Table 1 shows the characteristics of the formal context for the three families and the combined set.

[illegible] gene	[illegible]	[illegible]	[illegible]	1501	[illegible]	[illegible]	7155	8544	7605	5585	7397	8015	5586	5587	30198	5194	8285	9405	7517	5588	0004	5578	5594
[illegible]	.6	.6	[illegible]	1	.2																		
[illegible] 1						.6	.6																
[illegible] 1	.6	.6					.6																
[illegible] 1	[illegible]	.6																					
[illegible]	.6	.6	.2	1	.2			1	.6	.2													
[illegible]	.6	.6									.2	1	1										
4A1	[illegible]													.2									
4A2	[illegible]													1	.4								
[illegible]	.6						.6		1			1		1		.6	1	1					
[illegible]	.4						.6							.4	.4	.6							
5A3	.6	.6					.6				1								1	1			
9A1	.4						.6				1										.2	.6	1
9A2	.6			1																		.6	1

Fig. 1. Fuzzy formal context for the Collagen Family

Table 1. Characteristics of Fuzzy Formal Contexts for Gene Annotation files

Family	# of Objects	# of Attributes	Fill %
Collagen	13	23	20
Myotubularin	7	14	35
Receptor precursor	7	35	33
Three combined	27	63 (9 common)	10

3.2 Threshold vs. Fuzzy Closure

For all four test data sets, $\alpha \geq 0.2$ is used. Column 1 of Table 2 gives the test data set. Columns 2 and 3 give the number of concepts produced by the respective approaches. Notice the significant difference in the sizes of the concept lattices. Column 4 gives the number of α-cut concepts exactly matching a fuzzy closure concept. The extent of the α-cut concept must agree with that of the fuzzy closure concept both with respect to the set of objects and their membership degrees. The intent of the α-cut concept, however, only matches the set of attributes in the intent of the fuzzy closure concept. Very few exact concept matches are found. Column 5 gives the number of concepts from the fuzzy closure approach that match an α-cut concept ignoring the membership degrees. Column 6 gives the number of concepts in the alpha-cut approach having exactly the same extent including matching degrees. Column 7 gives the number of extents from the fuzzy closure approach that match to an α-cut extent, ignoring the membership degrees. Column 8 gives the number of intents from the fuzzy closure approach that match to an α-cut intent, ignoring the membership degrees.

Table 2. Threshold vs. Closure Summary

Test Cases	# of concepts		# concepts Exact matches	# concepts Matches	# Extent Exact matches	# Extent matches	# Intent matches
	Threshold	Closure					
Collagen	26	1160	2	12	26	1160	1160
Myotubularin	13	470	1	5	13	470	470
Receptor	17	913	1	19	17	913	913
Combined	62	3495	4	30	62	3495	3495

Columns 2 and 6 have identical values. That is, each extent produced by the threshold approach maps exactly to an extent in one of the fuzzy closure concepts. The set of extents from the threshold approach are a subset of those from the fuzzy closure approach. Secondly, column 3 values are identical to that of column 7 and column 8. Ignoring degrees, the extents produced by the closure approach result in the same set of extents from the threshold approach. A large number of extents produced by the closure approach are identical in the set of objects they contain. For example, for the 1160 extents of the collagen data, 26 sets of extents are unique and repeatedly used with varying membership degrees and similarly, for the intents of the collagen data. The fuzzy closure approach generates a vast number of fuzzy concepts, many of which are very close or similar to already existing fuzzy concepts within the lattice. The threshold approach using the smallest membership degree in the fuzzy formal context as the α-cut value is much simpler and produces all the extent sets and intent sets of the fuzzy closure approach ignoring membership degrees. The fuzzy closure approach provides additional information with respect to the intents membership degrees. The fuzzy closure operator approach generates a vast number of fuzzy concepts, many of which are very close or similar to already existing fuzzy concepts with small differences between the memberships degrees in the extent and intents going from one parent concept to its children. Examining such large fuzzy concept lattices to uncover information hidden in the fuzzy formal context could be very challenging.

3.3 Fuzzy Factor Analysis vs. Fuzzy Concept Lattice Creation

Table 3 compares the factor analysis and threshold approach. For factor analysis the Łukasiewicz implication and the min tnorm were used. Column 2 and column 3 give the number of concepts produced by the respective approaches. Column 4 gives the number of α-cut concepts that exactly match (object and its degree) to a factor in factor analysis. The intent of the α-cut concept, however, only has to agree with the set of attributes in the intent of the factor. Column 5 lists the number of concepts in the α-cut approach that match to a factor with respect to the set of objects in the extent and the set of attributes in the intent. Column 6 lists the number of concepts in the factor analysis that have exactly the same extent, (object and its degree) as a concept in the α-cut approach. Column 7 lists the number of factor extents that match to an α-cut extent, ignoring the degrees. Column 8 gives how many factor intents are matched to an α-cut intent, ignoring the membership degrees of the factor intent.

Table 3. Threshold vs. Factor Analysis Summary

Test Cases	# of concepts		# Factors Exact matches	# Factors Matches	# Extent Exact matches	# Extent matches	# Intent matches
	Threshold	Factor					
Collagen	26	13	0	7	2	13	1
Myotubularin	13	8	0	2	0	8	0
Receptor	17	9	0	6	0	9	0
Combined	62	32	1	16	2	32	3

Column 4 shows that for all cases except the combined, no exact matches exist between the two. Column 5, however, shows except for the myotubularin data that over 50% of the factors produced agree with a threshold concept with respect to its set of objects in the extent and the set of attributes in the intent. Column 6 shows that very few or none of the threshold fuzzy extent sets correspond to that of a factor extent set. Column 7 being identical to column 3, however, indicates that all the extent sets of factors correspond to an extent set of a concept from the threshold approach when degrees are ignored. The set of extents for the factor analysis is a subset of that of the threshold approach when membership degrees are ignored. Although the threshold approach has not been run giving the attributes preference, it seems highly likely that all the intents of these factors would also corresponds to intent of a fuzzy concept for the threshold approach if degrees are not considered.

4 Conclusions and Future Work

A major difference between the two approaches to creating a fuzzy concept lattice is the number of fuzzy concepts produced. Since the threshold approach transforms the fuzzy formal context into a crisp context, it simplifies the process and produces a much fewer number of fuzzy concepts. The set of fuzzy extents produced by the threshold method is always a subset of that of the fuzzy closure approach. If membership degree is ignored, the set of extents are identical. All the extent sets of the factors correspond to an extent set of a threshold fuzzy concept when degrees are ignored. Although not discussed in the above experimental section, experiments with implication operators for the fuzzy closure approach revealed that the Łukasiewicz implication always produced a significantly larger number of concepts than the Gödel implication, i.e., in the thousands versus in the hundreds. The combination of the Łukasiewicz implication with the min tnorm in factor analysis produced factors that more quickly covered a larger percentage of the formal context matrix. However, the factors produced by the Gödel implication with the min tnorm produces smaller concepts, i.e. with respect to the sizes of the intents and extents.

Future plans are to experiment with other real world application data and develop visualization features to browse a fuzzy concept lattice and the factors to allow users to see their coverage of the fuzzy formal context entries.

References

1. da Costa, P.C.G., d'Amato, C., Fanizzi, N., Laskey, K.B., Laskey, K.J., Lukasiewicz, T., Nickles, M., Pool, M. (eds.): URSW 2005 - 2007. LNCS (LNAI), vol. 5327. Springer, Heidelberg (2008)
2. Sanchez, E. (ed.): Fuzzy Logic and the Semantic Web. Elsevier Science, Amsterdam (2006)
3. Tho, Q.T., Hui, S.C., Fong, A.C.M., Cao, T.H.: Automatic Fuzzy Ontology Generation for the Semantic Web. IEEE Transactions on Knowledge and Data Engineering 18(6), 842–856 (2006)
4. Fenza, G., Loia, V., Senatore, S.: Concept Mining of Semantic Web Services by Means of Fuzzy Formal Concept Analysis (FFCA). In: IEEE International Conference on Systems, Man, and Cybernetics (IEEE SMC), Singapore, pp. 12–15 (October 2008)
5. De Maio, C., Fenza, L.V., Senatore, S.: Towards Automatic Fuzzy Ontology Generation. In: Proceedings of the 2009 IEEE International Conference on Fuzzy Systems, Jeju Island, Korea, August 20-24, pp. 1044–1049 (2009)
6. Belohlavek, R., De Baets, B., Outrata, B., Vychodil, J.: Computing the lattice of all fixpoints of a fuzzy closure operator. IEEE Trans. on Fuzzy systems (2010), doi: 10.1109/TFUZZ.2010.2041006
7. Belohlavek, R., Vychodil, V.: Factor Analysis of Incidence Data via Novel Decomposition of Matrices. In: Ferré, S., Rudolph, S. (eds.) ICFCA 2009. LNCS, vol. 5548, pp. 83–97. Springer, Heidelberg (2009)
8. Cross, V., Kandasamy, M., Yi, W.: Fuzzy Concept Lattices: Examples using the Gene Ontology. In: Proceedings of the North American Fuzzy Information Processing Society (NAFIPS), Toronto, CA, July 12-14 (2010)
9. Cross, V., Kandasamy, M., Yi, W.: Comparing Two Approaches to Creating Fuzzy Concept Lattices. In: Proc.of the North American Fuzzy Information Processing Society, El Paso, TX, March 18-19 (2011)
10. Gene Ontology Consortium. Gene Ontology: Tool for the Unification of Biology. Nature Genetics 25, 25–29 (2000)
11. Popescu, M., Keller, J., Mitchell, J.: Fuzzy Measures on the Gene Ontology for Gene Product Similarity. In: IEEE/ACM Transactions on Computational Biology and Bioinformatics, vol. 3(3), pp. 263–274 (July/September 2006)

Position Paper: Pragmatics in Fuzzy Theory

Karl Erich Wolff

Mathematics and Science Faculty
Darmstadt University of Applied Sciences
Schoefferstr. 3, D-64295 Darmstadt, Germany
karl.erich.wolff@t-online.de

Abstract. This position paper presents the main problems in classical and modern Fuzzy Theory and gives solutions in Formal Concept Analysis for many of these problems. To support the successful cooperation between scientists from the communities of Fuzzy Theory and Formal Concept Analysis the author starts with this position paper an initiative, called "Pragmatics in Fuzzy Theory".

Keywords: Fuzzy Sets, Formal Concept Analysis, Conceptual Scaling, Distributed Objects.

1 Fuzzy Theory and Formal Concept Analysis

We assume that the reader is familiar with Fuzzy Theory [12,13,1] and Formal Concept Analysis [7,4].

1.1 Fuzzy Concepts

The first paper combining Fuzzy Theory and Formal Concept Analysis was written by Burusco and Fuentes-Gonzáles [3]. A slightly different version of the same idea is the definition of a fuzzy concept lattice by Pollandt [6,5] and, independently, Belohlávek (see 3.3 in [2]). In 2002, Belohlávek published his nice book [1] on Fuzzy Relational Systems. A good overview over several constructions of certain fuzzy concept lattices is given in [2].

1.2 How Formal Concept Analysis Met Fuzzy Theory

As a member of the research group on Formal Concept Analysis in Darmstadt I can tell that we all had been surprised that there is a mathematically clear theory generalizing our conceptual approach. It was a good luck, that Silke Pollandt (née: Umbreit) joined our research group in Darmstadt in about 1995 when she had finished her doctoral thesis [6,5]. Her mathematical approach was fine, but we saw several problems in the applications of fuzzy concept lattices. At the International Conference on Conceptual Knowledge Processing in 1996 in Darmstadt Ana Burusco and R. Fuentes-Gonzáles gave a talk about their work; the next talk was "Conceptual Processing of Fuzzy Data" by the author. That was the early phase of a long lasting discussion about Fuzzy Theory and

S.O. Kuznetsov et al. (Eds.): RSFDGrC 2011, LNAI 6743, pp. 135–138, 2011.

Formal Concept Analysis. The middle phase of this discussion happened from 1997 - 2000 at several EUFIT-conferences with Lotfi Zadeh. The latest phase of this discussion started at the Third International Conference on Formal Concept Analysis (ICFCA) in Lens (France) 2005, where Radim Belohlávek and his group the first time joined an FCA-conference. From that time on we regularly met the very active group of Radim Belohlávek at the ICFCA and at his international conference on Concept Lattices and their Applications (CLA).

1.3 Problems in Fuzzy Theory

Now we mention the main problems in classical and modern Fuzzy Theory.

Problem 1.1 (Fuzzy Implications): What is the meaning of a fuzzy implication "If m_1 is v_1, then m_2 is v_2 ", for example "If the pressure is high, then the weather is fine"? In classical Fuzzy Theory one tries to represent the meaning of such an implication by the construction of a fuzzy relation $R \in [0,1]^{(X_1 \times X_2)}$ which depends only on the membership functions of v_1 and v_2. Examples are the Kleene-Dienes implication, the Lukasiewicz implication, the stochastic implication, the Goguen implication, the Gödel implication, the Sharp implication and the Mamdani implication - and they all have some disadvantages (see [8,9]).

Problem 1.2 (Linguistic Variables): Zadeh [13] has introduced the notion of a linguistic variable without defining it in a mathematical precise way and without a definition of the direct product of two linguistic variables. Such a definition is necessary for the meaningful representation of fuzzy implications. The problem lies in the fact that any product of two linguistic variables over the real unit interval $[0,1]$ has to be a linguistic variable over $[0,1] \times [0,1]$, but this square did not fit into the methodology of fuzzy sets. I believe, that this was the reason for the many different trials for the introduction of implications mentioned in Problem 1.1.

Problem 2.1 (Meaning of Residuated Lattices): This problem has two main aspects: the order theoretic aspect and the algebraic one. In all applications mentioned in the cited literature the ordinal aspect is used similarly as in conceptual scaling theory. The (usually truncated) algebraic operations are not connected to some meaning in the applications - and that yields problems in the interpretation of the fuzzy concept lattices which depend on the algebraic structure. For example, the differences in the meaning of the Lukasiewicz, Gödel, and the product algebra for the fuzzy concept lattices generated from them seem to be not well-understood (see [1], chapter 2, p. 239.)

Problem 3.1 (Fuzzy Concept Lattices): fuzzy concept lattices are too big. It was observed very early that a fuzzy concept lattice often has much more concepts than a concept lattice of the same data.

Problem 3.2 (Interpretation of Fuzzy Concepts): Fuzzy concept lattices are difficult to interpret. The reason is that there is no easily readable rule corresponding to the reading rule in FCA. For more details the reader is referred to the main theorem of fuzzy concept lattices (see [1], p. 262). As a consequence, all diagrams of fuzzy concept lattices in [1] are drawn without the usual labeling of the object and attribute concepts. This labeling is the main tool for the visualising power of line diagrams in FCA.

2 Solutions and Hints

The two main problems (Problem 1.1 and 1.2) in classical Fuzzy Theory have been solved by the author in [8,9] by introducing a Fuzzy Scaling Theory using (realized) linguistic variables (over an ordered set!) and their product which is a linguistic variable over the direct product of the corresponding ordered sets. That leads to a clear understanding of the question why fuzzy implications can not be treated meaningfully over linear logics: the product of two intervals is not an interval. It also leads to an understanding of the different roles of the set G of measured objects and the set X of values of the given measurement. The object distribution on the concept lattice of the derived context of the direct product of the linguistic variables does not only show the fuzzy implications as implications of a formal context, it also shows partial implications (which hold up to some exceptions).

Problem 2.1 should be treated in practice by using complete residuated lattices only if the algebraic structure of the residuated lattice corresponds to the practical problem - but that will happen quite seldom. I emphasize to compare applications of complete residuated lattices with applications treated by usual conceptual scaling theory. The investigation of the interrelation of conceptual and algebraic structures should be forced.

Problem 3.1 can be understood as follows: it is well-known (see [5], p. 34) that a fuzzy context (G, M, R) over a complete residuated lattice L where $R \in L^{G \times M}$ can be viewed as a many-valued context and conceptually scaled with the ordinal scale $(L, L, \geq)$ in such a way that the concept lattice of the derived formal context is isomorphic to the join-subsemilattice of the fuzzy concept lattice generated by the crisp formal objects $(g, 1)$ for $g \in G$. Since the fuzzy context can be reconstructed even from this (small) concept lattice of the derived context the usually much larger fuzzy concept lattice is not necessary for this reconstruction of the given information; and in that sense it is too big!

Problem 3.2 should be solved from a general point of view: In colloquial language we use terms like "this cloud" to denote a cloud without describing it in all details. The formal description should therefore also have the possibility to adapt the granularity of the description to the given purpose. I have the feeling that fuzzy concept lattices have lost some flexibility for a suitable adaptation. One of the main reasons seems to be the fact that the formal objects (x, a) with $x \in X$ and $a \in L$ (for some L-Fuzzy context (X, Y, R)) do not have a counterpart in colloquial language.

In the next section we give a short alternative conceptual description of fuzzyness without using membership functions.

3 Representation of Fuzzyness by Distributed Objects

To be short, the main idea for the representation of fuzzyness without using membership functions is described by an example: imagine a beautiful sunset with fuzzy clouds. To represent it, I take a digital photo. Assume that for each of the millions of pixels we have its place, given by two coordinates, and we have three numbers describing the color of this pixel. For a conceptual knowledge processing of these data I use the pixels as formal objects of a data table, formally described as a Conceptual Semantic System (CSS) as introduced by the author (see [11]). I also use the notion of an object in a CSS and the notion of the trace of an object. An object is just a tupel of values, for example a triple of color values. The trace of such a triple is then the set of places of those pixels which have the given color triple. These traces can be arbitrarily fuzzy. For a mathematical introduction the reader is referred to [11].

Concluding remark: I am looking forward to have valuable discussions concerning “Pragmatics in Fuzzy Theory”.

References

1. Belohlávek, R.: Fuzzy Relational Systems: Foundations and Principles. Kluwer Academic/Plenum Press, Norwell, USA (2002)
2. Belohlávek, R., Vychodil, V.: What is a fuzzy concept lattice? In: Proceedings of the 3rd International Conference on Concept Lattices and Their Applications 2005 (CLA 2005), pp. 34–45 (2005)
3. Burusco Juandeaburre, A., Fuentes-Gonzáles, R.: The study of the L-fuzzy concept lattice. Mathware & Soft Computing 3, 209–218 (1994)
4. Ganter, B., Wille, R.: Formal Concept Analysis: mathematical foundations. Springer, Heidelberg (1999); German version: Springer, Heidelberg (1996)
5. Pollandt, S.: Fuzzy-Begriffe: Formale Begriffsanalyse unscharfer Daten. Springer, Berlin (1997)
6. Umbreit, S.: Formale Begriffsanalyse mit unscharfen Begriffen. Dissertation, Martin-Luther-Universität Halle-Wittenberg (1995)
7. Wille, R.: Restructuring lattice theory: an approach based on hierarchies of concepts. In: Ferré, S., Rudolph, S. (eds.) ICFCA 2009. LNCS (LNAI), vol. 5548, pp. 314–339. Springer, Heidelberg (2009)
8. Wolff, K.E.: Conceptual Interpretation of Fuzzy Theory. In: Zimmermann, H.J. (ed.) EUFIT 1998, vol. I, pp. 555–562. Aachen (1998)
9. Wolff, K.E.: Concepts in Fuzzy Scaling Theory: Order and Granularity. Fuzzy Sets and Systems 132, 63–75 (2002)
10. Wolff, K.E.: States of Distributed Objects in Conceptual Semantic Systems. In: Dau, F., Mugnier, M.-L., Stumme, G. (eds.) ICCS 2005. LNCS (LNAI), vol. 3596, pp. 250–266. Springer, Heidelberg (2005)
11. Wolff, K.E.: Applications of Temporal Conceptual Semantic Systems. In: Wolff, K.E., et al. (eds.) KONT/KPP 2007. LNCS(LNAI), vol. 6581, pp. 59–78. Springer, Heidelberg (2011)
12. Zadeh, L.A.: Fuzzy Sets. Information and Control 8, 338–353 (1965)
13. Zadeh, L.A.: The concept of a linguistic variable and its application to approximate reasoning. Part I: Inf. Science 8, pp. 199–249; Part II: Inf. Science 8, pp. 301–357; Part III: Inf. Science 9, pp. 43–80 (1975)

Regularization of Fuzzy Cognitive Maps for Hybrid Decision Support System

Alexey N. Averkin and Sergei A. Kaunov

Computer center of RAS
Vavilova 40, 119991 Moscow, Russia
Averkin2003@inbox.Ru
SKaunov@gmail.com
http://www.CCAS.Ru

Abstract. In this paper an aspect of collaborative construction of decision support systems based on fuzzy cognitive maps (FCM) is considered. We propose a way for cooperation in developing process of this systems by different experts and tuning developed systems to given conditions. These goals are attained by employing regularization methods, available since FCM is considered as a neural network. Interpretation and motivation of such approach are described. On the base of fuzzy cognitive map and fuzzy hierarchy model the new approach of Fuzzy Hierarchical Modeling is introduced. Advantages of the method are described. A novel approach to overcoming inherent limitations of Hierarchical Methods by exploiting cognitive maps and multiple distributed information repositories is proposed.

Keywords: Fuzzy Cognitive Map, hybrid decision support, regularization.

1 Introduction

The work with experts implies a lot of problems. Some of them are closely examined in [1], where many aspects of FCM hybrid DSS are discussed. In this work we concentrate on two problems. The first one is problem of expert inaccessibility for tuning system to given environment when it vary a little from specific part of the expert experience. And the other one is allowing experts to work together on one system or allow their models to collaborate. Solution for these problems comes through generalization of models; we employ methods of regularization to realize it in custom hybrid decision support system.

2 Regularization

Regularization was introduced for solving ill-posed problems by substitution of original problem with finding of special function that is very close to original but differs a little to treat it as well-posed. Firstly it made an assumption about solution type than the function of this type included into the special function

S.O. Kuznetsov et al. (Eds.): RSFDGrC 2011, LNAI 6743, pp. 139–146, 2011.

with *regularizational parameter* which determining balance between solution distance from original problem and estimate function. Great parameter value leads solution to be the same function you intoduced and there is no solution at all when parameter equals zero. Regularization often addressed as Occam's razor application because it solve the problem by some degree of restriction on possible solution.

2.1 Neural Networks Regularization

Statistical methods of artificial intelligence employs regularization as one of major tool. Primarily it lets to control over-fitting by introducing some additional information to the model (eg. introducing of prior distribution). Usually this information is a penalty for complexity (eg. restrictions for smoothness or bounds on the vector space norm). Usually neural networks regularization is error functional minimization. Most often error functional is sum of bias squares with depending on function smoothness addition. Choose of this restrictions will reflects in structural changes of the network optimizing its size and work time.

2.2 Regularization of FCM

Regularization was not effectively applied to FCMs though some attempts were made (eg. [2]). A very nice approach described in [3]. They used FCM for core of their image clustering algorithm. The procedure is that rasterized image first presented as a network where each pixel becomes a node of the cognitive map. After some processing of training set—number of clusters is given—the FCM works as reasoner deciding what cluster should each pixel correspond by computing its degree of membership. The decision depends on linear combination of two coefficients, called "causal weights". The first is actual *regularization* and the other is addressed as "contextual" coefficient—which promote relation of nearby pixel–nodes to the same cluster—they used with some constant weight by themselves. This last define the balance of context and regularization for the decision. Actual regularization done as iterative process computing measure between estimate membership degree and its value at current moment (iteration) of the concept (pixel–node). So it means that we shift the membership degree to count not from zero but from its statistical estimate.

But notable efforts are made for developing different ways for FCMs refinement and adjustment. A lot of methods of artificial intelligence employed for this task. Neural nets learning is one of successful among them. This approach efficiently used in [4]. The essence of this approach is to let FCM behave as a fuzzy neural net where concepts become nodes and their causal edges become synapses. With this change we get the ability to improve the net, eg. to apply some learning algorithm or regularize it.

Hansen and Rassmussen evolved research on adaptive regularization (introduced to neural networks). Adaptive regularization is a way to take into account of imposed restrictions particular environment. This becomes possible through modifying regularizing functional. It is changed so it could react to few parameters of interest. This approach to regularization provide different ways of pruning weights. In their work [5] they demonstrated two methods for pruning with some sort of a threshold (which differ in the approach depth the first one is plain Bayesian classifier, the other based on minimizing of generalization error).

Since we switched from pure causal view on the net (FCM) to fuzzy neural perception of it this kind of processing seems very natural for its structure. Demand in such transformation of ready FCM comes from task of tuning the system in the same time preserving knowledge retrieved from expert. Applying adaptive regularization with different parameters required level of generalization could be obtained [6].

3 Interpretation of FCM Regularization

After appropriate pruning of weights in a FCM, it becomes consciously editable to the experts working with it. The task of placing FCM to optimal cognitive level of expert perception induced by requirement of negative emotions minimizing. *Here we define emotions as value of cognitive dissonance occurring at the given moment.*

Solving the problem of collaborative decision making include dealing with task of cooperative decision support systems construction. Each expert has very own experience and therefore representations of subject area slightly differs one from another. FCM approach helps to handle this contradictions by finding common ground for developing collaborative system accepted by all participating contributors. This common ground is a proper generalization of given individual structures. Generalization done as removing "redundant" connections between nodes being aware of eventually growing number of mistakes in model. The edges to remove computes by adaptive regularization of model. This let to customize given model to given conditions.

4 Hybrid Decision Support Model

4.1 Fuzzy Hierarchy Model

The background of the proposed method for comparative evaluation of alternatives (variants of the decisions made) is provided by the well-known Saati hierarchical model [7] modified in [8]. This model is a weighted oriented acyclic graph G that consists of a tree T, oriented from the root to leaves $l_1, ..., l_p$ and the set $A = \{a_1, ..., a_M\}$ of nodes-alternatives, each of which is connected by oriented arcs (l_i, a_j) with all leaves.

The tree T is the hierarchy of goals known in decision-making theory. The general goal of the situation control corresponds to the root of the tree. The hierarchy is constructed using the expert up-down method. The general goal is decomposed into the subgoals (criteria) that characterize it. In turn, the subgoals are divided into criteria of a more particular character. The process of structural decomposition of the control goal is terminated when all criteria that cannot be further decomposed $l_1, ..., l_p$ have been determined. Such criteria are called leaf criteria. When formulating criteria, the "rule of positive relation" should be valid. This means that if the estimate of attainability of the goal increases, then the estimate of attainability of the goal for an adjacent criterion of a higher level also increases.

The level of a tree node is defined as usual. The root has level 0. A node has level h if the length of the route from the root to it is h. All nodes (including the leaf ones) are numbered beginning with 1, and the number of a node of level h is always less than the number of any node of a lower level. Nonleaf nodes are also denoted by k_{hi}, where h is the level number and i is the number of the node. The expressions $(k_{hi}, k_{h+1,j})$ and (i, j) are equivalent and describe the same arc, and we always have $i < j$. The leaf nodes (which can be located at different levels), as above, are always denoted by $l_1, ..., l_p$.

All weights of arcs and nodes belong to the segment $[0, 1]$. For the weights of arcs of the tree T, we have the following *normalization condition*: for any node k_{hi}, the sum of weights of all arcs going from it to nodes of level $h+1$ is equal to one. The weight ω_{ij} of the arc $(k_{hi}, k_{h+1,j})$ means the relative importance of the criterion $k_{h+1,j}$ for the criterion of a higher level k_{hi}. Particular values of the weights of arcs are determined by receiving from experts pairwise comparisons of the relative importance of criteria and their subsequent processing with account of the normalization conditions.

Under the condition that the weights of arcs are known, the weights of tree nodes are up-down determined recurrently as follows. The weight u_0 of the root is equal to one. The weight u_j of the node $k_{h+1,j}$ adjacent to the node k_{hi} is determined by the formula

$$u_j = \omega_{ij} u_i \ . \tag{1}$$

By the normalization condition, the sum of weights of all nodes of level $h+1$ adjacent to the same node k_{ri} is equal to its weight, i.e., u_i.

For each alternative $a_j \in A$, its estimate $F(a_j)$ of the attainability of the general goal is determined by the formula

$$F(a_j) = \sum_{i=1}^{p} \omega_{ij} u_i \ , \tag{2}$$

where ω_{ij} is the weight of the arc (l_i, a_j) and u_i is the weight of the leaf criterion l_i. The alternative a_q is the best one if it has the maximum estimate among all other alternatives.

4.2 Fuzzy Cognitive Model

A cognitive map is a weighted oriented graph with n nodes. Nodes v_i are called factors (in English-language literature, they more often are called concepts). A variable $y_i(t)$ corresponds to a factor v_i $(i = 1, ..., n)$. This variable can take linguistic values. The set of such values is linearly ordered and generates a linguistic scale $Z_i = \{z_{i1}, z_{i2}, ..., z_{ir}\}$, where z_{i1} and z_{ir} are the minimum and maximum elements of the set. In general, each factor has its own linguistic scale the number of elements of which is determined by the experts. The vector

$Y(t) = (y_1(t), ..., y_n(t))$ is called the *situation state* at the time instant t. The weights of arcs are numbers and belong to the interval [-1, 1]. They are given by the adjacency matrix $W = ||w_{ij}||$; the value w_{ij} characterizes the effect of the factor v_i on the factor v_j, and the sign characterizes the character of the effect.

For convenience of computations, we define a mapping ϕ of linguistic scales of factors on the number interval [0, 1] as follows. For a linguistic scale $Z_i = \{z_{i1}, z_{i2}, ..., z_{ir}\}$, we divide the interval [0, 1] into r equal intervals, whose boundaries are denoted in the increasing order $b_0 = 0, b_1, ..., b_{r-1}, b_r = 1$. We set $\phi(z_{ik} = \frac{b_{k-1}-b_k}{2}$ (the element z_{ik} is mapped to the center of the kth interval). The mapping $\phi : Z_i \longrightarrow [0,1]$ allows us to make the algorithms of the model numerical. The inverse mapping $\phi^{-1} : [0,1] \longrightarrow Z_i$ is a homomorphism, all points that belong to the interval (b_{k-1}, b_k) are mapped into a single point z_{ik}. With the help of the mapping ϕ, the situation state can be represented in a numerical form $X(t) = \phi(Y(t)) = (\phi(y_1(t)), ..., \phi(y_n(t)))$. In what follows, we will work with the numerical representation of the situation state $X(t)$.

An *increment* of the factor v_i at the time instant $t + 1$ is the value $p_i(t+1) = x_i(t + 1)x_i(t)$. Since increments may be negative, the values of pi belong to the segment [-1, 1]. Assume that $v_{j1}, ..., v_{jk}$ is the set of all factors that are input factors for the factor v_i (i.e., the initial nodes of the arcs incoming v_i). Then, in general, the value x_i at the time instant $t + 1$ depends on the values of the incoming factors at the time instant t and the weights of arcs connecting these factors with the factor v_i

$$x_i(t + 1) = f_i(x_{j1}(t), ..., x_{jk}(t), w_{j1,i}, ..., w_{jk,i}), \tag{3}$$

and the functions f_i attached to different factors are different in general. In applications, a simpler case is considered when these functions for all factors are the same. This allows one to use matrix methods for solving problems of situation analysis. The functions depend only on the increments of the values of factors and do not depend on their state.

Obtainment of a prognosis of the situation development. The problem of prognosis is formulated as follows. Given: a cognitive map $G(V, W)$, where V is the set of nodes (situation factors), and W is the adjacency matrix; a set $Z_1, \ldots, Z_n$ of scales of all factors of the situation; an initial situation state $X(0) = (x_1(0), \ldots, x_n(0))$; and an initial vector of the increments of the situation factors $P(0) = (p_1(0), \ldots, p_n(0))$, it is necessary to find the situation state $X(1), \ldots, X(n)$ and the vectors of increments $P(1), \ldots, P(n)$ at consecutive discrete time instants $1, \ldots, n$, where n (the number of nodes) is chosen so that the influence of the initial perturbation can reach all nodes. The prognosis of situation development is determined by a matrix relation

$$P(t + 1) = P(t) \circ W, \tag{4}$$

where $\circ$ is the max-product rule, $p_i(t + 1) = \max_j(|p_j(t)w_{ji}|)$.

The scheme for dealing with the model is as follows. An expert specifies in linguistic values an initial state of the situation $Y(0)$ and the next state $Y(1)$, which arises after control actions. Based on these data, the numerical initial increment $P(0) = \phi(Y(1)) - \phi(Y(0)) = X(1) - X(0)$ is computed. The subsequent computations are numerical. Using operation 4, the increments at the consecutive time instants $t = 1, ..., n$ are computed, and the situation state is determined from the relation

$$X(t+1) = X(t) + P(t+1) . \tag{5}$$

To interpret the prognosis and to give the results to the expert, we use the inverse mapping ϕ^{-1} from numerical values to linguistic ones. When obtaining a prognosis, together with computation of the vector $P(t+1)$, we compute the vector $C = \{c_1(t+1), ..., c_n(t+1)\}$. The value $c_i(t+1)$ is called a *consonance* of the factor v_i and is determined as follows. Denote by $p_i^+(t+1)$ the maximum of positive increments that are input into the factor v_i, i.e., $p_i^+(t+1) = \max_j(p_j(t)w_{ji})$, $p_j(t)w_{ji} \geq 0$. Analogously, $p_i^-(t+1)$ is the maximum of the absolute values of negative increments that are input to the factor v_i; then, we have

$$c_i(t+1) = \frac{|p_i^+(t+1) + p_i^-(t+1)|}{|p_i^+(t+1)| + |p_i^-(t+1)|} . \tag{6}$$

The consonance $c_i(t+1)$ characterizes the degree of certainty of the prognosis at the time instant $t+1$.

4.3 Integrated Model

The main idea of the integration consists in (1) establishing a correspondence between a certain subset of factors of the cognitive map and leaf criteria of the estimation model and constructing a mapping of the values of factors to the values of leaf criteria of the estimation model; and (2) representing the alternative decisions proposed for the estimation in the hierarchical model, i.e., control actions, as the vectors of increments of control factors, i.e., the factor of the cognitive map which the DMP can directly affect. (Note that it is supposed that the estimation model does not change in the course of situation analysis.)

At the first stage of constructing the integrated model, a general goal is formulated, the hierarchy of criteria is constructed, the level of leaf criteria is selected, and the weights of leaf criteria are determined. At the second stage, a cognitive map that satisfies the following condition is constructed. Each leaf criterion of the estimation model corresponds to a certain factor of the cognitive map. The set of such factors that cover the whole set of leaf criteria is denoted by V^*. To construct a cognitive map, we suggest to use the methodology based on structural and functional decomposition of the situation. This methodology was described in detail in [9]. This methodology considers the situation in two aspects, structural and functional. The structural aspect is based on the decomposition (up-down) of the application domain into components (elements of the situation) connected by the relation "part-the whole object." For each element, the factors that determine it are selected. Among the set of factors that characterize all elements of the situation, the factors are selected that are close in meaning to the leaf criteria of the estimation hierarchy, which generate the set V^*.

Both structural decomposition and functional description (relations between the factors and their weights) are performed by experts. Therefore, the maps designed by different experts can be different, since they reflect their subjective understanding of the problem to some extent. However, the requirement to select the set V^* guarantees that all variants of cognitive maps contain the same factors formulated in the same language, in terms of the leaf criteria of the estimation model. The third stage of designing the integrated model consists in constructing leaf criteria based on the scale of factors of the set V^*.

The procedure of formation of alternative decisions and their estimation in the integrated model are as follows:
experts form the source alternatives;
each alternative a_i is represented in the form of a vector $P_i(0)$ of increments of all factors of the cognitive map;
by formulas 4 and 5, a prognosis of development of the situation under the initial increment $P_i(0)$ is computed, i.e., the vector $X_i(n)$ of the values of situation factors at the step n;
the values $x^*_{ij}(n)$ factors from the set V^* are mapped in the values of the corresponding leaf criteria, i.e., the value $\psi_j(x^*_{ij}(n))$;
the alternative is estimated in the estimation model by the formula

$$F(a_i) = \sum \psi_j(x^*_{ij})u_j \ , \qquad (7)$$

where the sum is over all factors from the set V^*;
as the best alternative, the alternative with the maximum estimate is chosen $F(a_q) = \max_{j=1,\ldots,M} F(a_j)$.

5 Conclusion

We showed conjunction of two fuzzy approaches to decision support to one hybrid system allowing both to act alternatively, basing on results of each other. The first part of system is fuzzy hierarchy model, the other one is fuzzy cognitive map. It process results of the hierarchy model and directs it work. For constructing of such complex system for a big subject area was proposed applying regularization to FCM component for its generalization and allowing collaboration on it. Also at CCAS A. N. Averkin with team research genetic programming approach to FCM training. Currently a method developed which allows to train FCMs based on fuzzy relational equations. This part of the work is at the stage of preparation for publication, and interested researchers will be able to learn about it soon.

Acknowledgments. This paper is partially supported by RFBR projects No. 10-01-00851 and 11-01-00959.

References

1. Averkin, A.N., Agrafonova, T.V., Titova, N.V.: System of Decision Making Support Based on Fuzzy Models. Journal of Computer and Systems Sciences International 48, 89–100 (2009)
2. Sahbi, H., Boujemaa, N.: Fuzzy Clustering: Consistency of Entropy Regularization. In: International Conference on Computational Intelligence (Special Session on Fuzzy Clustering), Dortmund, Germany (2004)
3. Pajares, G., Guijarro, M., Herrera, P.J., Ruz, J.J., de la Cruz, J.M.: Fuzzy Cognitive Maps Applied to Computer Vision Tasks. In: Glykas, M. (ed.) Fuzzy Cognitive Maps: Advances in Theory, Methodologies, Tools and Applications. STUDFUZZ, vol. 247, pp. 270–300. Springer, Heidelberg (2010)
4. Carlsson, C., Fuller, R.: Adaptive Fuzzy Cognitive Maps for Hyperknowledge Representation in Strategy Formation Process. In: Proceedings of International Panel Conference on Soft and Intelligent Computing, pp. 43–50. Technical University of Budapest (1996)
5. Hansen, L.K., Rasmussen, C.E.: Pruning from Adaptive Regularization. Neural Computation 6(6), 1222–1231 (1994)
6. Goutte, C., Hansen, L.K.: Regularization with a pruning prior. Neural Networks 10(6), 1053–1059 (1997)
7. Saati, T.: Decision Making: A Method for Analysis of Hierarchies. Radio i Svyaz, Moscow (1993) (in Russian)
8. Makeev, S.P., Shakhnov, I.F.: Arrangement of Objects in Hierarchical Systems. Izv. Akad. Nauk SSSR, Tekh. Kibern.. 3, 29–46 (1991)
9. Kulinich, A.A.: The Methodology of Cognitive Modeling of Complex Ill-Determined Situations. In: Proceedings of 2nd International Conference on Control Problems, Moscow, vol. 2 (2003) (in Russian)

On Designing of Flexible Neuro-Fuzzy Systems for Nonlinear Modelling

Krzysztof Cpałka[1,2], Olga Rebrova[3], Robert Nowicki[1,2] and Leszek Rutkowski[1,2]

[1] Czestochowa University of Technology, Department of Computer Engineering, Poland
[2] Academy of Management (SWSPiZ), Institute of Information Technology, Poland
[3] The Russian State Medical University, Institute of Pharmaeconomics, Russia
krzysztof.cpalka@kik.pcz.pl, o.yu.rebrova@gmail.com,
robert.nowicki@kik.pcz.pl, leszek.rutkowski@kik.pcz.pl

Abstract. In the paper the evolutionary strategy is used for learning of neuro-fuzzy structures of a Mamdani type applied to modelling of nonlinear systems. In the process of evolution we determine parameters of fuzzy membership functions, specific t-norm in a fuzzy inference, specific t-norm for aggregation of antecedents in each rule, and specific t-conorm describing an aggregation operator. The method is tested using well known approximation benchmarks.

1 Introduction

Neuro-fuzzy systems combine the natural language description of fuzzy systems and the learning properties of neural networks. In literature various structures of such systems and corresponding learning methods have been proposed (see e.g. [1]-[3], [6], [8], [10]). Recently several algorithms have been developed to increase interpretability and accuracy of these systems. For various methods of designing fuzzy rule-based systems the reader is referred e.g. to [2]-[4], [9]-[11], [13].

In our previous works ([2] [11]) we developed the idea of flexible neuro-fuzzy systems for classification. In such systems the type of triangular norms was found in the process of learning using the back propagation method. In this paper we extend that idea, using the evolutionary strategy (μ, λ) (see e.g. [5]), for modelling of nonlinear systems. In the process of evolution we determine parameters of fuzzy membership functions, specific t-norm in an a fuzzy inference, specific t-norm for aggregation of antecedents in each rule, and specific t-conorm describing an aggregation operator. It should be emphasized, that performance of fuzzy systems depends on applied triangular norms, therefore their proper choice is very important. Our method is tested using well known approximation benchmarks.

2 Description of Neuro-Fuzzy System of a Mamdani Type

We consider multi-input, multi-output neuro-fuzzy system mapping $\mathbf{X} \to \mathbf{Y}$, where $\mathbf{X} \subset \mathbf{R}^n$ and $\mathbf{Y} \subset \mathbf{R}^m$. The fuzzifier performs a mapping from the

S.O. Kuznetsov et al. (Eds.): RSFDGrC 2011, LNAI 6743, pp. 147–154, 2011.

observed crisp input space $\mathbf{X} \subset \mathbf{R}^n$ to the fuzzy sets defined in $\mathbf{X}$. The most commonly used fuzzifier is the singleton fuzzifier which maps $\bar{\mathbf{x}} = [\bar{x}_1, \ldots, \bar{x}_n] \in \mathbf{X}$ into a fuzzy set $A' \subseteq \mathbf{X}$ characterized by the membership function

$$\mu_{A'}(\mathbf{x}) = \begin{cases} 1 \text{ if } \mathbf{x} = \bar{\mathbf{x}} \\ 0 \text{ if } \mathbf{x} \neq \bar{\mathbf{x}} \end{cases}. \tag{1}$$

The fuzzy rule base consists of a collection of N fuzzy IF THEN rules in the form R^k :

$$\left[\text{IF } x_1 \text{ is } A_1^k \text{ AND} \ldots \text{AND } x_n \text{ is } A_n^k \text{ THEN } y_1 \text{ is } B_1^k \text{AND} \ldots \text{AND } y_m \text{ is } B_m^k\right], \tag{2}$$

where $\mathbf{x} = [x_1, \ldots, x_n] \in \mathbf{X}$, $\mathbf{y} = [y_1, \ldots, y_m] \in \mathbf{Y}$, $A_1^k, A_2^k, \ldots, A_n^k$ are fuzzy sets characterized by membership functions $\mu_{A_i^k}(x_i)$, $i = 1, \ldots, n$, $k = 1, \ldots, N$, whereas B_j^k are fuzzy sets characterized by membership functions $\mu_{B_j^k}(y_j)$, $j = 1, \ldots, m$, $k = 1, \ldots, N$. The fuzzy inference determines a mapping from the fuzzy sets in the input space $\mathbf{X}$ to the fuzzy sets in the output space $\mathbf{Y}$. Let $\mathbf{A}^k = A_1^k \times A_2^k \times \ldots \times A_n^k$. Each of N rules (2) determines fuzzy sets $\bar{B}_j^k \subset \mathbf{Y}$ given by the compositional rule of inference

$$\bar{B}_j^k = A' \circ \left(\mathbf{A}^k \to B_j^k\right) \tag{3}$$

characterized by membership functions

$$\mu_{\bar{B}_j^k}(y_j) = \mu_{A_1^k \times \ldots \times A_n^k \to B_j^k}(\bar{\mathbf{x}}, y_j) = \mu_{\mathbf{A}^k \to B_j^k}(\bar{\mathbf{x}}, y_j) = T\left\{\tau_k(\bar{\mathbf{x}}), \mu_{B_j^k}(y_j)\right\}, \tag{4}$$

where $T\{\cdot\}$ is a t-norm and $\tau_k(\bar{\mathbf{x}})$ is a firing strength of the k-th rule, k=1, …, N, defined as follows:

$$\tau_k(\bar{\mathbf{x}}) = \mu_{\mathbf{A}^k}(\bar{\mathbf{x}}) = \mathop{T}_{i=1}^{n}\left\{\mu_{A_i^k}(\bar{x}_i)\right\}. \tag{5}$$

The aggregation operator, applied in order to obtain the fuzzy set B_j' based on fuzzy sets $\bar{B}_j^k$, $k = 1, \ldots, N$, is the t-konorm operator. The aggregation is carried out by

$$B_{j'} = \bigcup_{k=1}^{N} \bar{B}_j^k. \tag{6}$$

The membership function of B_j' is determined by the uses of a t-norm, i.e.:

$$\mu_{B_j'}(y_j) = \mathop{S}_{k=1}^{N}\left\{\mu_{\bar{B}_j^k}(y_j)\right\}. \tag{7}$$

The defuzzifier performs a mapping from the fuzzy sets B_j' to a crisp point $\bar{y}_j$, $j = 1, \ldots, m$, in $\mathbf{Y} \subset \mathbf{R}$. The COA (center of area) method is defined by the following formula:

$$\bar{y}_j = \frac{\int_{\mathbf{Y}} y_j \cdot \mu_{B_j'}(y_j)\, dy_j}{\int_{\mathbf{Y}} \mu_{B_j'}(y_j)\, dy_j} \tag{8}$$

or by

$$\bar{y}_j = \frac{\sum\limits_{r=1}^{N} \bar{y}_{j,r}^{B} \cdot \mu_{B_j'}\left(\bar{y}_{j,r}^{B}\right)}{\sum\limits_{r=1}^{N} \mu_{B_j'}\left(\bar{y}_{j,r}^{B}\right)} \tag{9}$$

in the discrete form, where $\bar{y}_{j,r}^{B}$ are centres of the membership functions $\mu_{B_j^r}(y)$, i.e. for $j = 1, \ldots, m$ and $r = 1, \ldots, N$, we have:

$$\mu_{B_j^r}\left(\bar{y}_{j,r}^{B}\right) = \max_{y_j \in \mathbf{Y}} \left\{\mu_{B_j^r}(y_j)\right\}. \tag{10}$$

Consequently, in Mamdani approach formula (9) takes the following form:

$$\bar{y}_j = f(\bar{\mathbf{x}}) = \frac{\sum\limits_{r=1}^{N} \bar{y}_j^r \mathop{S}\limits_{k=1}^{N} \left\{ T\left\{ \mathop{T}\limits_{i=1}^{n} \left\{\mu_{A_i^k}(\bar{x}_i)\right\}, \mu_{B_j^k}\left(\bar{y}_j^r\right) \right\}\right\}}{\sum\limits_{r=1}^{N} \mathop{S}\limits_{k=1}^{N} \left\{ T\left\{ \mathop{T}\limits_{i=1}^{n} \left\{\mu_{A_i^k}(\bar{x}_i)\right\}, \mu_{B_j^k}\left(\bar{y}_j^r\right) \right\}\right\}}. \tag{11}$$

3 Neuro-Fuzzy Systems with Various Types of Triangular Norms

In our study we use non-parameterized (Zadeh, algebraic, bounded, drastic) and parameterised (Aczel-Alsina, Dombi, Frank, Schweizer-Sklar, Yager) triangular norms (see e.g. [7]). The hyperplanes corresponding to parameterized families of t-norms and t-conorms can be adjusted in the process of evolutionary learning of an appropriate parameter. In order to design a specific neuro-fuzzy system one should specify input and output membership functions and specify triangular norms: t-conorm for aggregation described by formula (7), t-norm for generation of inferences described by formula (4), and t-norm for aggregation of antecedents in each rule, described by formula (5). In this paper a new learning algorithm for evolution of flexible neuro-fuzzy inference systems, described by formula (11), is proposed. In the process of evolution we find type of triangular norms and all parameters described in the previous section.

4 Designing and Learning of Neuro-Fuzzy Systems with Various Types of Triangular Norms

The structure of the neuro-fuzzy system described by formula (11), its parameters and type of adjustable triangular norms used for aggregation of rules, connections of antecedents and consequences and aggregation of antecedents were found using the evolutionary strategy (μ, λ). Moreover, the learning algorithm has the following components:

- A repair procedure of the temporary population for the adjustable triangular norms described as follows:

$$\left\{p_k^{\mathcal{T}} < \varepsilon_0 \Rightarrow p_k^{\mathcal{T}} = \varepsilon_0, p_k^{\mathrm{I}} < \varepsilon_0 \Rightarrow p_k^{\mathrm{I}} = \varepsilon_0, p^{\mathrm{agr}} < \varepsilon_0 \Rightarrow p^{\mathrm{agr}} = \varepsilon_0\right\}, \qquad (12)$$

  where ε_0 is a small positive number chosen before the evolution process.
- Fitness function calculated as the average of the classification mistakes in the learning sequence.
- Crossover and mutation operations for chromosomes $\mathbf{X}_j^{\mathrm{str}}$ and $\mathbf{X}_j^{\mathrm{alg}}$ analogous to the classical genetic algorithm with obvious modifications (see e.g. [12]).

4.1 Evolution of Structure

In our study a structure of neuro-fuzzy systems is found based on the evolution of chromosomes $\mathbf{X}_j^{\mathrm{str}}$ given by

$$\mathbf{X}_j^{\mathrm{str}} = \begin{pmatrix} \text{type of triangular norm of aggregation of antecedents,} \\ \text{type of triangular norm of inference operator,} \\ \text{type of triangular norm of aggregation of rules} \end{pmatrix}, \qquad (13)$$
$$= \left(X_{j,1}^{\mathrm{str}}, X_{j,2}^{\mathrm{str}}, X_{j,3}^{\mathrm{str}}\right)$$

where $j = 1, \ldots, \mu$ for parent population or $j = 1, \ldots, \lambda$ for temporary population. The gene named "type of triangular norm of aggregation of antecedents", "type of triangular norm of inference operator" and "type of triangular norm of aggregation of rules" takes values from the set 0 for Zadeh triangular norm, 1 for algebraic triangular norm, 2 for bounded triangular norm, 3 for drastic triangular norm, 4 for Aczel-Alsina triangular norm, 5 for Dombi triangular norm, 6 for Frank triangular norm, 7 for Schweizer-Sklar triangular norm, 8 for Yager triangular norm.

For temporary population we use the recombination (crossover) and the mutation operations:

- Single-point crossover, with probability p_c chosen before the evolution process, analogous to the classical genetic algorithm.
- Mutation, with probability p_m chosen before the evolution process, analogous to the classical genetic algorithm with obvious modifications. For example, mutation of gene coding type of triangular norm is performed by a random selection of its value from the set $\{0, 1, 2, 3, 4, 5, 6, 7, 8\}$. Choosing value of probability p_m we should realize that chromosomes $\mathbf{X}_j^{\mathrm{str}}$, $j = 1, \ldots, \lambda$, before and after mutation are independent each of other. Therefore, if p_m increases, then mutation resembles random sampling.

4.2 Evolution of Parameters

In the system described by formula (11) there are $L = 2N(n+m+1)+1$ parameters to be found in the evolution process: parameters $\{\bar{x}_{i,k}^{A}, \sigma_{i,k}^{A}\}$ of Gaussian membership functions $\mu_{A_i^k}(\bar{x}_i)$, parameters $\{\bar{y}_{j,k}^{B}, \sigma_{j,k}^{B}\}$ of Gaussian membership functions $\mu_{B_j^r}(\bar{x}_i)$, parameters $\{p_k^{\tau}\}$ of parameterized t-norm for aggregation of antecedents in each rule, parameters $\{p_k^{\mathrm{I}}\}$ of parameterized t-norm for generation of inferences described by formula (4), parameter $\{p^{\mathrm{agr}}\}$ of parameterized t-conorm for aggregation described by formula (7). We apply evolutionary strategy (μ, λ) for learning all parameters taking into account restrictions imposed on particular parameters. In a single chromosome, according to the Pittsburgh approach, a complete linguistic model is coded in the following way:

$$\mathbf{X}_j^{\mathrm{par}} = \begin{pmatrix} \bar{x}_{1,1}^{A}, \sigma_{1,1}^{A}, \ldots, \bar{x}_{n,1}^{A}, \sigma_{n,1}^{A}, \bar{y}_{1,1}^{B}, \sigma_{1,1}^{B}, \ldots, \bar{y}_{m,1}^{B}, \sigma_{m,1}^{B}, p_1^{\tau}, p_1^{\mathrm{I}}, \ldots \\ \bar{x}_{1,N}^{A}, \sigma_{1,N}^{A}, \ldots, \bar{x}_{n,N}^{A}, \sigma_{n,N}^{A}, \bar{y}_{1,N}^{B}, \sigma_{1,N}^{B}, \ldots, \bar{y}_{m,N}^{B}, \sigma_{m,N}^{B}, p_N^{\tau}, p_N^{\mathrm{I}}, \\ p^{\mathrm{agr}} \end{pmatrix}, \\ = \left(X_{j,1}^{\mathrm{par}}, X_{j,2}^{\mathrm{par}}, \ldots, X_{j,L}^{\mathrm{par}}\right) \tag{14}$$

where $j = 1, \ldots, \mu$ for parent population or $j = 1, \ldots, \lambda$ for the temporary population. The self-adaptive feature of the algorithm is realized by assigning to each gene a separate mutation range described by the standard deviation

$$\sigma_j^{\mathrm{par}} = \left(\sigma_{j,1}^{\mathrm{par}}, \sigma_{j,2}^{\mathrm{par}}, \ldots, \sigma_{j,L}^{\mathrm{par}}\right), \tag{15}$$

where $j = 1, \ldots, \mu$ for the parent population or $j = 1, \ldots, \lambda$ for the temporary population.

For temporary population we use the recombination (crossover) and the mutation operations:

- Crossover with averaging the values of the genes:

$$X_{j1,g}^{\mathrm{par}'} = \tfrac{1}{2}\left(X_{j1,g}^{\mathrm{par}} + X_{j2,g}^{\mathrm{par}}\right), X_{j2,g}^{\mathrm{par}'} = X_{j1,g}^{\mathrm{par}'}, \tag{16}$$

  and

$$\sigma_{j1,g}^{\mathrm{par}'} = \tfrac{1}{2}\left(\sigma_{j1,g}^{\mathrm{par}} + \sigma_{j2,g}^{\mathrm{par}}\right), \sigma_{j2,g}^{\mathrm{par}'} = \sigma_{j1,g}^{\mathrm{par}'}, \tag{17}$$

  where $g = 1, \ldots, L$.
- Mutation:

$$\sigma_{j,g}^{\mathrm{par}'} = \sigma_{j,g}^{\mathrm{par}} \exp\left(\tau' N(0,1) + \tau N_{j,g}(0,1)\right), \tag{18}$$

  and

$$X_{j,g}^{\mathrm{par}'} = X_{j,g}^{\mathrm{par}} + \sigma_{j,g}^{\mathrm{par}'} N_{j,g}(0,1), \tag{19}$$

  where $\sigma_{j,g}^{\mathrm{par}}$ denotes current value of the mutation range of the j-th chromosome, $j = 1, \ldots, \lambda$, of the g-th gene, $g = 1, \ldots, L$, $\sigma_{j,g}^{\mathrm{par}'}$ denotes a new value of the mutation range, $N(0,1)$ is the number drawn from the normal distribution, $N_{j,g}(0,1)$ is the number drawn the normal distribution of the j-th chromosome, $j = 1,\ldots,\lambda$, of the g-th gene, $g = 1,\ldots,L$, and τ', τ denote constants chosen before the evolution process. The following formulas can be found in literature (see e.g. [5]):

$$\tau' = \frac{C}{\sqrt{2L}}, \tag{20}$$

and

$$\tau = \frac{C}{\sqrt{2\sqrt{L}}}, \tag{21}$$

where C takes value 1 the most frequently. In order to avoid convergence of the mutation range to 0, we use the following formula:

$$\sigma_{j,g}^{\mathrm{par}'} < \varepsilon_0 \Rightarrow \sigma_{j,g}^{\mathrm{par}'} = \varepsilon_0, \tag{22}$$

where ε_0 is a small positive number chosen before the evolution process.

Each individual of the parental and temporary populations is represented by sequence of chromosomes $\langle \mathbf{X}_j^{\mathrm{str}}, \mathbf{X}_j^{\mathrm{par}}, \sigma_j^{\mathrm{par}} \rangle$, given by formulas (13)-(15). The genes of the first chromosome take integer values, whereas the genes of the two last chromosomes take real values.

5 Simulation Results

The neuro-fuzzy system (11) is simulated on the chemical plant problem and modeling of static nonlinear function (HANG problem) (see e.g. [13]).

The evolution process is characterized by the following parameters: $\mu = 10$, $\lambda = 500$, $p_m = 0.077$, $p_c = 0.770$, $\psi = 0.01$, $C = 1.2$, number of generations $= 250$, and $\varepsilon_0 = 0.01$. For the problems we found the type of triangular norm of aggregation of antecedents, type of triangular norm of implication, and type of triangular norm of aggregation of rules.

5.1 Chemical Plant Problem

We deal with a model of an operator's control of a chemical plant. The plant produces polymers by polymerisating some monomers. Since the start up of the plant is very complicated, men have to perform the manual operations at the plant. Three continuous inputs are chosen for controlling the system: monomer concentration, change of monomer concentration and monomer flow rate. The output is the set point for the monomer flow rate.

The best result ($RMSE = 0.0085$) is obtained for the neuro-fuzzy system (11) with $m = 1$, $N = 6$, Dombi t-norm in aggregation of antecedents, algebraic triangular norm in implication, and algebraic triangular norm in aggregation of rules (see Table 1).

Table 1. Simulation results

No	Name of gene	Problem	
		Chemical Plant	HANG
1	type of tr. norm of aggregation of antecedents	Dombi	Yager
2	type of tr. norm of implication	algebraic	Zadeh
3	type of tr. norm of aggregation of rules	algebraic	algebraic

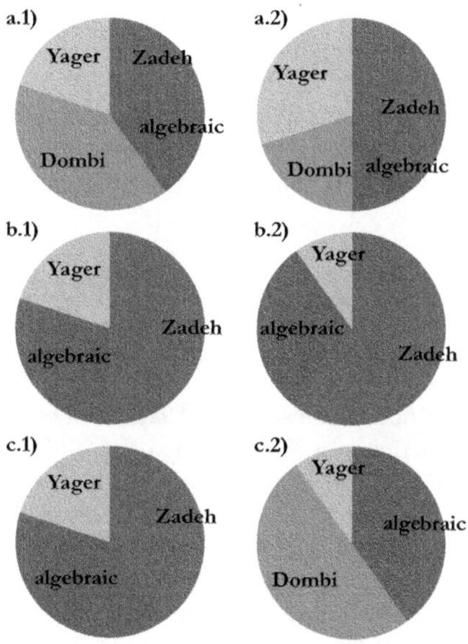

Fig. 1. The types of triangular norms encoded in the genes of chromosomes of the population μ after the last generation of learning of neuro-fuzzy system (11) in: a) aggregation of antecedents, b) implication, c) aggregation of rules for chemical plant problem (a.1, b.1, c.1) and HANG problem (a.2, b.2, c.2)

5.2 HANG Problem

The problem is to approximate a nonlinear function given by

$$y(x_1, x_2) = \left(1 + x_1^{-2} + x_2^{-1.5}\right)^2. \tag{23}$$

We obtained 50 input output data by sampling the input range $x_1, x_2 \in [1, 5]$.

The best result ($RMSE = 0.0691$) is obtained for the neuro-fuzzy system (11) with $m = 1$, $N = 5$, Yager t-norm in aggregation of antecedents, Zadeh triangular norm of implication, and algebraic triangular norm in aggregation of rules (see Table 1).

In Fig. 1 we present the types of triangular norms encoded in the genes of chromosomes of the population μ after the last generation of learning of neuro-fuzzy system (11). One can see that in our simulations, which illustrate approximation problems, nonparametric triangular norms were selected to build aggregation operators and inference operators, whereas parametric triangular norms were selected to build operators of aggregation of premises of rules.

6 Conclusions

In the paper we presented a novel method for designing and learning of flexible neuro-fuzzy systems for nonlinear approximation. The method is based on the evolutionary strategy (μ, λ) and allows to find both the structure and parameters of the system in the process of evolution.

Acknowledgements. This work was partly supported by the Polish Ministry of Science and Higher Education (Polish Singapore Research Project 2008-2011), and the Foundation for Polish Science TEAM project 2010-2014.

References

1. Casillas, J., Cordon, O., Herrera, F., Magdalena, L. (eds.): Interpretability Issues in Fuzzy Modeling. Springer, Heidelberg (2003)
2. Cpałka, K.: A New Method for Design and Reduction of Neuro-Fuzzy Classification Systems. IEEE Transactions on Neural Networks 20(4), 701–714 (2009)
3. Cpałka, K.: On evolutionary designing and learning of flexible neuro-fuzzy structures for nonlinear classification. In: Nonlinear Analysis Series A: Theory, Methods & Applications, vol. 71. Elsevier, Amsterdam (2009)
4. Czogała, E., Łęski, J.: Fuzzy and Neuro-Fuzzy Intelligent Systems. Physica-Verlag. A Springer Company, Heidelberg, New York (2000)
5. Fogel, D.B.: Evolutionary Computation: Toward a New Philosophy of Machine Intelligence, 3rd edn. IEEE Press, Piscataway (2006)
6. Gabryel, M., Rutkowski, L.: Evolutionary Learning of Mamdani-Type Neuro-fuzzy Systems. In: Rutkowski, L., Tadeusiewicz, R., Zadeh, L.A., Żurada, J.M. (eds.) ICAISC 2006. LNCS (LNAI), vol. 4029, pp. 354–359. Springer, Heidelberg (2006)
7. Klement, E.P., Mesiar, R., Pap, E.: Triangular Norms. Kluwer Academic Publishers, Dordrecht (2000)
8. Kumar, M., Stoll, R., Stoll, N.: A robust design criterion for interpretable fuzzy models with uncertain data. IEEE Trans. Fuzzy Syst. 14(2), 314–328 (2006)
9. Łęski, J.: A Fuzzy If-Then Rule-Based Nonlinear Classifier. Int. J. Appl. Math. Comput. Sci. 13(2), 215–223 (2003)
10. Rutkowski, L.: Computational Intelligence. Springer, Heidelberg (2007)
11. Rutkowski, L., Cpałka, K.: Flexible neuro-fuzzy systems. IEEE Trans. Neural Networks 14(3), 554–574 (2003)
12. Sivanandam, S.N., Deepa, S.N.: Introduction to Genetic Algorithms. Springer, Heidelberg (2008)
13. Sugeno, M., Yasukawa, T.: A fuzzy logic based approach to qualitative modeling. IEEE Trans. on Fuzzy Systems 1, 7–31 (1993)

Time Series Processing and Forecasting Using Soft Computing Tools

Nadezhda Yarushkina[1], Irina Perfilieva[2], Tatiana Afanasieva[1], Andrew Igonin[1], Anton Romanov[1], and Valeria Shishkina[1]

[1] Ulyanovsk State Technical University, Severny Venec, 32, 432027 Ulyanovsk, Russia
[2] Institute for Research and Applications of Fuzzy Modeling University of Ostrava 30. dubna 22, 701 03 Ostrava 1, Czech Republic
jng@ultsu.ru, Irina.Perfilieva@osu.cz, tv.afanaseva@mail.ru, agigonin@gmail.com, romanov73@gmail.com, shvv@ulstu.ru

Abstract. The aim of this contribution is to show that the combination of F-transform, fuzzy relations, neural networks and genetic algorithms can be successfully used in analysis and forecast of short time series encountered in financial analysis of a small enterprize. We propose to represent a time series trend by the direct F-transform components and to model it by one of three different models that are based on a linear autoregressive equation, neural network or fuzzy relation autoregressive equation. An optimal model of the trend will be chosen by a genetic algorithm. In comparison with other time series techniques the proposed one is simple and effective in computation and forecast.

In the application part, we present a description of a new software system that has been elaborated on the basis of the proposed theory. It includes analysis of time series and their tendencies in linguistic terms.

Keywords: F-transform, time series processing, fuzzy tendencies, soft. computing

1 Introduction

The analysis and forecasting of time series is based on its suitable decomposition. There are two approaches to the decomposition of a time series. The first one (called *Box-Jenkins methodology* [1]) uses a combination of an autoregressive model and a moving average model, so that a time series is represented by its values in previous time moments. The second approach uses aggregative decomposition components such as trend, cycle, etc., so that a time series is represented by an additive or multiplicative combination of these components contemporarily. In our contribution, we propose a new combination of the two above described approaches, where the aggregative decomposition components are modeled by respective autoregressive models. The advantage of our methodology over those from which it stems, consists in its simplicity and thus, higher efficiency.

S.O. Kuznetsov et al. (Eds.): RSFDGrC 2011, LNAI 6743, pp. 155–162, 2011.

Because the notion of trend has no precise definition, we use two methods for the trend modeling: the F-transform [4] and the method of local tendencies (see [8]), see details in Section 4.

The notion of a fuzzy transform (*F-transform*, for short) see, e.g., [4,5] turned out to be very useful in combination with other soft computing techniques. In [6,3], we successfully applied the inverse F-transform and perception-based logical deduction to time series analysis and prediction. In this contribution, we propose a novel approach which consists in using the direct F-transform in combination with fuzzy relation processing tool, neural networks and genetic algorithms. We show that time series analysis and processing can be successfully performed on the basis of this powerful combination.

The term *fuzzy tendency of a fuzzy time series* has been introduced in [8] with the purpose to characterize dynamics (systematic movement) of a TS. In the second part of our contribution, we explain how time series decomposition and forecast can be made on the basis the F-transform together with other soft computing tools. An advantage of the presented method is its applicability to the analysis and forecast of short time series which cannot be processed by statistical methods due to their insufficient length. Last, an application to express analysis of short time series will be demonstrated.

2 Discrete F-Transform and Its Matrix Representation

2.1 Fuzzy Partition and Its Matrix Representation

Generally, the F-transform of a function $f: P \longrightarrow \mathbb{R}$ is a vector whose components can be considered as weighted local mean values of f. Throughout this paper we will assume that $\mathbb{R}$ is the set of real numbers, $[a,b] \subseteq \mathbb{R}$, and $P = \{p_1,\dots,p_l\}$, $n < l$, is a finite set of points such that $P \subseteq [a,b]$. Function $f: P \longrightarrow \mathbb{R}$ defined on the set P is called *discrete*.

Below, we will remind basic facts about the F-transform as they were presented in [4], and then introduce its relaxed version and matrix representation.

Let $f: P \longrightarrow \mathbb{R}$ be an arbitrary function on P. The first step in the definition of the F-transform of f is a selection of a *fuzzy partition* of the interval $[a,b]$ by a finite number $n: 2 \leq n \leq l-2$ of fuzzy sets $A_1,\dots,A_n : [a,b] \to [0,1]$, identified with their membership functions. There are three axioms which characterize a fuzzy partition:

1. (locality) - for every $k = 1,\dots,n$, $A_k(x) = 0$ if $x \in [a,b] \setminus [x_{k-1},x_{k+1}]$;
2. (continuity) - for every $k = 1,\dots,n$, A_k is continuous on $[x_{k-1},x_{k+1}]$;
3. (density) - $\sum_{j=1}^{l} A_k(p_j) > 0$, $k = 1,\dots,n$.

The membership functions $A_1,\dots,A_n$ in the fuzzy partition are called *basic functions*. We say that the basic function A_k *covers* a point p_j if $A_k(p_j) > 0$.

In the subsequent text, we will fix an interval $[a,b]$, a finite set of points $P \subseteq [a,b]$ and a fuzzy partition $A_1,\dots,A_n$ of $[a,b]$. Denote $a_{kj} = A_k(p_j)$ and consider $n \times l$ matrix A with elements a_{kj}. We will say that A is a *partition matrix* of P.

2.2 Discrete F-Transform

Once the basic functions $A_1,\ldots,A_n$ are selected, we define (see [4]) the (direct) *F-transform* of a function $f : P \longrightarrow \mathbb{R}$ as a vector $(F_1,\ldots,F_n)$ where the k-th *component* F_k is equal to

$$F_k = \frac{\sum_{j=1}^{l} f(p_j)\cdot A_k(p_j)}{\sum_{j=1}^{l} A_k(p_j)}, \quad k = 1,\ldots,n. \tag{1}$$

Let us identify the function $f : P \longrightarrow \mathbb{R}$ with the row vector $\mathbf{f} = (f_1,\ldots,f_l)$ of its values on P so that $f_j = f(p_j)$, $j = 1,\ldots,l$. Moreover, let partition $A_1,\ldots,A_n$ be represented by the matrix A. Then we will say that the row vector $F_n(\mathbf{f}) = (F_1,\ldots,F_n)$ is the F-transform of f if

$$(F_1,\ldots,F_n) = \left(\frac{(A\mathbf{f}^T)_1}{a_1},\ldots,\frac{(A\mathbf{f}^T)_n}{a_n}\right) \tag{2}$$

where $(A\mathbf{f}^T)_k$ is the k-th component of the product $A\mathbf{f}$, $a_k = \sum_{j=1}^{l} a_{kj}$, $k = 1,\ldots,n$. The following properties characterize the $F_n(\mathbf{f})$:

P1. The mapping $F_n : \mathbb{R}^l \to \mathbb{R}^n$ such that $F_n : \mathbf{f} \to F_n(\mathbf{f})$ is linear.
P2. Components $F_1,\ldots,F_n$ of $F_n(\mathbf{f})$ minimize the following function

$$\Phi(y_1,\ldots,y_n) = \sum_{k=1}^{n}\sum_{j=1}^{l}(f_j - y_k)^2 a_{kj}.$$

3 Time Series Decomposition

Assume that y_t, $t = 1,\ldots,T$, $T \geq 3$, is a time series. We consider it as a discrete function which is defined on the set $P_T = \{1,\ldots,T\}$ of time moments. Let $A_1,\ldots,A_n$, $n < T$, be basic functions which constitute a fuzzy partition of the interval $[1,T]$. Denote P_k, $k = 1,\ldots,n$, a subset of P_T consisting of points covered by A_k. Note that due to the density condition on a fuzzy partition, every P_k is not empty.

Let $F_n(\mathbf{y}) = (Y_1,\ldots,Y_n)$ be the F-transform of time series y_t with respect to $A_1,\ldots,A_n$. We say that $\{y_t - Y_k \mid t \in P_k\}$ is the k-th *residual vector* of y_t with respect to A_k, $k = 1,\ldots,n$. For $t = 1,\ldots,T$, $k = 1,\ldots,n$ we denote

$$r_{tk} = \begin{cases} y_t - Y_k, & \text{if } t \in P_k, \\ -\infty, & \text{otherwise} \end{cases}$$

so that $R = (r_{tk})$ is a $T \times n$ matrix of residua.

In the following two propositions we show an estimation of the F-transform of a residual vector and a decomposition of a time series y_t.

Proposition 1. *Let $R = (r_{tk})$ be the $T \times n$ matrix of residua of a time series y_t with respect to fuzzy partition $A_1,\ldots,A_n$ of $[1,T]$. Let A be the $n \times T$ partition matrix of the set P_T. Then*

- $AR = 0$,
- $F_n(\mathbf{r_k}) = 0$, $k = 1,\ldots n$.

Proposition 2. *Let y_t be a time series on $[1,T]$ and $A_1,\ldots,A_n$ a fuzzy partition of $[1,T]$. If $F_n(\mathbf{y})$ is the F-transform of y_t and $R=(r_{tk})$ is the $T\times n$ matrix comprised of residual vectors then y_t can be represented as follows:*

$$y_t = \bigvee_{k=1}^{n} (Y_k + r_{tk}). \tag{3}$$

Remark. In the decomposition (3), the F-transform components are considered as *trend* components of the time series y_t.

4 Models of Trend

In this section, we will use decomposition (3) of a time series and propose models of thus obtained parts. These models will be then used for forecast. Because we use similar approaches for modeling trend components and residual vectors, we will describe only the first one.

Let the following sequence of trend components $Y_1,\ldots,Y_n$ be an autoregressive of order q, $1\leq q\leq 5$, so that

$$Y_{k+q} = \varphi(Y_k,Y_{k+1},\ldots,Y_{k+q-1}), \quad k=1,\ldots,n-q-1. \tag{4}$$

In our approach, we investigate three general models of φ: linear, neural network and fuzzy relation. Let us briefly characterize each of them.

(a) Formally, a linear model of (4) is characterized by

$$Y_{k+q} = \alpha_1 Y_k + \alpha_2 Y_{k+1} + \cdots + \alpha_q Y_{k+q-1}, \quad k=1,\ldots,n-q-1. \tag{5}$$

Obviously, (5) is a system of linear equations with unknown $\alpha_1,\alpha_2,\ldots,\alpha_q$. We propose to solve this system numerically.

(b) Neural network approximation of (4) is realized by a group of three models that differ in a way of representation of arguments in (4). The neural network is configured as a multilayer perceptron.

(c) Fuzzy relation model of (4) has been proposed in [7]. Without going into specific details, we can reproduce it by the following equation:

$$\tilde{Y}_{k+q} = (\tilde{Y}_k,\ldots,\tilde{Y}_{k+q-1})\circ R, \quad k=1,\ldots,n-q-1, \tag{6}$$

where $(\tilde{Y}_k,\ldots,\tilde{Y}_{k+q-1},\tilde{Y}_{k+q})$ are respective fuzzy sets on a range Y of a time series, $\circ$ is a max – min-composition, and R is a fuzzy relation on Y^{q+1}. We extend the group of fuzzy relation models (6) by the following two:

$$\Delta\tilde{Y}_{k+q} = (\Delta\tilde{Y}_k,\ldots,\Delta\tilde{Y}_{k+q-2})\circ R, \quad k=1,\ldots,n-q-1, \tag{7}$$

and

$$\tilde{Y}_{k+q} = (\tilde{Y}_k,\ldots,\tilde{Y}_{k+q-1},\Delta\tilde{Y}_k,\ldots,\Delta\tilde{Y}_{k+q-2})\circ R, \quad k=1,\ldots,n-q-1, \tag{8}$$

where $\Delta\tilde{Y}_k$ is a fuzzy set that characterizes the first difference $Y_{k+1}-Y_k$. It is worth noticing that $\Delta\tilde{Y}_k$ is an instance of a *local tendency* of a time series (see [8] for detailed theory). Expressions (6)–(8) represent systems of fuzzy relation equations with respect to unknown fuzzy relations R. Under certain conditions, these systems are solvable. A solution R is a required model of (4).

5 Short Time Series Forecast

In this section, we will apply the time series decomposition (3) to forecast short time series. Due to short lengths, they can be hardly processed by statistical methods. Since our decomposition (3) consists of two parts, we will forecast both of them, i.e. a trend component and its residual counterpart. We will explain our approach in more details below.

Our forecast will be based on the model (4) which contains two a priori unknown parameters: q and φ. They will be identified by the following standard procedure: the last component Y_n is not used in (4) and is forecasted by each of three specified above models. The best model minimizes the criterion MAPE:

$$MAPE = \frac{1}{n}\sum_{k=1}^{n}\left|\frac{|Y_k - \widehat{Y_k}|}{Y_k}\right| 100\%. \tag{9}$$

It will be used for the forecast of an unknown value of Y_{n+1} (and similarly, $\mathbf{r}_{n+1}$). Then obtained forecasts Y_{n+1} and $\mathbf{r}_{n+1}$ will be combined by (3) and used for the forecast of the next time series value.

In (Figure 1) we illustrate our method on the example of one chosen time series of the length 25 with real economical data. Four different partitions of $[1,T]$ with triangular shaped basic functions are considered. In all cases, we produce forecasts of the last F-transform component and of the respective portion of the time series.

We see that the MAPE value of the trend forecast does not directly depend on a robustness of a partition. Thus other parameters (e.g., the above discussed) should be included into the optimization procedure. More details are in the next section.

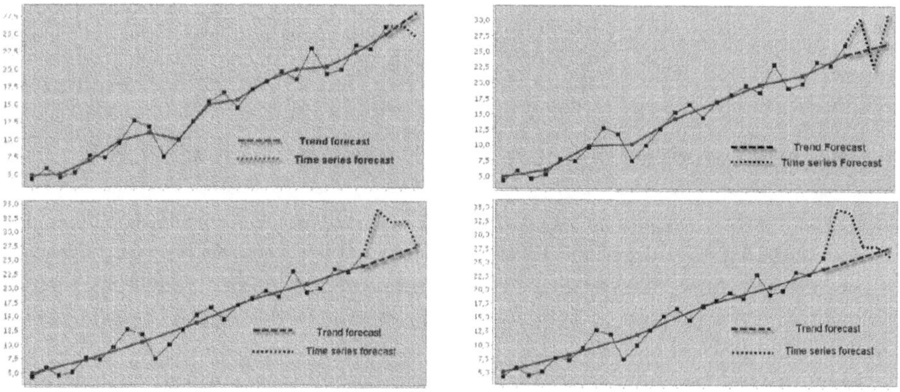

Fig. 1. Four forecasts of the same time series: *Upper left.* The forecast of the last 2 values. Each basic function covers 5 points. Trend MAPE=0,21. Time series MAPE=7.32. *Upper right.* The forecast of the last 3 values. Each basic function covers 7 points. Trend MAPE=4,16. *Lower left.* The forecast of the last 4 values. Each basic function covers 9 points. Trend MAPE=15,75. *Lower right.* The forecast of the last 5 values. Each basic function covers 11 points. Trend MAPE=4,068. Time series MAPE=15,08.

The same time series as in the example above, has been processed by the ForecastPro software package which combines a majority of recommended statistical methods such as ARMA, ARIMA, Box-Jenkins, etc. Because the given time series is too short, the exponential smoothing method was applied as the only possible one. We illustrate the result of the ForecastPro in Figure 2 below where a forecast of the last 5 values is shown. It is seen that comparable (by the number of forecasted values) forecasts are in Figure 1 (Lower right) and Figure 2. Both forecasts have almost equal MAPE values.

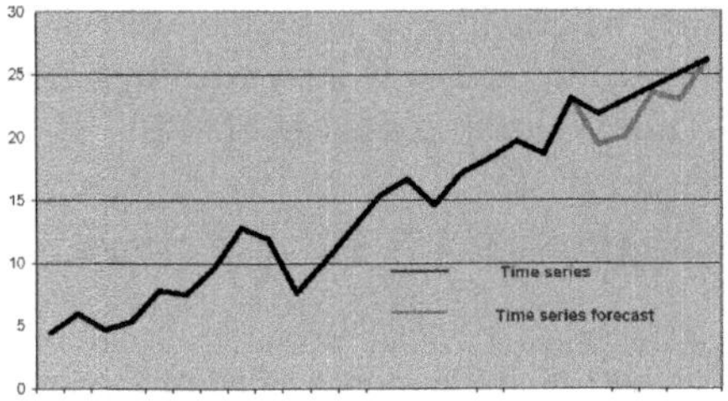

Fig. 2. Time series of the length 25, and the forecast of the last 5 values made by the ForecastPro. Time series MAPE=14,51.

At the end of this section, we show that on the benchmark "Alabama", which contains data on university entrants, our forecast is significantly better than known statistical and fuzzy methods. The below given table contains the values of MAPE of four known methods used for a forecast of one future value.

Table 1. Comparison of TS modeling for University of Alabama TS

ARIMA	Song [7]	Huarng [2]	The proposed method
5,49	3,11	1,5294	0, 96

6 Express Analysis of Short Time Series and Web-service

In this section, we present a demonstration of the software system that has been elaborated on the basis of the proposed theory. The system is focused on express analysis of short time series. Moreover, it is a part of the Internet service which has been developed with the purpose to help small companies in verification of their economic stability.

More than 40 real time series have been processed by the software system. In the below given table we show models that have been chosen after optimization as well as a characteristic (number of covered points (Number CP)) of an optimal partition. The choice was among two models of (4): linear (LN), and neural network (NN) (see Section 4).

A brief analysis of optimal methods and their parameters shows that robust partitions are preferable to finer ones.

Table 2. Optimal methods for forecast of time series trends and their MAPE

TS (code)	Length of TS	Number CP	Method	MAPE (trend)%
2-6	40	11	NN	4,95
2-8	43	13	NN	3,7
2-9	42	7	LN	2,3
2-10	90	11	LN	5,99
2-11	57	9	NN	16,77
2-12	150	5	LN	0,85
2-13	25	11	LN	2,04
3-15	13	9	NN	11,10
3-16	13	9	NN	3,30
3-17	13	7	LN	1,98
3-24	12	5	NN	0,21
3-25	12	9	NN	17,39

6.1 Genetic Optimization

In this section, we will describe the optimization algorithm that helps to reduce a dimension of a search space. By this we mean the space which we characterized in the previous section and which is connected with the following parameters: number of points that are covered by basic functions (Number CP), order of the regression model q, degree of seasonality, structure of an input vector in the neural network. The last parameter determines whether our regression model depends on previous trend values, or it depends on previous trend values and their first differences, or it depends on first differences of previous trend values. The optimization criterion is a value of MAPE.

The chosen optimization algorithm is a genetic algorithm with a classical structure. We restrict the number of epochs and by this exclude those combinations of parameters which cannot be optimal for a certain time series. It is worth noticing that genetic optimization is the first step of our software system. It does not solve the problem of global optimization, but significantly reduces the search space. The final optimal choice of parameters that are used for a final forecast is made on the basis of a full overview.

7 Conclusion

In this contribution, we have shown that the combination of F-transform, fuzzy relations, neural networks and genetic algorithms can be successfully used in analysis and forecast of short time series encountered in financial analysis of a small enterprize. We described how a time series can be decomposed into a sequence of trend values and a sequence of residual vectors, and proposed the algorithm which forecasts both components.

We outlined the new software system that has been elaborated on the basis of the proposed theory. Besides the F-transform, it includes analysis of time series and their tendencies in linguistic terms, time series processing with the help of neural networks, optimization of forecast by genetic algorithms. The elaborated software system is focused on express analysis of short time series which cannot be processed by statistical methods.It is shown that on some benchmarks, the proposed software system works better that known statistical and fuzzy methods. This software system is a part of the Internet service which has been developed with the purpose to help small companies in verification of their economic stability.

Acknowledgments. The research has been supported by the grant 1M0572 of the MŠMT ČR.

References

1. Box, G., Jenkins, G.: Time Series Analysis: Forecasting and Control. Holden-Day, San Francisco (1970)
2. Huarng, K.: The application of neural networks to forecast fuzzy time series. Physica 336, 481–491 (2006)
3. Novák, V., Štěpnička, M., Dvořák, A., Perfilieva, I., Pavliska, V., Vavříčková, L.: Analysis of seasonal time series using fuzzy approach. Int. Journ. of General Systems 39, 305–328 (2010)
4. Perfilieva, I.: Fuzzy transforms: Theory and applications. Fuzzy Sets and Systems 157, 993–1023 (2006)
5. Perfilieva, I.: Fuzzy transforms: A challenge to conventional transforms. In: Hawkes, P.W. (ed.) Advances in Images and Electron Physics, vol. 147, pp. 137–196. Elsevier Academic Press, San Diego (2007)
6. Perfilieva, I., Novák, V., Pavliska, V., Dvořák, A., Štěpnička, M.: Analysis and prediction of time series using fuzzy transform. In: Proc. of IEEE Int. Conf. on Neural Networks WCCI 2008, Hong Kong, pp. 3875–3879 (2008)
7. Song, Q., Chissom, B.S.: Fuzzy time series and its models. Fuzzy Sets and Systems 54, 269–277 (1993)
8. Yarushkina, N.G.: Osnovy teorii nechetkih i gibridnyh system. Finansy i Statistika, Moskva (2004)

Fuzzy Linear Programming – Foreign Exchange Market

Biljana R. Petreska[1], Tatjana D. Kolemisevska-Gugulovska[1], and Georgi M. Dimirovski[2]

[1] Faculty of EE and IT, SS Cyril and Methodius University, Skopje, R. Macedonia
bpetreska@gmail.com, tanjakg@feit.ukim.edu.mk
[2] Department of Computer Engineering, Dogus University of Istanbul, Turkey
gdimirovski@dogus.edu.tr

Abstract. Strategic asset management is concerned with a complex process that usually involves a variety decision making situations. But managers, decision makers, and experts dealing with optimization problems often have a lack of information on the exact values of some parameters and in the same time are unable to put into the model important aspects that couldn't be defined as crisp. To deal with this kind of imprecise data, Fuzzy Sets provide a powerful tool to model and solve these problems. A method for solving Fuzzy Linear Programming problems is provided, and this method is applied to a concrete real problem for foreign exchange market. With proposed fuzzy model in this work, asset managers obtain a tool for making better decisions for FX transactions when exchange rates are not known.

Keywords: Fuzzy Linear Programming, Foreign Exchange Market.

1 Introduction

By its very nature, financial dynamics of assets markets concerns largely volatile market systems that simultaneously exhibit phenomena of randomness, uncertainty, and fuzziness [1,2,3]. Hence this system and other social systems currently are being revisited using computational intelligence models and techniques. As a humanistic science, economics should thus have been one of the early prime targets for utilizing Fuzzy Set theory. Applications of Fuzzy Sets within the field of intelligent decision making have, for the most part, consisted of fuzzifications of the classical theories of decision making.

The main motivation for this paper is to give an example of application of Fuzzy Theory and latest achievements in researches on a particular problem. The formulated scenario in this paper, for foreign exchange market is impossible to solve with classical or crisp mathematical models, especially fuzzyness in right-hand site coefficients. The importance of elaborated application in this paper is significant, having in mind that the cash management separately and generally as important part of the asset management is conducted in very unstable environment and benefits of such non-crisp applications can lead to better results in middle and long term time periods.

S.O. Kuznetsov et al. (Eds.): RSFDGrC 2011, LNAI 6743, pp. 163–166, 2011.

2 Fuzzy Linear Programming

A linear programming problem (LPP) in general may be defined as the problem of maximizing or minimizing a linear function subject to linear constraints. Fuzzy Linear Programming is used when optimization occurs under uncertainty, when any vague quantity appears in LPP, in coefficients or constraints [4,5,6,7,8,9,12].

This offers many non-crisp definable possibilities. First we may not want to actually maximize or minimize the objective function. Secondly, the constraints might be vague. The $\leq$ sign might not be defined in the traditional sense to the degree that smaller violations may be acceptable.

Most general form of Fuzzy LPP where all coefficients we assume are Fuzzy Numbers can be written as:

$$\begin{aligned} &\max\ \ \mathbf{c_1}x_1 + ... + \mathbf{c_n}x_n \quad \text{subject to}: \\ &\mathbf{a_{i1}}x_1 + ... + \mathbf{a_{in}}x_n \leq_g \mathbf{b_i}, i = 1, ...m; x_j \geq 0, j = 1, ..., n\ . \end{aligned} \tag{1}$$

The methods for solving Fuzzy LPP are well established [10,12,13]. In this work, the elaborated problem is stated as Fuzzy Linear Programming problem with technological and right-hand-side coefficients. Most widely known is the method proposed by Bellman and Zadeh, known as Symmetric method [7,13]. Mathematically the Fuzzy Set for the solution of Fuzzy LPP can be written as:

$$D = [(\bigcap_{i=1}^{m} C_i)\bigcap G](X) \tag{2}$$

where C_i are defined Fuzzy Sets for the constraints, and G is Fuzzy Set of fuzzy optimal values.

In order to develop the more concrete form for this solution, some assumptions should be accepted. That is,$\mathbf{a_{ij}}$ and $\mathbf{b_i}$ are Fuzzy Numbers with the membership functions of this form:

$$\mu_v(x) = \begin{cases} 1\ , x < v \\ \frac{v+u-x}{u}\ , v \leq x < v+u \\ 0\ , x \geq v+u\ . \end{cases} \tag{3}$$

Here $x \in R$, so the Fuzzy Numbers in Fuzzy LPP can be written as $\mathbf{a_{ij}}$=[a_{ij} a_{ij}+d_{ij}] and $\mathbf{b_i}$=[b_i b_i+p_i].

Defuzzification of the problem is based on the calculation of the lower and upper bounds of the optimal values. Objective function takes values between z_l and z_u while technological coefficients take values between a_{ij} and a_{ij}+d_{ij} and the right-hand side coefficients take values between b_i and b_i+p_i. Defuzzified form of (2) becomes:

$$\begin{aligned} &\max \lambda \\ &\lambda(z_u - z_l) - \sum_{j=1}^{n} c_j x_j + z_l \leq 0 \\ &\sum_{j=1}^{n} (a_{ij} + \lambda d_{ij})x_j + \lambda p_i - b_i \leq 0,\ 1 \leq i \leq m; x_j \geq 0,\ 0 \leq \lambda \leq 1\ . \end{aligned} \tag{4}$$

3 Foreign Exchange Market and Illustrative Example

Foreign exchange market exists everywhere where one currency is traded for another. For interest in this work are a long term planned transactions where the exact exchange rate is unknown in the time when the trade will take place.

For purpose in this work we will assume that the total cost is a linear function of transferred amount, i.e. $cost_{ij}=k_{ij}*x_{ij}$. Optimization problem arises to determine amounts x_{ij} that should be transferred from account i to account j in way the function of cost be minimized. x_{ij} represents the amounts expressed in the currency B_j.

$$
\begin{aligned}
&\min f_{cost} = \sum_{i=1,j=1}^{k,l} k_{ij}x_{ij} \\
&\sum_{j=1}^{m} x_{ij} \leq a_i,\ i = 1..n,\ n \in N \\
&\sum_{i=1}^{n} \mathbf{er_{ij}} x_{ij} \geq \mathbf{b_j},\ j = 1..m,\ m \in N;\ x_{ij} \geq 0
\end{aligned}
\qquad (5)
$$

where $\mathbf{er_{ij}}$ are the Fuzzy Numbers for exchange rates, a_i available amount of currency A_i,$\mathbf{b_j}$ minimum but not exact amount we should indemnify of currency B_j that is different from A_i.

Illustrative example: A Bank on the account in a Brazilian bank has 100.000 Brazilian Reals. Main goal of the Bank is to indemnify minimum 10.000 USD in USA and 15.000 CAD in Canada. The transfer costs between the banks are 5% and 4% from transferred money respectively.

With respect to the past values of exchange rate between BRL and USD and BRL and CAD, appropriate Fuzzy Numbers are defined as $\mathbf{er_{11}}$=[0,58 0,6], $\mathbf{er_{21}}$=[0,56 0,62], $\mathbf{b_1}$=[10000 20000], $\mathbf{b_2}$=[15000 35000].

On this definition, the symmetric method proposed by Bellman and Zadeh is applied.

After solving 16 crisp LPP using Optimization toolbox in MATLAB, results are z_l=-4224,1 and z_{up}=-1801.1. In solving non convex problem (4) special approach is required (see e.g. [10]). Solving these problems with iterations is quite precise, but generally it takes much more iterations to obtain wanted precision. In this paper problem is solved with iteration.

The value of λ=0,51050 is the maximum value for which the problem (5) has a feasible solution. For that value of λ, the cost would be 2987,16 BLR. This result can be interpreted as, at certainty level 0,51050 the cost will be minimized for the stated problem. So if the exchange rates are lower than 0,59021 and 0,59063 we can revoke maximum 15105 USD and 25210 CAD at minimum cost. For better exchange rate it is recommended to redesign the problem.

4 Conclusion

We showed how the concepts of Fuzzy Set Theory can be applied to the real problem. We studied how to incorporate unpredictable values of exchange rates and aspiration levels, which are human-formulated, into the solution model. The defined model represents a valuable tool for better decision making in volatile environment. This paper shows one more time that economics is a field where applications of Fuzzy Sets Theory can be very successful.

In order to find the maximum value of λ in further research, one may apply an adaptive intelligent controller. According to the past values and recent trends that affect exchange rates, it should model the membership functions for exchange rates.

References

1. Ingersoll, J.E.: Theory of Financial Decision Making, pp. 65–140. Rowman & Littlefield Publishers Inc., Baltimore (1987)
2. Campbell, J.Y., Lo, A.W., MacKinley, A.C.: The Econometrics of Finance Markets. Princeton University Press, Princeton (1997)
3. Dimirovski, G.M., Dinibütün, A.T., Kile, F., Neck, R., Stahre, J., Vlacic, L.: Control system approaches for susstainable development in globalization age. Annual Reviews in Control 30(1), 103–115 (2006)
4. Zimmermann, H.J.: Fuzzy programming and linear programming with several objective functions. Fuzzy Sets & Systems 1, 45–55 (1978)
5. Bellman, R.E., Zadeh, L.A.: Decision making in fuzzy environment. Management Science 17, 209–215 (1970)
6. Rogers F., Neggers J., Younbae J.: Method for optimizing linear problems with fuzzy constraints. In: International Mathematical Forum, vol. 3(23) (2008)
7. Klir, G.J., Yuan, B.: Fuzzy Sets and Fuzzy Logic: Theory and Applications. Prentice-Hall, Upper Saddle River (1995)
8. Bruckley, J.J., Feruing, T.: Evolution algorithm solution to fuzzy problems: Fuzzy linear programming. Fuzzy Sets & Systems 109, 35–53 (2000)
9. Wang, L.X.: A Course in Fuzzy Systems and Controls. Prentice-Hall, Englewood Cliffs
10. Gasimov R.N., Yenilmez K.: Solving Fuzzy Linear Programming Problems with Linear Membership Functions, Internet published paper `citeseerx.ist.psu.edu/viewdoc/download?doi=10.1.1.95.8155\&rep=rep1\&type=pdf`
11. Rogers F., Neggers J., Jun Y.: Method for optimizing linear problems with fuzzy constraints. In: International Mathematical Forum, vol. 3(23) (2008)
12. Cadenas, J.M., Verdegay, J.L.: Using fuzzy numbers in Linear Programming. IEEE Transactions on SMC-part B: Cybernetics 27(6)
13. Fuller R., Zimmerman H.J.: Approximate Reasoning for Solving Fuzzy Linear programming Problems, Internet published paper `citeseerx.ist.psu.edu/viewdoc/download?doi=10.1.1.49.7492\&rep=rep1\&type=pdf`
14. Royal Forex, FOREX, Study book for successful foreign exchange dealing, Internet publications `www.earnforex.com/forex-e-books/beginner-forex-trading/Study_Book_for_Successful_Foreign_Exchange_Dealing.pdf`

Fuzzy Optimal Solution of Fuzzy Transportation Problems with Transshipments

Amit Kumar, Amarpreet Kaur and Manjot Kaur

School of Mathematics and Computer Applications
Thapar University, Patiala-147 004, India
amit_rs_iitr@yahoo.com, amanpreettoor@gmail.com,
manjot.thaparian@gmail.com

Abstract. In this paper a new method, named as Mehar's method, is proposed for solving fuzzy transportation problems with transshipments. Also, it is shown that it is better to use Mehar's method as compared to the existing method.

1 Introduction

In conventional transportation and transshipment problems [1] it is assumed that decision maker is sure about the precise values of transportation cost, availability and demand of the product. In real world applications, all these parameters may not be known precisely due to uncontrollable factors. To deal with such situations several authors have represented the different parameters by fuzzy numbers [18] and proposed different methods for solving fuzzy transportation problems [2,9,11,12,13,15,16,17] and fuzzy transshipment problems [3,4,5,6,7,8,10,14].

2 Proposed Method

Kumar et al. [10] proposed fuzzy linear programming approach for finding the fuzzy optimal solution of fuzzy transportation problems with transshipment. In the existing method [10] the fuzzy linear programming formulation of the chosen fuzzy transportation problem with transshipment is converted into four crisp linear programming formulations of crisp transportation problems with transshipment and then all the obtained crisp linear programming problems are solved by Simplex method [1]. But in the literature, it is pointed out that it is better to use modified distribution method [1] for finding the solution of crisp transportation problems as compared to Simplex method.

Due to the same reason, in this section a new method, named as Mehar's method, based on modified distribution method, is proposed for finding the fuzzy optimal solution of same type of problems.

The steps of the proposed method are as follows:

Step 1. Split Table 3 [10] into four crisp transportation tables i.e., Table 1, Table 2, Table 3 and Table 4 respectively.

Step 2. Find the optimal solution a_{ij}; $b_{ij} - a_{ij}$; $c_{ij} - b_{ij}$ and $d_{ij} - c_{ij}$ by solving

S.O. Kuznetsov et al. (Eds.): RSFDGrC 2011, LNAI 6743, pp. 167–170, 2011.

the crisp transportation problems, shown by Table 1; Table 2; Table 3 and Table 4 respectively, by using modified distribution method.
where, $\lambda_{ij} = \frac{a'_{ij}+b'_{ij}+c'_{ij}+d'_{ij}}{4}$, $\rho_{ij} = \frac{b'_{ij}+c'_{ij}+d'_{ij}}{4}$, $\delta_{ij} = \frac{c'_{ij}+d'_{ij}}{4}$, and $\xi_{ij} = \frac{d'_{ij}}{4}$.

Step 3. Find the values of a_{ij}, b_{ij}, c_{ij} and d_{ij} by solving the equations obtained in Step 2 and also find $\tilde{x}_{ij} = (a_{ij}, b_{ij}, c_{ij}, d_{ij})$.

Step 4. Find the minimum total fuzzy transportation cost by putting the values of $\tilde{x}_{ij}$ in $\sum_{i=1}^{m+n}\sum_{j=1}^{m+n} \tilde{c}_{ij} \otimes \tilde{x}_{ij}$.

Table 1. First crisp transportation table

	S_1	S_2	$\cdots$	S_m	D_1	D_2	$\cdots$	D_n	
S_1	0	λ_{12}	$\cdots$	λ_{1m}	$\lambda_{1(m+1)}$	$\lambda_{1(m+2)}$	$\cdots$	$\lambda_{1(m+n)}$	q_1
S_2	λ_{21}	0	$\cdots$	λ_{2m}	$\lambda_{2(m+1)}$	$\lambda_{2(m+2)}$	$\cdots$	$\lambda_{2(m+n)}$	q_2
$\vdots$	$\vdots$	$\vdots$	$\vdots$	$\vdots$	$\vdots$	$\vdots$	$\vdots$	$\vdots$	$\vdots$
S_m	λ_{m1}	λ_{m2}	$\cdots$	0	$\lambda_{m(m+1)}$	$\lambda_{m(m+2)}$	$\cdots$	$\lambda_{m(m+n)}$	q_m
D_1	$\vdots$	$\vdots$	$\vdots$	$\vdots$	$\vdots$	$\vdots$	$\vdots$	$\vdots$	p_1
D_2	$\vdots$	$\vdots$	$\vdots$	$\vdots$	$\vdots$	$\vdots$	$\vdots$	$\vdots$	p_1
$\vdots$	$\vdots$	$\vdots$	$\vdots$	$\vdots$	$\vdots$	$\vdots$	$\vdots$	$\vdots$	$\vdots$
D_n	$\lambda_{(m+n)1}$	$\lambda_{(m+n)2}$	$\cdots$	$\lambda_{(m+n)m}$	$\lambda_{(m+n)(m+1)}$	$\lambda_{(m+n)(m+2)}$	$\cdots$	0	p_1
	p_1	p_1	$\cdots$	p_1	q'_1	q'_2	$\cdots$	q'_n	

Table 2. Second crisp transportation table

	S_1	S_2	$\cdots$	S_m	D_1	D_2	$\cdots$	D_n	
S_1	0	ρ_{12}	$\cdots$	ρ_{1m}	$\rho_{1(m+1)}$	$\rho_{1(m+2)}$	$\cdots$	$\rho_{1(m+n)}$	$r_1 - q_1$
S_2	ρ_{21}	0	$\cdots$	ρ_{2m}	$\rho_{2(m+1)}$	$\rho_{2(m+2)}$	$\cdots$	$\rho_{2(m+n)}$	$r_2 - q_2$
$\vdots$	$\vdots$	$\vdots$	$\vdots$	$\vdots$	$\vdots$	$\vdots$	$\vdots$	$\vdots$	$\vdots$
S_m	ρ_{m1}	ρ_{m2}	$\cdots$	0	$\rho_{m(m+1)}$	$\rho_{m(m+2)}$	$\cdots$	$\rho_{m(m+n)}$	$r_m - q_m$
D_1	$\vdots$	$\vdots$	$\vdots$	$\vdots$	$\vdots$	$\vdots$	$\vdots$	$\vdots$	$p_2 - p_1$
D_2	$\vdots$	$\vdots$	$\vdots$	$\vdots$	$\vdots$	$\vdots$	$\vdots$	$\vdots$	$p_2 - p_1$
$\vdots$	$\vdots$	$\vdots$	$\vdots$	$\vdots$	$\vdots$	$\vdots$	$\vdots$	$\vdots$	$\vdots$
D_n	$\rho_{(m+n)1}$	$\rho_{(m+n)2}$	$\cdots$	$\rho_{(m+n)m}$	$\rho_{(m+n)(m+1)}$	$\rho_{(m+n)(m+2)}$	$\cdots$	0	$p_2 - p_1$
	$p_2 - p_1$	$p_2 - p_1$	$\cdots$	$p_2 - p_1$	$r'_1 - q'_1$	$r'_2 - q'_2$	$\cdots$	$r'_n - q'_n$	

Table 3. Third crisp transportation table

	S_1	S_2	$\cdots$	S_m	D_1	D_2	$\cdots$	D_n	
S_1	0	δ_{12}	$\cdots$	δ_{1m}	$\delta_{1(m+1)}$	$\delta_{1(m+2)}$	$\cdots$	$\delta_{1(m+n)}$	$s_1 - r_1$
S_2	δ_{21}	0	$\cdots$	δ_{2m}	$\delta_{2(m+1)}$	$\delta_{2(m+2)}$	$\cdots$	$\delta_{2(m+n)}$	$s_2 - r_2$
$\vdots$	$\vdots$	$\vdots$	$\vdots$	$\vdots$	$\vdots$	$\vdots$	$\vdots$	$\vdots$	$\vdots$
S_m	δ_{m1}	δ_{m2}	$\cdots$	0	$\delta_{m(m+1)}$	$\delta_{m(m+2)}$	$\cdots$	$\delta_{m(m+n)}$	$s_m - r_m$
D_1	$\vdots$	$\vdots$	$\vdots$	$\vdots$	$\vdots$	$\vdots$	$\vdots$	$\vdots$	$p_3 - p_2$
D_2	$\vdots$	$\vdots$	$\vdots$	$\vdots$	$\vdots$	$\vdots$	$\vdots$	$\vdots$	$p_3 - p_2$
$\vdots$	$\vdots$	$\vdots$	$\vdots$	$\vdots$	$\vdots$	$\vdots$	$\vdots$	$\vdots$	$\vdots$
D_n	$\delta_{(m+n)1}$	$\delta_{(m+n)2}$	$\cdots$	$\delta_{(m+n)m}$	$\delta_{(m+n)(m+1)}$	$\delta_{(m+n)(m+2)}$	$\cdots$	0	$p_2 - p_1$
	$p_3 - p_2$	$p_3 - p_2$	$\cdots$	$p_3 - p_2$	$s'_1 - r'_1$	$s'_2 - r'_2$	$\cdots$	$s'_n - r'_n$	

Table 4. Fourth crisp transportation table

	S_1	S_2	$\cdots$	S_m	D_1	D_2	$\cdots$	D_n	
S_1	0	ξ_{12}	$\cdots$	ξ_{1m}	$\xi_{1(m+1)}$	$\xi_{1(m+2)}$	$\cdots$	$\xi_{1(m+n)}$	$t_1 - s_1$
S_2	ξ_{21}	0	$\cdots$	ξ_{2m}	$\xi_{2(m+1)}$	$\xi_{2(m+2)}$	$\cdots$	$\xi_{2(m+n)}$	$t_2 - s_2$
$\vdots$	$\vdots$	$\vdots$	$\vdots$	$\vdots$	$\vdots$	$\vdots$	$\vdots$	$\vdots$	$\vdots$
S_m	ξ_{m1}	ξ_{m2}	$\cdots$	0	$\xi_{m(m+1)}$	$\xi_{m(m+2)}$	$\cdots$	$\xi_{m(m+n)}$	$t_m - s_m$
D_1	$\vdots$	$\vdots$	$\vdots$	$\vdots$	$\vdots$	$\vdots$	$\vdots$	$\vdots$	$p_4 - p_3$
D_2	$\vdots$	$\vdots$	$\vdots$	$\vdots$	$\vdots$	$\vdots$	$\vdots$	$\vdots$	$p_4 - p_3$
$\vdots$	$\vdots$	$\vdots$	$\vdots$	$\vdots$	$\vdots$	$\vdots$	$\vdots$	$\vdots$	$\vdots$
D_n	$\xi_{(m+n)1}$	$\xi_{(m+n)2}$	$\cdots$	$\xi_{(m+n)m}$	$\xi_{(m+n)(m+1)}$	$\xi_{(m+n)(m+2)}$	$\cdots$	0	$p_4 - p_3$
	$p_4 - p_3$	$p_4 - p_3$	$\cdots$	$p_4 - p_3$	$t'_1 - s'_1$	$t'_2 - s'_2$	$\cdots$	$t'_n - s'_n$	

3 Advantages of the Proposed Method

To show the advantage of the proposed method over existing method [10] a fuzzy transportation problem with transshipment, [Example 5.1, 10], is solved by using the proposed method and it is shown that the obtained results are same while it is easy to apply the proposed method as compared to the existing method [10].

3.1 Results

On solving the fuzzy transshipment problem [10], the obtained fuzzy optimal solution and minimum total fuzzy transportation cost is $\tilde{x}_{11} = (16, 30, 44, 56)$, $\tilde{x}_{13} = (6, 8, 10, 20)$, $\tilde{x}_{14} = (4, 10, 12, 14)$, $\tilde{x}_{16} = (0, 2, 8, 8)$, $\tilde{x}_{21} = (0, 0, 0, 2)$, $\tilde{x}_{22} = (16, 30, 44, 58)$, $\tilde{x}_{26} = (0, 4, 8, 10)$, $\tilde{x}_{33} = (16, 30, 44, 58)$, $\tilde{x}_{44} = (16, 30, 44, 58)$, $\tilde{x}_{54} = (6, 6, 6, 6)$, $\tilde{x}_{55} = (16, 30, 44, 58)$, $\tilde{x}_{66} = (16, 30, 44, 58)$and remaining are $(0, 0, 0, 0)$ and $(8, 38, 90, 166)$.

3.2 Discussion

It can be easily seen that the results of the fuzzy transportation problems with transshipment, obtained by using the existing method [10] and the proposed method are same but as discussed in, Section 2, it is easy to use the proposed method as compared to existing method [10].

4 Conclusion

The shortcoming of an existing method [10] for finding the fuzzy optimal solution of fuzzy transportation problem with transshipment are pointed out and to overcome the shortcoming of the existing method a new method, named as Mehar's method, is proposed for solving the same type of problem.

Acknowledgements. The authors would like to thank to the anonymous reviewers for their suggestions. Special thanks go to Ms. Mehar Kaur.

References

1. Dantzig, G.B., Thapa, M.N.: Linear programming: 2: theory and extensions. Princeton University Press, New Jersey (1963)
2. Gani, A., Razak, K.A.: Two stage fuzzy transportation problem. Journal of Physical Sciences 10, 63–69 (2006)
3. Ghatee, M., Hashemi, S.M.: Ranking function-based solutions of fully fuzzified minimal cost flow problem. Information Sciences 177, 4271–4294 (2007)
4. Ghatee, M., Hashemi, S.M.: Generalized minimal cost flow problem in fuzzy nature: An application in bus network planning problem. Applied Mathematical Modelling 32, 2490–2508 (2008)
5. Ghatee, M., Hashemi, S.M.: Application of fuzzy minimum cost flow problems to network design under uncertainty. Fuzzy Sets and Systems 160, 3263–3289 (2009)
6. Ghatee, M., Hashemi, S.M.: Optimal network design and storage management in petroleum distribution network under uncertainty. Engineering Applications of Artificial Intelligence 22, 796–807 (2009)
7. Ghatee, M., Hashemi, S.M., Hashemi, B., Dehghan, M.: The solution and duality of imprecise network problems. Computers and Mathematics with Applications 55, 2767–2790 (2008)
8. Ghatee, M., Hashemi, S.M., Zarepisheh, M., Khorram, E.: Preemp-tive priority-based algorithms for fuzzy minimal cost flow problem: An application in hazardous materials transportation. Computers and Industrial Engineering 57, 341–354 (2009)
9. Gupta, P., Mehlawat, M.K.: An algorithm for a fuzzy transportation problem to select a new type of coal for a steel manufacturing unit. TOP 15, 114–137 (2007)
10. Kumar, A., Kaur, A., Gupta, A.: Fuzzy linear programming approach for solving fuzzy transportation problems with transshipment. Journal of Mathematical Modelling and Algorithms (2010), doi: 10.1007/s10852-010-9147-8
11. Li, L., Huang, Z., Da, Q., and Hu, J.: A new method based on goal programming for solving transportation problem with fuzzy cost. In: International Symposiums on Information Processing, 3–8 (2008)
12. Lin, F.T.: Solving the transportation problem with fuzzy coefficients using genetic algorithms. In: IEEE International Conference on Fuzzy Systems, pp. 1468–1473 (2009)
13. Liu, S.T., Kao, C.: Solving fuzzy transportation problems based on extension principle. European Journal of Operational Research 153, 661–674 (2004)
14. Liu, S.T., Kao, C.: Network flow problems with fuzzy arc lengths. IEEE Transactions on Systems, Man and Cybernetics-Part B: Cybernetics 34, 765–769 (2004)
15. Oheigeartaigh, M.: A fuzzy transportation algorithm. Fuzzy Sets and Systems 8, 235–243 (1982)
16. Pandian, P., Natarajan, G.: A new algorithm for finding a fuzzy optimal solution for fuzzy transportation problems. Applied Mathematical Sciences 4, 79–90 (2010)
17. Stephen Dinagar, S., Palanivel, K.: The transportation problem in fuzzy environment. International Journal of Algorithms, Computing and Mathematics 2, 65–71 (2009)
18. Zadeh, L.A.: Fuzzy sets. Information and Control 8, 338–353 (1965)

Fuzzy Optimal Solution of Fully Fuzzy Project Crashing Problems with New Representation of *LR* Flat Fuzzy Numbers

Amit Kumar, Parmpreet Kaur, and Jagdeep Kaur

School of Mathematics and Computer Applications
Thapar University, Patiala-147 004, India
amit_rs_iitr@yahoo.com, parmpreetsidhu@gmail.com,
sidhu.deepi87@gmail.com

Abstract. In this paper, a new method, named as Mehar's method, is proposed for solving fully fuzzy project crashing problems and a new representation of LR flat fuzzy numbers, named as JMD representation of LR flat fuzzy numbers, are introduced. Also, it is shown that it is better to use JMD representation of LR flat fuzzy numbers as compared to the existing representation of LR flat fuzzy numbers.

1 Introduction

Management of complex projects that consists of a large number of interrelated activities poses problems involved in planning, scheduling, and control, especially when the project activities have to be performed in a specified technological sequence. But in real-world applications, the time required to complete the various activities in a research and development project may be known only approximately due to insufficient information. To deal quantitatively with imprecise information, the concepts and techniques of probability could be employed. However, probability distribution requires a priori predictable regularity or a posteriori frequency determination to construct. As an alternative, uncertain values can be represented by fuzzy sets.

For finding the fuzzy critical path and fuzzy project crashing, several approaches and algorithms are proposed over the past years [5,2,1,4].

2 *JMD* Representation of *LR* Flat Fuzzy Numbers

Kumar and Kaur [3] proposed JMD representation of triangular fuzzy numbers and shown that it is better to use JMD representation of triangular fuzzy numbers as compared to existing representation of triangular fuzzy numbers. On the same direction in this section, a new representation of LR flat fuzzy numbers, named as JMD representation of LR flat fuzzy numbers, are introduced.

Definition 1. *Let* $(m, n, \alpha^L, \alpha^R)_{LR}$ *be an LR flat fuzzy number then its JMD representation is* $(x, \alpha^L, \alpha^M, \alpha^R)_{LR}^{JMD}$*, where* $x = m - \alpha^L$, $\alpha^M = n - m$.

S.O. Kuznetsov et al. (Eds.): RSFDGrC 2011, LNAI 6743, pp. 171–174, 2011.

Definition 2. *A JMD LR flat fuzzy number* $\widetilde{A} = (x, \alpha^L, \alpha^M, \alpha^R)_{LR}^{JMD}$ *is said to be non-negative JMD LR flat fuzzy number iff* $x \geq 0$.

Definition 3. *Let* $(x, \alpha^L, \alpha^M, \alpha^R)_{LR}^{JMD}$ *be a JMD LR flat fuzzy number then* $\Re(x, \alpha^L, \alpha^M, \alpha^R)_{LR}^{JMD} = x + \frac{3(\alpha^L)+2(\alpha^M)+\alpha^R}{4}$ *for* $L(x) = R(x) = max\,\{0, 1-x\}$.

3 Mehar's Method with *JMD LR* Flat Fuzzy Numbers

In this section, a new method, named as Mehar's method, is proposed to find the fuzzy optimal solution of the fully fuzzy project crashing (FFPC) problems. The steps of Mehar's method are as follows:

Step 1. If all the parameters $\widetilde{C}_{ij}$, $\widetilde{D}_{ij}$, $\widetilde{Y}_{ij}$, $\widetilde{t}_{ij}$, $\widetilde{x}_j$ and $\widetilde{T}$ are represented by *JMD LR* flat fuzzy numbers $(c_{ij}, \alpha_{ij}^L, \alpha_{ij}^M, \alpha_{ij}^R)_{LR}^{JMD}$, $(d_{ij}, \delta_{ij}^L, \delta_{ij}^M, \delta_{ij}^R)_{LR}^{JMD}$, $(y_{ij}, \beta_{ij}^L, \beta_{ij}^M, \beta_{ij}^R)_{LR}^{JMD}$, $(t_{ij}, \eta_{ij}^L, \eta_{ij}^M, \eta_{ij}^R)_{LR}^{JMD}$, $(x_j, \gamma_j^L, \gamma_j^M, \gamma_j^R)_{LR}^{JMD}$ and $(t, \psi^L, \psi^M, \psi^R)_{LR}^{JMD}$ respectively, then the linear programming formulation of project crashing problem [6] can be written as:
Minimize $\sum_{(i,j)\in A} (c_{ij}, \alpha_{ij}^L, \alpha_{ij}^M, \alpha_{ij}^R)_{LR}^{JMD} \otimes (y_{ij}, \beta_{ij}^L, \beta_{ij}^M, \beta_{ij}^R)_{LR}^{JMD}$
subject to $(x_j, \gamma_j^L, \gamma_j^M, \gamma_j^R)_{LR}^{JMD} \oplus (y_{ij}, \beta_{ij}^L, \beta_{ij}^M, \beta_{ij}^R)_{LR}^{JMD} \succeq (x_i, \gamma_i^L, \gamma_i^M, \gamma_i^R)_{LR}^{JMD} \oplus (d_{ij}, \delta_{ij}^L, \delta_{ij}^M, \delta_{ij}^R)_{LR}^{JMD}$ $\forall (i,j) \in A$; $(y_{ij}, \beta_{ij}^L, \beta_{ij}^M, \beta_{ij}^R)_{LR}^{JMD} \preceq (t_{ij}, \eta_{ij}^L, \eta_{ij}^M, \eta_{ij}^R)_{LR}^{JMD} \forall (i,j) \in A$; $(x_1, \gamma_1^L, \gamma_1^M, \gamma_1^R)_{LR}^{JMD} = (0,0,0,0)_{LR}^{JMD}$; $(x_n, \gamma_n^L, \gamma_n^M, \gamma_n^R)_{LR}^{JMD} \preceq (t, \psi^L, \psi^M, \psi^R)_{LR}^{JMD}$; $(y_{ij}, \beta_{ij}^L, \beta_{ij}^M, \beta_{ij}^R)_{LR}^{JMD}$ is a non-negative *JMD LR* flat fuzzy number $\forall (i,j) \in A$; $(x_j, \gamma_j^L, \gamma_j^M, \gamma_j^R)_{LR}^{JMD}$ is an unrestricted *JMD LR* flat fuzzy number $\forall\, j$ where, A = set of all activities (i,j), $\widetilde{D}_{ij}$ = fuzzy normal time for (i,j), $\widetilde{t}_{ij}$ = maximum allowable fuzzy crash time for (i,j), $\widetilde{C}_{ij}$ = incremental fuzzy crashing costs for (i,j), $\widetilde{Y}_{ij}$ = number of fuzzy time units by which duration of (i,j) is crashed, $\widetilde{x}_i$ = earliest fuzzy time for event i, $\widetilde{x}_1$ = project fuzzy start time, $\widetilde{x}_n$ = project fuzzy completion time, $\widetilde{T}$ = specified project fuzzy completion time.

Step 2. Using Definition 2 and Definition 3, convert the fuzzy linear programming (FLP) problem, obtained in Step 1, into the following crisp linear programming (CLP) problem:
Minimize $\Re(\sum_{(i,j)\in A} (c_{ij}, \alpha_{ij}^L, \alpha_{ij}^M, \alpha_{ij}^R)_{LR}^{JMD} \otimes (y_{ij}, \beta_{ij}^L, \beta_{ij}^M, \beta_{ij}^R)_{LR}^{JMD})$
subject to $\Re((x_j, \gamma_j^L, \gamma_j^M, \gamma_j^R)_{LR}^{JMD}) + \Re((y_{ij}, \beta_{ij}^L, \beta_{ij}^M, \beta_{ij}^R)_{LR}^{JMD}) \geq \Re((x_i, \gamma_i^L, \gamma_i^M, \gamma_i^R)_{LR}^{JMD}) + \Re((d_{ij}, \delta_{ij}^L, \delta_{ij}^M, \delta_{ij}^R)_{LR}^{JMD}) \forall (i,j) \in A$; $\Re((y_{ij}, \beta_{ij}^L, \beta_{ij}^M, \beta_{ij}^R)_{LR}^{JMD}) \leq \Re(t_{ij}, \eta_{ij}^L, \eta_{ij}^M, \eta_{ij}^R)_{LR}^{JMD}) \forall\ (i,j) \in A$; $\Re((x_1, \gamma_1^L, \gamma_1^M, \gamma_1^R)_{LR}^{JMD}) = \Re((0,0,0,0)_{LR}^{JMD})$; $\Re((x_n, \gamma_n^L, \gamma_n^M, \gamma_n^R)_{LR}^{JMD}) \leq \Re((t, \psi^L, \psi^M, \psi^R)_{LR}^{JMD})$; $\gamma_j^L, \gamma_j^M, \gamma_j^R, \geq 0\ \forall\, j$; $y_{ij}, \beta_{ij}^L, \beta_{ij}^M, \beta_{ij}^R \geq 0\ \forall\, (i,j) \in A$; x_j is a real number $\forall\, j$.

Step 3. Solve the CLP problem, obtained in Step 2, to find the minimum fuzzy crashing cost for completing the project within specified fuzzy time by putting the values of $\widetilde{Y}_{ij} = (y_{ij}, \beta^L_{ij}, \beta^M_{ij}, \beta^R_{ij})$ in $\sum_{(i,j)\in A} \widetilde{C}_{ij} \otimes \widetilde{Y}_{ij}$.

4 Advantages of *JMD LR* Flat Fuzzy Numbers

In this section, the results of the same problem obtained by using existing and *JMD LR* flat fuzzy numbers are compared in Table 1.

Table 1. Comparative study

Details of crisp constraints and crisp variables in the CLP formulation obtained by using Mehar's method with existing representation of LR flat fuzzy numbers $((m, n, \alpha^L, \alpha^R)_{LR})$	Details of crisp constraints and crisp variables in the CLP formulation obtained by using Mehar's method with JMD representation of LR flat fuzzy numbers $((x, \alpha^L, \alpha^M, \alpha^R)^{JMD}_{LR})$
Number of crisp constraints corresponding to N fuzzy constraints $= N$	Number of crisp constraints corresponding to N fuzzy constraints $= N$
Crisp variables corresponding to one non-negative LR flat fuzzy variables $((m, n, \alpha^L, \alpha^R)_{LR})$ are $m, n, \alpha^L, \alpha^R, m - \alpha^L, n - m \geq 0$,	Crisp variables corresponding to one non-negative JMD LR flat fuzzy variables $((x, \alpha^L, \alpha^M, \alpha^R)^{JMD}_{LR})$ are $\alpha^L, \alpha^M, \alpha^R, x \geq 0$
Crisp variables corresponding to one unrestricted LR flat fuzzy variables $((m, n, \alpha^L, \alpha^R)_{LR})$ are $\alpha^L, \alpha^R, n - m \geq 0$ and m, n are real numbers	Crisp variables corresponding to one unrestricted JMD LR flat fuzzy variables $((x, \alpha^L, \alpha^M, \alpha^R)^{JMD}_{LR})$are $\alpha^L, \alpha^M, \alpha^R \geq 0$ and x is a real number
Number of crisp variables corresponding to one non-negative LR flat fuzzy variables $((m, n, \alpha^L, \alpha^R)_{LR})$ are 6	Number of crisp variables corresponding to one non-negative JMD LR flat fuzzy variables $((x, \alpha^L, \alpha^M, \alpha^R)^{JMD}_{LR})$ are 4
Number of non-negative crisp variables corresponding to (u) non-negative LR flat fuzzy variables $= 6 \times u$	Number of non-negative crisp variables corresponding to (u) non-negative JMD LR flat fuzzy variables $= 4 \times u$
Number of unrestricted crisp variables corresponding to (v) unrestricted LR flat fuzzy variables $= 5 \times v$	Number of unrestricted crisp variables corresponding to (v) unrestricted JMD LR flat fuzzy variables $= 4 \times v$
Total number of crisp constraints $= N + 6 \times u + 5 \times v$	Total number of crisp constraints $= N + 4 \times u + 4 \times v$

To show the advantages of JMD representation of LR flat fuzzy numbers over existing representation of LR flat fuzzy numbers, a FFPC problems, chosen in Example 1, is solved by using Mehar's method with existing representation of LR flat fuzzy numbers and Mehar's method with JMD representation of LR flat fuzzy numbers and it is shown that it is better to use Mehar's method with JMD representation of LR flat fuzzy numbers over Mehar's method with existing representation of LR flat fuzzy numbers.

Table 2. $\widetilde{D}_{ij}$, $\widetilde{t}_{ij}$, $\widetilde{C}_{ij}$ for each activity

Activity	Activity name	$\widetilde{D}_{ij}$(days)	$\widetilde{t}_{ij}$(days)	$\widetilde{C}_{ij}$($)
(1, 2)	Build foundation	$(3, 6, 2, 4)_{LR}$	$(1, 3, 1, 1)_{LR}$	$(20, 40, 10, 10)_{LR}$
(2, 3)	Build walls and ceilings	$(6, 9, 1, 3)_{LR}$	$(2, 4, 1, 1)_{LR}$	$(10, 15, 5, 15)_{LR}$
(3, 5)	Build roofs	$(9, 12, 5, 3)_{LR}$	$(0.5, 1.5, 0.5, 0.5)_{LR}$	$(15, 25, 10, 10)_{LR}$
(3, 6)	Do electrical wiring	$(4, 6, 3, 3)_{LR}$	$(1.5, 2.5, 0.5, 0.5)_{LR}$	$(80, 100, 60, 20)_{LR}$
(3, 4)	Put in windows	$(3, 5, 1, 1)_{LR}$	$(1, 3, 1, 1)_{LR}$	$(20, 23, 8, 2)_{LR}$
(4, 5)	Put on siding	$(5, 6, 2, 4)_{LR}$	$(2, 4, 1, 1)_{LR}$	$(15, 30, 5, 35)_{LR}$
(5, 6)	Paint house	$(2, 3.5, 0.5, 1.5)_{LR}$	$(0.5, 1.5, 0.5, 0.5)_{LR}$	$(30, 50, 20, 20)_{LR}$

Example 1. Find the minimum fuzzy crashing cost for completing the project within $(15, 25, 10, 10)_{LR}$ days. The fuzzy normal time $\widetilde{D}_{ij}$, incremental fuzzy crashing cost $\widetilde{C}_{ij}$ and the maximum allowable fuzzy crash time $\widetilde{t}_{ij}$ for each activity are given in Table 2, where, $L(x) = R(x) = \max\ \{0, 1 - x\}$.

4.1 Results

For solving the FFPC problem, chosen in Example 1, by using Mehar's method with existing representation of LR flat fuzzy numbers and Mehar's method with JMD representation of LR flat fuzzy numbers there is need to solve CLP problems having 88 and 68 constraints respectively while the minimum fuzzy crashing cost obtained by using both the methods are same.

4.2 Physical Interpretation of Results

Using Mehar's method with JMD representation of LR flat fuzzy numbers the fuzzy crashing cost is $(45, 55, 75, 85)_{LR}^{JMD} = (100, 175, 55, 85)_{LR}$ which can be physically interpreted as follow:

The least amount of minimum crashing cost is 45 \$, the most possible amount of minimum crashing cost lies between 100 \$ and 175 \$ and the greatest amount of minimum crashing cost is 260 \$.

5 Conclusion

A new method, named as Mehar's method, for finding the fuzzy optimal solution of fully fuzzy project crashing problems and a new representation of LR flat fuzzy numbers, named as JMD representation of LR flat fuzzy numbers, are introduced.

Acknowledgements. The authors would like to thank to the anonymous reviewers for their suggestions. Special thanks go to Ms. Mehar.

References

1. Chen, S.P., Hsueh, Y.J.: A simple approach to fuzzy critical path analysis in project networks. Applied Mathematical Modelling 32, 1289–1297 (2008)
2. Guang, J.C., Shang, J.Z., Yan, L., Min, Z.Y., Dong, H.Z.: Research on the fully fuzzy time-cost trade-off based on genetic algorithms. Journal of Marine Science and Application 4, 18–23 (2005)
3. Kumar, A., Kaur, P.: A New Method for Fuzzy Critical Path Analysis in Project Networks with a New Representation of Triangular Fuzzy Numbers. Applications and Applied Mathematics: An International Journal 5, 1442–1466 (2010)
4. Lin, F.T.: Fuzzy crashing problem on project management based on confidence-interval estimates. In: Eighth International Conference on Intelligent Systems Design and Applications, vol. 2, pp. 164–169 (2008)
5. Liu, S.T.: Fuzzy activity times in critical path and project crashing problems. Cybernetics and Systems 34, 161–172 (2003); 73, 227–234 (1995)
6. Winston, W.L.: Operations Research: Applications and Algorithms, Singapore (2003)

A Prototype System for Rule Generation in Lipski's Incomplete Information Databases

Hiroshi Sakai[1], Michinori Nakata[2], and Dominik Ślęzak[3,4]

[1] Mathematical Sciences Section, Department of Basic Sciences,
Faculty of Engineering, Kyushu Institute of Technology
Tobata, Kitakyushu 804, Japan
sakai@mns.kyutech.ac.jp
[2] Faculty of Management and Information Science,
Josai International University
Gumyo, Togane, Chiba 283, Japan
nakatam@ieee.org
[3] Institute of Mathematics, University of Warsaw
Banacha 2, 02-097 Warsaw, Poland
[4] Infobright Inc., Poland
Krzywickiego 34 pok. 219, 02-078 Warsaw, Poland
slezak@infobright.com

Abstract. This paper advances rule generation in Lipski's incomplete information databases, and develops a software tool for rule generation. We focus on three kinds of information incompleteness. The first is non-deterministic information, the second is missing values, and the third is intervals. For intervals, we introduce the concept of a resolution. Three kinds of information incompleteness are uniformly handled by *NIS-Apriori* algorithm. An overview of a prototype system in Prolog is presented.

Keywords: Lipski's incomplete information databases, Rule generation, Apriori algorithm, Rough sets, Prolog.

1 Introduction

In our previous research, we coped with rule generation in *Non-deterministic Information Systems* (*NISs*) [9]. In contrast to *Deterministic Information Systems* (*DISs*) [8,12], *NISs* were proposed by Pawlak [8] and Orłowska [7] to better handle information incompleteness in data. Recently, we focused on Lipski's *Incomplete Information Databases* (*IIDs*) [5,6], and proposed rule generation in *IIDs* [11]. We treat the obtained methodology as a step toward more general rule-based data analysis, where both data values and descriptors take various forms of incompleteness, vagueness or non-determinism.

In this paper, we advance the previous rule generation in *IIDs*, and develop a prototype system, which can handle three kinds of information incompleteness. The first kind of information incompleteness is non-deterministic information [8,7], the second is missing values [3,4], and the third is intervals.

S.O. Kuznetsov et al. (Eds.): RSFDGrC 2011, LNAI 6743, pp. 175–182, 2011.

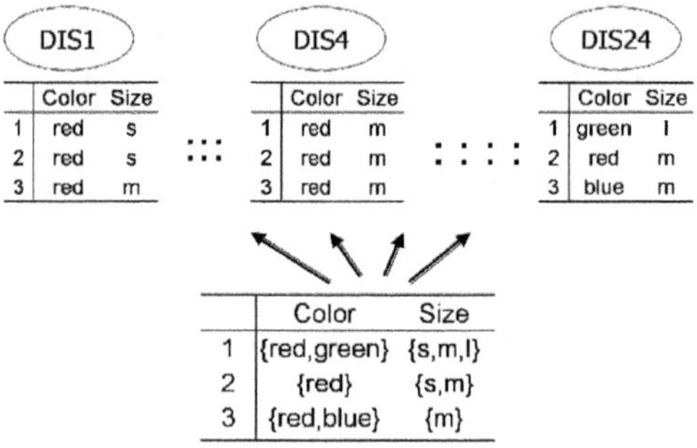

Fig. 1. A *NIS* and 24 derived *DISs*. The number of derived *DISs* is finite. However, it usually increases in the exponential order with respect to the level of incompleteness of *NIS's* values.

The paper is organized as follows: Section 2 recalls data representation and rule generation in *DISs* and *NISs*. Section 3 introduces the same for *IIDs*, and presents implementation and execution. Section 4 concludes the paper.

2 Rule Generation in DISs and NISs

We omit formal definitions of *DISs* and *NISs*. Instead, we show an example in Figure 1. We identify a *DIS* with a standard table. In a *NIS*, each attribute value is a set. If the value is a singleton, there is no incompleteness. Otherwise, we interpret it as a set of possible values, i.e., each set includes the actual value but we do not know which of them is the actual one.

A rule (more correctly, a candidate for a rule) is an implication τ in the form of *Condition_part* $\Rightarrow$ *Decision_part*. We employ $support(\tau)$ and $accuracy(\tau)$ to express the rule's appropriateness as follows [1,8] (see also Figure 2):

Specification of the rule generation task in a ***DIS***
For threshold values α and β ($0 < \alpha, \beta \leq 1$), find each implication τ satisfying $support(\tau) \geq \alpha$ and $accuracy(\tau) \geq \beta$.

In *NISs*, the same τ may be generated by different tuples, so we use notation τ^x to express that τ is generated by an object x. Let $DD(\tau^x)$ denote $\{\psi \mid \psi$ is a derived *DISs* and τ^x occurs in $\psi\}$, and we define the next task.

Specification of the rule generation task in a ***NIS***
(The lower system) Find each implication τ such that $support(\tau^x) \geq \alpha$ and $accuracy(\tau^x) \geq \beta$ (for an object x) hold in each $\psi \in DD(\tau^x)$.
(The upper system) Find each implication τ such that $support(\tau^x) \geq \alpha$ and $accuracy(\tau^x) \geq \beta$ (for an object x) hold in some $\psi \in DD(\tau^x)$.

Both above systems depend on $|DD(\tau^x)|$. In [10], we proved some simplifying results illustrated by Figure 3. We also showed how to effectively compute $support(\tau^x)$ and $accuracy(\tau^x)$ for ψ_{min} and ψ_{max} independently from $|DD(\tau^x)|$.

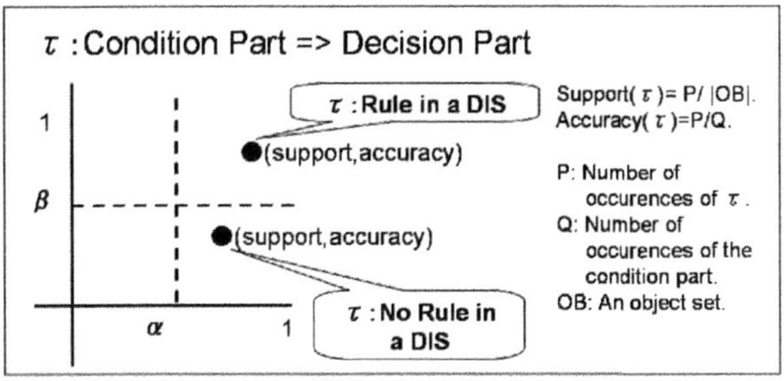

Fig. 2. A pair (*support,accuracy*) corresponding to the implication τ

Due to Figure 3, we have the following equivalent specification.

Equivalent specification of the rule generation task in a NIS

(The lower system). Find each implication τ such that $minsupp(\tau^x) \geq \alpha$ and $minacc(\tau^x) \geq \beta$ for an object x (see Figure 3).

(The upper system). Find each implication τ such that $maxsupp(\tau) \geq \alpha$ and $maxacc(\tau) \geq \beta$ for an object x (see Figure 3).

In [10], we extended rule generation onto $NISs$ and implemented a software tool called *NIS-Apriori*. *NIS-Apriori* does not depend upon the number of derived $DISs$. This paper is extending this software tool to Lipski's *Incomplete Information Databases*.

3 Rule Generation in Incomplete Information Databases

Now, we advance from $NISs$ to $IIDs$. We introduce an example of an IID, and consider it. The formal definitions of an IID are in [11].

3.1 An Example of an Incomplete Information Database

In Table 1, we have $Domain_{Age}$=$\{20, 21, ..., 70\}$, $Domain_{Sex}$=$\{male, female\}$, $Domain_{Department}$=$\{dp1, dp2, dp3\}$ and $Domain_{Salary}$=$\{400, 401, 402, ..., 2000\}$. For handling information incompleteness, the attribute values of *Age* and *Salary* are intervals, and the attribute values of *Sex* and *Department* are either a value, a subset of the domain or a missing value $*$. Missing values $*$ and intervals are often employed for handling information incompleteness.

3.2 Non-deterministic Information and Missing Values

In Table 1, we have two missing values, i.e., two $*$ symbols. In rough sets, the domain DOM is usually a finite set, therefore we identify $*$ with non-deterministic

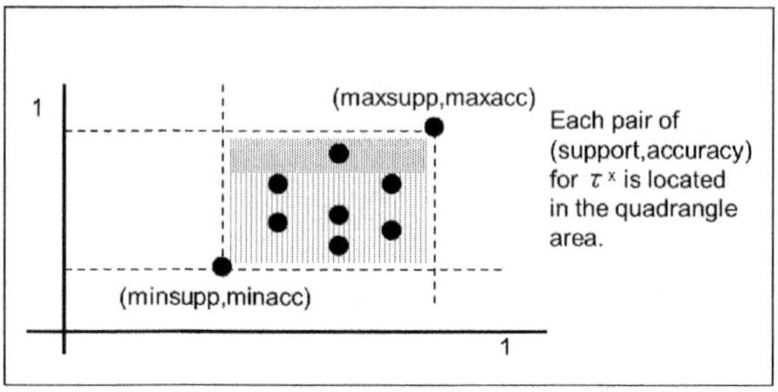

Fig. 3. A distribution of pairs (*support*,*accuracy*) for τ^x. There exists $\psi_{min} \in DD(\tau^x)$ which makes both $support(\tau^x)$ and $accuracy(\tau^x)$ the minimum. There exists $\psi_{max} \in DD(\tau^x)$ which makes both $support(\tau^x)$ and $accuracy(\tau^x)$ the maximum. We denote such quantities as *minsupp*, *minacc*, *maxsupp* and *maxacc*, respectively.

information DOM. Namely, we replace two $*$ symbols with $\{dp1, dp2, dp3\}$ and $\{male, female\}$, and we obtain a NIS for a set of attributes $\{Sex, Department\}$. Thus, we consider 96 ($=2^5 \times 3$) derived $DISs$ like in Figure 1, and we see that an actual DIS exists within 96 derived $DISs$.

3.3 Information Incompleteness about Intervals and Derived DISs

Now, we consider information incompleteness for intervals. We usually interpret an interval $[lower, upper]$ as that the actual value is between *lower* and *upper*. Information incompleteness for intervals is a relative concept. For example, let us consider number π. The interval $[3.14, 3.15]$ will be enough for students, but it will be too simple for researcher. This example is also related to granularity and granular computing in general [13]. Consider the following definition.

Definition 1. *[11] For an attribute A whose values are intervals, let us fix a threshold value* $\gamma_A > 0$. *We say that an interval* $[lower, upper]$ *is "*$definite$*", if its length (upper − lower) is not higher than* γ_A. *Otherwise, we say that it is "*$indefinite$*". We call* γ_A *a resolution of* VAL_A.

Example 1. In Table 1, consider γ_{Age}=1. Then information about x_4, x_6 and x_8 is definite, and information about other objects is indefinite. For x_1, there are three possible intervals: $[22, 23]$, $[23, 24]$, $[24, 25]$. For γ_{Age}=10, information about all objects is definite, and there is no information incompleteness.

According to Definition 1 and Example 1, we can re-define derived $DISs$ (depending upon the resolution) for intervals. We can also consider a figure Figure 1 for Table 1. However, the number of derived $DISs$ may not be finite. For example, for an interval $[0, 1]=\{x : real\_number | 0 \leq x \leq 1\}$ and γ=0.1, the number of definite intervals is infinite.

Table 1. A example of Incomplete Information Database (IID)

OB	Age	Sex	$Department$	$Salary$
x_1	$[22, 25]$	$female$	$*$	$[400, 500]$
x_2	$[20, 25]$	$female$	$\{dp2, dp3\}$	$[500, 600]$
x_3	$[25, 30]$	$male$	$\{dp1, dp2\}$	$[400, 700]$
x_4	$[36, 36]$	$*$	$dp2$	$[700, 750]$
x_5	$[37, 40]$	$male$	$\{dp1, dp2\}$	$[500, 800]$
x_6	$[43, 43]$	$female$	$\{dp2, dp3\}$	$[600, 800]$
x_7	$[45, 50]$	$male$	$dp3$	$[700, 900]$
x_8	$[52, 52]$	$male$	$dp3$	$[800, 900]$
x_9	$[53, 57]$	$male$	$dp3$	$[1000, 1500]$
x_{10}	$[60, 70]$	$male$	$dp3$	$[1100, 2000]$

3.4 Descriptors and Rule Generation in $IIDs$

In a DIS, we consider each implication τ from a table. If τ satisfies $support(\tau) \geq \alpha$ and $accuracy(\tau) \geq \beta$, we pick up this τ as a candidate of rule. In a NIS, we followed this strategy, and defined $DD(\tau^x)$ in Section 2. For handling categorical data in rough sets, we usually suppose that each domain of attribute values is finite. So, we implicitly handled finite number of descriptors, and we did not specify any descriptor for rule generation.

However, we may need to specify descriptors in an IID, because there may be infinite number of them. Also, each rule is expressed by descriptors, so the selection of descriptors is very important. We see this is the next important issue for rule generation in $IIDs$. In the current prototype system in Prolog, we explicitly specify each descriptor in a data set.

Our rule generation basically depends upon the consistency in rough sets, and we also need to consider the *Dominance based Rough Sets Approach* ($DRSA$) [2]. By using the property of the ordered set, we will be able to generate a software with more general functionality. This is the next important issue, too. The following is the tentative rule generation task in $IIDs$.

Specification of the tentative rule generation task in an IID
(Assumption). Descriptors are given, and each implication τ is defined by given descriptors. Each $DD(\tau^x)$ is a set of derived $DISs$ with definite intervals.
(The lower system). The same definition in $NISs$.
(The upper system). The same definition in $NISs$.

3.5 Data Expression and Equivalence Classes

The following is the real data for Table 1. The prototype system in Prolog can handle any data set in the following syntax.

```
object(10,4). /* #object=10, #attribute=4 */
support(0.2). /* constraint: support is more than 0.2 */
accuracy(0.5). /* constraint: accuracy is more than 0.5 */
decision(4). /* decision attribute */
attrib(1,age,5,[[25,30],[30,40],[40,50],[50,60],[60,100]]).
resolution(1,interval,5). /* resolution of age */
attrib(2,sex,2,[male,female]).
resolution(2,set,1).
attrib(3,department,3,[dp1,dp2,dp3]).
resolution(3,set,1).
attrib(4,salary,4,[[300,600],[600,800],[800,1000],[1000,2000]]).
resolution(4,interval,100).

data(1,[[22,25],female,nil,[400,500]]).
data(2,[[20,25],female,[dp2,dp3],[500,600]]).
data(3,[[25,30],male,[dp1,dp2],[400,700]]).
data(4,[[36,36],nil,dp2,[700,750]]).
data(5,[[37,40],male,[dp1,dp2],[500,800]]).
data(6,[[43,43],female,[dp2,dp3],[600,800]]).
data(7,[[45,50],male,dp3,[700,900]]).
data(8,[[52,52],male,dp3,[800,900]]).
data(9,[[53,57],male,dp3,[1000,1500]]).
data(10,[[60,70],male,dp3,[1100,2000]]).
```

In this data set, five descriptors for an attribute *Age* and four descriptors for an attribute *Salary* are specified. For attributes *Sex* and *Department*, $[sex, male]$, $[sex, female]$, $[department, dp1]$, $[department, dp2]$ and $[department,\ dp3]$ are specified. According to the values of *support* and *resolution*, this data set is at first translated to the internal data. The following is a part of it:

```
upper(3,1,[department,dp1],[],[1,3,5]).
upper(3,2,[department,dp2],[4],[1,2,3,4,5,6]).
lower(3,3,[department,dp3],[7,8,9,10],[1,2,6,7,8,9,10]).
lower(4,1,[salary,[300.0,600.0]],[1,2],[1,2,3,5]).
lower(4,2,[salary,[600.0,800.0]],[4,6],[3,4,5,6,7]).
upper(4,3,[salary,[800.0,1000.0]],[8],[7,8]).
lower(4,4,[salary,[1000.0,2000.0]],[9,10],[9,10]).
```

The fourth and fifth arguments mean the minimum equivalence class and the maximum equivalence class for a descriptor. For *Sex* and *Department*, if attribute value of an object x is definite, x is added to fourth and fifth arguments of the descriptor. If attribute value is indefinite, x is added to the fifth argument of the related descriptors. For *Age* and *Salary*, we suppose the intervals INT_x of an object x and INT_{desc} of a descriptor. If $INT_x \subseteq INT_{desc}$, x is added to fourth and fifth arguments of the descriptor. Otherwise, if $[lower, upper]{=}INT_x \cap INT_{desc} \neq \emptyset$ and $upper - lower \geq \gamma$, x is added to the fifth argument of the related descriptors. By using the fourth argument inf and the fifth argument sup, we can easily obtain $minsupp(\tau^x)$, $minacc(\tau^x)$, *maxsupp*(τ^x) and *maxacc*(τ^x), and we may apply *NIS-Apriori* algorithm by using these four criterion values [10] to rule generation in Table 1.

3.6 Execution for Table 1

Now, we show the example of real execution for Table 1. By *step*1 command, we obtain rules in the form of $[Attribute_A, val_A] \Rightarrow [Salary, val_{Salary}]$. In the lower system, we obtained a rule (*minsupp*(τ)=0.2, *minacc*(τ)=0.5), which we call *certain rule*. This rule τ satisfies the constraints of *support* and *accuracy* in each derived *DIS*, where τ occurs. Object 1 and 2 support this τ. In the upper system, we obtained 11 rules, which we call *possible rules*.

```
--- 1st STEP --------------------------
File=tlip.pl, Support=0.2, Accuracy=0.5
===== Lower System =====================================================
[13] [sex,female]=>[salary,[300,600]] (0.2,0.5) [1,2]
The Rest Candidates:[[[2,1],[4,4]],[[3,3],[4,4]]]
(Next Candidates are Remained)
===== Upper System =====================================================
[2] [age,[30,40]]=>[salary,[600,800]] (0.2,1.0) [4,5] IGC [5]
[6] [age,[40,50]]=>[salary,[600,800]] (0.2,1.0) [6,7] IGC [7]
[14] [sex,male]=>[salary,[600,800]] (0.4,0.5714285714) [3,4,5,7]
[17] [sex,female]=>[salary,[300,600]] (0.2,0.6666666667) [1,2]
[18] [sex,female]=>[salary,[600,800]] (0.2,0.5) [4,6] IGC [4]
         :          :          :
[32] [department,dp3]=>[salary,[1000,2000]] (0.2,0.5) [9,10]
The Rest Candidates:[[[2,1],[4,1]],[[2,1],[4,3]],[[2,1],[4,4]],:::
(Next Candidates are Remained)
EXEC_TIME=0.0(sec)
```

In order to obtain rules in the form of $[Attribute_A, val_A] \wedge [Attribute_B, val_B] \Rightarrow [Salary, val_{Salary}]$, we execute *step*2, and we have the following:

```
--- 2nd STEP --------------------------
===== Lower System =====================================================
[1] [sex,male]&[department,dp3]=>[salary,[1000,2000]] (0.2,0.5) [9,10]
The Rest Candidates:[]
(Lower System Terminated)
===== Upper System =====================================================
[3] [sex,male]&[dep::,dp3]=>[salary,[800,1000]] (0.2,0.5) [7,8] IGC [7]
[4] [sex,male]&[department,dp3]=>[salary,[1000,2000]] (0.2,0.5) [9,10]
The Rest Candidates:[]
(Upper System Terminated)
EXEC_TIME=0.0(sec)
```

4 Concluding Remarks

In this paper, we proposed how to formulate and solve the rule generation problem for Incomplete Information Databases. Our prototype was examined for an exemplary practical data set (mammographic.csv, the object size is 150, the attribute size is 6, the number of derived $DISs$ is about 10^{46}).

Acknowledgment. The first author is supported by the Grant-in-Aid for Scientific Research (C) (No.18500214, No.22500204), Japan Society for the Promotion of Science. The third author is supported by the grant N N516 077837 from the Ministry of Science and Higher Education of the Republic of Poland.

References

1. Agrawal, R., Srikant, R.: Fast Algorithms for Mining Association Rules. In: Proc. of VLDB, pp. 487–499 (1994)
2. Dembczyński, K., Greco, S., Słowiński, R.: Rough Set Approach to Multiple Criteria Classification with Imprecise Evaluations and Assignments. European J. Operational Research 198, 626–636 (2009)
3. Grzymała-Busse, J.: Data with Missing Attribute Values: Generalization of Indiscernibility Relation and Rule Induction. Transactions on Rough Sets 1, 78–95 (2004)
4. Kryszkiewicz, M.: Rules in Incomplete Information Systems. Information Sciences 113, 271–292 (1999)
5. Lipski, W.: On Semantic Issues Connected with Incomplete Information Data Base. ACM Trans. DBS. 4, 269–296 (1979)
6. Lipski, W.: On Databases with Incomplete Information. Journal of the ACM 28, 41–70 (1981)
7. Orłowska, E., Pawlak, Z.: Representation of Nondeterministic Information. Theoretical Computer Science 29, 27–39 (1984)
8. Pawlak, Z.: Rough Sets. Kluwer Academic Publishers, Dordrecht (1991)
9. Sakai, H., Okuma, A.: Basic Algorithms and Tools for Rough Non-deterministic Information Analysis. Transactions on Rough Sets 1, 209–231 (2004)
10. Sakai, H., Ishibashi, R., Nakata, M.: On Rules and Apriori Algorithm in Non-deterministic Information Systems. Transactions on Rough Sets 9, 328–350 (2008)
11. Sakai, H., Nakata, M., Ślęzak, D.: Rule Generation in Lipski's Incomplete Information Databases. In: Szczuka, M., Kryszkiewicz, M., Ramanna, S., Jensen, R., Hu, Q. (eds.) RSCTC 2010. LNCS, vol. 6086, pp. 376–385. Springer, Heidelberg (2010)
12. Skowron, A., Rauszer, C.: The Discernibility Matrices and Functions in Information Systems. In: Intelligent Decision Support - Handbook of Advances and Applications of the Rough Set Theory, pp. 331–362. Kluwer Academic Publishers, Dordrecht (1992)
13. Zadeh, L.A.: Toward a Theory of Fuzzy Information Granulation and its Centrality in Human Reasoning and Fuzzy Logic. Fuzzy Sets and Systems 90, 111–127 (1997)

How to Reconstruct the System's Dynamics by Differentiating Interval-Valued and Set-Valued Functions

Karen Villaverde[1] and Olga Kosheleva[2]

[1] New Mexico State University, Las Cruces, NM 88003, USA
kvillave@cs.nmsu.edu
[2] University of Texas at El Paso, El Paso, TX 79968, USA
olgak@utep.edu

Abstract. To predict the future state of a physical system, we must know the differential equations $\dot{x} = f(x)$ that describe how this state changes with time. In many practical situations, we can observe individual trajectories $x(t)$. By differentiating these trajectories with respect to time, we can determine the values of $f(x)$ for different states x; if we observe many such trajectories, we can reconstruct the function $f(x)$. However, in many other cases, we do not observe individual systems, we observe a set X of such systems. We can observe how this set X changes, but not how individual states change. In such situations, we need to reconstruct the function $f(x)$ based on the observations of such "set trajectories" $X(t)$. In this paper, we show how to extend the standard differentiation techniques of reconstructing $f(x)$ from vector-valued trajectories $x(t)$ to general set-valued trajectories $X(t)$.

Keywords: prediction under uncertainty, differentiation of interval-valued and set-valued functions.

1 Formulation of the Problem

One of the main objectives of science and engineering: a brief reminder. One of the main objectives of *science* is to predict the future state of different systems. We want to predict the future weather, we want to predict the future trajectories of celestial bodies, etc. To make these predictions, we need to know the current state of the system, and we need to know how the state evolves with time.

For *engineering*, the main objective is to produce a design that satisfies the given properties, a control that leads the object into the given location, etc. In all such problems, we also need to be able to predict the future behavior of the designed and/or controlled system. The state of a physical object (system) can be characterized by the values $x = (x_1, \ldots, x_n)$ of different physical characteristics $x_1, \ldots, x_n$ of this object. For a celestial object, these characteristics include its mass, its location, its velocity, its angular velocity relative to different axes, its brightness and reflectivity at different places, etc. For the atmosphere, these

S.O. Kuznetsov et al. (Eds.): RSFDGrC 2011, LNAI 6743, pp. 183–190, 2011.

characteristics include temperature, atmospheric pressure, wind speed, etc., at different locations. The evolution of macro-objects is usually reasonably well described by a (deterministic) ordinary differential equation $\dot{x} = f(x)$, where $\dot{x} \stackrel{\text{def}}{=} \dfrac{dx}{dt}$ is the time derivative.

Need for empirical differentiation. In many cases, the mapping $f(x)$ that describes the system's dynamics is known. For example, a point object can be characterized by its location r and its velocity v: $x = (r, v)$. Newton's equations $m\ddot{r} = F(r)$ describe the dynamics of this object in the force field $F(r)$. These equations can be described in the desired form $\dot{x} = f(x)$ as follows: $\dot{x} = v$, $\dot{v} = F(r)$, i.e., $f(r, v) = (v, F(r))$.

However, often, we do not know the exact dynamics $f(x)$. In such situations, we need to reconstruct the values $f(x)$ based on the observed trajectories of a system, i.e., on the values $x(t_i)$ measured for different values $t_1 < t_2 < \ldots < t_m$. When the observations are close in time, we can approximately describe the corresponding time derivatives as $\dot{x}(t_i) \approx \dfrac{x(t_i) - x(t_{i-1})}{t_i - t_{i-1}}$, and then reconstruct $f(x)$ from the condition that $\dot{x}(t_i) = f(x(t_i))$ for all observation moments t_i – and for all observed objects $x(t)$.

Need for interval-valued and set-valued functions: case of uncertainty. The above description is based on the ideal case when we observe a single object and its trajectory. In practice, often, instead of a *single* object, we observe the whole *group* of objects, a group in which it is very difficult to distinguish between individual objects.

For example, in biology, we can analyze the spread of the bacteria or viruses by tracing the corresponding epidemics, but it is practically impossible to trace individual bacteria or viruses. In meteorology, we can trace, e.g., how water goes from one state into another, from clouds to rain to rivers to evaporation etc., but it is impossible to trace individual molecules. In such cases, at any given moment of time t, instead of a single state $x(t)$, we observe the *collection* (*set*) $X(t)$ of different states.

Need to extend differentiation techniques to interval-valued and set-valued functions. In order to make predictions, we need to know the dynamics $f(x)$. Thus, we need to be able to reconstruct the dynamics from the observed sets $X(t_1)$, $X(t_2)$, ..., $X(t_n)$. For the case of exactly known states, when each set $X(t_i)$ consists of a single state $X(t_i) = \{x(t_i)\}$, this reconstruction is based on the differentiation. Thus, it is reasonable to call the process of reconstructing the dynamics $f(x)$ from the observed sets "differentiations" of the corresponding set-valued function $X(t)$.

Differentiation of set-valued functions: what is known. There have been many generalizations of differentiation to set-valued functions. Many such generalizations appeared in *rough set* theory; see, e.g., [19]. The main idea behind rough sets is that instead of the exact set X of possible states, we only store its lower and upper approximations $\underline{X}$ and $\overline{X}$: the set $\underline{X} \subseteq X$ is formed by all the granules that are fully contained in X, and the set $\overline{X} \supseteq X$ is formed by all the granules that may have common elements with X. When the situation changes, both sets change. To describe the rate of such change, Pawlak described differentiation of rough sets [19,20]; see also [14,15,18,22,23,24,25].

Similar set-valued differentiation procedures [1,4,8,9,10,11,12,21] have been defined in within a *set-valued analysis* [2,3], when we need, e.g., to find the optimal shape (set) of a design (see, e.g. [16]). Several papers describe the application of these techniques to the important problem of find the range of the solution to a differential equation under uncertain initial conditions and uncertain values of the parameters; see, e.g., [5,6,7].

What we do. None of the existing set differentiation procedures directly solves our problem – of reconstructing the function $f(x)$ from the observed set trajectories $X(t)$. We show, however, that by properly modifying the known differentiation techniques, we can extract the dynamics $f(x)$ from the observed behavior $X(t_i)$. This extraction uses techniques that generalize the standard differentiation techniques from the case of vector-valued trajectories $x(t)$ to the more general case of set-valued trajectories $X(t)$.

In this paper, we first consider a 1-D (interval-valued) case in Section 2, then a fuzzy case in Section 3, and finally, the general multi-D case in Section 4.

2 Case of Interval-Valued Functions

Formulation of the case. Let us start with the simplest case in which the state of a system is characterized by a single variable x, i.e., when $x = x_1$ and $n = 1$. In this case, each state is a point on a real line, and thus, for each moment of time t, the set $X(t)$ of all observed states is a subset of the real line. In general, the set $X(t)$ of observed states can be disconnected (in the standard topological sense). However, in this case, we would be able to individually trace every connected component separately. So, for our purpose, it makes sense to consider the case when the set $X(t)$ of possible states is connected. It also makes sense to consider the case when this set is bounded – since in practice, most observed collections are bounded. On the real line, the only bounded connected sets are intervals. Thus, we can conclude that for every t, we observe the corresponding interval $X(t) = [\underline{x}(t), \overline{x}(t)]$.

Analysis of the problem. Let $f(x)$ be a function that describes the system's dynamics. This means that once at some moment t, we have a state $x(t)$, then at the next moment of time $t+\Delta t$, we have the state $x(t+\Delta t) \approx x(t)+f(x(t))\cdot\Delta t$. Different values $x \in [\underline{x}(t), \overline{x}(t)]$ lead, in general, to different values $x + f(x) \cdot \Delta t$. Thus, to find the upper endpoint $\overline{x}(t + \Delta t)$ of the interval

$$X(t + \Delta t) = [\underline{x}(t + \Delta t), \overline{x}(t + \Delta t)],$$

we need to find the largest possible value of the expression $x + f(x) \cdot \Delta t$ when $x \in [\underline{x}(t), \overline{x}(t)]$.

In physics, in most dynamical equations, the transformation $f(x)$ is usually smooth (differentiable). Thus, it is reasonable to assume that $f(x)$ is differentiable and thus, that the mapping $x \to x + f(x) \cdot \Delta t$ is also differentiable. The derivative of this mapping with respect to x is equal to $1 + f'(x) \cdot \Delta t$. When the time step Δt is sufficiently small, we have $|f'(x) \cdot \Delta t| \ll 1$ and thus, $1 + f'(x) \cdot \Delta t > 0$. Hence, on this interval, the transformation $x \to x + f(x) \cdot \Delta t$ is strictly increasing. Thus, the largest value $\overline{x}(t+\Delta t)$ of the expression $x(t+\Delta t) = x(t) + f(x(t)) \cdot \Delta t$ is attained when $x(t)$ attains its largest value, i.e., when $x(t) = \overline{x}(t)$. In other words, $\overline{x}(t + \Delta t) \approx \overline{x}(t) + f(\overline{x}(t)) \cdot \Delta t$. Thus, we have $\dot{\overline{x}} = f(\overline{x})$.

Similarly, the smallest possible value $\underline{x}(t + \Delta t)$ of the expression $x(t + \Delta t) = x(t) + f(x(t)) \cdot \Delta t$ is attained when $x(t)$ attains its smallest value, i.e., when $x(t) = \underline{x}(t)$. In other words, $\underline{x}(t + \Delta t) \approx \underline{x}(t) + f(\underline{x}(t)) \cdot \Delta t$. Thus, we have $\dot{\underline{x}} = f(\underline{x})$. Hence, we arrive at the following conclusion.

Conclusion: how to reconstruct the dynamics from the interval-valued observations. If, for each moment of time t_i, we know the interval $X(t_i) = [\underline{x}(t_i), \overline{x}(t_i)]$, then we can reconstruct the dynamics $f(x)$ as follows. First, we estimate

$$\dot{\overline{x}}(t_i) \approx \frac{\overline{x}(t_i) - \overline{x}(t_{i-1})}{t_i - t_{i-1}}; \quad \dot{\underline{x}}(t_i) \approx \frac{\overline{x}(t_i) - \overline{x}(t_{i-1})}{t_i - t_{i-1}}.$$

Then, we reconstruct $f(x)$ from the conditions that $\dot{\overline{x}}(t_i) = f(\overline{x}(t_i))$ and $\dot{\underline{x}}(t_i) = f(\underline{x}(t_i))$ for all observation moments t_i – and for all observed interval-valued trajectories $X(t)$.

Example. For radioactive decay, $\dot{x} = -k \cdot x$, so $x(t) = x(0) \cdot \exp(-k \cdot t)$. Thus, if we start with an interval $X(0) = [1, 2]$, we get $X(t) = [\exp(-k \cdot t), 2 \cdot \exp(-k \cdot t)]$. By differentiating the lower endpoint, we conclude that for every t, we have $f(\exp(-k \cdot t)) = -k \cdot \exp(-k \cdot t)$, i.e., that indeed $f(x) = -k \cdot x$.

3 Fuzzy Case: Observation

Formulation of the problem. Instead of observing the crisp interval $X(t)$, we can be observing a *fuzzy* interval. In other words, in addition to the interval $[\underline{x}(t), \overline{x}(t)]$ that is guaranteed to contain all the observed objects, for every degree α from the interval $(0, 1)$, we also have narrower intervals $[\underline{x}_\alpha(t), \overline{x}_\alpha(t)]$ (alpha-cuts of the corresponding fuzzy sets) that contain $x(t)$ with certainty α.

Analysis of the problem. It is known that for every bounded continuous transformation, the alpha-cut of the result is equal to the result of applying this transformation to the original alpha-cut; see, e.g., [13,17]. Thus, for each α, the corresponding α-cut intervals $[\underline{x}_\alpha(t_i), \overline{x}_\alpha(t_i)]$ form a sequence from which we can extract $f(x)$ – by using the interval-based techniques described in the previous section. As a result, we arrive at the following technique.

Conclusion: how to reconstruct the dynamics from the fuzzy-valued observations. Let us assume that for each moment of time t_i, we know the fuzzy value $X(t_i)$.

In other words, we assume that for every moment of time t_i and for every degree $\alpha \in (0,1]$, we know the interval $[\underline{x}_\alpha(t_i), \overline{x}_\alpha(t_i)]$. Then we can reconstruct the dynamics $f(x)$ as follows. First, we estimate

$$\dot{\overline{x}}_\alpha(t_i) \approx \frac{\overline{x}_\alpha(t_i) - \overline{x}_\alpha(t_{i-1})}{t_i - t_{i-1}}; \quad \dot{\underline{x}}_\alpha(t_i) \approx \frac{\overline{x}_\alpha(t_i) - \overline{x}_\alpha(t_{i-1})}{t_i - t_{i-1}}.$$

Then, we reconstruct $f(x)$ from the conditions that $\dot{\overline{x}}_\alpha(t_i) = f(\overline{x}_\alpha(t_i))$ and $\dot{\underline{x}}_\alpha(t_i) = f(\underline{x}_\alpha(t_i))$ for all observation moments t_i, for all degrees α, and for all observed interval-valued trajectories $X(t)$.

4 General Multi-D Case

Formulation of the problem. In the multi-dimensional case, at different moments of time t, we observe the set $X(t)$ of states. For an individual state $x(t) \in X(t)$, we do not know what will be the corresponding state $x(t + \Delta t)$ at the next moment of time $t + \Delta t$, we only know that this unknown state $x(t + \Delta t)$ belongs to the observed set $X(t + \Delta t)$. We also know that all the states from the set $X(t + \Delta t)$ are obtained from the states of the set $X(t)$ by the corresponding evolution $\dot{x} = f(x)$. We may observe several different evolving sets $X^{(1)}(t)$, $X^{(2)}(t)$, ... Our objective is, based on this information, to reconstruct the dynamics $f(x)$.

Definitions and the main result. Let us formulate our main result in precise terms. By a *dynamical system*, we mean a smooth function $f : \mathbb{R}^n \to \mathbb{R}^n$. By a *smooth set* X, we mean a simply connected open set whose boundary ∂X is a smooth surface. For every dynamical system f and for every smooth set X, by a *set trajectory*, we mean a function that maps each positive real number t into the set $X(t) = \{x(t) : x(0) \in X \,\&\, \dot{x} = f(x)\}$. Let us denote the class of all set trajectories corresponding to a system f by $T(f)$.

Our main result is that is that a dynamical system is uniquely determined by the class of its set trajectories, i.e., if $T(f) = T(f')$ then $f = f'$.

Comment. As we will see from the proof, in order to uniquely determine f, it is not necessary to know *all* set trajectories, it is sufficient to have a class of set trajectories for which, for every point $x \in \mathbb{R}^n$, we have n set trajectories $X^{(i)}(t)$ and moments of time t_i at which $x \in \partial X^{(i)}(t_i)$ and at which the n normal vectors $N^{(i)} \perp \partial X^{(i)}(t_i)$ are linearly independent.

Analysis of the problem. Let x_0 be any point on the border $\partial X(t)$, and let N be a normal vector, i.e., the unit vector orthogonal to $\partial X(t)$ at the point x_0 (i.e., orthogonal to the tangent plane to $X(t)$).

As the states evolve, each state $x \in X(t)$ changes into the next state $x + f(x) \cdot \Delta t$. Locally, when x is close to x_0, the value $f(x)$ is close to $f(x_0)$ and thus, the whole plane shifts by the vector $\Delta x \stackrel{\text{def}}{=} f(x_0) \cdot \Delta t$. Let us first consider the situation when the vector $f(x_0)$ is in the tangent plane. In this situation, while each state changes, the plane itself does not change. Thus, since we only observe the set $X(t)$ – i.e., in effect, its boundary $\partial X(t)$ – locally, we will not observe any difference. A general shift Δx can be represented as a linear combination of two shifts:

- a shift in the direction from the plane – which we cannot observe, and
- a shift in the direction orthogonal to the plane, i.e., in the direction parallel to the normal vector; this shift we can observe.

In geometric terms, the shift in the direction of N can be represented as $(\Delta x, N) \cdot N$, where $(a, b) \stackrel{\text{def}}{=} \sum\limits_i a_i \cdot b_i$ is a scalar (dot) product of two vectors, and the shift in the direction from the plane has the form $\Delta x - (\Delta x, N) \cdot N$. The value $(\Delta x, N)$ is actually equal to the distance between the two tangent planes:

- the tangent plane to the border set $\partial X(t)$ at the point x_0, and
- the tangent plane to the border set $\partial X(t + \Delta t)$ at the point closest to x_0.

So, by measuring this distance, we can find the scalar product $(\Delta x, N) = (f(x_0), N) \cdot \Delta t$, and thus, we can find the scalar product $(f(x_0), N)$. If we have several families of sets $X^{(1)}(t)$, ..., $X^{(k)}(t)$, then, in general, the normal vectors $N^{(j)}$ corresponding to the moment when the boundaries of the corresponding sets pass through (or close to) the point x_0, are different. Thus, we can get the scalar products $(f(x_0), N^{(j)})$ corresponding to different vectors $N^{(j)}$. Once we know a sufficient number of such products, we can thus uniquely reconstruct the vector $f(x_0)$ – i.e., all n coordinates $f_i(x_0)$ of this vector – from the corresponding system of linear equations $\sum\limits_{i=1}^{n} f_i(x_0) \cdot N_i^{(j)} = (f(x_0), N^{(j)})$. As a result, we arrive at the following conclusion.

Conclusion: how to reconstruct the dynamics from the set-valued observations. Let us assume that we have k dynamically changing sets. For each such set $X^{(j)}$, $j = 1, \ldots, k$, for different moments of times $t_1^{(j)} < t_2^{(j)} < \ldots < t_k^{(j)} < \ldots$ we observe the sets $X^{(j)}(t_k^{(j)})$ of possible states. As before, we assume that the consecutive moments are close to each other, i.e., $t_{k+1}^{(j)} \approx t_k^{(j)}$. Then, we can reconstruct the function $f(x)$ as follows.

For each family j, for each moment $t_k^{(j)}$, and for each point x from the boundary $\partial X^{(j)}(t_k^{(j)})$ of the set $X^{(j)}(t_k^{(j)})$, we compute the distance $\Delta\rho_k^{(j)}(x)$ between the following two planes:

- the plane P tangent to $\partial X^{(j)}(t_k^{(j)})$ at the point x, and
- the plane tangent to $\partial X^{(j)}(t_{k+1}^{(j)})$ at a point which is the closest to x.

Then, we compute the ratio $\dfrac{\Delta\rho_k^{(j)}(x)}{t_{k+1}^{(j)} - t_k^{(j)}}$. We also compute the unit vector $N_k^{(j)}(x)$ which is orthogonal to the plane P. Then, we conclude that the value $f(x)$ of the desired dynamical function $f(x)$ satisfies the equation $(f(x), N_k^{(j)}(x)) =$

$\frac{\Delta\rho_k^{(j)}(x)}{t_{k+1}^{(j)} - t_k^{(j)}}$. After that, we reconstruct the desired function $f(x)$ from the fact that these equations – with known values of the right-hand side – must be satisfied for all the points x on all the boundaries $\partial X^{(j)}(t_k^{(j)})$. Specifically, to find the value $f(x_0)$ for a given x_0, we collect all such equations for close values $x \approx x_0$. Since $x_0 \approx x$, we conclude that $f(x_0) \approx f(x)$ and thus, that $(f(x_0), N_k^{(j)}(x)) \approx (f(x), N_k^{(j)}(x)) = \frac{\Delta\rho_k^{(j)}(x)}{t_{k+1}^{(j)} - t_k^{(j)}}$. In general, for these equations, the normal vectors $N_k^{(j)}(x)$ will be different, so we have sufficiently many linear equations of the type $(f(x_0), N_k^{(j)}(x)) \approx \frac{\Delta\rho_k^{(j)}(x)}{t_{k+1}^{(j)} - t_k^{(j)}}$. from which we can uniquely reconstruct the vector $f(x_0)$.

5 Conclusions

In order to predict the evolution of a system, we need to know the differential equations that describe how its state changes with time. These equations can be determined from observations when we observe several trajectories of individual systems. However, in many practical situations, we do not observe individual trajectories, we observe the whole set of systems that evolve together, we observe how this set changes, but not how individual trajectories change. In this paper, we show that based on several such set observations, we can also uniquely reconstruct the differential equations that describe the system's dynamics.

Acknowledgments. The authors are thankful to the anonymous referees for valuable suggestions.

References

1. Assev, S.M.: Quasilinear operators and their application in the theory of multivalued mappings. Proceedings of the Steklov Institute of Mathematics 2, 23–52 (1986)
2. Aubin, J.P., Franskowska, H.: Set-Valued Analysis. Birkhäuser, Basel (1990)
3. Aubin, J.P., Franskowska, H.: Introduction: Set-valued analysis in control theory. Set-Valued Analysis 8, 1–9 (2000)
4. Banks, H.T., Jacobs, M.Q.: A differential calculus for multifunctions. Journal of Mathematical Analysis and Applications 29, 246–272 (1970)
5. Bede, B.: Note on Numerical solutions of fuzzy differential equations by predictor-corrector method. Information Sciences 178, 1917–1922 (2008)
6. Bede, B., Gal, S.G.: Generalizations of the differentiability of fuzzy number valued functions with applications to fuzzy differential equation. Fuzzy Sets and Systems 151, 581–599 (2005)
7. Bede, B., Rudas, I.J., Bencsik, A.L.: First order linear fuzzy differential equations under generalized differentiability. Information Sciences 177, 1648–1662 (2007)

8. Chalco-Cano, Y., Román-Flores, H.: On the new solution of fuzzy differential equations. Chaos, Solitons & Fractals 38, 112–119 (2008)
9. Chalco-Cano, Y., Román-Flores, H.: Comparation between some approaches to solve fuzzy differential equations. Fuzzy Sets and Systems 160, 1517–1527 (2008)
10. De Blasi, F.S.: On the differentiability of multifunctions. Pacific Journal of Mathematics 66, 67–81 (1976)
11. Hukuhara, M.: Integration des applications mesurables dont la valeur est un compact convexe. Funkcialaj Ekvacioj 10, 205–223 (1967)
12. Ibrahim, A.-G.M.: On the differentiability of set-valued functions defined on a Banach space and mean value theorem. Applied Mathematics and Computers 74, 76–94 (1996)
13. Klir, G.J., Yuan, B.: Fuzzy Sets and Fuzzy Logic: Theory and Applications. Prentice-Hall, Upper Saddle River (1995)
14. Li, D.G., Chen, X.E.: Improvement of definitions of one-element rough function and binary rough function and investigation of their mathematical analysis properties. Journal of Shanxi University 23(4), 318–321 (2000)
15. Li, D.G., Hu, G.R.: Definitions of n-element rough function and investigation of its mathematical analysis properties. Journal of Shanxi University 24(4), 299–302 (2001)
16. Nguyen, H.T., Kreinovich, V.: How to divide a territory? a new simple differential formalism for optimization of set functions. International Journal of Intelligent Systems 14(3), 223–251 (1999)
17. Nguyen, H.T., Walker, E.A.: A First Course in Fuzzy Logic. Chapman & Hall/CRC, Boca Raton (2006)
18. Pawlak, Z.: Rough functions. Bulletin of the Polish Academy of Sciences, Technical Series PAS Tech. Ser. 355(5-6), 249–251 (1997)
19. Pawlak, Z.: Rough sets, rough function and rough calculus. In: Pal, S.K., Skowron, A. (eds.) Rough Fuzzy Hybridization, pp. 99–109. Springer, Berlin (1999)
20. Pawlak, Z., Pal, S.K., Skowron, A.: Rough-Fuzzy Hybridization: A New Trend in Decision-Making. Springer, New York (1999)
21. Stefanini, L., Bede, B.: Generalized Hukuhara differentiability of interval-valued functions and interval differential equations. Nonlinear Analysis 71, 1311–1328 (2009)
22. Wang, Y., Guan, Y.Y., Wang, H.: Rough derivatives in rough function model and their application. In: Proc. of the 4th International Conference on Fuzzy Systems and Knowledge Discovery, vol. 3, pp. 193–197 (2007)
23. Wang, Y., Wang, J.M., Guan, Y.Y.: The theory and application of rough integration in rough function model. In: Proc. of the 4th International Conference on Fuzzy Systems and Knowledge Discovery, vol. 3, pp. 224–228 (2007)
24. Wang, Y., Xu, X., Yu, Z.: Notes for rough derivatives and rough continuity in rough function model. In: Proceedings of the 7th International Conference on Fuzzy Systems and Knowledge Discovery FSKD 2010, Yantai, China, August 10-12, pp. 245–247. IEEE Press, Los Alamitos (2010)
25. Zhuo, Z.Q.: Improvement of definition of rough function and proof of rough derivatives properties in rough set theory. Journal of Huaibei Coal Normal University 23(3), 10–15 (2002)

Symbolic Galois Lattices with Pattern Structures

P. Agarwal[1], M. Kaytoue[1], S.O. Kuznetsov,[2], A. Napoli[1], and G. Polaillon[3]

[1] LORIA – Campus Scientifique, B.P. 70239 – Vandœuvre-lès-Nancy – France
[2] HSE – Pokrovskiy Bd. 11 – 109028 Moscow – Russia
[3] E3S Supélec – 3 rue Joliot-Curie – 91192 Gif sur Yvette – France

Abstract. Concept lattices are mathematical structures useful for many tasks in knowledge discovery and management. A concept lattice is basically obtained from binary data encoding the membership of some attributes to some objects. Dealing with complex data brings the important problem of discretization and the associated loss of information. To avoid discretization, (i) pattern structures and (ii) symbolic data analysis provide means to analyze such complex data directly. We compare both these approaches and show how they are mutually beneficial.

Keywords: Concept analysis, Symbolic data analysis, Pattern structures.

1 Introduction

Many classification problems can be formalized by means of a *formal context*, a binary relation between an object set and an attribute set indicating whether an object has or does not have an attribute [4]. According to the so-called *Galois connection*, one may classify within *formal concepts* a set of objects sharing a same maximal set of attributes, and vice-versa. Concepts are ordered in a lattice structure called *concept lattice* within the Formal Concept Analysis (FCA) framework [4]. FCA can be used for a number of purposes like knowledge formalization and acquisition, ontology design, and data mining. To handle complex data in FCA, *pattern structures* have been proposed as a generalization of formal contexts to complex data [9,8]. On the other hand, Symbolic Data Analysis (SDA [1]) aims at analyzing data such as numbers, intervals, sets of discrete values, etc. An object is described by a vector of values with each dimension corresponding to a variable, and each variable may be of different type. [2,3] addressed the problem of building concept lattices by formalizing "symbolic objects" in SDA and properly defined Galois connections between these symbolic individuals and their descriptions. The links between the FCA and SDA approaches still remain unclear. Both methods show the same behaviour when working on the same data, but the goal of this paper is to argue how the SDA formalism for building concept lattices can be taken into account in FCA in a universal way, to facilitate comprehension and future extension.

The paper is organized as follows. Section 2, 3 respectively present SDA, and pattern structures. Both approaches are compared and discussed in Section 4. Limited by space, we assume that the reader is familiar with FCA [4].

S.O. Kuznetsov et al. (Eds.): RSFDGrC 2011, LNAI 6743, pp. 191–198, 2011.

2 Symbolic Galois Lattices in Symbolic Data Analysis

Symbolic Data Analysis takes its roots in data analysis and as well in probability theory and statistics [1]. Let Ω be a set of individuals, $\mathcal{D}$ be a set of descriptions, and Y be a mapping between Ω and $\mathcal{D}$. $\mathcal{D}$ may include any kind of descriptions, e.g. numbers, symbolic values, intervals, propositional variables, etc. The mapping Y associates with each $\omega \in \Omega$ a description $d \in \mathcal{D}$ having the form of a vector of values. For example, considering Table 1 below, we have: $Y(\omega_1) = (y_1(\omega_1), y_2(\omega_1)) = ([75, 80], [1, 2])$. The description of an individual or a class of individuals is called an *intensional description.*

Description of two individuals can be compared w.r.t. the type of values, thanks to comparison operators in $\{=, \subseteq, \supseteq, \in, \ni, \longrightarrow, \ldots\}$. For example, $Y(\omega_4) \subseteq Y(\omega_1)$ in Table 1, means that $Y(\omega_4) = (y_1(\omega_4), y_2(\omega_4)) \subseteq Y(\omega_1) = (y_1(\omega_1), y_2(\omega_1))$. Now, given Ω and $\mathcal{D}$, we precisely define the notion of a *symbolic object.*

Symbolic object – Let O_i be the "description space" for variable y_i and $\mathcal{D} = O_1 \times \ldots \times O_p$, with $y_i(\omega) = d_i \in O_i$ for an individual ω in Ω. d_i is the description of attribute $y_i(\omega) \in Y(\omega)$. A symbolic object s is a triplet $s = (a, R, d)$ where a is a mapping between Ω and $\{0, 1\}$ (or $\{false, true\}$) called *extension*, R is a comparison operator between descriptions, and $d = (d_1, \ldots, d_p)$ is a description vector.

For example, in Table 1, $\Omega = \{\omega_1, \omega_2, \omega_3, \omega_4\}$, $Y = \{y_1, y_2\}$, with $Y(\omega_1) = (y_1(\omega_1), y_2(\omega_1)) = ([75, 80], [1, 2])$. Then, $s = (a, =, ([75, 80], [1, 2]))$ is a symbolic object describing individual ω_1 as $y_1(\omega_1) = [75, 80]$ and $y_2(\omega_1) = [1, 2]$. A straightforward extension of the notion of symbolic object is given by *assertion objects.*

Table 1. A data table

	y_1	y_2
ω_1	[75, 80]	[1, 2]
ω_2	[60, 80]	[1, 1]
ω_3	[50, 70]	[2, 2]
ω_4	[72, 73]	[1, 2]

Assertion object – Let us define an assertion object with the following mapping:

$$\psi\colon \omega \longrightarrow [y_1(\omega) R d_1] \wedge \ldots \wedge [y_p(\omega) R d_p]$$

such that $\psi(\omega) = 1$ iff $y_i(\omega) R d_i \; \forall i = 1, \ldots, p$ holds, $\wedge$ being the logical conjunction. Intuitively, $\psi(\omega)$ defines a conjunction of events which is true iff every event is itself true. For example, an assertion object ψ_1 is given by $(y_1 \subseteq [60, 80]) \wedge (y_2 \subseteq [1, 2])$, with $\psi_1(\omega_1) = \psi_1(\omega_2) = \psi_1(\omega_4) = 1$. Accordingly, the *extent* of an assertion object ψ w.r.t. Ω is given by $ext_\Omega(\psi) = \{\omega \in \Omega \mid \psi(\omega) = 1\}$. The extent of ψ is the set of individuals whose description fulfills the description of the assertion object. For example, $ext_\Omega(\psi_1) = \{\omega_1, \omega_2, \omega_4\}$. Moreover, $ext_\Omega([y_1 \supseteq [72, 73]] \wedge [y_2 \subseteq [1, 2]]) = \{\omega_2, \omega_4\}$.

Symbolic order – Let A denote the set of assertion objects. A partial ordering is defined on A as follows: $\forall \alpha, \beta \in A, \; \alpha \leq_\Omega \beta \Rightarrow ext_\Omega(\alpha) \subseteq ext_\Omega(\beta)$. Further, Galois connections can be defined between $\wp(\Omega)$ and A depending on the choice of a "generalization operator" for building the upper bound of two assertions objects (see [2,10]).

Hereon, we consider only the interval data for sake of simplicity and for comparison with the FCA formalism. We introduce two Galois connections, one based on generalization by *union* and the other *by intersection.*

Generalization by union – The common description of a set of assertion objects is based on the *union* of the description of each assertion object. More precisely, let us consider the two dual derivation operators f and g:

$$\begin{aligned}
&f : (\wp(\Omega), \subseteq) \longrightarrow (A, \leq_\Omega)\\
&\quad X \longrightarrow \psi = \wedge_j [y_j(\omega)\ R \bigcup_{\omega_i \in X} \{y_j(\omega_i)\}]\\
&g : (A, \leq_\Omega) \longrightarrow (\wp(\Omega), \subseteq)\\
&\quad \psi = \wedge_j [y_j(\omega) R W_j] \longrightarrow ext_\Omega(\psi) = \{\omega \mid y_j(\omega)\ R\ W_j = 1, j = 1 \ldots p\}\\
&\quad where\ W_j = \{y_j(\omega)\ R \bigcup_{\omega_i \in X} \{y_j(\omega_i)\}
\end{aligned}$$

For example, in Table 1:

$$\begin{aligned}
f(\{\omega_1, \omega_2\}) &= [y_1 \subseteq ([75, 80] \cup [60, 80])] \wedge [y_2 \subseteq ([1, 2] \cup [1, 1])]\\
&= [y_1 \subseteq [60, 80]] \wedge [y_2 \subseteq [1, 2]]\\
g(f(\{\omega_1, \omega_2\})) &= \{\omega_1, \omega_2, \omega_4\}
\end{aligned}$$

The compositions of derivation operators f and g, i.e. $f \circ g$ and $g \circ f$, are closure operators as in classical FCA [2,10]. In sequence, the pair $(\{\omega_1, \omega_2, \omega_4\}, ([y_1 \subseteq [60, 80]] \wedge [y_2 \subseteq [1, 2]]))$ defines a *concept* w.r.t. generalization by union.
Generalization by intersection – Generalization by intersection is defined in the same way as generalization by union, replacing the union operation $\cup$ with the intersection operation $\cap$, and inclusion $\subseteq$ by reverse inclusion $\supseteq$. For example, from Table 1:

$$\begin{aligned}
f(\{\omega_1, \omega_2\}) &= [y_1 \supseteq ([75, 80] \cap [60, 80])] \wedge [y_2 \supseteq ([1, 2] \cap [1, 1])]\\
&= [y_1 \supseteq [75, 80]] \wedge [y_2 \supseteq [1, 1]]\\
g(f(\{\omega_1, \omega_2\})) &= \{\omega_1, \omega_2\}
\end{aligned}$$

Accordingly, the pair $(\{\omega_1, \omega_2\}, ([y_1 \supseteq [75, 80]] \wedge [y_2 \supseteq [1, 1]]))$ defines a *concept* w.r.t. generalization by intersection.

Given a Galois connection (f, g), the closure operators $f \circ g$ and $g \circ f$, a concept is a pair with an extension which is a closed set (of objects) for $g \circ f$ and an intension which is a closed set (of attributes) for $f \circ g$. A lattice of assertion objects can be built just as a concept lattice in FCA called *symbolic Galois lattice* (but algorithims are not exactly the same and are inefficient). We also discuss the relation between the formalism of symbolic/assertion objects and FCA. As we can see, the formalism of symbolic/assertion objects is powerful and very interesting because it can be applied to various and very heterogeneous data. However, this formalism looks rather complicated to comprehend and explain. Moreover, the algorithmic aspects have not been considered important unlike their treatment in FCA. Additionaly, some extensions of symbolic/assertion objects, e.g. to graphs, are not straightforward.

In the next section, we introduce pattern structures, an extension of FCA for working with complex data and in particular intervals, which in our opinion, subsume the symbolic/assertion object formalism in most of its dimensions.

3 Pattern Concept Lattices

Pattern structures are introduced [5] in full compliance with FCA and can be thought of as a "generalization" of formal contexts to complex data from which a concept lattice can be built without any *a priori* scaling.

In classical FCA, the operators of the Galois connection put in correspondence, the elements of the lattices $(2^G, \subseteq)$ of objects and $(2^M, \subseteq)$ of attributes and vice-versa. These lattices are partially ordered sets. This means that if one needs to build concept lattices where objects are not described by binary attributes but by complex descriptions (graphs, intervals, ...), one has to define a partial order on object descriptions. This is the main idea of *pattern structures* formalizing objects from G and their descriptions called *patterns* from a set D where patterns are ordered in a meet-semi-lattice $(D, \sqcap)$ [5]. In classical FCA, if we consider the lattice of attributes $(2^M, \subseteq)$, it is straightforward that $\forall N, O \subseteq M$, $N \subseteq O \Leftrightarrow N \cap O = N$, e.g. with $M = \{a, b, c\}$, $\{a, b\} \subseteq \{a, b, c\} \Leftrightarrow \{a, b\} \cap \{a, b, c\} = \{a, b\}$. The set-intersection operator $\cap$ has the properties of a meet operator in a semi-lattice. This is the underlying idea for ordering patterns with a subsumption relation $\sqsubseteq$: given two patterns $c, d \in D$, $c \sqsubseteq d \Leftrightarrow c \sqcap d = c$. Then, building the concept lattice is in full compliance with FCA theory.

Formally, let G be a set of objects, $(D, \sqcap)$ be a semi-lattice of object descriptions, and $\delta : G \rightarrow D$ be a mapping. $(G, (D, \sqcap), \delta)$ is called a *pattern structure*. Elements of D are called *patterns* and are ordered by ordering relation $\sqsubseteq$: given $c, d \in D$ one has $c \sqsubseteq d \Longleftrightarrow c \sqcap d = c$. We use the operator $(.)^\square$ to derive two sets as follows:

$$A^\square = \textstyle\bigsqcap_{g \in A} \delta(g), \text{ for any } A \subseteq G,$$
$$d^\square = \{g \in G \mid d \sqsubseteq \delta(g)\}, \text{ for any } d \in (D, \sqcap)$$

These operators form a Galois connection between $(\mathfrak{P}(G), \subseteq)$ and $(D, \sqsubseteq)$. $(.)^{\square\square}$ is a closure operator. *Pattern concepts* of $(G, (D, \sqcap), \delta)$ are pairs of the form (A, d), $A \subseteq G$, $d \in D$, such that $A^\square = d$ and $A = d^\square$, and d is called a *pattern intent* while A is a *pattern extent*. When partially ordered by $(A_1, d_1) \leq (A_2, d_2) \Leftrightarrow A_1 \subseteq A_2$ $(\Leftrightarrow d_2 \sqsubseteq d_1)$, the set of all pattern concepts forms a complete lattice called a *pattern concept lattice*. An example is given in the next section. Standard FCA algorithms need slight modification to compute the pattern concept lattice, see e.g. [5,8].

4 Symbolic Galois Lattices with Pattern Structures

4.1 Handling Heterogeneous Variables with Pattern Structures

SDA works on data tables where each column corresponds to a variable y_i which has a range with different types of values. Pattern structures consider a partially ordered set of descriptions $(D, \sqcap)$, corresponding to one variable in terms of SDA. However, we can consider a semi-lattice $(D_{y_i}, \sqcap_{y_i})$ for each variable $y_i \in Y$. The direct product of all these semi-lattices gives a semi-lattice $(D, \sqcap)$ with all possible descriptions of objects and sets of objects. Therefore given a set $Y = \{y_1, ..., y_p\}$ of p variables, we define

$$(D, \sqcap) = (D_{y_1}, \sqcap_{y_1}) \times ... \times (D_{y_p}, \sqcap_{y_p})$$

where each $(D_{y_i}, \sqcap_{y_i})$, $i \in [1, p]$ is the poset of descriptions for variable $y_i \in Y$. $(D, \sqcap)$ corresponds to the set $\mathcal{D}$ in SDA, i.e. the description space, provided with a partial ordering $\sqsubseteq$ such that, for any $c, d \in D$, $c \sqcap d = c \iff c \sqsubseteq d$.

We call a pattern $d \in D$, as defined above, a *pattern vector*. Writing d_i as the ith component of pattern vector $d \in D$, we have for any $c, d \in D$:

$$c \sqcap d = (c_1 \sqcap_{y_1} d_1, ..., c_p \sqcap_{y_p} d_p) \text{ and } c \sqsubseteq d \Leftrightarrow c_i \sqsubseteq_{y_i} d_i \ \forall i = 1...p$$

Table 2. A data table

	y_1	y_2	y_3
g_1	[75,80]	[1,2]	{a,b}
g_2	[60,80]	[1,1]	{d,e}
g_3	[50,70]	[2,2]	{a,c}
g_4	[72,73]	[1,2]	{a}

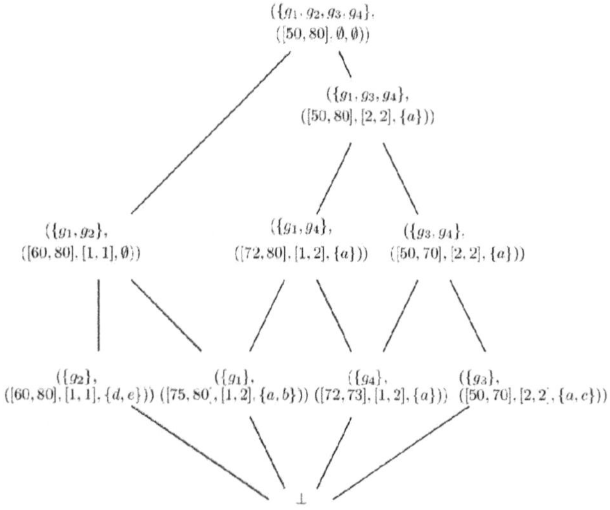

Fig. 1. Pattern concept lattice raised from Table 2

Each dimension i of a pattern vector corresponds to a variable y_i. Each variable y_i may have a different type for which a semi-lattice operation $\sqcap_{y_i}$ has to be defined according to the current data analysis and goals. Consider variable y_1 in Table 2. As in SDA, we can define $\sqcap_{y_1}$ as either interval convexification or intersection. In case of interval convexification, with $a_1, b_1, a_2, b_2 \in \mathbb{R}$, we have:

$$[a_1, b_1] \sqcap_y [a_2, b_2] = [min(a_1, a_2), max(b_1, b_2)]$$
$$[a_1, b_1] \sqsubseteq_y [a_2, b_2] \Leftrightarrow [a_1, b_1] \supseteq [a_2, b_2]$$

When $\sqcap_{y_i}$ corresponds to interval intersection, we have:

$$[a_1, b_1] \sqcap_y [a_2, b_2] = [max(a_1, a_2), min(b_1, b_2)]$$
$$[a_1, b_1] \sqsubseteq_y [a_2, b_2] \Leftrightarrow [a_1, b_1] \subseteq [a_2, b_2]$$

Now that object descriptions are partially ordered, the general Galois connection defined for pattern structures allows us to directly compute pattern concepts and lattices from heterogeneous data. Consider the example in Table 2. It can be represented by a pattern structure $(G, (D, \sqcap), \delta)$ where $G = \{g_1, g_2, g_3, g_4\}$ and $\delta(g_1) = ([75, 80], [1, 2], \{a, b\})$. Descriptions contain two inter-valued variables - y_1, y_2 and one categorical variable - y_3. We choose the following semi-lattice operation for each variable – interval convexification for $\sqcap_{y_1}$, interval intersection for $\sqcap_{y_2}$ and set intersection for $\sqcap_{y_3}$. The general Galois connection is illustrated as follows:

$$\begin{aligned}
\{g_1, g_3\}^{\square} &= \delta(g_1) \sqcap \delta(g_3) \\
&= ([75, 80], [1, 2], \{a, b\}) \sqcap ([50, 70], [2, 2], \{a, c\}) \\
&= ([75, 80] \sqcap_{y_1} [50, 70], [1, 2] \sqcap_{y_2} [2, 2], \{a, b\} \sqcap_{y_3} \{a, c\}) \\
&= ([50, 80], [2, 2], \{a\})
\end{aligned}$$

$$\begin{aligned}
\{g_1, g_3\}^{\square\square} &= ([50, 80], [2, 2], \{a\})^{\square} \\
&= \{g \in G \mid ([50, 80], [2, 2], \{a\}) \sqsubseteq \delta(g)\} \\
&= \{g_1, g_3, g_4\}
\end{aligned}$$

The first operator $(.)^{\square}$ gives the infimum of descriptions $\delta(g_1)$ and $\delta(g_3)$ in $(D, \sqcap)$. The second operator $(.)^{\square}$ gives the set of objects "sharing" a given description. Hence, $(\{g_1, g_3, g_4\}, ([50, 80], [2, 2], \{a\})$ is a pattern concept of $(G, (D, \sqcap), \delta)$. The set of all pattern concepts gives rise to a pattern concept lattice given in Figure 1. Each node gives a pattern concept, while each line denotes ordering relation $\leq$ on concepts. Accordingly, the higher a concept is, the more objects it has in its extent. Due to the choice of semi-lattice operation for each variable, the higher a concept is, the larger are the intervals for variable y_1 (convexification), the smaller are the intervals and attribute sets for variable y_2 and y_3 respectively (intersection). The links with SDA formalism are then natural. We provide an example of pattern concept and its equivalent concept in SDA terms:

$$\begin{aligned}
&(\{g_1, g_4\}, ([72, 80], [1, 2], \{a\})) \\
&\equiv (\{g_1, g_4\}, ([y_1 \subseteq [72, 80]] \wedge [y_2 \supseteq [1, 2]] \wedge [y_3 \supseteq \{a\}]))
\end{aligned}$$

While SDA first requires to represent complex objects as symbolic/assertion objects (conjunction of events), pattern structures consider object descriptions in their original form. SDA proposes two kinds of Galois connection depending on a generalization operation (intersection or union). In pattern structures we have to define a semi-lattice operation inducing a partial ordering and can use the general Galois connection (and associated FCA algorithms) to build concept lattices from heterogeneous data. Generalization with union or intersection are particular cases of the $\sqcap$ operation in $(D, \sqcap)$.

4.2 Handling Missing Values

It often happens that objects are partially described and, therefore, have missing values. In SDA, two kinds of missing values are proposed: (i) the object has a value, but we do not know it, and (ii) the object has no value for this variable and should be considered as such. Given a variable y_i and its description space O_{y_i}, SDA describes such missing values as (i) its definition domain with $[y \subseteq O_{y_i}]$, and (ii) the empty set with $[y_i \supseteq \emptyset]$.

These two kinds of missing values can also be considered with pattern structures. These values should belong to the poset $(D, \sqcap)$ and be ordered w.r.t. $\sqsubseteq$. Now, we denote the missing values (i) as $*$ and (ii) as ?, and $*, ? \in D$. We simply have $d \sqsubseteq *$ and $? \sqsubseteq d$ for any $d \in D$. Accordingly, the description space is no more a semi-lattice $(D, \sqcap)$ but rather a lattice $(D, \sqcap, \sqcup)$, such that $c \sqcup d = d \iff c \sqsubseteq d$. Actually, both missing values correspond to the smallest and the largest element in this lattice, respectively. For an interval-valued variable y_i provided with convexification as infimum $\sqcap_{y_i}$, the element $*$ corresponds to the largest interval defined on the variable domain, while the element ? corresponds to the empty interval.

4.3 Links with Conceptual Scaling

With an appropriate discretization procedure, a pattern structure can be reduced to a formal context such that concepts in both data representations are in a *one*-to-*one* correspondence [5]. Processing the former may be more efficient and obvious than processing the latter, but this should be investigated for each kind of complex data and semi-lattice operation [5,8].

For example, consider the pattern structure $(G, (D, \sqcap), \delta)$ where $G = \{g_1, g_2\}$, $D = \{[1,1], [2,2], [1,2]\}$, $\delta(g_1) = [1,1]$ and $\delta(g_2) = [2,2]$. Here we have a single attribute. Let $\sqcap$ be defined as interval convexification. One has $\{g_1, g_2\}^\square = [1,2]$. Now, consider the formal context (G, M, I) with same set of objects and the set of attributes $M = \{y \leq w, y \geq w, \forall w \in \{1,2\}\}$, i.e. $M = \{y \leq 1, y \leq 2, y \geq 1, y \geq 2\}$. $\{1,2\}$ is the set of all interval end-points of object descriptions. A binary attribute $m \in M$ is a constraint on $\mathbb{R}$, e.g. "$y \leq 1$" $\in M$. Then, $(g, m) \in I$ holds if the value describing object g in the above pattern structure respects the constraint m. Therefore, $g_1' = \{y \leq 1, y \geq 1, y \leq 2\}$, while $g_2' = \{y \geq 1, y \leq 2, y \geq 2\}$. We get $\{g_1, g_2\}' = g_1' \cap g_2' = \{y \geq 1, y \leq 2\}$ whose interpretation in $\mathbb{R}$ exactly corresponds to $\{g_1, g_2\}^\square = [1,2]$.

The operation that builds (G, M, I) from $(G, (D, \sqcap), \delta)$ in this example is called *interordinal scaling* [4]. In this case, [8] showed that pattern concepts and formal concepts in both data representations are in *one*-to-*one* correspondence and equivalent w.r.t. to their interpretation in $\mathbb{R}$. Therefore, one can consider either pattern structures or so-called representation contexts, since both data representations highlight different computational properties which depend on original data (size, distribution, etc). SDA does not suggest to investigate representation contexts.

5 Conclusion

Pattern structures allow to directly consider complex data, avoiding to represent descriptions as symbolic/assertion objects. One general Galois connection is sufficient to consider several data-types, hence it is not required to define a new Galois connection for different data-types and description generalization operations (with union and intersection in SDA). Indeed, the main core of pattern structures lies in defining an appropriate semi-lattice operation inducing a partial order of descriptions. This is rather simple with numerical and categorical data as illustrated in this paper, but much more difficult with graph data, as discussed in [5].

Avoiding discretization and loss of information, generally leads to a great amount of concepts, as noticed both in SDA and pattern structures [3,6,8,7]. In SDA, the authors of [2,3] showed how to reduce concept lattices to simpler hierarchies. These reduction techniques are based on quality criteria defined in SDA, but require that the concept lattice is already computed, a bottleneck for very large databases. On the other hand, pattern structures propose to project object descriptions with "simpler ones" before the computation, allowing to reduce the number of concepts. This gives interesting perspectives of research to consider well studied SDA quality criteria within pattern structures.

References

1. Bock, H.H., Diday, E. (eds.): Analysis of Symbolic Data – Exploratory Methods for Extracting Statistical Information from Complex Data. Springer, Heidelberg (2000)
2. Brito, P.: Hierarchical and Pyramidal Clustering with Complete Symbolic Objects. In: Bock, H.H., Diday, E. (eds.) [1], pp. 312–324
3. Brito, P., Polaillon, G.: Structuring probabilistic data by Galois lattices. Mathématiques et sciences humaines 169 (2005)
4. Ganter, B., Wille, R.: Formal Concept Analysis. Springer, Heidelberg (1999)
5. Ganter, B., Kuznetsov, S.O.: Pattern structures and their projections. In: Delugach, H.S., Stumme, G. (eds.) ICCS 2001. LNCS (LNAI), vol. 2120, pp. 129–142. Springer, Heidelberg (2001)
6. Kaytoue, M., Assaghir, Z., Messai, N., Napoli, A.: Two complementary classification methods for designing a concept lattice from interval data. In: Link, S., Prade, H. (eds.) FoIKS 2010. LNCS, vol. 5956, pp. 345–362. Springer, Heidelberg (2010)
7. Kaytoue, M., Assaghir, Z., Napoli, A., Kuznetsov, S.O.: Embedding tolerance relations in formal concept analysis: an application in information fusion. In: Huang, J., Koudas, N., Jones, G., Wu, X., Collins-Thompson, K., An, A. (eds.) CIKM, pp. 1689–1692. ACM, New York (2010)
8. Kaytoue, M., Kuznetsov, S.O., Napoli, A., Duplessis, S.: Mining gene expression data with pattern structures in formal concept analysis. Information Science (2010) (in press)
9. Kuznetsov, S.O.: Pattern structures for analyzing complex data. In: Sakai, H., Chakraborty, M.K., Hassanien, A.E., Ślęzak, D., Zhu, W. (eds.) RSFDGrC 2009. LNCS, vol. 5908, pp. 33–44. Springer, Heidelberg (2009)
10. Polaillon, G.: Pyramidal Classification for Interval Data using Galois Lattice Redutcion. In: Bock, H.H., Diday, E. (eds.) [1], pp. 324–341

Multiargument Relationships in Fuzzy Databases with Attributes Represented by Interval-Valued Possibility Distributions

Krzysztof Myszkorowski

Institute of Information Technology,
Technical University of Łódź, Wólczańska 215, 90-924 Łódź, Poland
kamysz@ics.p.lodz.pl

Abstract. The paper contains an analysis of integrity constraints for multiargument relationships in fuzzy databases. It is assumed that attribute values are represented by means of interval-valued possibility distributions. The analysis is carried out using the theory of functional dependencies. The notion of functional dependency has been appropriately extended according to the representation of fuzzy data. The paper formulates the rules to which (n-1)-ary relationships embedded in the n-ary relationship must be subordinated.

Keywords: Fuzzy databases, interval-valued fuzzy sets, fuzzy functional dependencies, fuzzy normal forms, possibility distribution, n-ary relationships.

1 Introduction

Fuzzy database models have been created for managing and retrieving imperfect information [1,2]. There are two major approaches concerning fuzzy data representation, namely, the similarity-based approach [3] and the possibility-based approach [4]. In the present paper it is assumed that attribute values are represented by means of interval-valued possibility distributions. The idea of this concept addresses the problem of applying imprecise possibility measures. The interval-valued possibility distribution is defined using the notion of an interval-valued fuzzy set [5,6]. Functional dependencies which exist between attributes reflect integrity constraints and should be studied during the design process. The notion of functional dependency has to be extended according to the representation of fuzzy data [7,8,9,10]. In further considerations it will be applied the definition using an interval-valued fuzzy implicator [11,12].

In database models, usually binary relationships occur between entity sets. When designing, it may be necessary to define n-ary relationships. Furthermore, within such connections there may exist relationships comprising fewer than n sets. However, there is no complete arbitrariness. The relationships "embedded" in n-ary relationships are subjected to certain restrictions. This issue for ternary relationships was presented in [13,14]. For various types of ternary relationships

S.O. Kuznetsov et al. (Eds.): RSFDGrC 2011, LNAI 6743, pp. 199–206, 2011.

the authors formulated the rules to which the binary relationships between pairs of sets are subjected. This analysis may be also carried out using the theory of functional dependencies (FDs). A multiargument relationship may be formally presented using the relational notation: $R(X_1, X_2, \dots, X_n)$, where R is the name of the relationship and attributes X_i denote keys of entity sets which participate in it. The functional dependencies

$$U - \{X_i\} \rightarrow X_i, \quad i = 1, 2, \dots, n \tag{1}$$

describe the integrity constraints for R and must not be infringed. They constitute a restriction for (n-1)-ary relationships. The FDs describing the relationships between (n-1) attributes "embedded" in the n-ary relationship may be presented as follows:

$$U - \{X_i, X_j\} \rightarrow X_i, \quad i \neq j, \quad i = 1, 2, \dots, n \quad . \tag{2}$$

The imposition of the FD (2) is possible if X_i does not belong to any candidate key of R.

The paper extends the idea presented in [15]. The aim is to analyze multiargument relationships in fuzzy databases with attributes represented by interval-valued possibility distributions. The paper is organized as follows. Section 2 presents the basic notions related to interval-valued possibility distributions. Section 3 contains the definition of a fuzzy functional dependency, extended Armstrong's rules and extended normal forms. These notions are used in section 4 in analysis of fuzzy multiargument relationships.

2 Interval-Valued possibility distributions

The idea of an interval-valued fuzzy set (IVF) extends a traditional notion of the Zadeh fuzzy set [16,17]. Membership degrees assigned to elements in a given IVF are expressed by means of closed subintervals of [0,1].

Definition 1. *Let U be a universe of discourse. An interval-valued fuzzy set F in U is a set of ordered pairs:*

$$F = \{< x, \mu_F(x) >: x \in U, \quad \mu_F(x) : U \rightarrow Int([0, 1])\}, \tag{3}$$

where $\mu_F(x) = [\mu_{F_L}(x),\ \mu_{F_U}(x)]$, *Int([0,1]) stands for the set of all closed subintervals of [0,1].*

The interval $\mu_F(x)$ approximates the correct value of the membership degree. Values of $\mu_{F_L}(x)$ and $\mu_{F_U}(x)$ are interpreted as the lower and upper membership functions, respectively. Interval-valued fuzzy sets F and G are equal iff $\forall x\ \mu_F(x) = \mu_G(x)$. In order to define a closeness measure between IVFs one has to establish an order relation for intervals [18]. In further considerations it will be assumed the following partial order [11]:

$$[a_L, a_U] \leq [b_L, b_U] \Leftrightarrow a_L \leq b_L \ \ and \ \ a_U \leq b_U. \tag{4}$$

An IVF A is included in an IVF B, $A \subseteq B$, if and only if $\forall x\ \mu_{A_L}(x) \leq \mu_{B_L}(x)$ and $\mu_{A_U}(x) \leq \mu_{B_U}(x)$ [19]. The degree of inclusion $\subseteq (A, B)$ can be obtained by means of an interval-valued fuzzy implicator I: Int([0,1]) × Int([0,1]) → Int([0,1]):

$$\subseteq (A, B)_L = \inf_x I(\mu_A(x), \mu_B(x))_L, \quad \subseteq (A, B)_U = \inf_x I(\mu_A(x), \mu_B(x))_U. \quad (5)$$

A closeness measure $=_c (A, B)$ between IVFs A and B is a mapping $IF(U) \times IF(U) \rightarrow$ Int([0,1]), where $IF(U)$ denotes a set of all IVFs in U. The bounds of $=_c (A, B)$ can be expressed by the following formulas:

$$\begin{aligned} =_c (A, B)_L &= \min(\inf_x I(\mu_A(x), \mu_B(x))_L,\ \inf_x I(\mu_A(x), \mu_B(x))_L), \\ =_c (A, B)_U &= \min(\inf_x I(\mu_A(x), \mu_B(x))_U,\ \inf_x I(\mu_A(x), \mu_B(x))_U). \end{aligned} \quad (6)$$

In further considerations we will apply an extension of the Gödel-Brouwer implicator [11]:

$$I([a, b], [c, d]) = \begin{cases} [c, d] & \text{if } a > c \text{ and } b > d \\ [c, 1] & \text{if } a > c \text{ and } b \leq d \\ [1, 1] & \text{if } a \leq c \text{ and } b \leq d \\ [d, d] & \text{if } a \leq c \text{ and } b > d \end{cases} \quad (7)$$

Based on the concept of IVFs one can define the concept of interval-valued possibility distributions.

Definition 2. *Let U be a universe of discourse, X be a variable on U and F be an interval-valued fuzzy set with $\mu_F(x) = [\mu_{F_L}(x),\ \mu_{F_U}(x)]$. The interval-valued possibility distribution of X with respect to F is defined as*

$$\Pi_X = \{\pi_X(x)/x : x \in U, \quad \pi_X(x) = \mu_F(x)\}, \quad (8)$$

where $\pi_X(x) = [\pi_{X_L}(x),\ \pi_{X_U}(x)]$ is a closed subinterval of [0,1].

A closeness measure $=_c (\Pi_X,\ \Pi_Y)$ between two interval-valued possibility distributions Π_X and Π_Y is defined as the possibility that $\Pi_X = \Pi_X$ [10]:

$$\begin{aligned} =_c (\Pi_X, \Pi_Y)_L &= \sup_x \min(\pi_{X_L}(x), \pi_{Y_L}(x)), \\ =_c (\Pi_X, \Pi_Y)_U &= \sup_x \min(\pi_{X_U}(x), \pi_{Y_U}(x)). \end{aligned} \quad (9)$$

This measure can be extended for tuple closeness within the possibilistic database framework. A tuple t of relation $R(X_1, X_2, \ldots, X_n)$ is of the form: $t = (\Pi_{X_1}, \Pi_{X_2}, \ldots, \Pi_{X_n})$. Let t and t' be two tuples of relation R. The degree that $t = t'$ is expressed by the following formulas:

$$=_c (t, t')_L = \min_i(=_c (\Pi_{X_i}, \Pi'_{X_i})_L), \quad =_c (t, t')_U = \min_i(=_c (\Pi_{X_i}, \Pi'_{X_i})_U). \quad (10)$$

3 Functional Dependencies in Fuzzy Databases

The existence of a classical FD $X \rightarrow Y$ between attributes X and Y means that the values of X uniquely determine the values of Y. In fuzzy databases this interpretation is not sufficient because of imprecise values of attributes. Therefore the classical definition has to be extended. In [10] Chen has introduced the definition of the fuzzy functional dependency (FFD) for the possibilistic fuzzy data model with the use of a fuzzy implicator. If attribute values are allowed to be interval-valued possibility distributions this definition has to be modified. We will name this type of FFD an interval fuzzy functional dependency (IFFD).

Definition 3. *Let $R(U)$ be a relation scheme and let X and Y be subsets of U: $X, Y \subseteq U$, where $U = \{X_1, X_2, \dots, X_n\}$. An interval fuzzy functional dependency $X \rightarrow_\theta Y$ is said to exist in $\theta = [\theta_L, \theta_U]$ degree if and only if for every relation r of R the following conditions are met:*

$$\begin{aligned} &\min_{t_1,t_2 \in r} I(t_1(X) =_c t_2(X), t_1(Y) =_c t_2(Y))_L \geq \theta_L, \\ &\min_{t_1,t_2 \in r} I(t_1(X) =_c t_2(X), t_1(Y) =_c t_2(Y))_U \geq \theta_U, \end{aligned} \tag{11}$$

where θ_L, θ_U, $\in$ [0,1], $=_c$ is a closeness measure (9) *and I is an interval-valued fuzzy implicator* (7).

A dependency $X \rightarrow_\theta Y$ is partial if there exists a set $X' \subset X$ such that $X' \rightarrow_\theta Y$. In the opposite case $X \rightarrow_\theta Y$ fully. The set of attributes $K \subseteq U$ such that $K \rightarrow_\theta U$ fully is a θ-key of $R(U)$. Its elements are called θ-prime-attributes.

Example 1. Consider the relationship $PES(P, E, S)$ between the post held (P), education (E) and salary (S) with the following IFFDs: $PE \rightarrow_{[0.7,\ 0.8]} S$ and $SE \rightarrow_{[0.5,\ 0.6]} P$. The scheme PES has two candidate keys: PE - [0.7, 0.8]-key and SE - [0.5, 0.6]-key.

Like in classical relational databases there are the following inference rules known as extended Armstrong's axioms:

A1: $Y \subseteq X \Rightarrow X \rightarrow_\theta Y$ for all θ
A2: $X \rightarrow_\theta Y \Rightarrow XZ \rightarrow_\theta YZ$
A3: $X \rightarrow_\alpha Y \wedge Y \rightarrow_\beta Z \Rightarrow X \rightarrow_\gamma Y, \quad \gamma = [\min(\alpha_L, \beta_L), \min(\alpha_U, \beta_U)]$

From A1, A2 and A3 the following inference rules can be derived:

D1: $X \rightarrow_\alpha Y \wedge X \rightarrow_\beta Z \Rightarrow X \rightarrow_\lambda YZ, \quad \gamma = [\min(\alpha_L, \beta_L), \min(\alpha_U, \beta_U)]$
D2: $X \rightarrow_\alpha Y \wedge WY \rightarrow_\beta Z \Rightarrow XW \rightarrow_\lambda Z, \quad \gamma = [\min(\alpha_L, \beta_L), \min(\alpha_U, \beta_U)]$
D3: $X \rightarrow_\alpha Y \wedge Z \subseteq Y \Rightarrow X \rightarrow_\alpha Z$
D4: $X \rightarrow_\alpha Y \wedge X \rightarrow_\beta Y$ for $\alpha \leq \beta$

Based on the notion of IFFD the definitions of classical normal forms can be appropriately extended. The first fuzzy normal form (F1NF) specifies the structure of relations. It is required that any attribute value of relations in F1NF is represented by an excluding possibility distribution which means that all the elements of the attribute domain are mutually exclusive. The further normal forms are based on IFFDs which exist between the key and nonkey attributes.

Definition 4. *Scheme R $(X_1, X_2, \ldots, X_n)$ is in θ-fuzzy second normal form (θ-F2NF) if every θ-nonprime-attribute is fully functionally dependent on a θ-key in α degree, where $\alpha = [\alpha_L, \alpha_U]$.*

Example 2. Scheme PES from the previous example is in [0.5, 0.6]-F3NF. Let us augment it by attribute A determining age and assume that its values are connected with values of attribute P. Let us assume that the relationship between the post and age is expressed by $P \rightarrow_{\phi} A$, $\phi = [\phi_L, \phi_U]$. In result of such modification PE is a θ-key with $\theta_L = \min(\phi_L, 0.5)$ and $\theta_U = \min(\phi_U, 0.6)$. However because of the introduced dependency the modified scheme is not in [0.5, 0.6]-F3NF. There is a non-prime-attribute dependent on a part of the θ-key.

Definition 4 excludes the possibility of the occurrence of θ-key subset, which would determine a θ-nonprime-attribute.

Definition 5. *Scheme R (U), $U = \{X_1, X_2, \ldots, X_n\}$, is in θ-fuzzy third normal form (θ-F3NF) if for every IFFD $X \rightarrow_{\phi} Y$, where $X, Y \subseteq U$, $Y \not\subseteq X$, X contains a θ-key of R or Y is a θ-prime-attribute.*

The definition of θ-F3NF eliminates the possible occurrence of transitive dependencies between the attributes.

Example 3. Let us consider the relation EES with attributes Em - employee, E - education and S - salary. Let us assume that between its attributes there are IFFDs: $Em \rightarrow_{[0.6,\ 0.8]} E$ and $E \rightarrow_{[0.7,\ 0.9]} S$. Basing on transitivity axiom they yield the dependency $Em \rightarrow_{[0.6,\ 0.8]} S$. The key of EES is Em. It is the [0.6, 0.8]-key. Attributes E and S disturb the conditions of definition 5, because E is not a θ-key and S is not a θ-prime-attribute. The θ-fuzzy third normal form is obtained in result of the decomposition into relations with schemes: $EE(Em, E)$ - [0.6, 0.8]-F3NF and $ES(E, S)$ - [0.7, 0.9]-F3NF. This decomposition maintains the dependencies.

Eliminating from definition 5 the possibility that attribute Y in $X \rightarrow_{\theta} Y$ is θ-prime leads to a stronger definition. This is a definition of θ-fuzzy Boyce-Codd normal form.

Definition 6. *Scheme R (U), $U = \{X_1, X_2, \ldots, X_n\}$, is in θ-fuzzy Boyce-Codd normal form (θ-FBCNF) if for every IFFD $X \rightarrow_{\phi} Y$, where $X, Y \subseteq U$, $Y \not\subseteq X$, X contains a θ-key of R.*

4 Fuzzy Multiargument Relationships

Let us consider the ternary relationship $R(X,Y,Z)$ with the following IFFDs:

$$XY \rightarrow_{\alpha} Z, \quad XZ \rightarrow_{\beta} Y, \quad YZ \rightarrow_{\gamma} X, \tag{12}$$

where $\alpha = [\alpha_L, \alpha_U]$, $\beta = [\beta_L, \beta_U]$, $\gamma = [\gamma_L, \gamma_U]$. The scheme R has three candidate θ-keys. Let us define the following interval-valued fuzzy sets of attributes:

$$\mathcal{L} = \{\gamma^c/X,\ \beta^c/Y,\ \alpha^c/Z\}, \quad \mathcal{B} = \{\alpha/X,\ \beta/Y,\ \gamma/Z\}. \tag{13}$$

where $\alpha^c = [1 - \alpha_U,\ 1 - \alpha_L]$, $\beta^c = [1 - \beta_U,\ 1 - \beta_L]$, $\gamma^c = [1 - \gamma_U,\ 1 - \gamma_L]$. Membership grades of attributes depend on levels of suitable IFFDs. If an attribute does not occur on the right side of any dependency (12) its degrees of membership to $\mathcal{L}$ and $\mathcal{B}$ are equal to [1, 1] and [0, 0] respectively.

The possibility to impose fuzzy binary relationships is limited by intervals α, β and γ. Let us consider the imposition of the dependency $Y \rightarrow_\phi X$, where $\alpha = [\phi_L,\ \phi_U]$. Its consequence is the dependency $YZ \rightarrow_\phi X$. If $\phi > \gamma$ the intervals determining the membership of X in sets $\mathcal{L}$ and $\mathcal{B}$ change, which means the disturbance of the integrity conditions. Hence the following theorem can be formulated:

Theorem 1. *In the fuzzy ternary relationship $R(X,Y,Z)$ with IFFDs* (12) *there may exist binary relationships determined by IFFDs of the form $V \rightarrow_\phi W$, where $V, W \in \{X, Y, Z\}$, if $\phi \leq \alpha$ for $W = Z$, $\phi \leq \beta$ for $W = Y$, $\phi \leq \gamma$ for $W = X$.*

Remark 1. Let us notice that the imposition of $Y \rightarrow_\phi X$ introduces $Y \rightarrow_\lambda Z$, where $\lambda = [\lambda_L,\ \lambda_U]$. For we have: $Y \rightarrow_\phi X \wedge XY \rightarrow_\alpha Z \Rightarrow Y \rightarrow_\lambda Z$, where $\lambda_L = \min(\phi_L,\ \alpha_L)$, $\lambda_U = \min(\phi_U,\ \alpha_U)$.

Example 4. The sets $\mathcal{L}$ and $\mathcal{B}$ for relation PES from example 1 are as follows: $\mathcal{L} = \{[0.4,\ 0.5]/P,\ [1,\ 1]/E,\ [0.2,\ 0.3]/S\}$ and $\mathcal{B} = \{[0.5,\ 0.6]/P,\ [0.7,\ 0.8]/S\}$. Let us consider the imposition of the dependency $E \rightarrow_{[0.3,\ 0.4]} P$. This imposition creates another dependency: $E \rightarrow_{[0.3,\ 0.4]} S$ (rule D2). Thus a new θ-key (attribute E) with $\theta = [0.3,\ 0.4]$ has been created. Scheme PES is in [0.3, 0.4]-FBCNF. Membership grades in sets $\mathcal{L}$ and $\mathcal{B}$ remain the same. The imposition is admissible. If the level ϕ of the imposed dependency were greater then [0.5, 0.6] the integrity constraints would be disturbed.

The obtained result can generalized for fuzzy n-ary relationships. Let there exist in the relationship $R(X_1,\ X_2,\ \dots,\ X_n)$ the following IFFDs:

$$U - \{X_i\} \rightarrow_{\alpha_i} X_i, \quad \text{where } \alpha_i = [\alpha_{i_L},\ \alpha_{i_U}],\ i = 1, 2,\ \dots, m, \quad m \leq n. \tag{14}$$

The scheme $R(X_1,\ X_2,\ \dots,\ X_n)$ has m θ-keys in the form $U - \{X_i\}$. From dependencies (14) the following sets $\mathcal{L}$ and $\mathcal{B}$ result:

$$\mathcal{L} = \{\alpha_1^c/X_1,\ \dots, \alpha_m^c/X_m, 1/X_{m+1},\ \dots, 1/X_n\}, \tag{15}$$

$$\mathcal{B} = \{\alpha_1/X_1,\ \dots, \alpha_m/X_m\}. \tag{16}$$

where $\alpha_i^c = [1 - \alpha_{i_U},\ 1 - \alpha_{i_L}]$. Attributes $X_{m+1}, \dots, X_n$ fully belong to $\mathcal{L}$. Their number equals $n - m$. The imposition of (n-1)-ary relationships cannot disturb integrity constraints determined by dependencies (14). Therefore the levels γ_i of the imposed IFFDs:

$$U - \{X_i, X_j\} \rightarrow_{\gamma_i} X_i, \quad \text{where } i \neq j, \quad i, j = 1, 2, \dots, n\ ,\ \gamma_i = [\gamma_{i_L},\ \gamma_{i_U}] \tag{17}$$

must satisfy the following condition: $\gamma_i \leq \alpha_i$, where α_i denotes the level of the relevant dependency (14). Due to the Armstrong's rules: $U - \{X_i, X_j\} \rightarrow_{\gamma_i} X_i \Rightarrow$

$U-\{X_i\} \rightarrow_{\gamma_i} X_i$. If $\gamma_i \leq \alpha_i$, then $U-\{X_i\} \rightarrow_{\alpha_i} X_i \Rightarrow U-\{X_i\} \rightarrow_{\gamma_i} X_i$. Thus the obtained dependency is not contradictory to (14). Otherwise the imposition is not allowed. Its consequence is a change of the membership intervals for X_i in sets $\mathcal{L}$ and $\mathcal{B}$.

Theorem 2. *In the fuzzy n-ary relationship* $R\ (X_1,\ X_2,\ \ldots,\ X_n)$ *with IFFDs* (14)*, there may exist* (*n-1*)*-ary relationships determined by IFFDs* (17)*, in which* $\gamma_i \leq \alpha_i$.

Scheme $R(X_1,\ X_2,\ \ldots,\ X_n)$ with IFFDs (14) occurs in θ-FBCNF, where $\theta = [\min_i(\alpha_{i_L}),\ \min_i(\alpha_{i_U})]$. After having introduced IFFDs (17) the conditions of definition 6 may be disturbed. Let us examine the consequences of imposing the dependency $U-\{X_i, X_j\} \rightarrow_{\gamma_i} X_i$, where $i \neq j$, $i = 1, 2, \ldots, m$, $j = 1, 2, \ldots, n$ and $\gamma_i \leq \alpha_i$. Let us assume that $m > 1$. If $m < n$ there are attributes fully belonging to $\mathcal{L}$. They cannot occur on the right side of any IFFD (14). If $j \leq m$ from $U-\{X_j\} \rightarrow_{\alpha_j} X_j$ the attribute X_i can be eliminated. Basing on rule D2 we obtain: $U-\{X_i, X_j\} \rightarrow_{\gamma_i} X_i \wedge U-\{X_j\} \rightarrow_{\alpha_j} X_j \Rightarrow U-\{X_i, X_j\} \rightarrow_{\lambda_{i,j}} X_j$, where $\lambda_{i,j} = [\min(\gamma_{i_L},\ \alpha_{j_L}),\ \min(\gamma_{i_U},\ \alpha_{j_U})]$. A new key arises. This is formed by attributes occurring on the left side of the introduced dependency. The conditions of the definition of θ-FBCNF have not been disturbed. If $j > m$, i.e. X_j fully belongs to $\mathcal{L}$, no new key will be formed. Scheme $R(X_1,\ X_2,\ \ldots,\ X_n)$ will not occur in θ-FBCNF. The left side of the dependency $U-\{X_i, X_j\} \rightarrow_{\gamma_i} X_i$ does not contain the key. However, it will remain in θ-F3NF, because X_i is θ-prime. If $(m = 1)$ there exists only one admissible dependency (7). Its imposition introduces a partial dependency of attribute X_i on the θ-key which means a disturbance in the conditions defining the θ-fuzzy second normal form.

5 Conclusions

The paper analysis fuzzy multiargument relationships with attributes represented by means of interval-valued possibility distributions. The starting point are interval fuzzy functional dependencies (14) existing between all attributes of the relationship $R(X_1,\ X_2,\ \ldots,\ X_n)$. Their levels are determined by means of subintervals of [0,1]. They determine the integrity constraints for R. The subject of considerations was the possible occurrence - within the n-ary relationship - of interval fuzzy functional dependencies (17) describing the relationships between (n-1) attributes. They cannot disturb the integrity constraints. Such dependencies are admissible if their levels do not exceed in the sense of order (4) the levels of relevant dependencies (14). The imposition of them may disturb the normal form of scheme R.

References

1. Galindo, J., Urrutia, A., Piattini, M.: Fuzzy Databases. In: Modeling, Design and Implementation. Idea Group Publishing, London (2005)
2. Petry, F.: Fuzzy Databases: Principles and Applications. Kluwer Academic Publishers, Boston (1996)

3. Buckles, B.P., Petry, F.E.: A Fuzzy Representation of Data for Relational Database. Fuzzy Sets and Systems 7, 213–226 (1982)
4. Prade, H., Testemale, C.: Generalizing Database Relational Algebra for the Treatment of Incomplete or Uncertain Information and Vague Queries. Information Science 34, 115–143 (1984)
5. Sambuc, R.: Founctions Φ-floues. Application á l'aide au diagnostic en pathologie thyroidienne. PhD thesis, University de Marseillé, France (1975)
6. Karnik, N.N., Mendel, J.M.: An Introduction to Type-2 Fuzzy Logic Systems. University of Southern California, Los Angeles (1998)
7. Cubero, J.C., Vila, M.A.: A new definitions of fuzzy functional dependency in fuzzy relational databases. International Journal for Intelligent Systems 9, 441–448 (1994)
8. Sozat, M.I., Yazici, A.: A Complete Approximation for Fuzzy Functional and Multivalued Dependencies in Fuzzy Database Relations. Fuzzy Sets and Systems 117, 161–181 (2001)
9. Tyagi, B.K., Sharfuddin, A., Dutta, R.N., Devendra, K.T.: A complete axiomatization of fuzzy functional dependencies using fuzzy function. Fuzzy Sets and Systems 151, 363–379 (2005)
10. Chen, G.: Fuzzy Logic in Data Modeling - semantics, constraints and database design. Kluwer Academic Publishers, Boston (1998)
11. Alcade, C., Burusco, A., Fuentes-Gonzales, R.: A constructive method for the definition of interval-valued fuzzy implication operators. Fuzzy Sets and Systems 153, 211–227 (2005)
12. Deschrijver, G., Kerre, E.E.: Implicators based on binary aggregation operators in interval valued fuzzy set theory. Fuzzy Sets and Systems 153, 229–248 (2005)
13. Dullea, J., Song, I., Lamprou, I.: An Analysis of Structural Validity in Entity-Relationship Modeling. Data and Knowledge Engineering 47, 167–205 (2001)
14. Jones, T., Song, I.: Analysis of Binary/Ternary Relationships Cardinality Combinations in Entity-Relationship Modeling. Data and Knowledge Engineering 19, 39–64 (1996)
15. Myszkorowski, K.: Fuzzy Functional Dependencies in Multiarguments Relationships. In: Rutkowski, L., Scherer, R., Tadeusiewicz, R., Zadeh, L.A., Zurada, J.M. (eds.) ICAISC 2010. LNCS, vol. 6113, pp. 152–159. Springer, Heidelberg (2010)
16. Zadeh, L.A.: Fuzzy sets. Information and Control 8, 338–353 (1965)
17. Zadeh, L.A.: Fuzzy sets as a basis for a theory of possibility. Fuzzy Sets and Systems 1, 3–28 (1978)
18. Sengupta, A., Pal, T.K., Chakraborty, D.: Interpretation of inequality constraints involving interval coefficients and a solution to interval linear programming. Fuzzy Sets and Systems 119, 129–138 (2001)
19. Mondal, T.K., Samanta, S.K.: Topology of interval-valued intuitionistic fuzzy sets. Fuzzy Sets and Systems 119, 483–494 (2001)

Disjunctive Set-Valued Ordered Information Systems Based on Variable Precision Dominance Relation

Guoyin Wang, Qingshan Yang, and Qinghua Zhang

Institute of Computer Science & Technology, Chongqing University of Posts and Telecommunications, Chongqing, 400065, P.R. China
wanggy@cqupt.edu.cn

Abstract. Through analyzing limitations of four existing dominance relations in disjunctive set-valued information systems, a variable precision dominance relation is proposed, and an extended rough set model based on the variable precision dominance relation is defined. In order to derive much simpler attribute representation, a discernibility matrix is defined, and an attribute reduction method based on the discernibility matrix is developed.

Keywords: disjunctive set-valued information system; rough set; variable precision dominance relation; attribute reduction.

1 Introduction

Classical rough set theory is based on the indiscernibility relation, and mainly studies the complete information system. But in many practical issues, some attribute values in an information system may be unknown or multi-values. Furthermore, attributes are sometimes with preference-ordered domains, and the ordering of properties of attributes plays a crucial role. To solve these problems, Zhang et al. [6] proposed the concept of set-valued information system and used it to process incomplete information. Greco et al. [1, 2] proposed the dominance-based rough set approach to take account of the ordering properties of attribute. Qian et al. [4] introduced four possible dominance relations to the disjunctive set-valued information systems.

The four dominance relations in [4] are based on comparing the minimum or the maximum value in the value domain of one object with that of another object, but do not consider attribute values between the minimum value and the maximum value. Therefore, the four existing dominance relations cannot perfectly characterize the degree that one object dominates another one, and lack of adaptability of noise data. In addition, all the objects in disjunctive set-valued information systems can only get one value in the value domain, so it may happen that the condition of the dominance relation cannot meet, just because the minimum or the maximum value in the value domain is not available. For these reasons, we propose a variable precision dominance relation based on the probability theory, in order to depict certain degree of the dominance relation between two objects, and enhance adaptability of noisy data.

The rest of the paper is organized as follows. In Section 2, a variable precision dominance relation is proposed, and a rough set model based on the variable precision

S.O. Kuznetsov et al. (Eds.): RSFDGrC 2011, LNAI 6743, pp. 207–210, 2011.

dominance relation is established. In Section 3, an attribute reduction method is presented. Finally, the paper is summarized in Section 4.

2 Rough Set Model Based on Variable Precision Dominance Relation

Definition 1. Set-valued Information System(SvIS) can be defined as 4-tuple $S=(U,A,V,f)$, where $U=\{x_1,x_2,\ldots,x_n\}$ is a non-empty finite set of objects, $A=\{a_1,a_2,\ldots,a_m\}$ is a non-empty finite set of attributes, and V is the set of attribute values. f: $U\times A\rightarrow 2^V$ is a set-valued mapping such that $\forall_{x\in U, a\in A} |f(x,a)|\geq 1$, where $|\cdot|$ denotes the cardinality of a set.

Definition 2. Let $S=(U,A,V,f)$ be SvIS, if the domain of condition attributes is ordered according to the decreasing or increasing preference, then S is called Set-valued Ordered Information System(SvOIS).

It is assumed that the domain of attribute $a\in A$ is completely pre-ordered, let $\succeq_a$ denote, then $y\succeq_a x$ means that y is at least as good as x with respect to attribute a, and $y\succeq_a x \Leftrightarrow f(y,a)\geq f(x,a)$ (the domain of attribute a is increasing preference).

In disjunctive SvIS, some objects may have more than one value for an attribute. Furthermore, for any $x\in U$, $a\in A$, the value of each attribute is only one of value domain. In this paper, we assume that the probability that every object gets each value in the value domain is equal. For any x_i, $x_j\in U$, $a\in A$, several kinds of possible relationships between $f(x_i,a)$ and $f(x_j,a)$ are shown in Fig.1.

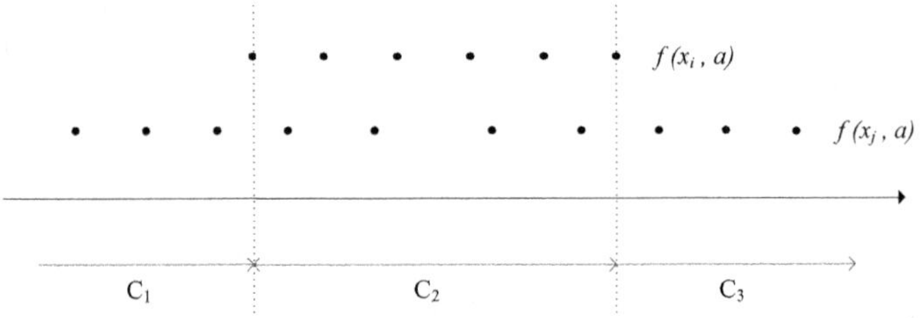

Fig. 1. Possible relationships of arbitrary two attribute set-values

In Fig.1, C_1 denotes the situation of $f(x_j,a)<\min f(x_i,a)$, C_2 stands for the situation of $\min f(x_i,a)\leq f(x_j,a)<\max f(x_i,a)$, and C_3 depicts the situation of $f(x_j,a)\geq \max f(x_i,a)$. In the following, we give the definition of minimum probability that x_j dominates x_i with respect to attribute a, denoted by P_{ji}^a.

$$P_{ji}^a=\frac{|C_2|}{|f(x_i,a)|\cdot|f(x_j,a)|}+\frac{|C_3|}{|f(x_j,a)|}=\frac{|f(x_j,a)|-|C_1|-|C_3|}{|f(x_i,a)|\cdot|f(x_j,a)|}+\frac{|C_3|}{|f(x_j,a)|}. \tag{1}$$

where $|\cdot|$ denotes the cardinality of a set, $P_{ji}^a\in[0,1]$ shows the degree that x_j dominates x_i, and the larger the value of P_{ji}^a, the greater the degree.

Definition 3. Let $S=(U,A,V,f)$ be a disjunctive SvOIS, $B \subseteq A$, for a given variable precision $k \in (0,1]$, the variable precision dominance relation $R_B^{\geq k}$ and the corresponding dominance class $[x_i]_B^{\geq k}$ can be respectively defined as follows:

$$R_B^{\geq k} = \{(x_j, x_i) \in U \times U \mid \forall_{a \in B}\ P_{ji}^a \geq k\},\ [x_i]_B^{\geq k} = \{\ x_j \in U \mid (x_j, x_i) \in R_B^{\geq k}\ \}. \quad (2)$$

Definition 4. Let $S=(U,A,V,f)$ be a disjunctive SvOIS, for any $X \subseteq U$ and $B \subseteq A$, the lower approximation $\underline{R_B^{\geq k}}(X)$, the upper approximation $\overline{R_B^{\geq k}}(X)$ of X are respectively defined as follows:

$$\underline{R_B^{\geq k}}(X) = \{x \in U \mid (x \cup [x]_B^{\geq k}) \subseteq X\},\ \overline{R_B^{\geq k}}(X) = \{x \in U \mid (x \cup [x]_B^{\geq k}) \cap X \neq \varnothing\}. \ (3)$$

From definition 3 and definition 4, we can easily get the following properties:

Property 1. Let $R_B^{\geq k}$ be the variable precision dominance relation, then

(1) if $\forall_{a \in B} P_{ii}^a \geq k$, then $R_B^{\geq k}$ is reflexive, but not symmetric and transitive;

(2) if $\exists_{a \in B} P_{ii}^a < k$, then $R_B^{\geq k}$ is not reflexive, symmetric and transitive;

(3) if $C \subseteq B \subseteq A$, then $R_C^{\geq k} \supseteq R_B^{\geq k} \supseteq R_A^{\geq k}$, $[x_i]_C^{\geq k} \supseteq [x_i]_B^{\geq k} \supseteq [x_i]_A^{\geq k}$;

(4) if $0 \leq k_1 \leq k_2 \leq 1$, then $R_B^{\geq k_1} \supseteq R_B^{\geq k_2}$, $[x_i]_B^{\geq k_1} \supseteq [x_i]_B^{\geq k_2}$;

(5) if $0 \leq k_1 \leq k_2 \leq 1$, then $\underline{R_B^{\geq k_1}}(X) \subseteq \underline{R_B^{\geq k_2}}(X) \subseteq X$, $X \subseteq \overline{R_B^{\geq k_2}}(X) \subseteq \overline{R_B^{\geq k_1}}(X)$;

(6) if $X \subseteq Y \subseteq U$, then $\underline{R_B^{\geq k}}(X) \subseteq \underline{R_B^{\geq k}}(Y)$, $\overline{R_B^{\geq k}}(X) \subseteq \overline{R_B^{\geq k}}(Y)$.

3 Attribute Reduction in Disjunctive SvOIS

In OIS, some attributes can be reduced from the original attribute set, as long as the ordering of objects can be preserved in terms of a given dominance relation [3,4].

Definition 5. Let $S=(U,A,V,f)$ be a disjunctive SvOIS, $B \subseteq A$, $R_B^{\geq k}$ is the variable precision dominance relation. If $R_B^{\geq k} = R_A^{\geq k}$, and for any $b \in B$, $R_{B \setminus \{b\}}^{\geq k} \neq R_A^{\geq k}$, then B is called attribute reduction of S.

Obviously, an attribute reduction is a minimum subset of attributes satisfying $R_B^{\geq k} = R_A^{\geq k}$ in a disjunctive SvOIS.

Definition 6. Let $S=(U,A,V,f)$ be a disjunctive SvOIS, if $B=\{B_1, B_2, \ldots, B_m\}$ is the set of all attribute reduction, then $core = \bigcap_{1 \leq l \leq m} B_l$ is the core attribute set, $K = \bigcup_{1 \leq l \leq m} B_l \setminus core$ is the relatively indispensable attribute set, and $I = U \setminus \bigcup_{1 \leq l \leq m} B_l = U \setminus (core \cup K)$ is the dispensable attribute set. Where '\' denotes set difference.

Definition 7. Let $S=(U,A,V,f)$ be a disjunctive SvOIS. For any $x_i, x_j \in U$, if $Dis^k(x_j,x_i) = \{a \in B \mid (x_j, x_i) \notin R_{\{a\}}^{\geq k}\}$, then $Dis^k(x_j,x_i)$ is called discernibility attribute set between x_i and x_j with respect to B , and $DIS^k = (Dis^k(x_j,x_i) \mid x_i , x_j \in U)$ is the discernibility matrix.

Afer the discernibility matrix is established, we can continue to use the discernibility function in [5] to simplify the disjunctive SvOIS.

Property 2. Suppose that $S=(U,A,V,f)$ is a disjunctive SvOIS. If $Dis^k(x_j,x_i)$ is the discernibility attribute set of S with respect to $R_B^{\geq k}$, then $R_B^{\geq k} = R_A^{\geq k}$ if and only if $B \cap Dis^k(x_j,x_i) \neq \varnothing$, where, $Dis^k(x_j,x_i) \neq \varnothing$.

4 Conclusion

Set-valued information systems are generalized models of single-valued information systems. They could be divided into two categories: conjunctive and disjunctive set-valued information systems. In this paper, we proposed a variable precision dominance relation based on the probability theory in disjunctive set-valued information systems. And then, we established a rough set model based on the variable precision dominance relation. Finally, we defined the discernibility matrix, and developed an attribute reduction method based on the discernibility matrix.

Acknowledgments

This work is supported by the National Natural Science Foundation of China (No. 61073146) and the Science Fund for Distinguished Young Scholars of Chongqing of China (No.2008BA2041).

References

1. Greco, S., Matarazzo, B., Slowinski, R.: Rough Approximation of A Preference Relation by Dominance Relations. European Journal of Operational Research. 117(1), 63–83 (1999)
2. Greco, S., Matarazzo, B., Slowinski, R.: Rough Approximation by Dominance Relation. International Journal of Intelligent Systems 17(2), 153–171 (2002)
3. Inuiguchi, M., Yoshioka, Y., Kusunoki, Y.: Variable-precision Dominance-based Rough Set Approach and Attribute Reduction. International Journal of Approximate Reasoning 50(8), 1199–1214 (2009)
4. Qian, Y.H., Liang, J.Y., Song, P., Dang, C.Y.: On Dominance Relations in Disjunctive Set-valued Ordered Information Systems. International Journal of Information Technology & Decision Making 9(1), 9–33 (2010)
5. Skowron, A., Rauszer, C.: The Discernibility Matrices and Functions in Information Systems. In: Sowiski, R. (ed.) Intelligent Decision Support. Handbook of Applications and Advances of the Rough Sets Theory, pp. 331–362. Kluwer, Dordrecht (1992)
6. Zhang, W.X., Wu, W.Z., Liang, J.Y., Li, D.Y.: Theory and Method of Rough Sets. Science Press, Beijing (2001) (in Chinese)

An Interval-Valued Fuzzy Soft Set Approach for Normal Parameter Reduction

Xiuqin Ma and Norrozila Sulaiman

Faculty of Computer Systems and Software Engineering Universiti Malaysia Pahang
Lebuh Raya Tun Razak, Gambang 26300, Kuantan, Malaysia
xueener@gmail.com, norrozila@ump.edu.my

Abstract. Soft set theory in combination with the interval-valued fuzzy set has been proposed as the concept of the interval-valued fuzzy soft set. However, up to the present, few documents have focused on parameter reduction of the interval-valued fuzzy soft sets. In this paper, we propose a definition of normal parameter reduction of interval-valued fuzzy soft sets, which considers the problems of sub-optimal choice and added parameters. Then, a heuristic algorithm of normal parameter reduction for interval-valued fuzzy soft sets is presented. Finally, an illustrative example is employed to show our contribution.

Keywords: Fuzzy soft sets, Soft sets, Interval-valued fuzzy soft sets, Reduction, Normal parameter reduction.

1 Introduction

Soft set theory was firstly proposed by a Russian Mathematician Molodtsov [1] in 1999. It is a new mathematical tool for dealing with uncertainties. In recent years, there has been a rapid growth in interest in soft set theory and its applications[2], and great progresses of study on soft set theory have been made. Furthermore soft set models in combination with the interval-valued fuzzy set have been proposed as the concept of the interval-valued fuzzy soft set [3] by Yang et al. And it is worthwhile to mention that some effort has been done to such issues concerning reduction [4] of soft sets. However, up to the present, few documents have focused on parameter reduction of the interval-valued fuzzy soft sets. So, in this paper, we propose a definition of normal parameter reduction of interval-valued fuzzy soft sets and give a heuristic algorithm to achieve the normal parameter reduction of interval-valued fuzzy soft sets.

This paper is organized as follows. Section 2 reviews the basic notions of interval-valued fuzzy soft sets. Section 3 gives a definition of normal parameter reduction of interval-valued fuzzy soft sets and a related heuristic algorithm which is illustrated by an example. Finally, section 4 presents the conclusion from our work.

S.O. Kuznetsov et al. (Eds.): RSFDGrC 2011, LNAI 6743, pp. 211–214, 2011.

2 Essential Rudiments

In this section, we review some definitions with regard to soft sets, interval-valued fuzzy soft sets and the related application.

Let U be a non-empty initial universe of objects, E be a set of parameters in relation to objects in U, $P(U)$be the power set of U, and $A \subset E$.The definition of soft set is given as follows.

Definition 1. *(See [1]).A pair (F,A)is called a soft set over U if and only if F is a mapping of E into the set of all subsets of the set U.*

That is, a soft set over U is a parameterized family of subsets of the universe U.

Definition 2. *(See [5]).An interval-valued fuzzy set $\hat{X}$ on an universe U is a mapping such that*

$$\hat{X} : U \rightarrow Int([0,1]) \tag{1}$$

where Int([0,1])represents the set of all closed subintervals of [0,1], the set of all interval-valued fuzzy sets on U is denoted by $\widetilde{\psi}(U)$. Let $\hat{X} \in \widetilde{\psi}(U)$. For every $x \in U$, $\mu^-_{\hat{X}}(U)$ and $\mu^+_{\hat{X}}(U)$ are regarded as the lower and upper degrees of membership x to $\hat{X}$ $(0 \leq \mu^-_{\hat{X}}(U) \leq \mu^+_{\hat{X}}(U) \leq 1)$, respectively. $\mu_{\hat{X}}(x) = [\mu^-_{\hat{X}}(U), \mu^+_{\hat{X}}(U)]$ is referred to as the degree of membership an element x to $\hat{X}$.

Definition 3. *(See [3]).Let U be an initial universe of objects and E be a set of parameters in relation to objects in U. A pair ($\widetilde{\omega}$,E)is called an interval-valued fuzzy soft set over $\widetilde{\psi}(U)$, where $\widetilde{\omega}$ is a mapping given by*

$$\widetilde{\omega} : E \rightarrow \widetilde{\psi}(U) \tag{2}$$

That is, an interval-valued fuzzy soft set is a parameterized family of interval-valued fuzzy subsets of U. Hence, its universe is the set of all interval-valued fuzzy sets of U, i.e$\widetilde{\psi}(U)$.

Algorithm 1. Yang et al. [3] presented the algorithm to solve fuzzy decision making problems based on interval-valued fuzzy soft sets in the following. More details can be found in [3].

3 Normal Parameter Reduction of Interval-Valued Fuzzy Soft Sets

In this section, we give a definition of normal parameter reduction of interval-valued fuzzy soft sets and present a heuristic algorithm to achieve it.

Definition 4. *For interval-valued fuzzy soft set $(\widetilde{S}, E)$,$U = \{h_1, h_2, ..., h_n\}$, $E = \{e_1, e_2, ..., e_n\}$,$\mu_{\widetilde{S}(e_j)}(h_i) = [\mu^-_{\widetilde{S}(e_j)}(h_i), \mu^+_{\widetilde{S}(e_j)}(h_i)]$ is the degree of membership an element h_i to $\widetilde{S}(e_j)$. We denote $d_{\widetilde{S}(e_j)}(h_i)$ as score of membership degrees for e_j,where it is formulated as*

$$d_{\widetilde{S}(e_j)}(h_i) = \sum_{k=1}^{n}(\mu^-_{\widetilde{S}(e_j)}(h_i) - \mu^-_{\widetilde{S}(e_j)}(h_k)) + \sum_{k=1}^{n}(\mu^+_{\widetilde{S}(e_j)}(h_i) - \mu^+_{\widetilde{S}(e_j)}(h_k)) \quad (3)$$

Definition 5. *For interval-valued fuzzy soft set $(\widetilde{S}, E)$,$U = \{h_1, h_2, ..., h_n\}$, $E = \{e_1, e_2, ..., e_n\}$,if there exists a subset $A = \left\{e_1', e_2', ..., e_p'\right\} \subset E$ satisfying $\sum_{e_k \in A} d_{\widetilde{S}(e_k)}(h_1) = \sum_{e_k \in A} d_{\widetilde{S}(e_k)}(h_2) = ... = \sum_{e_k \in A} d_{\widetilde{S}(e_k)}(h_n)$, then A is dispensable,otherwise, A is indispensable. $B \subset E$ is defined as a normal parameter reduction of E, if the two conditions as follows are satisfied*
(1) B is indispensable
(2) $\sum_{e_k \in E-B} d_{\widetilde{S}(e_k)}(h_1) = \sum_{e_k \in E-B} d_{\widetilde{S}(e_k)}(h_2) = ... = \sum_{e_k \in E-B} d_{\widetilde{S}(e_k)}(h_n)$

Algorithm 2. Normal parameter reduction of interval-valued fuzzy soft sets

(1) Input interval-valued fuzzy soft sets $(\widetilde{S}$, E)and the parameter set E;
(2) Compute score of membership degrees $d_{\widetilde{S}(e_j)}(h_i)$, for $1 \leq i \leq n$,$1 \leq j \leq m$;
(3) Check A, where $A = \left\{e_1', e_2', ..., e_p'\right\} \subset E$, if
$\sum_{e_k \in A} d_{\widetilde{S}(e_k)}(h_1) = \sum_{e_k \in A} d_{\widetilde{S}(e_k)}(h_2) = ... = \sum_{e_k \in A} d_{\widetilde{S}(e_k)}(h_n)$,
and then A is put into a candidate parameter reduction set.
(4) Find the maximum cardinality of A in the candidate parameter reduction set and get E-A as the optimal normal parameter reduction.

Example 1. Let $(\widetilde{S}$, E) be an interval-valued fuzzy soft set with the tabular representation displayed in Table 1. Suppose that $U = \{h_1, h_2, h_3, h_4, h_5, h_6\}$, $E = \{e_1, e_2, e_3, e_4, e_5, e_6\}$.

Table 1. An interval-valued fuzzy soft set $(\widetilde{S}$, E)

U/E	e_1	e_2	e_3	e_4	e_5	e_6	c_i	r_i
h_1	[0.5,0.8]	[0.1,0.3]	[0.3,0.5]	[0.4,0.5]	[0.7,0.9]	[0.5,0.8]	[2.5,3.8]	0.6
h_2	[0.3,0.4]	[0.5,0.6]	[0.6,0.8]	[0.6,0.8]	[0.4,0.6]	[0.2,0.3]	[2.6,3.5]	-0.6
h_3	[0.4,0.6]	[0.6,0.7]	[0.5,0.6]	[0.3,0.4]	[0.6,0.7]	[0.1,0.3]	[2.5,3.3]	-2.4
h_4	[0.7,0.9]	[0.2,0.3]	[0.2,0.3]	[0.0,0.1]	[0.9,1.0]	[0.1,0.2]	[2.1,2.8]	-7.8
h_5	[0.1,0.3]	[0.5,0.7]	[0.8,0.9]	[0.5,0.7]	[0.2,0.5]	[0.6,0.7]	[2.7,3.8]	1.8
h_6	[0.1,0.2]	[0.7,0.9]	[0.8,1.0]	[0.9,1.0]	[0.2,0.4]	[0.6,0.8]	[3.3,4.3]	8.4

We can obtain score of membership degrees for$(\widetilde{S}$, E) by the formula of (3), which is given in Table 2.

Table 2. The score of membership degrees for ($\widetilde{S}$, E)

U/E	e_1	e_2	e_3	e_4	e_5	e_6	r_i
h_1	2.5	-3.7	-2.5	-0.8	2.5	2.6	0.6
h_2	-1.1	0.5	1.1	2.2	-1.1	-2.2	-0.6
h_3	0.7	1.7	-0.7	-2.0	0.7	-2.8	-2.4
h_4	4.3	-3.1	-4.3	-5.6	4.3	-3.4	-7.8
h_5	-2.9	1.1	2.9	1.0	-2.9	2.6	1.8
h_6	-3.5	3.5	3.5	5.2	-3.5	3.2	8.4

From the Table 2, we can obtain$\{e_1, e_3\}$ and $\{e_3, e_5\}$ satisfying

$$\sum_{e_k \in A} d_{\widetilde{S}(e_k)}(h_1) = \sum_{e_k \in A} d_{\widetilde{S}(e_k)}(h_2) = ... = \sum_{e_k \in A} d_{\widetilde{S}(e_k)}(h_n) = 0$$

Thus $\{e_2, e_4, e_5, e_6\}$ and $\{e_1, e_2, e_4, e_6\}$ are the normal reduction of the interval-valued fuzzy soft set($\widetilde{S}$, E).

4 Conclusion

Pioneering work on the interval-valued fuzzy soft sets has been done by Yang et al. However, up to the present, few documents have focused on parameter reduction of the interval-valued fuzzy soft sets. In this paper, we have proposed a definition and the related property of normal parameter reduction of interval-valued fuzzy soft sets and give a heuristic algorithm to achieve the normal parameter reduction of interval-valued fuzzy soft sets, which considers the problems of sub-optimal choice and added parameters. Finally, an example illustrates our contribution.

References

1. Molodtsov, D.: Soft set theory-First results. Computers and Mathematics with Applications 37(4/5), 19–31 (1999)
2. Herawan, T., Mat Deris, M.: A Soft Set Approach for Association Rules Mining. Knowledge Based Systems (2010), doi: 10.1016/j.knosys.2010.08.005
3. Yang, X., Lin, T.Y., Yang, J., Dongjun, Y.L.A.: Combination of interval-valued fuzzy set and soft set. Computers and Mathematics with Applications 58, 521–527 (2009)
4. Kong, Z., Gao, L., Wang, L., Li, S.: The normal parameter reduction of soft sets and its algorithm. Computers and Mathematics with Applications 56(12), 3029–3037 (2008)
5. Gorzalzany, M.B.: A method of inference in approximate reasoning based on interval-valued fuzzy sets. Fuzzy Sets and Systems 21, 1–17 (1987)

Incorporating Game Theory in Feature Selection for Text Categorization

Nouman Azam and JingTao Yao

Department of Computer Science, University of Regina, Regina, Saskatchewan, Canada S4S 0A2
{azam200n,jtyao}@cs.uregina.ca

Abstract. Feature selection remains as one of effective and efficient techniques in text categorization. Selecting important features is crucial for effective performance in case of high imbalance in data. We introduced a method which incorporates game theory to feature selection with the aim of dealing with high imbalance situations for text categorization. In particular, a game is formed between negative and positive categories to identify the suitability of features for their respective categories. Demonstrative example suggests that this method may be useful for feature selection in text categorization problems involving high imbalance.

1 Introduction

Feature selection is a process which selects a subset of features, that are considered as important [1,2,3,4]. Text categorization systems are often faced with high class imbalance. For example, in binary classification setting where the category of interest is taken as positive class while the union of all other categories are taken as negative [5]. The number of positive examples is fewer than negative ones in such cases. Previous studies suggest that efficiently tailoring feature selection can be useful in such cases for increased performance [6,7].

There are two major feature selection approaches in text categorization, namely, one sided and two sided approaches [7]. Methods in these approaches assign positive or negative values to features. These values indicate feature's importance level or utilities to be used for classification. Features representing positive category are assigned positive values, while features representing negative category are assigned negative values. Methods in one sided approach select features with high positive values. Methods in two sided approach selects features with high absolute value, implicitly [5]. Both approaches have their limitations. Methods in one sided approach ignore the importance of negative features. It is suggested that the presence of negative features are necessary for higher classification rate as they can increase confidence in rejecting irrelevant examples [6,7]. Similarly, methods in two sided (implicit combination) approach, based on their definitions, will assign higher absolute values to positive features compared to negative ones [7]. In reality, when signs are ignored, fewer negative features are selected. To overcome the limitations, an explicit combinational approach was

S.O. Kuznetsov et al. (Eds.): RSFDGrC 2011, LNAI 6743, pp. 215–222, 2011.

Table 1. Probabilities of words

	w_1	w_2	w_3	w_4	w_5	w_6	w_7	w_8
Probabilities in positive category	0.7	0.5	0.4	0	0	0	0.5	0.6
Probabilities in negative category	0	0	0	0.7	0.5	0.3	0.6	0.5

introduced [7]. Highly positive and negative features generated by a particular one sided method were selected with this approach. It is suggested that the explicit combinational approach may provide better results than the above two approaches [7]. In addition, a comparative study suggested that the explicit combinational approach has superior performance than implicit approaches [5].

Most of the existing approaches favor features that are indicative of either positive or negative category. However, there might be features that are indicative of both categories. It is plausible to include such features in some applications. The challenge remains on how to effectively select these features. In this article, we employ game theory for such purpose. A game is setup with positive and negative category players. The goal of players is to reach a dominant position over others or to cooperate with others in order to reach an optimal position.

2 Feature Selection in High Class Imbalance

Existing feature selection approaches are not suitable for selecting features that are representative of both positive and negative categories. We will examine four feature selection methods. Correlation coefficient (CC) and GSS coefficient (GSS) [7] are chosen for one sided approach while chi square (CHI) [8] and gini index (GINI) [9] for two sided approach.

For text categorization, we demonstrate the incompetence of these approaches with an example. Let us consider an imbalanced data with 10 documents in positive and 100 in negative categories. There are eight words in these documents represented as $w_1, w_2, ..., w_8$. Table 1 shows the probabilities of these words in documents of the two categories. The probability of a word in a category, refers to the fraction of documents from that category containing the word. For example, the probability of 0.7 for w_1 in positive category suggest that 7 out of 10 positive documents contains w_1. The scores of the words in respective methods are summarized in Table 2.

Table 3 presents the results of feature rankings. The rankings for explicit approach were obtained from methods in one sided approach. Words with highest scores receive the highest ranks in respective methods. We note that w_7 and w_8 are not considered as important by any method. If we are interested in top three features, they will be ignored for all the methods.

Let us consider w_7 of Table 1. The word has similar level of probabilities in both categories. The existing methods consider w_7 less important, as it does not differentiate these two categories. However, the same word may be useful, if we

Table 2. Scores of features corresponding to different methods

		w_1	w_2	w_3	w_4	w_5	w_6	w_7	w_8
One sided metrics	CC	8.244	6.91	6.143	-4.183	-2.887	-1.937	-0.585	0.575
	GSS	0.058	0.042	0.033	-0.058	-0.041	-0.025	-0.008	0.008
Implicit combinational metrics	CHI	67.96	47.62	37.74	17.50	8.333	3.75	0.342	0.331
	GINI	0.49	0.25	0.16	0	0	0	0.002	0.004

Table 3. Features rankings for different approaches

		Feature rankings							
		1	2	3	4	5	6	7	8
One sided	CC and GSS	w_1	w_2	w_3	w_8	w_7	w_6	w_5	w_4
Implicit combination	CHI	w_1	w_2	w_3	w_4	w_5	w_6	w_8	w_7
	GINI	w_1	w_2	w_3	w_8	w_7	$(w_4 w_5 w_6)$		
Explicit combination	ranking for positive	w_1	w_2	w_3	w_8	w_7	w_6	w_5	w_4
	ranking for negative	w_4	w_5	w_6	w_7	w_8	w_3	w_2	w_1

consider the times it appear in documents of the two categories. If the times of occurrence of w_7 in positive documents is much large than the times in negative documents, we may consider w_7 as an important feature for positive category.

Suppose, that we use an explicit combinational approach, the rankings in categories may be obtained by sorting the words based on their probabilities in respective categories. The new ranking of the words with their probabilities in Table 1 would be

- $w_1, w_8, (w_7, w_2), w_3, (w_4, w_5, w_6)$, for positive and
- $w_4, w_7, (w_8, w_5), w_6, (w_1, w_2, w_3)$, for negative category.

Words in brackets are of equal probabilities. We may select features based on the rankings. Suppose that we consider positive category to be twice as important as negative category. We may select two positive features for every negative feature. Therefore, w_1, w_8 and w_4 would be selected in this case. Three types of features, that is, those indicative of positive category (i.e., w_1), negative category (i.e., w_4), and both of them (i.e., w_8) are selected.

A particular feature may be considered as good for positive category, negative category, both of them or neither of them. We try to find a systematic method, which can be used to find the best choice among the four decisions.

3 Feature Selection with Game Theory

Game theory attempts to mathematically assess situations, where an individual success in making a decision among choices depends on the decision choices of others [10,11]. It has been utilized in many areas such as economics, machine learning and rough sets [12,13,14,15].

A game can be formally defined as $G = \{P, S, F\}$, where P represents a set of players, S a set of strategies, and F a set of payoff functions. We follow the following four steps for a feature selection game.

Identifying the Player Set: Two players representing positive category and negative category, are considered in this study. The set of players is denoted as $P = \{C^+, C^-\}$, where C^+ represents positive category and C^- the negative category. Players will try to determine the features utility for their respective categories using suitable payoff functions. The final decision is based on each player's achieving maximum payoff.

Identifying the Strategy Set: To participate in a game, each player should have a set of strategies. Individual strategies may be realized as actions. Hence, we have two sets of strategies with actions corresponding to the two players. Setting the strategy sets for C^+ as S^+ and C^- as S^-, the two sets are denoted as $S^+ = \{a_1, a_2, ..., a_n\}$ and $S^- = \{a_1, a_2, ..., a_n\}$. This means that there are n actions available for each player. We simplify two actions here, i.e., action a_1 to keep the feature and a_2 to discard the feature. To differentiate the actions of the two players, we use notation a_i^+ to denote the actions of C^+ and a_i^- for C^- where i represents the action.

Determining the Payoff Functions: A payoff or utility function is a result of a player's action. More formally, the payoff of player i, performing action a_j can be measured as $u_{i,j} = u(a_j)$. The set of all u functions acting in game G can be represented by a set F. In this game $F = \{u_{C^+}, u_{C^-}\}$, where u_{C^+} represents the set of payoff functions for the player C^+ and u_{C^-} the player C^-. Given an action of the opponent, a player can choose from two possible actions. Hence, we need four payoff functions for each player. We denote the payoff of player i, performing an action j, given action k by the second player as $u_{i(j|k)}$. The payoff set for each player can be denoted as,

$$u_{C^+} = \{u_{C^+(a_1^+|a_1^-)},\quad u_{C^+(a_1^+|a_2^-)},\quad u_{C^+(a_2^+|a_1^-)},\quad u_{C^+(a_2^+|a_2^-)}\},$$

$$u_{C^-} = \{u_{C^-(a_1^-|a_1^+)},\quad u_{C^-(a_1^-|a_2^+)},\quad u_{C^-(a_2^-|a_1^+)},\quad u_{C^-(a_2^-|a_2^+)}\}.$$

We now define individual payoff functions. Let us denote the positive category as cat and the negative $\overline{cat}$. For a particular word w, we denote the number of documents from cat and $\overline{cat}$ containing w as A and B respectively. Similarly, the number of documents from cat and $\overline{cat}$ that does not contain w are denoted by C and D respectively. The conditional probabilities of a particular word w's presence or absence in category cat or $\overline{cat}$ are defined as,

$$P(w|cat) = \frac{A}{\sum_{d \in cat} d}, \qquad P(w|\overline{cat}) = \frac{B}{\sum_{d \in \overline{cat}} d},$$

$$P(\overline{w}|cat) = \frac{C}{\sum_{d \in cat} d}, \text{ and } \qquad P(\overline{w}|\overline{cat}) = \frac{D}{\sum_{d \in \overline{cat}} d}.$$

Table 4. Action scenarios for players

Player	Action	Objective	Required Characteristic
C^+	a_1^+	Include feature	Present in many and absent in few positive examples i.e. high value for $P(w\|cat)$
	a_2^+	Do not include feature	Absent in many and present in few positive examples i.e. high value for $P(\overline{w}\|cat)$
C^-	a_1^-	Include feature	Present in many and absent in few negative examples i.e. high value for $P(w\|\overline{cat})$
	a_2^-	Do not include feature	Absent in many and present in few negative examples i.e. high value for $P(\overline{w}\|\overline{cat})$

When both players decide to keep a feature, high probability is desired in both categories (i.e., higher value of $P(w|cat)$ and $P(w|\overline{cat})$). The average function can be useful to represent utility of players in this case. The average will be relatively low if any of the two probabilities is low. Higher average value can be expected, when both probabilities are high. The payoffs for the players are calculated as $\{P(w|cat) + P(w|\overline{cat})\}/2$ in this case. Alternatively, both players may decide to discard a feature. The desired property in this case is high probability of features absence in both categories. The payoff for players are calculated similarly as $\{P(\overline{w}|cat) + P(\overline{w}|\overline{cat})\}/2$. The player C^+ may decides to keep a feature while C^- decides to discard, the payoffs are calculated as $P(w|cat)$ and $P(\overline{w}|\overline{cat})$, respectively. Finally, C^+ may choose to discard and C^- to keep a feature. The corresponding payoffs are calculated as $P(\overline{w}|cat)$ and $P(w|\overline{cat})$, respectively. Table 4 summarize the action scenarios for both players.

Implementing Competition: Finally, we need to express the game as a competition or cooperation between players in a payoff table. This is presented in Table 5. In order to find actions of players, we determine Nash equilibrium [11] within the payoff table. Intuitively, this means that none of the players can be benefited by changing his or her strategy, given the other player's chosen action.

The actions of the players in the game mutually define the usefulness of a feature belonging to positive or negative categories. If we define sets, FS^+ and FS^- as sets of features representing positive and negative categories, the game will determine the inclusion or exclusion of features in these sets. In particular, a feature is included in FS^+, when the players actions are a_1^+ and a_2^-, while in

Table 5. The payoff matrix for C^+ and C^-

		C^-	
		a_1^-	a_2^-
C^+	a_1^+	$u_{C^+(a_1^+\|a_1^-)}, u_{C^-(a_1^-\|a_1^+)}$	$u_{C^+(a_1^+\|a_2^-)}, u_{C^-(a_2^-\|a_1^+)}$
	a_2^+	$u_{C^+(a_2^+\|a_1^-)}, u_{C^-(a_1^-\|a_2^+)}$	$u_{C^+(a_2^+\|a_2^-)}, u_{C^-(a_2^-\|a_2^+)}$

FS^-, when actions are a_2^+ and a_1^-. A feature may be included in both sets when actions of the players are a_1^+ and a_1^-. Finally, we may discard a feature when the actions are a_2^+ and a_2^-. A game may be implemented for all features in this way. The final features set can be obtained as $FS = FS^+ \bigcup FS^-$.

4 A Demonstrative Example

Let us demonstrate the application of game theory using the earlier example. Based on probabilities in Table 1, one can obtain the values of payoff functions. For example, the payoff of C^+ for taking action a_1^+ given action a_1^- of C^- is,

$$u_{C^+(a_1^+|a_1^-)} = \{P(w|cat) + P(w|\overline{cat})\}/2 = \{0.7 + 0\}/2 = 0.35.$$

Table 6 shows the payoff table for w_1. The cells with bold numbers represent the Nash equilibrium. The actions of players in the state of equilibrium are a_1^+ for C^+ and a_2^- for C^-. In this state, none of the players can achieve a higher payoff, given the other player's chosen action. For example, changing the action of C^+ from a_2^+ to a_1^+ will decrease the payoff from 0.7 to 0.65. The actions of players mutually decide to include w_1 in FS^+.

Table 6. The payoff table for w_1

		C^-	
		a_1^-	a_2^-
C^+	a_1^+	0.35, 0.35	**0.70, 1.0**
	a_2^+	0.30, 0.0	0.65, 0.65

Table 7. The payoff table for w_2

		C^-	
		a_1^-	a_2^-
C^+	a_1^+	0.25, 0.25	0.50, 1.0
	a_2^+	0.50, 0.0	**0.75, 0.75**

For the other two words i.e., w_2 and w_3, the equilibrium in their respective payoff tables results in the final decision of discarding them. This is presented in Table 7 and Table 8.

Table 8. The payoff table for w_3

		C^-	
		a_1^-	a_2^-
C^+	a_1^+	0.20, 0.20	0.40, 1.0
	a_2^+	0.60, 0.0	**0.80, 0.80**

Table 9. The payoff table for w_4

		C^-	
		a_1^-	a_2^-
C^+	a_1^+	0.35, 0.35	0.0, 0.30
	a_2^+	**1.0, 0.7**	0.65, 0.65

We now consider w_4's payoff as shown in Table 9. The actions of players in equilibrium are a_2^+ for C^+ and a_1^- for C^-. We note that both of players have maximized their payoffs, given the chosen action of another player. The chosen actions of players in this case, corresponds to the decision of including w_4 in FS^-. Similarly, for w_5 and w_6, the actions in equilibrium corresponds to the decision of discarding. This is shown in Table 10 and Table 11.

Table 10. The payoff table for w_5

		C^-	
		a_1^-	a_2^-
C^+	a_1^+	0.25, 0.25	0.0, 0.50
	a_2^+	1.0, 0.50	**0.75, 0.75**

Table 11. The payoff table for w_6

		C^-	
		a_1^-	a_2^-
C^+	a_1^+	0.15, 0.15	0.0, 0.70
	a_2^+	1.0, 0.30	**0.85, 0.85**

For the last two words, w_7 and w_8, the payoff tables are shown in Table 12 and Table 13. For both words, the actions for C^+ and C^- in equilibrium are a_1^+ and a_1^- respectively. In other words, both players agree that the feature is important and thus should be included in both FS^+ and FS^-.

Table 12. The payoff table for w_7

		C^-	
		a_1^-	a_2^-
C^+	a_1^+	**0.55, 0.55**	0.50, 0.40
	a_2^+	0.50, 0.60	0.45, 0.45

Table 13. The payoff table for w_8

		C^-	
		a_1^-	a_2^-
C^+	a_1^+	**0.55, 0.55**	0.60, 0.50
	a_2^+	0.40, 0.50	0.45, 0.45

After implementing the game for all of the words, we have $FS^+ = \{w_1, w_7, w_8\}$ and $FS^- = \{w_4, w_7, w_8\}$. The final features set $FS = \{w_1, w_4, w_7, w_8\}$, contains all the three types of features. We observed that the words w_7 and w_8, which represent both categories, are included in the final features set. Furthermore, the method decides the ratio between positive and negative features implicitly. In reality, one can modify the payoff functions, which may result different ratio, consistent with importance assigned to categories. It is suggested that the game theory approach may help us in selecting important features that are indicative of both positive and negative categories.

5 Conclusion

Existing feature selection approaches give preference to features indicating either positive or negative categories. These approaches may not be suitable for selecting features that indicate both categories. This article presents a feature selection method incorporated with game theory, for text categorization problems involving high class imbalance. In particular, a game between two players, positive and negative category is implemented. The players find the utilities of features for their respective category and take appropriate actions. The importance of the method is that it is able to include features indicating either positive category, negative category or both of them. Demonstrative example suggests that it may be useful in text categorization applications involving high imbalance.

References

1. Liu, H., Yu, L.: Toward integrating feature selection algorithms for classification and clustering. IEEE Transactions on Knowledge and Data Engineering 17(4), 491–502 (2005)
2. Yao, J.T., Zhang, M.: Feature selection with adjustable criteria. In: Ślęzak, D., Wang, G., Szczuka, M.S., Düntsch, I., Yao, Y. (eds.) RSFDGrC 2005. LNCS (LNAI), vol. 3641, pp. 204–213. Springer, Heidelberg (2005)
3. Zhong, N., Dong, J., Ohsuga, S.: Using rough sets with heuristics for feature selection. Journal of Intelligent Information Systems 16(3), 199–214 (2001)
4. Swiniarski, R.W., Skowron, A.: Rough set methods in feature selection and recognition. Pattern Recognition Letters 24(6), 833–849 (2003)
5. Ogura, H., Amano, H., Kondo, M.: Comparison of metrics for feature selection in imbalanced text classification. Expert Systems with Applications 38(5), 4978–4989 (2011)
6. Forman, G.: An extensive empirical study of feature selection metrics for text classification. Journal of Machine Learning Research 3, 1289–1305 (2003)
7. Zheng, Z., Wu, X., Srihari, R.: Feature selection for text categorization on imbalanced data. ACM SIGKDD Explorations Newsletter 6(1), 80–89 (2004)
8. Yang, Y., Pedersen, J.O.: A comparative study on feature selection in text categorization. In: Proceedings of the 14th International Conference on Machine Learning, pp. 412–420 (1997)
9. Shang, W., Huang, H., Zhu, H., Lin, Y., Qu, Y., Wang, Z.: A novel feature selection algorithm for text categorization. Expert Systems with Applications 33(1), 1–5 (2007)
10. Myerson, R.B.: Game Theory: Analysis of Conflict. Harvard University Press, Cambridge (1991)
11. Neumann, J.V., Morgenstern, O.: Theory of Games and Economic Behavior. Princeton University Press, Princeton (1944)
12. Nash, J.: The bargaining problem. Econometrica 18, 155–162 (1950)
13. Sako, M., Helper, S.: Determinants of trust in supplier relations: Evidence from the automotive industry in japan and the united states. Journal of Economic Behavior and Organization 34(3), 387–417 (1998)
14. Herbert, J., Yao, J.T.: A game-theoretic approach to competitive learning in self-organizing maps. In: Wang, L., Chen, K., S. Ong, Y. (eds.) ICNC 2005. LNCS, vol. 3610, pp. 129–138. Springer, Heidelberg (2005)
15. Herbert, J., Yao, J.T.: Game-theoretic rough sets. Fundamenta Informaticae 108(3-4), 267–286 (2011)

Attribute Reduction in Random Information Systems with Fuzzy Decisions

Wei-Zhi Wu and You-Hong Xu

School of Mathematics, Physics and Information Science,
Zhejiang Ocean University, Zhoushan, Zhejiang, 316004, P.R. China
wuwz@zjou.edu.cn (W.-Z. Wu), xyh@zjou.edu.cn (Y.-H. Xu)

Abstract. Knowledge reduction is one of the main problems in the study of rough set theory. This paper deals with knowledge reduction in the sense of reducing attributes in random information systems with fuzzy decisions based on the Dempster-Shafer theory of evidence. The concepts of lower approximation reducts, upper approximation reducts, random belief reducts and random plausibility reducts in random fuzzy decision systems are introduced. The relationships among these reducts are examined.

Keywords: Belief functions, fuzzy decisions, knowledge reduction, random information systems, rough sets.

1 Introduction

Imprecision and uncertainty are two important aspects of incompleteness of information. One theory for the study of insufficient and incomplete information in intelligent systems is rough set theory [2]. The primitive notions of rough set theory are a dual pair of lower and upper approximations induced from an approximation space. Another important method used to deal with uncertainty in information systems is the Dempster-Shafer theory of evidence [3]. The fundamental numeric measures are a dual pair of belief and plausibility functions derived from the belief structure.

There are strong connections between rough set theory and Dempster-Shafer theory of evidence. It has been demonstrated that various belief structures are associated with various approximation spaces such that the different dual pairs of lower and upper approximation operators induced by approximation spaces may be used to interpret the corresponding dual pairs of belief and plausibility functions induced by belief structures [4,7,8,10]. Based on this observation, the Dempster-Shafer theory of evidence may be used to analyze knowledge reduction and knowledge acquisition in information systems (see e.g. [1,5,6,9,12]).

In the traditional rough set approach, the values of attributes are assumed to be nominal data, i.e. symbols. In many applications, however, the decision attribute-values can be linguistic terms (i.e. fuzzy sets). The traditional rough set approach would treat these values as symbols, thereby some important information included in these values such as the partial ordering and membership

S.O. Kuznetsov et al. (Eds.): RSFDGrC 2011, LNAI 6743, pp. 223–230, 2011.

degrees is ignored, which means that the traditional rough set approach cannot effectively deal with fuzzy initial data (e.g. linguistic terms). On the other hand, the available database may be obtained by a randomization method. Thus a new rough set model is needed to deal with such data. In this paper, we will propose the concept of random information systems with fuzzy decisions and discuss the issue of attribute reduction in such systems by using the Dempster-Shafer theory of evidence.

2 Fuzzy Evidence Theory Induced by a Crisp Belief Structure

Throughout this paper, U will be a nonempty finite set called the universe of discourse. The class of all subsets (fuzzy subsets, respectively) of U will be denoted by $\mathcal{P}(U)$ (by $\mathcal{F}(U)$, respectively). For a fuzzy set $X \in \mathcal{F}(U)$, if there exists an $x \in U$ such that $X(x) = 1$, then X is referred to as a normalized fuzzy set. The cardinality of a fuzzy set X is denoted by $|X| = \sum_{x \in U} X(x)$. We use the symbols $\vee$ and $\wedge$ to denote the maximum and minimum, respectively. We state that if X is a fuzzy set of U and P is a probability measure on U, then the probability of the fuzzy set X, denoted by $\mathrm{P}(X)$, is defined, in the sense of Zadeh [11], by

$$\mathrm{P}(X) = \sum_{x \in U} X(x)\mathrm{P}(\{x\}). \tag{1}$$

Notice that if we define $\mathrm{P}(x) =: \mathrm{P}(\{x\}) = 1/|U|$ for all $x \in U$ and $\mathrm{P}(X) = |X|/|U|$ for all $X \in \mathcal{P}(U)$, then $\mathrm{P} : \mathcal{P}(U) \to [0, 1]$ is a probability measure on U.

The Dempster-Shafer theory of evidence, also called the "evidence theory" or the "belief function theory", is treated as a promising method of dealing with uncertainty in intelligence systems. The basic representational structure in the Dempster-Shafer theory of evidence is a belief structure [3].

Definition 1. *Let U be a non-empty finite set, a set function $m : \mathcal{P}(U) \to [0, 1]$ is referred to as a crisp basic probability assignment if it satisfies axioms* (M1) *and* (M2)*:*

$$\text{(M1)}\ m(\emptyset) = 0, \quad \text{(M2)} \sum_{A \subseteq U} m(A) = 1.$$

A set $X \in \mathcal{P}(U)$ with nonzero basic probability assignment is referred to as a focal element. We denote by $\mathcal{M}$ the family of all focal elements of m. The pair $(\mathcal{M}, m)$ is called a belief structure on U.

Associated with each belief structure, a pair of fuzzy belief and plausibility functions can be defined [8].

Definition 2. *Let $(\mathcal{M}, m)$ be a crisp belief structure on U. A fuzzy set function* $\mathrm{Bel} : \mathcal{F}(U) \to [0, 1]$ *is referred to as a CF-belief function induced from $(\mathcal{M}, m)$ on U if*

$$\mathrm{Bel}(X) = \sum_{\{Y : Y \in \mathcal{M}\}} m(Y) \bigwedge_{u \in Y} X(u) \quad \forall X \in \mathcal{F}(U). \tag{2}$$

A fuzzy set function $\mathrm{Pl}: \mathcal{F}(U) \to [0,1]$ *is referred to as a CF-plausibility function induced from* $(\mathcal{M}, m)$ *on* U *if*

$$\mathrm{Pl}(X) = \sum_{\{Y: Y \in \mathcal{M}\}} m(Y) \bigvee_{u \in Y} X(u) \quad \forall X \in \mathcal{F}(U). \tag{3}$$

It can be verified that

$$\mathrm{Bel}(X) \leq \mathrm{Pl}(X) \quad \forall X \in \mathcal{F}(U). \tag{4}$$

Moreover, the CF-belief and CF-plausibility functions based on the same belief structure are connected by the dual property

$$\mathrm{Pl}(X) = 1 - \mathrm{Bel}(\sim X) \quad \forall X \in \mathcal{F}(U), \tag{5}$$

where $\sim X$ is the complement of the fuzzy set X. Similar to the crisp belief and plausibility functions, it can be proved that the CF-belief function is a fuzzy monotone Choquet capacity on U, i.e., it satisfies the following properties [7]:

(FMC1) $\mathrm{Bel}(\emptyset) = 0$,
(FMC2) $\mathrm{Bel}(U) = 1$,
(FMC3) for all $X_i \in \mathcal{F}(U), i = 1, 2, \ldots, k$,
$\mathrm{Bel}(\bigcup_{i=1}^{k} X_i) \geq \sum_{\emptyset \neq J \subseteq \{1,2,\ldots,k\}} (-1)^{|J|+1} \mathrm{Bel}(\bigcap_{i \in J} X_i)$.

And the CF-plausibility function is a fuzzy alternating Choquet capacity on U, i.e., it satisfies the following properties:

(FAC1) $\mathrm{Pl}(\emptyset) = 0$,
(FAC2) $\mathrm{Pl}(U) = 1$,
(FAC3) for all $X_i \in \mathcal{F}(U), i = 1, 2, \ldots, k$,
$\mathrm{Pl}(\bigcap_{i=1}^{k} X_i) \leq \sum_{\emptyset \neq J \subseteq \{1,2,\ldots,k\}} (-1)^{|J|+1} \mathrm{Pl}(\bigcup_{i \in J} X_i)$.

3 Random Information Systems with Fuzzy Decisions and Rough Fuzzy Approximations

The notion of information systems provides a convenient tool for the representation of objects in terms of their attribute values. An information system (IS for short) is a pair (U, AT), where $U = \{x_1, x_2, \ldots, x_n\}$ is a non-empty, finite set of objects called the universe of discourse and $AT = \{a_1, a_2, \ldots, a_m\}$ is a non-empty, finite set of attributes, such that $a: U \to V_a$ for any $a \in AT$, where V_a is called the domain of a.

Each non-empty subset $B \subseteq AT$ determines an indiscernibility relation as follows:

$$R_B = \{(x, y) \in U \times U : a(x) = a(y) \; \forall a \in B\}. \tag{6}$$

Since R_B is an equivalence relation on U, it forms a partition $U/R_B = \{[x]_B : x \in U\}$ of U, where $[x]_B$ denotes the equivalence class determined by x with respect to (wrt) B, i.e., $[x]_B = \{y \in U : (x,y) \in R_B\}$.

A decision system (sometimes called a decision table) is a pair $(U, C \cup \{d\})$ where (U, C) is an IS, and $d \notin C$ is a distinguished attribute called the decision, in this case C is called the conditional attribute set, d is a mapping $d : U \to V_d$ from the universe U into the value set V_d, we assume, without any loss of generality, that $V_d = \{d_1, d_2, \ldots, d_r\}$. d is called a fuzzy decision if, for each $x \in U$, $d(x)$ is a fuzzy subset of V_d, i.e., $d : U \to \mathcal{F}(V_d)$, with no lose of generality, we represent d as follows:

$$d(x_i) = d_{i1}/d_1 + d_{i2}/d_2 + \cdots + d_{ir}/d_r, i = 1, 2, \ldots, n, \tag{7}$$

where $d_{ij} \in [0, 1]$. In this case, $(U, C \cup \{d\})$ is called an IS with fuzzy decisions. For the fuzzy decision d, we define a fuzzy indiscernibility binary relation R_d on U as follows:

$$R_d(x_i, x_k) = \min\{1 - |d_{ij} - d_{kj}| : j = 1, 2, \ldots, r\}, i, k = 1, 2, \ldots, n. \tag{8}$$

Then, we obtain a fuzzy similarity class $S_d(x)$ of $x \in U$ in the system $S = (U, C \cup \{d\})$ as follows:

$$S_d(x)(y) = R_d(x, y), \quad y \in U. \tag{9}$$

Since $S_d(x)(x) = R_d(x, x) = 1$, we see that $S_d(x) : U \to [0, 1]$ is a normalized fuzzy set of U. Denote by U/R_d the fuzzy similarity classes induced by the fuzzy decision d, i.e.

$$U/R_d = \{S_d(x) : x \in U\}. \tag{10}$$

If P is a normalized probability measure on U, that is, $\mathrm{P}(x) > 0$ for all $x \in U$ and $\sum_{x \in U} \mathrm{P}(x) = 1$, then the triple (U, P, AT) is referred to as a random IS. Likewise, $(U, \mathrm{P}, C \cup \{d\})$ is called a random decision system with fuzzy decisions where $(U, C \cup \{d\})$ is a decision system with fuzzy decisions. It should be noted that an IS may be treated as a random IS with a special probability $\mathrm{P}(x) = 1/|U|$ for all $x \in U$.

Definition 3. *Let $S = (U, C \cup \{d\})$ be an IS with fuzzy decisions. For $B \subseteq C$ and $X \in \mathcal{F}(U)$, we define the lower and upper approximations of X wrt (U, R_B) as follows:*

$$\underline{R_B}(X)(x) = \bigwedge_{y \in [x]_B} X(y), \quad \overline{R_B}(X)(x) = \bigvee_{y \in [x]_B} X(y), \quad x \in U. \tag{11}$$

According to Definition 3, the following Property 1 can be easily concluded.

Property 1. Let $S = (U, C \cup \{d\})$ be an IS with fuzzy decisions. If $A \subseteq B \subseteq C$ and $X \in \mathcal{F}(U)$, then

$$\underline{R_A}(X) \subseteq \underline{R_B}(X) \subseteq X \subseteq \overline{R_B}(X) \subseteq \overline{R_A}(X). \tag{12}$$

Theorem 1. *Let $S = (U, \mathrm{P}, C \cup \{d\})$ be a random IS with fuzzy decisions. For $B \subseteq C$ and $X \in \mathcal{F}(U)$, if $\underline{R_B}(X)$ and $\overline{R_B}(X)$ are, respectively, the lower and upper approximations of X wrt (U, R_B) defined by Definition 3, denote*

$$\begin{aligned} \mathrm{Bel}_B(X) &= \mathrm{P}(\underline{R_B}(X)) = \sum_{x \in U} \underline{R_B}(X)(x)\mathrm{P}(x), \\ \mathrm{Pl}_B(X) &= \mathrm{P}(\overline{R_B}(X)) = \sum_{x \in U} \overline{R_B}(X)(x)\mathrm{P}(x), \end{aligned} \tag{13}$$

then $\mathrm{Bel}_B : \mathcal{F}(U) \to [0,1]$ and $\mathrm{Pl}_B : \mathcal{F}(U) \to [0,1]$ are, respectively, a CF-belief function and a CF-plausibility function on U, and the corresponding basic probability assignment m_B is

$$m_B(Y) = \begin{cases} \mathrm{P}(Y), & \text{if } Y \in U/R_B, \\ 0, & \text{otherwise.} \end{cases} \tag{14}$$

By Property 1 and Theorem 1, we can easily conclude following

Property 2. Let $S = (U, \mathrm{P}, C \cup \{d\})$ be a random IS with fuzzy decisions. If $A \subseteq B \subseteq C$ and $X \in \mathcal{F}(U)$, then

$$\mathrm{Bel}_A(X) \leq \mathrm{Bel}_B(X) \leq \mathrm{P}(X) \leq \mathrm{Pl}_B(X) \leq \mathrm{Pl}_A(X). \tag{15}$$

4 Attribute Reducts in Random Information Systems with Fuzzy Decisions

In this section, we discuss attribute reducts in a random IS with fuzzy decisions.

Definition 4. *Let $S = (U, \mathrm{P}, C \cup \{d\})$ be a random IS with fuzzy decisions and $B \subseteq C$. Then*

(1) *B is referred to as a lower approximation consistent set of S if*

$$\underline{R_B}(S_d(x)) = \underline{R_C}(S_d(x)) \ \forall x \in U. \tag{16}$$

If B is a lower approximation consistent set of S and no proper subset of B is a lower approximation consistent set of S, then B is referred to as a lower approximation reduct of S.

(2) *B is referred to as an upper approximation consistent set of S if*

$$\overline{R_B}(S_d(x)) = \overline{R_C}(S_d(x)) \ \forall x \in U. \tag{17}$$

If B is an upper approximation consistent set of S and no proper subset of B is an upper approximation consistent set of S, then B is referred to as an upper approximation reduct of S.

(3) *B is referred to as a random belief consistent set of S if*

$$\mathrm{Bel}_B(S_d(x)) = \mathrm{Bel}_C(S_d(x)) \ \forall x \in U. \tag{18}$$

If B is a random belief consistent set of S and no proper set of B is a random belief consistent set of S, then B is referred to as a random belief reduct of S.

(4) *B is referred to as a random plausibility consistent set of S if*

$$\mathrm{Pl}_B(S_d(x)) = \mathrm{Pl}_C(S_d(x)) \ \forall x \in U. \tag{19}$$

If B is a random plausibility consistent set of S and no proper subset of B is a random plausibility consistent set of S, then B is referred to as a random plausibility reduct of S.

Theorem 2. *Let $S = (U, \mathrm{P}, C \cup \{d\})$ be a random IS with fuzzy decisions and $B \subseteq C$, then*

(1) *B is a lower approximation consistent set of S iff B is a random belief consistent set of S.*

(2) *B is a lower approximation reduct of S iff B is a random belief reduct of S.*

Proof. (1) "$\Rightarrow$" If B is a lower approximation consistent set of S, that is, $\underline{R_B}(S_d(x)) = \underline{R_C}(S_d(x))$ for all $S_d(x) \in U/R_d$. Then

$$\sum_{y \in U} \underline{R_B}(S_d(x))(y)\mathrm{P}(y) = \sum_{y \in U} \underline{R_C}(S_d(x))(y)\mathrm{P}(y) \ \forall S_d(x) \in U/R_d. \tag{20}$$

That is,

$$\mathrm{P}(\underline{R_B}(S_d(x)) = \mathrm{P}(\underline{R_C}(S_d(x)) \ \forall S_d(x) \in U/R_d. \tag{21}$$

Consequently

$$\mathrm{Bel}_B(S_d(x)) = \mathrm{Bel}_C(S_d(x)) \ \forall S_d(x) \in U/R_d. \tag{22}$$

Thus, B is a random belief consistent set of S.

"$\Leftarrow$" If B is a random belief consistent set of S, for any $x \in U$, by the definition, we have $\mathrm{Bel}_B(S_d(x)) = \mathrm{Bel}_C(S_d(x))$, that is,

$$\mathrm{P}(\underline{R_B}(S_d(x)) = \mathrm{P}(\underline{R_C}(S_d(x)). \tag{23}$$

Equivalently,

$$\sum_{y \in U} \underline{R_B}(S_d(x))(y)\mathrm{P}(y) = \sum_{y \in U} \underline{R_C}(S_d(x))(y)\mathrm{P}(y). \tag{24}$$

Since $B \subseteq C$, by Eq. (12), we have

$$\underline{R_B}(S_d(x))(y) \le \underline{R_C}(S_d(x))(y) \ \forall y \in U. \tag{25}$$

Notice that P is a normalized probability on U, i.e., $\mathrm{P}(y) > 0$ for all $y \in U$, then, combining Eqs. (24) and (25), we must have

$$\underline{R_B}(S_d(x))(y) = \underline{R_C}(S_d(x))(y) \ \forall y \in U. \tag{26}$$

Therefore

$$\underline{R_B}(S_d(x)) = \underline{R_C}(S_d(x)). \tag{27}$$

Thus, we have proved that B is a lower approximation consistent set of S.

(2) It follows immediately from (1).

Similar to Theorem 2, the following Theorem 3 implies that in a random IS with fuzzy decisions the concepts of upper approximation reduct and plausibility reduct are equivalent.

Theorem 3. *Let $S = (U, \mathrm{P}, C \cup \{d\})$ be a random IS with fuzzy decisions and $B \subseteq C$, then*

(1) B is an upper approximation consistent set of S iff B is a random plausibility consistent set of S.

(2) B is an upper approximation reduct of S iff B is a random plausibility reduct of S.

Proof. It is similar to the proof of Theorem 2.

According to Eq. (15) and Theorems 2 and 3, we can conclude following

Theorem 4. *Let $S = (U, \mathrm{P}, C \cup \{d\})$ be a random IS with fuzzy decisions and $B \subseteq C$. Then*

(1) B is a random belief consistent set of S iff

$$\sum_{D \in U/R_d} Bel_B(D) = \sum_{D \in U/R_d} Bel_C(D). \tag{28}$$

(2) B is a random plausibility consistent set of S iff

$$\sum_{D \in U/R_d} Pl_B(D) = \sum_{D \in U/R_d} Pl_C(D). \tag{29}$$

(3) B is a random belief reduct of S iff

$$\sum_{D \in U/R_d} Bel_B(D) = \sum_{D \in U/R_d} Bel_C(D) \tag{30}$$

and for each $B' \subset B$,

$$\sum_{D \in U/R_d} Bel_{B'}(D) < \sum_{D \in U/R_d} Bel_C(D). \tag{31}$$

(4) B is a random plausibility reduct of S iff

$$\sum_{D \in U/R_d} Pl_B(D) = \sum_{D \in U/R_d} Pl_C(D) \tag{32}$$

and for each $B' \subset B$,

$$\sum_{D \in U/R_d} Pl_{B'}(D) > \sum_{D \in U/R_d} Pl_C(D). \tag{33}$$

Theorem 4 implies that, in a random IS with fuzzy decisions, a random belief reduct (respectively, a random plausibility redcut) is a minimal attribute set to keep the sum of degrees of belief (respectively, plausibility) of fuzzy similarity decision classes generated by the full conditional attributes.

5 Conclusion

Rough set theory and Dempster-Shafer theory of evidence are two important ones to deal with imprecision and uncertainty. The lower and upper approximations and belief and plausibility measures of a set respectively characterize the non-numeric and numeric uncertain aspects of the available information. In this paper, we have introduced the notions of lower approximation reducts, upper approximation reducts, random belief reducts and random plausibility reducts in random information systems with fuzzy decisions. We have examined relationships among these reducts. We will investigate knowledge reduction and knowledge acquisition in random fuzzy decision systems in our further study.

Acknowledgement. This work was supported by grants from the National Natural Science Foundation of China (Nos. 61075120, 11071284, 60673096 and 60773174) and the Natural Science Foundation of Zhejiang Province in China (No. Y107262).

References

1. Lingras, P.J., Yao, Y.Y.: Data mining using extensions of the rough set model. Journal of the American Society for Information Science 49, 415–422 (1998)
2. Pawlak, Z.: Rough Sets: Theoretical Aspects of Reasoning about Data. Kluwer Academic Publishers, Boston (1991)
3. Shafer, G.: A Mathematical Theory of Evidence. Princeton University Press, Princeton (1976)
4. Skowron, A.: The rough sets theory and evidence theory. Fundamenta Informaticae 13, 245–262 (1990)
5. Wu, W.-Z.: A comparative study of belief and plausibility reducts in information systems with fuzzy decisions. In: Proceedings of 2010 International Conference on Machine Learning and Cybernetics (ICMLC 2010), Qingdao, China, July 11-14, pp. 552–557 (2010)
6. Wu, W.-Z.: Attribute reduction based on evidence theory in incomplete decision systems. Information Sciences 178, 1355–1371 (2008)
7. Wu, W.-Z., Leung, Y., Mi, J.-S.: On generalized fuzzy belief functions in infinite spaces. IEEE Transactions on Fuzzy Systems 17, 385–397 (2009)
8. Wu, W.-Z., Leung, Y., Zhang, W.-X.: Connections between rough set theory and Dempster-Shafer theory of evidence. International Journal of General Systems 31, 405–430 (2002)
9. Wu, W.-Z., Zhang, M., Li, H.-Z., Mi, J.-S.: Knowledge reduction in random information systems via Dempster-Shafer theory of evidence. Information Sciences 174, 143–164 (2005)
10. Yao, Y.Y.: Interpretations of belief functions in the theory of rough sets. Information Sciences 104, 81–106 (1998)
11. Zadeh, L.A.: Probability measures of fuzzy events. Journal of Mathematical Analysis and Applications 23, 421–427 (1968)
12. Zhang, M., Xu, L.D., Zhang, W.-X., Li, H.-Z.: A rough set approach to knowledge reduction based on inclusion degree and evidence reasoning theory. Expert Systems 20, 298–304 (2003)

Discernibility-Matrix Method Based on the Hybrid of Equivalence and Dominance Relations

Yan Li, Jin Zhao, Na-Xin Sun, Xi-Zhao Wang, and Jun-Hai Zhai

Faculty of Mathematics and Computer Science, Hebei University, Baoding 071002, Hebei Province, China
ly@hbu.cn,zhaojinjin111@126.com,wangxz@hbu.cn

Abstract. The attribute set of some information systems is composed of both regular attributes and criteria. In order to obtain information reduction of this type of information systems, equivalence relation should be defined on the regular attributes and dominance relation on the criteria. Firstly, suppose condition attributes are criteria and decision attributes are regular attributes, dominance-equivalence relation is introduced,and the Discernibility-Matrix (DM) method of reduct generation is developed and compared with the attribute significance method. Secondly, when condition attributes are the hybrid of regular attributes and criteria, equivalence-dominance relation is then defined and Discernibility-Matrix approach of reduction generation is also provided.The effectiveness of this method is shown by both theoretical proof and illustrative example.

Keywords: Equivalence relation, Dominance relation, Postitive domian Reduction, Discernibility matrix, Attribute significance.

1 Introduction

In rough set theory, the concept of reduct is introduced based on equivalence relation [1], especially for incomplete or inconsistent systems [2-4], which removes unnecessary information without losing classification ability of information systems. On the other hand, the information systems based on dominance relations [5] can deal with attributes with partial order, such as real number attributes and ordinal symbolic attributes. Different types of reductions are introduced based on dominance relations in [6-10]. However, sometimes there are both ordinal and non-ordinal attributes in one information system such as table 1. The values of color are light and dark, which are non-ordered; while the preference-order of price values are continuous and should not be ignored. The equivalence relation cannot take the ordered information into account and therefore the generated rules (knowledge) cannot reflect the information either. To deal with this type of information systems, Greco, Slowinski and Blaszczynski [11-16] developed the framework of dominance-based rough set approach(DRSA), where the ordinal attribute is called as criterion, and the non-ordinal attribute is called as regular attribute. In general, DRSA assumes that there are monotonic constraints

S.O. Kuznetsov et al. (Eds.): RSFDGrC 2011, LNAI 6743, pp. 231–239, 2011.

Table 1. Information system with the hybrid of criteria and regular attributes

NO.	color	size	year	price	choice
1	light	large	2010	117000	no
2	light	large	2003	35900	yes
3	dark	normal	2006	29900	yes
4	dark	medium	2004	19300	no
5	dark	small	2004	11400	no
6	light	small	2006	14200	yes
7	light	large	2009	106000	no
8	light	large	2004	420000	yes
9	dark	medium	2006	20410	no
10	light	normal	2007	29800	no
11	light	large	2010	83900	no
12	light	large	2010	83900	yes

between condition attributes and decision attributes. For the hybrid attribute set which contains both criteria and regular attributes, they suggest in [15][16]to transform all the regular attributes into criteria by duplicating these attributes respectively with increasing and decreasing order, for instance,color is replaced by color_1 and color_2, where the values of color_1 are increased, that is light<dark; while in color_2, it is considered that light>dark. Then the information system completely consists of criteria,and it can be handled by DRSA.However, the duplication of attributes makes the information systems more complicated, thus increase the computational load in reduction generation. Furthermore, considering a completely no-ordinal attribute(e.g., color takes values "red, blue, green") as criteria is difficult to understand in the sense of semantic meaning, and the generated rules often need post-processing, such as simplifying $\geq$ (or $\leq$) to $=$. To address this problem, we propose to introduce respectively equivalence relation to the regular attributes and dominance relation to criteria. For example, define equivalence relation to color and size, and define dominance relation to year and price, and the information system will be based on the equivalence-dominance relations. In order to compute reductions, equivalence classes and dominance classes are merged to equivalence-dominance classes, which is coverage of the universe. Then we directly define the concept of reduction and the discernibility-matrix method of reduction computation.

In the remainder of the paper, Section 2 briefly introduces a few basic concepts in DRSA. In Section 3, we consider information systems whose decision attributes are regular attributes and all the condition attributes are criteria, the dominance-equivalence relation is defined and the lower-approximation reduction is point out to be positive domain reduction in [8]. We then develop the discernibility-matrix method (called DM method) to the above reduction,and compare it with the attribute significance method. In Section 4, when the conditional attributes are the hybrid set of criteria and regular attributes, we define the equivalence-dominance relation and reduction and further give the DM method.Finally,conclusions are given in Section 5.

2 Basic Concepts in Dominance-Based Rough Set Approach

Definition 2.1 [6]. A 5-tuple$DS = (U, A, F, D, G)$ is referred to as a target information system, where(U, A, F) is an information system

$U = \{x_1, x_2, \cdots, x_n\}$ is a non-empty finite set of objects;
$A = \{a_1, a_2, \cdots, a_n\}$ is a finite set of condition attributes;
$D = \{d_1, d_2, \cdots, d_q\}$ is a finite set of decision attributes;
$F = \{f_k : U \to V_k, k \leq p\}$,$V_k$ is the finite domain of a_k;
$G = \{g_k : U \to V_k', k\prime \leq q\}$,$x \in U$, V_k' is the finite domain of d_k.

Definition 2.2 [6]. Let $I = (U, A, F, D, G)$ be a target information system, $B \subseteq A$, the dominance relation induce by B of I are respectively denoted as:

$$R_B^{\leq} = \{(x_i, x_j) \in U \times U : f_l(x_i) \leq f_l(x_j), \forall a_l \in B\}$$

B-dominance relation class of an object x can be defined as:

$$[x_i]_B^{\leq} = \{x_j \in U : (x_i, x_j) \in R_B^{\leq}\} = \{x_j \in U : f_l(x_i) \leq f_l(x_j), \forall a_l \in B\}$$

Definition 2.3 [9]. For any $X \subseteq U$, upper approximation and lower approximation of X respect to dominance relation $R_B^{\leq}$ can be defined as:

$$\overline{R_A^{\leq}}(X) = \{x_i \in U : [x_i]_A^{\leq} \bigcap X \neq \emptyset\}, \underline{R_A^{\leq}}(X) = \{x_i \in U : [x_i]_A^{\leq} \subseteq X\}$$

Definition 2.4 [8]. Assume that decision attributes be regular attribute. Based on dominance relation, the positive region of decision attribute set D relative to condition attribute (criteria) set A can be defined as:

$$POS_A(D) = \cup_{X \in U/D} \underline{R_A^{\leq}}(X)$$

Definition 2.5 [8]. Let $B \subseteq A$. If $POS_B(D) = POS_A(D)$, B is said to be a positive domain consistent set. If no proper subset of B is positive domain consistent set, then B is called a positive domain reduction.

[8] has proposed an algorithm for the computation of positive domain reduction based on attribute significance.

Definition 2.6. The significance of attribute subset $B(B' \subset B \subseteq A)$ can be denoted as:

$$SGF(B', B, D) = \frac{|POS_B(D)| - |POS_{B \setminus B'}(D)|}{|U|}$$

When $SGF(B', B, D) = 0, B'$ is not necessary in B; otherwise, B' is indispensable to B and cannot be removed.

3 Dominance-Equivalence Relation and DM Method

For target information systems whose decision attributes are regular attributes and all conditional attributes are criteria, the dominance relation $R_B^{\leq}$ is called dominance-equivalence relation in this paper. This is to distinguish this type of information systems with the completely dominance-based systems, in which both condition and decision attributes are criteria.[8] defines positive domain reduction and introduces condition attribute significance to obtain one positive domain reduction. In this section, we provide a DM method based on dominance-equivalence relation.

Definition 3.1. Let $I = (U, A, F, D, G)$ be a target information system, $B \subseteq A$, Denote $U/R_B^{\leqq} = \{[x_i]_B^{\leqq} : x_i \in U\}, U/R_D = \{D_1, D_2, \cdots, D_r\}$,

$$\eta_B = (\underline{R_B^{\leqq}}(D_1), \underline{R_B^{\leqq}}(D_2), \cdots, \underline{R_B^{\leqq}}(D_r))$$

where$[x_i]_B^{\leqq} = \{y \in U : (x, y) \in R_B^{\leqq}\}, \eta_B$ is the dominance domain function of U.

Proposition 3.1. Let $I = (U, A, F, D, G)$ be a target information system, $B \subseteq A$, B is said to be a positive domain consistent set if and only if for any $D_i \in U/R_D$, we have $\underline{R_B^{\leqq}}(D_i) = \underline{R_A^{\leqq}}(D_i)$, that is, $\eta_B = \eta_A$. If no proper subset of B is the positive domain consistent set, then B is called a positive domain reduction.

Theorem 3.1. Let $I = (U, A, F, D, G)$ be a target information system, $B \subseteq A$, B is said to be a positive domain consistent set if and only if for any $D_i \in U/R_D$, when $x \in \underline{R_B^{\leqq}}(D_i), y \notin \underline{R_B^{\leqq}}(D_i), \exists b \in B$ such that $f_b(x) > f_b(y)$

Proof : The proof is similar to that of theorem 1 in [7].

Definition 3.2. Let $I = (U, A, F, D, G)$ be a target information system,denote:

$$D_\eta^* = \{(x_i, x_j) : x_i \in \underline{R_A^{\leqq}}(D_i), x_j \notin \underline{R_A^{\leqq}}(D_i)\}$$

$$D_\eta(x_i, x_j) = \begin{cases} a_k \in A, f_{a_k}(x_i) > f_{a_k}(x_j), (x_i, x_j) \in D_\eta^*; \\ \emptyset, (x_i, x_j) \notin D_\eta^*. \end{cases}$$

where $D_\eta(x_i, x_j)$ is the positive domain discernibility attribute set of x_i and x_j. And$M_\eta = (D_\eta(x_i, x_j), x_i, x_j \in U)$ is referred as the positive domain discernibility matrix.

Using the defined discernibility matrix,multiple reductions can be obtained. Some set of rules are then extracted and can be used in the task of classification. We compare this decernibility-matrix method with the reduction computation algorithm based on attribute significance [8]. In our method, majority voting of the rules is used in classification. 15 data sets from the UCI machine learning repository are selected to conduct the experiments, which have regular decision attributes.Randomly select 50% samples as training set, and the rest as testing set.

Table 2 shows the results. $M\#$ is the number of attributes, "Correct recognition" is the proportion of the correctly classified samples, and "rejection rate" is the proportion of the samples that are not covered by the rules.We can see that the results of the two methods are similar for both correct recognition rate and rejection rate. Compared with attribute significance method, DM method has higher correct recognition rate and lower rejection rate on eight data sets. For other data sets, attribute significance method is slightly better. Therefore DM method can obtain multiple reductions without losing correct recognition rate.

Table 2. Discernibility-Matrix Method and Attribute Significance Method

Data set	M#	Attribute significance method		Discernibility matrix method	
		Correct recognition rate	Rejection rate	Correct recognition rate	Rejection rate
Auto-mpg	7	55.80%	25.00%	53.93%	24.57%
adult-stretch	8	70.91%	24.45%	72.27%	24.55%
balance-scale	5	73.51%	0.00%	75.64%	0.00%
bezdekIris	5	90.73%	3.07%	90.73%	3.47%
breast-cancer	10	91.60%	0.27%	90.17%	0.86%
bupa	7	64.19%	4.77%	65.84%	3.50%
Diabetes	8	73.22%	5.72%	70.87%	6.54%
sky	4	63.12%	7.50%	65.00%	8.75%
shuttle-landing	6	63.12%	36.77%	65.00%	35.00%
Heart-statlog	11	57.00%	24.52%	62.48%	22.11%
wine	13	47.22%	36.11%	45.00%	34.44%
Wine-quality	11	47.63%	20.79%	48.50%	31.59%
Ecoli	7	59.41%	11.71%	61.18%	9.06%
Iris	4	90.00%	4.13%	89.20%	4.93%
Pima	8	71.90%	6.64%	71.85%	6.15%
Average	-	67.96%	14.16%	68.51%	14.34%

4 Equivalence-Dominance Relation and DM Method

In some information systems, condition attributes is a hybrid set of regular attributes and criteria, and decision attributes are still regular attributes. We introduce the concept of equivalence-dominance relation, and correspondingly define some relevant concepts.

Definition 4.1. C is a hybrid set of criterion and regular attributes, $A \subset C$ is a regular attribute set, $B = C - A$ is a criterion set. The equivalence-dominance relation induced by C is denoted as:

$$R_C^{\approx\leq} = \{(x,y)|f_A(x) = f_A(y) \text{ and } f_B(x) \leq f_B(y)\}$$

Definition 4.2. Let A, B, C be the same as in Definition 4.1. C- equivalence-dominance relation class of an object x can be defined as follows:

$$[x_i]_c^{\approx\leq} = [x_i]_A \bigcap [x_i]_B^{\leq}$$

Definition 4.3. For any $X \subseteq U$, upper approximation and lower approximation of equivalence-dominance relation of $R^{\approx\leq}$ can be defined as follows:

$$\overline{R_C^{\approx\leq}}(X) = \{x_i \in U : [x_i]_C^{\approx\leq} \cap X \neq \emptyset\}, \underline{R_C^{\approx\leq}}(X) = \{x_i \in U : [x_i]_C^{\leq} \subseteq X\}$$

The relative positive region of decision attribute set D to C can be defined as: $POS_C^{\approx\leq}(D) = \cup\{x_j|[x_j]_C^{\approx\leq} \subseteq D_i\} = \cup_{X\in U/D}\underline{R_C^{\approx\leq}}(X)$, where $j = 1, 2, \cdots, n. D_i$ is the $i-th$ equivalence class defined on D.

Definition 4.4. $P \subseteq C$, if P can retain the positive region based on equivalence-dominance relation, and the subset of P cannot, then P is called a positive domain reduction of C.

Example 4.1. Considering the information system in Table 1 based on equivalence-dominance relation.

Denote $D = choice; a_1 = color, a_2 = size, a_3 = year, a_4 = price$

Let $A = \{a_1, a_2\}; B = \{a_3, a_4\}$.According to the definitions, we have:

$[x_1]_C^{\approx\leq} = \{x_1, x_{11}, x_{12}\}; [x_2]_C^{\approx\leq} = \{x_2\}; [x_3]_C^{\approx\leq} = \{x_3\}; [x_4]_C^{\approx\leq} = \{x_4\}$

$[x_5]_C^{\approx\leq} = \{x_5\}; [x_6]_C^{\approx\leq} = \{x_6\}; [x_7]_C^{\approx\leq} = \{x_7, x_{11}, x_{12}\}; [x_8]_C^{\approx\leq} = \{x_8\}$

$[x_9]_C^{\approx\leq} = \{x_9\}; [x_{10}]_C^{\approx\leq} = \{x_{10}\}; [x_{11}]_C^{\approx\leq} = [x_{12}]_C^{\approx\leq} = \{x_{11}, x_{12}\}$

$D_1 = \{x_1, x_4, x_5, x_7, x_9, x_{10}, x_{11}\}, D_2 = \{x_2, x_3, x_6, x_8, x_{12}\}$

Based on the equivalence-dominance relation,the lower approximations and positive region are:

$\underline{R_C^{\approx\leq}}(D_1) = \{x_2, x_3, x_6, x_8\}; \underline{R_C^{\approx\leq}}(D_2) = \{x_4, x_5, x_9, x_{10}\}$

$\underline{POS_C^{\approx\leq}}(D) = \{x_2, x_3, x_4, x_5, x_6, x_8, x_9, x_{10}\}$

According to the definition 4.3-4.4,$\{a_1, a_2\} \cap \{a_4\}$ is the positive domain reduction. Next,DM method of positive domain reduction based on equivalence-dominance relation is given.

Definition 4.5. Let $I = (U, A, F, D, G)$ be an information system based on equivalence-dominance relation, and $P \subseteq C$. Denote $U/R_P^{\approx\leq} = \{[x_i]_P^{\approx\leq} : x \in U\}, U/R_D = \{D_1, D_2, \cdots, D_r\}$,

$$\rho_P^{\approx\leq} = (\underline{R_P^{\approx\leq}}(D_1), \underline{R_P^{\approx\leq}}(D_2), \cdots, \underline{R_P^{\approx\leq}}(D_r))$$

where $[x_i]_P^{\approx\leq} = \{y \in U : (x, y) \in R_P^{\approx\leq}\}, \rho_P^{\approx\leq}$ is called dominance domain function of P in U.

Definition 4.6. Let $I = (U, A, F, D, G)$ be an information system based on equivalence-dominance relation, and $P \subseteq C$. If $\rho_P^{\approx\leq} = \rho_C^{\approx\leq}$, P is said to be a positive domain consistent set; If no proper subset of P is positive domain consistent set, P is also called a positive domain reduction.

Proposition 4.1. Let $I = (U, A, F, D, G)$ be an information system based on equivalence-dominance relation, and $P \subseteq C$. P is a positive domain consistent set if and only if for any $D_i \in U/R_D$,$\underline{R_P^{\approx\leq}}(D_i) = \underline{R_C^{\approx\leq}}(D_i)$ holds.

Then we give the judgment theorem of positive domain reduction.

Theorem 4.1. Let $I = (U, A, F, D, G)$ be an information system based on equivalence-dominance relation, $A \subset C$ is a regular attribute set, $B = C - A$ is a criterion set, and $P \subseteq C$. P is a positive domain consistent set if and only if for any $D_i \in U/R_D$, when $x \in \underline{R_C^{\approx\leq}}(D_i), y \notin \underline{R_C^{\approx\leq}}(D_i), \exists p \in P$ (1)If $p \in A$ then $f_p(x) \neq f_p(y)$(2)If $p \in B, then f_p(x) > f_p(y)$.

Proof "⇒"The proof is by contradiction.

There exist $D_i \in U/R_D$, assume when $x \in \underline{R_C^{\approx\leq}}(D_i), y \notin \underline{R_C^{\approx\leq}}(D_i), \exists p \in P$ (1)If $p \in A$ then $f_p(x) = f_p(y)$(2)If $p \in B, then f_p(x) \leq f_p(y)$. That is $\forall p \in P, f_p(x) \leq$

$f_p(y)$.So,$y \in [x]_P^{\approx\leq}$. On the other hand, B is a positive domain consistent set, for $\forall D_i \in U/R_D$, we have $\underline{R_P^{\approx\leq}}(D_i) = \underline{R_C^{\approx\leq}}(D_i)$. Because $x \in \underline{R_C^{\approx\leq}}(D_i), x \in \underline{R_P^{\approx\leq}}(D_i)$, that is, $[x]_P^{\approx\leq} \subseteq D_i$, and $y \in [x]_P^{\approx\leq}$, $[y]_P^{\approx\leq} \subseteq [x]_P^{\approx\leq}$, we have $[y]_P^{\approx\leq} \subseteq D_i$ so $y \in \underline{R_P^{\approx\leq}}(D_i)$. Therefore, $y \in \underline{R_C^{\approx\leq}}(D_i)$, it is in contradiction to $y \notin \underline{R_C^{\approx\leq}}(D_i)$.

"$\Leftarrow$"Assume that P is a positive domain consistent set, there exist $D_i \in U/R_D$, such that $\underline{R_P^{\approx\leq}}(D_i) \neq \underline{R_C^{\approx\leq}}(D_i)$. That is $\in x_0 \in \underline{R_C^{\approx\leq}}(D_i)$ and $x_0 \notin \underline{R_P^{\approx\leq}}(D_i)$, so $[x_0]_C^{\approx\leq} \subseteq D_i$, and $[x_0]_P^{\approx\leq} \nsubseteq D_i$. Because $[x_0]_C^{\approx\leq} \subseteq [x_0]_P^{\approx\leq}$, there exist $y_0 \in [x_0]_P^{\approx\leq}$, and $y_0 \notin D_i$, that is, $y_0 \notin \underline{R_C^{\approx\leq}}(D_i)$, which implies $x_0 \in \underline{R_C^{\approx\leq}}(D_i), y_0 \notin \underline{R_C^{\approx\leq}}(D_i)$. Therefore, (1)If $\exists p \in A$, we have $f_p(x_0) \neq f_p(y_0)$(2)If $\exists p \in B, f_p(x_0) > f_p(y_0)$holds. Obviously, it is in contradiction to $y_0 \in [x_0]_P^{\approx\leq}$. This completes the proof.

Definition 4.7. Discernibility matrix of positive domain reduction can be defined as:

$$M_{n\times n} = (c_{ij})_{n\times n} = \begin{pmatrix} c_{11} & c_{12} & \cdots & c_{1n} \\ c_{21} & c_{22} & \cdots & c_{2n} \\ \vdots & \vdots & \ddots & \vdots \\ c_{n1} & c_{n2} & \cdots & c_{nn} \end{pmatrix}$$

where $c_{ij} = c_{ij}\prime \cup c_{ij}\prime\prime . i,j = 1,2,\cdots,n$

$$c_{ij}\prime = \begin{cases} \{\alpha \in A, f_\alpha(x_i) \neq f_\alpha(x_j)\}, (x_i,x_j) \in D_\eta^*; \\ \emptyset, (x_i,x_j) \notin D_\eta^*. \end{cases}$$

$$c_{ij}\prime\prime = \begin{cases} \{a_k \in B, f_{a_k}(x_i) > f_{a_k}(x_j)\}, (x_i,x_j) \in D_\eta^*; \\ \emptyset, (x_i,x_j) \notin D_\eta^*. \end{cases}$$

$D_\eta^* = \{(x_i,x_j) : x_i \in \underline{R_C^{\approx\leq}}(D_i), x_j \notin \underline{R_C^{\approx\leq}}(D_i)\}$,$D_i \in U/R_D$

Example 4.2. Consider Table 1.We can compute positive domain discernibility matrix as follows, where i means the i-th attribute a_i.

∅	∅	∅	∅	∅	∅	∅	∅	∅	∅	∅	∅
4	∅	∅	1,2	1,2	∅	4	∅	1,2	2	4	4
1,2,4	∅	∅	2,3	2	∅	1,2,4	∅	2	1	1,2,4	1,2,4
1,2,4	1,2,3,4	2,4	∅	∅	1,2	1,2,4	1,2,4	∅	∅	1,2,4	1,2,4
1,2,4	1,2,3,4	2,4	∅	∅	1,2	1,2,4	1,2,4	∅	∅	1,2,4	1,2,4
2,4	∅	∅	1,2,3,4	2,3	∅	2,4	∅	1,2,4	2,4	2,4	2,4
∅	∅	∅	∅	∅	∅	∅	∅	∅	∅	∅	∅
4	∅	∅	1,2	1,2	2	4	∅	1,2	2	4	4
1,2,4	1,2,3,4	2,4	∅	∅	1,2	1,2,4	1,2,3,4	∅	∅	1,2,4	1,2,4
2,4	2,3,4	1,2,4	∅	∅	2,3	2,4	2,3	∅	∅	2,4	2,4
∅	∅	∅	∅	∅	∅	∅	∅	∅	∅	∅	∅
∅	∅	∅	∅	∅	∅	∅	∅	∅	∅	∅	∅

Consequently, we have

$$(a_1) \wedge (a_2) \wedge (a_4) \wedge (a_1 \vee a_2 \vee a_4) \vee (a_2 \vee a_4) \wedge (a_1 \vee a_3 \vee a_4) = a_1 \wedge a_2 \wedge a_3$$

This result is consistent with that using the definition of reduction in Example 4.1. Using this discernibility matrix, other reductions can be also obtained.

5 Conclusion

In order to extract the most important information from the complex data (whether it is ordinal or non-ordinal), reduction computation is necessary. We give a new discernibility-matrix method to obtain the positive domain reduction and compare it with the attribute significance method in [8]. Then a reduction method with more complex data is given based on equivalence-dominance relation. For this type of data,we also provide discernibility-matrix method to obtain the positive domain reduction. The effectiveness of this method is guaranteed by both theoretical proof and illustrative examples.

Acknowledgments. This work is supported by NSFC (No.60903088), Natural Science Foundation of Hebei Province(No.F2009000227,A2010000188,F2010000323), 100-Talent Programme of Hebei Province(CPRC002),key project of Educational Department of Hebei Province(ZD2010139), and Hong Kong PolyU grant A-PJ18.

References

1. Pawlak, Z.: Rough set: Theoretical aspects of reasoning about data. Kluwer Academic Publishers, Boston (1991)
2. Kryszkiewicz, M.: Rough Set Approach to Incomplete Information System. Information Sciences 112, 39–49 (1998)
3. Kryszkiewicz, M.: Comparative studies of alternative of knowledge reduction in inconsistent systems. Intelligent Systems 16(1), 105–120 (2001)
4. Zhang, W.X., Mi, J.S., Wu, W.Z.: Approaches to Knowledge Reductions in Inconsistent Systems. Chinese Journal of Computers 26, 12–18 (2003)
5. Greco, S., Matarazzo, B., Slowingski, R.: Intelligent Decision Support: Rough approximation of a preference relation by dominance relation. European Journal of Operation Research 117, 63–83 (1999)
6. Xu, W.H., Zhang, W.X.: Distribution reduction in inconsistent information systems based on dominance relations. Fuzzy Systems and Mathematics 21(4), 124–131 (2007) (in Chinese)
7. Xu, W.H., Zhang, X.Y., Zhang, W.X.: Lower approximation reduction in inconsistent information systems based on dominance relations. Computer Engineering and Applications 45(16), 66–68 (2009) (in Chinese)
8. Chen, J., Wang, G.Y., Hu, J.: Positive Domain Reduction Based on Dominance Relation in Inconsistent System. Computer Science 35(3), 216–218, 227 (2008) (in Chinese)

9. Zhang, W.X., Qiu, G.F.: Uncertain decision making based on rough set. Tsinghua University Press, Beijing (2005)
10. Shao, M.W., Zhang, X.Y.: Dominance relation and rules in an incomplete ordered information system. International Journal of Intelligent Systems 20, 13–27 (2005)
11. Greco, S., Matarazzo, B., Slowinski, R.: Rough sets methodology for sorting problems in presence of multiple attributes and criteria. European Journal of Operational Research 138, 247–259 (2002)
12. Blaszczynski, J., Greco, S., Slowinski, R.: Multi-criteria classification-A new scheme for application of dominance-based decision rules. European Journal of Operational Research 181, 1030–1044 (2007)
13. Greco, S., Matarazzo, B., Slowinski, R.: Dominance-Based Rough Set Approach as a Proper Way of Handling Graduality in Rough Set Theory. Transactions on Rough Sets 7, 36–52 (2007)
14. Blaszczynski, J., Greco, S., Slowinski, R., Szelag, M.: Monotonic variable consistency rough set approaches. International Journal of Approximate Reasoning 50, 979–999 (2009)
15. Blaszczynski, J., Greco, S., Slowinski, R.: Ordinal and non-ordinal classification using monotonic rules. In: 8th International Conference of Modeling and Simulation-MOSIM 2010 (2010)
16. Blaszczynski, J., Slowinski, R., Szelag, M.: Sequential covering rule induction algorithm for variable consistency rough set approaches. Information Sciences (in press)

Studies on an Effective Algorithm to Reduce the Decision Matrix

Takurou Nishimura[1], Yuichi Kato[1], and Tetsuro Saeki[2]

[1] Shimane University,
1060 Nishikawatsu-cho, Matsue city, Shimane 690-8504, Japan
ykato@cis.shimane-u.ac.jp
[2] Yamaguchi University,
2-16-1 Tokiwadai, Ube city, Yamaguchi 755-8611, Japan
tsaeki@yamaguchi-u.ac.jp

Abstract. In the conventional rough set theory, the decision matrix method is known as one of the method extracting the rules[1]. However, devising an efficient algorithm for the decision matrix method has seldom been reported to date. Consequently, this paper studies the process of reducing the decision matrix, finds several properties useful for the rule extraction and proposes an effective algorithm for the extraction. The algorithm is implemented in a piece of software and a simulation experiment is conducted to compare the reduced time of the software with that of LEM2[2][3]. As the results, the newly developed software is confirmed to perform exceptionally well under taxing conditions.

1 Consideration on the Decision Matrix

In order to study the decision matrix method proposed by Shan and Ziarko[1], we specify a decision table which has condition attributes $\{A(k) \mid k = 1, ..., K\}$ with their attribute values $A(k) = \{a(k, v) \mid v = 1, ..., V(k)\}$ and the corresponding decision attribute $Y = \{y1, y2\}$. The decision matrix to discern samples of $Y = y1$ from those of $Y = y2$ is expressed by the following two equations:

$$p(i) = \wedge_{j=1}^{J} D(i, j), \tag{1}$$

$$R1 = \vee_{i=1}^{I} p(i), \tag{2}$$

where $D(i, j) = \vee_{k=1}^{K} a(k, v(i))|_{a(k,v(i)) \neq a(k,v(j))}$ means that $a(k, v(i))$ which denotes a value of the k-th attribute of the i-th sample $s(i)$ of $Y = y1$, differs from that of the j-th sample $s(j)$ of $Y = y2$.

1.1 Consideration on Reducing Eq. (1)

With regard to (1), the following property is approved:

Proposition 1: $p(i)$ possibly includes only $a(1, v(i))$, $a(2, v(i))$, ..., $a(K, v(i))$ which construct $s(i)$.

S.O. Kuznetsov et al. (Eds.): RSFDGrC 2011, LNAI 6743, pp. 240–243, 2011.

Proposition 1 is evident by the property of $D(i,j)$. Accordingly, the reduced result of (1) can be once encoded by $a(k)$ ($k = 1, ..., K$) and replace them by the original values when decoded. One fast method of finding the rules permitted by (1) is the following procedure:

Step1: Transform $D(i,j)$ into the following bit pattern $\Delta(i,j)$:

$$\Delta(i,j) = (\delta(1,v(i)),...,\delta(k,v(i)),...,\delta(K,v(i))), \tag{3}$$

where $\delta(k,v(i)) = \begin{cases} 1 & (a(k,v(i)) \neq a(k,v(j))) \\ 0 & (a(k,v(i)) = a(k,v(j))) \end{cases}$.

Step2: Execute the conjunction of the corresponding bit between $\Delta(i,j)$ and $Bp(n)$, where $Bp(n) = (\delta(1), ..., \delta(k), ..., \delta(K))$, $n = \sum_{k=0}^{K-1} \delta(K-k)2^k$, and $\delta(k) = 1$ or 0 ($n = 1, ..., 2^K - 1$).
The bit pattern $Bp(n)$, which $\Delta(i,j)$, that is $D(i,j)$, permits, has the results of the conjunction that all of the bits are not zero. On the contrary, the prohibited bit pattern $Bp(n)$ has the result that all bits are 0.
Step3: Execute Step 2 for $j = 1$ to J.
Step4: Arrange the permitted $Bp(n)$ as $SBp = \{Bp(n(l)) \mid l = 1, ..., L\}$.

The set SBp should be arranged not to be redundant, that is, to be independent of each other in the members $Bp(n(l_1))$ and $Bp(n(l_2))$ ($1 \leq l_1 < l_2 \leq L$) by the following procedures (Arrange Strategy):

If $Bp(n(l_1)) \wedge Bp(n(l_2)) = Bp(n(l_1))$ then delete $Bp(n(l_2))$ from SBp ($\because$ $Bp(n(l_2))$ is contained in $Bp(n(l_1))$),
If $Bp(n(l_1)) \wedge Bp(n(l_2)) = Bp(n(l_2))$ then delete $Bp(n(l_1))$ from SBp ($\because$ $Bp(n(l_1))$ is contained in $Bp(n(l_2))$).

Suppose $SBp = \{Bp(m) \mid m = 1, ..., M\}$ arranged by the above strategy. The results of (1) is obtained by decoding $Bp(m)$ ($m = 1, ..., M$) by use of the attribute values of the i-th sample $s(i)$ ($= a(1,v(i))\ a(2,v(i))\ ...\ a(K,v(i))$) and denoted by $Q = \{q(m) \mid m = 1, ..., M\}$.

1.2 Consideration on Reducing Eq. (2)

We examine the way of reducing (2) based on the results of (1). The following two properties are approved in the processes of the reduction (due to the limitation of paper space, we omit the proofs of the below two propositions):

Proposition 2: The reduction by the type of the complementarity law: $xy \vee x\overline{y} = x$, never occurs in the Boolean operation of (2).

Proposition 3: The reduction by the type of the absorption law: $x \vee xy = x$, never occurs in the Boolean operation of (2).

Accordingly, only the type of the idempotent law: $x \vee x = x$, occurs in the reduction in (2) and the different terms derived from (1) are the results of (2), which largely omits the reduction operations in (2).

```
List No. Procedure DMM
1:   (input: sets Srow = {s(i)|i = 1,...,I} and Scolum = {s(j)|j = 1,...,J} ,
     output: a set τ)
2:   Initialize: I, J, K set; τ = φ; T = φ; T = {t(n_T)|n_T = 1,..,N_T}; N_T = 0;
3:   For (i = 1; i ≤ I; i++) {
       SBp = {sbp(n_SBp)|n_SBp = 1,...,N_SBp}; SBp = φ; N_SBp = 0;
4:     For (n = 1; n ≤ 2^K − 1; n++) {
5:       For (j = 1; j ≤ J; j++) { if Bp(n) ∧ Δ(i,j) == 0 then goto L1; }
         N_SBp++; sbp(N_SBp) ← Bp(n); // add Bp(n) to SBp
       L1: } // end of n-loop
6:     arrange SBp by arrange strategy → SBp = {Bp(m)|m = 1,...,M}
7:     decode SBp → Q = {q(m)|m = 1,...,M}; // results of (1)
8:     For (m = 1; m ≤ M; m++) {
9:       For (n_T = 1; n_T ≤ N_T; nT++) { if t(n_T) == q(m) then goto L2; }
10:      N_T++; t(N_T) ← q(m) // add q(m) to T
11:    L2: } // end of m-loop
12:  } // end of i-loop
13:  G ← Srow;
14:  While (G ≠ φ) {
15:  Select t(l) such that |t(l) ∩ G| is maximum; If a tie occurs, select the first t(l);
     // For set X, |X| denotes the cardinality of X.
     τ ← τ + t(l);
16:  G ← G − t(l) ∩ G; T ← T − t(l);
     } // end of while
```

Fig. 1. An algorithm of fast decision matrix method (FDMM)

2 An Effective Algorithm to Reduce the Decision Matrix

Figure 1 shows an algorithm for FDMM (Fast Decision Matrix Method) described in C language style based on the considerations in 1.1 and 1.2. In the figure, the input data sets Srow and Scolum are a decision table whose contradicted data are arranged by use of the lower or upper approximation method. The output data τ is a set of the conjunction terms derived from (1) and (2). This algorithm proceeds each sample $s(i)$, and hence the algorithm requires only a memory area of $I + J$ although a matrix generally requires that of $I \times J$. After reducing (1) and (2), procedures from line 13 to line 16 is added in order to obtain the minimum number of rules covering Srow since the result of reducing the decision matrix is not necessarily optimal.

3 Simulation Experiment — Comparison of the Execution Time with Existing Software

We developed the algorithm shown in Fig. 1 into a piece of software using C language, and implemented it in a personal computer to compare the execution time [sec] of extracting rules with that of LEM2 software[3] within the ROSE2 system[2].

The following two rules were specified in advance in order to generate samples and make their decision matrix:

Rule1: if $(A(1) = a(1,1)$ and $A(2) = a(2,1))$ or $(A(3) = a(3,1)$ and $A(4) = a(4,1))$ then $Y = y1$,
Rule2: if $(A(1) = a(1,2)$ and $A(2) = a(2,2))$ or $(A(3) = a(3,2)$ and $A(4) = a(4,2))$ then $Y = y2$.

Table 1. A comparison of the time on reducing rules between FDMM and LEM2 ($N_{actual} \rightarrow (t_{FDMM}, t_{LEM2})$, N_{actual}: actual data number, t_{FDMM} [s]: time of FDMM, t_{LEM2} [s]: time of LEM2)

(a) $(K, V(k)) = (5, 5)$		
$N = 1000$	$N = 5000$	$N = 10000$
(907)→(0, 3)	(3288)→(3, 85)	(4640)→(4, 98)
(912)→(0, 3)	(3322)→(3, 82)	(4660)→(3, 127)
(918)→(0, 5)	(3333)→(3, 86)	(4725)→(3, 81)
(b) $(K, V(k)) = (6, 6)$		
$N = 1000$	$N = 5000$	$N = 10000$
(993)→(1, 6)	(4846)→(21, 1158)	(9417)→(75, 13129)
(994)→(1, 5)	(4854)→(21, 1059)	(9432)→(84, 12319)
(995)→1, 6)	(4871)→(21, 1307)	(9442)→(77, 11383)

The sample data of the condition attribute values were generated randomly, and the corresponding decision value was decided according to Rule 1 and Rule 2. However, the decision value of Y was randomly assigned in the case where the generated set satisfied both of the specified rules and/or neither of the specified rules could be applied.

The specifications and conditions for the experiment were as follows: The number of the condition attributes K was 5 and 6, and the number of the attribute values $V(k)$ corresponding to the two cases was constantly 5 and 6, regardless of each respective condition attribute. We denote this specification as $(5, 5)$ and $(6, 6)$ respectively. The number of the generated data N was 1000, 5000 and 10000, and the lower approximation was used to make the consistent data set. The experiment was conducted three times for each case using a personal computer with an Intel Core2 CPU with 1.66 [GHz] clock speed and 2038 [MB] of RAM memory.

The results of the experiments are summarized in Table 1, where the time of extracting rules by FDMM t_{FDMM} and by LEM2 t_{LEM2} are shown corresponding to the actual data number N_{actual}, deleting the redundant data in the form: $N_{actual} \rightarrow (t_{FDMM}, t_{LEM2})$, since the data was randomly generated and included the same data. The times shown in the table are truncated to the nearest second. The difference between FDMM and LEM2 grew to more than 100 times at $N = 10000$ and $(K, V(k)) = (6, 6)$. The results in Table 1 show that FDMM performs exceptionally well under taxing conditions such as with a large number of data and growing number of condition attributes and their values.

References

1. Shan, N., Ziarko, V.: Data-based acquisition and incremental modification of classification rules. Computational Intelligence 11(2), 357–370 (1995)
2. Laboratory of Intelligent Decision Support System (IDSS), `http://www.idss.cs.put.poznan.pl/site/idss-en.html`
3. Grzymala-Busse, J.W.: LERS — A system for learning from examples based on rough sets. In: Słowiński, R. (ed.) Intelligent Decision Support, Handbook of Applications and Advances of the Rough Sets Theory, pp. 3–18. Kluwer Academic Publishers, Dordrecht (1992)

Accumulated Cost Based Test-Cost-Sensitive Attribute Reduction

Huaping He[1] and Fan Min[2]

[1] School of Computer Science,
Sichuan University of Science and Engineering, Zigong 643000, China
hehuaping_001@163.com
[2] Key Lab of Granular Computing,
Zhangzhou Normal University, Zhangzhou 363000, China
minfanphd@163.com

Abstract. As a generalization of the classical reduct problem, test-cost-sensitive attribute reduction aims at finding a minimal test-cost reduct. The performance of an existing algorithm is not satisfactory, partly because that the test-cost of an attribute is not appropriate to adjust the attribute significance. In this paper, we propose to use the test-cost sum of selected attributes instead and obtain a new attribute significance function, with which a new algorithm is designed. Experimental results on the Zoo dataset with various test-cost settings show performance improvement of the new algorithm over the existing one.

Keywords: Cost-sensitive learning, attribute reduction test-cost, accumulated cost.

1 Introduction

In many data mining applications, redundant data make the mining task rather difficult. Removing them can facilitate the mining task and make the data more visible. Attribute reduction [1][2][3] is a successful technique for this purpose. Recently we have indicated the test-cost-sensitive attribute reduction problem in [4] and defined it formally in [5]. It focuses on the test-cost instead of the classification accuracy. We argue that this problem is important since data are not free [6]. The algorithm framework in [5] is devoted to this problem. Specifically, we designed an attribute significance function based on the test-cost of each attribute and a user-specified factor λ, and obtain a substantial algorithm. The performance of the algorithm is, however, not satisfactory.

This paper proposes a new attribute significance function based on the accumulated test-cost, namely the total test-costs of selected attributes. Respective heuristic algorithm is designed with this function. We compare the new algorithm with the existing one using some exemplary data, and found that neither algorithm always wins. This phenomenon coincides with no-free-lunch theorems (www.no-free-lunch.org/). Fortunately, the new algorithm is stably better than the existing one from the statistical point of view. This claim is validated through experimental on four UCI [7] datasets.

S.O. Kuznetsov et al. (Eds.): RSFDGrC 2011, LNAI 6743, pp. 244–247, 2011.

2 Problem Definition

We consider the simplest while most widely used test-cost-sensitive decision systems as follows.

Definition 1. *[4] A* test-cost-independent decision system *(TCI-DS) S is the 6-tuple:*

$$S = (U, C, D, \{V_a | a \in C \cup D\}, \{I_a | a \in C \cup D\}, c), \tag{1}$$

where U is a finite set of objects called the universe, C is the set of conditional attributes, D is the set of decision attributes, V_a is the set of values for each $a \in C \cup D$, $I_a : U \to V_a$ is an information function for each $a \in C \cup D$, $c : C \to \mathbb{R}^+ \cup \{0\}$ is the test-cost function. The test-cost of any $\emptyset \subset A \subseteq C$ is given by $c(A) = \sum_{a \in A} c(\{a\}) = \sum_{a \in A} c(a)$.

The concept of a reduct [1] is well known in the rough sets society. In this paper, we are interested in reducts with the minimal test-cost.

Definition 2. *Let $Red(S)$ be the set of all reducts of S. Any $R \in Red(S)$ where $c(R) = \min\{c(R') | R' \in Red(S)\}$ is called a* minimal test-cost reduct.

3 Accumulated Cost Based Heuristic Function

The heuristic function is to the central of a heuristic reduction algorithm. In [5] we proposed a function as follows:

$$f^{wei}(B, a_i, c) = f_e(B, a_i) c(a_i)^{\lambda}, \tag{2}$$

where λ is a non-positive number, and $f_e(B, a_i)$ is the information gain of a_i when attributes in B are selected. Here we propose a new function as follows:

$$f^{acc}(B, a_i, c) = f_e(B, a_i) c(B \cup \{a_i\})^{\lambda}. \tag{3}$$

That is, we use the total test-cost of $B \cup \{a_i\}$ instead of the test-cost of a_i to produce the weighting factor. Therefore the new function is called the *accumulated cost based attribute significance function*, and the new algorithm is called the *accumulated cost based test-cost-sensitive attribute reduction algorithm*. The influence of this revision will be discussed in detail in Section 4.

Note that superscripts *wei* and *acc* are employed to distinguish these two functions. Although computing $c^*(B \cup \{a_i\})$ in Equation (3) is slightly more complex than computing $c^*(a_i)$ in Equation (2), the time complexity is still less than that of computing $f_e(B, a_i)$. Hence this algorithm has the same time complexity as that of [5]. For brevity, in the following context algorithms based on f^{wei} and f^{acc} will be called WEI and ACC, respectively.

4 Experiments

We have undertaken experiments on 4 different datasets from the UCI library [7]. On all datasets the new algorithm is stably better than the existing one from statistical point of view. Due to the space limitation, we only list results on Zoo. Since there is no test-cost setting, we generate random numbers in [1, 100], and assign them as test-costs. Note that given different test-cost settings, the dataset is viewed as different. Therefore our experiments are far from one run.

There is a need to define a number of evaluation metrics to compare their performances [5]. Let the number of experiments be K, and the number of successfully finding an optimal reduct be k, the *finding optimal* factor (FOF) is defined as $op = \frac{k}{K}$. For a dataset with a particular test-cost setting, let R' be an optimal reduct. The *exceeding factor* of a reduct R is $ef(R) = \frac{c^*(R)-c^*(R')}{c^*(R')}$. The exceeding factor provides a quantitative metrics to evaluate the performance of a reduct. Let the number of experiments be K. In the i-th experiment ($1 \leq i \leq K$), the reduct computed by the algorithm be R_i. The *average exceeding factor* (AEF) is defined as $\frac{\sum_{i=1}^{K} ef(R_i)}{K}$.

4.1 Running Examples

Neither algorithm is better than the other one. We list one example here.

Example 1. Let the cost vector c = [88, 51, 33, 14, 56, 70, 85, 64, 87, 51, 53, 72, 15, 93, 52, 13], By running the exhaustive algorithm we know that the optimal reduct is {c, f, h, m, p}, and its test-cost is 195. Program executions of these two algorithms are listed in Table 1.

Table 1 indicates that ACC is better than WEI in Example 1. Note that counterexamples also exists, that is, WEI outperforms ACC in some other cases.

Table 1. An example where ACC is better than WEI

Algorithm		WEI	ACC
Operation	Step	Attribute subset	Attribute subset
Compute the core	1	{f, m}	{f, m}
Add attributes	2	{d, f, m}	{d, f, m}
	3	{d, f, m, p}	{d, f, m, p}
	4	{c, d, f, m, p}	{c, d, f, m, p}
	5	{c, d, f, k, m, p}	{c, d, f, h, m, p}
	6	{c, d, f, j, k, m, p}	
	7	{c, d, f, h, j, k, m, p}	
Remove attributes	8	{d, f, h, j, k, m, p}	{c, f, h, m, p}
	9	{d, f, h, k, m, p}	
The constructed reduct		{d, f, h, k, m, p}	{c, f, h, m, p}
The exceeding-factor		0.1744	0
Is optimal		NO	YES

4.2 Results

Fig. 1 (a) shows the results of FOF, where λ is from -1.5 to -7. The FOF of the new algorithm is about 1.86% better than that of the existing algorithm. Fig. 1 (b) shows the results of AEF. The AEF of the new algorithm is about 0.24% better than that of the existing algorithm.

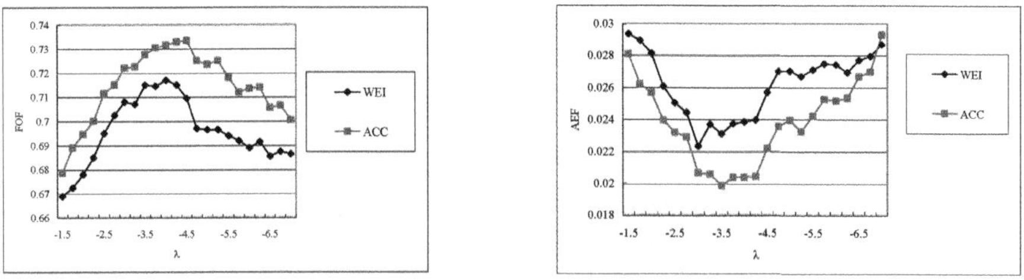

(a) (b)

Fig. 1. (a) Finding optimal factor; (b) Average exceeding factor

5 Conclusions

This paper proposes a new attribute significance function for the test-cost-sensitive reduction problem, and therefore obtain a respective algorithm ACC. ACC is not always better than the existing algorithm WEI while running particular examples. However, from the statistical point of view, it is stably better than WEI in terms of finding optimal factor and average exceeding factor.

Acknowledgements. This work is supported in part by National Natural Science Foundation of China, Grant No. 60873077.

References

1. Pawlak, Z.: Rough sets. International Journal of Computer and Information Sciences 11, 341–356 (1982)
2. Zhang, W., Mi, J., Wu, W.: Knowledge reductions in inconsistent information systems. Chinese Journal of Computers 26(1), 12–18 (2003)
3. Zhu, W., Wang, F.Y.: Reduction and axiomization of covering generalized rough sets. Information Sciences 152(1), 217–230 (2003)
4. Min, F., Liu, Q.: A hierarchical model for test-cost-sensitive decision systems. Information Sciences 179(14), 2442–2452 (2009)
5. Min, F., He, H., Qian, Y., Zhu, W.: Test-cost-sensitive attribute reduction. Submitted to Information Sciences (2011)
6. Turney, P.D.: Cost-sensitive classification: Empirical evaluation of a hybrid genetic decision tree induction algorithm. Journal of Artificial Intelligence Research 2, 369–409 (1995)
7. Blake, C.L., Merz, C.J.: UCI repository of machine learning databases (1998), `http://www.ics.uci.edu/~mlearn/mlrepository.html`

Approximate Bicluster and Tricluster Boxes in the Analysis of Binary Data

Boris G. Mirkin[1,2,*] and Andrey V. Kramarenko[1]

[1] National Research University – Higher School of Economics, Moscow, Russia
[2] Department of Computer Science and Information Systems, Birkbeck University of London, UK
bmirkin@hse.ru, mirkin@dcs.bbk.ac.uk

Abstract. A disjunctive model of box bicluster and tricluster analysis is considered. A least-squares locally-optimal one cluster method is proposed, oriented towards the analysis of binary data. The method involves a parameter, the scale shift, and is proven to lead to "contrast" box bi- and tri-clusters. An experimental study of the method is reported.

Keywords: box, bicluster, tricluster.

1 Introduction

The concept of bicluster emerges when the data relate two different sets of objects to each other so that highly related pairs of subsets, partitions or even hierarchies can be distinguished in each of the sets. This cluster structure was first made explicit by J.Hartigan[2,3] and dubbed as biclustering by B.Mirkin[7]. The concept and corresponding methods gained popularity in several applied areas of which probably the most effective is bioinformatics (see, for example, Madeira and Oliveira[5], Prelic et al.[10]). A somewhat more conservative and mathematically driven approach to establishing relations between a set of objects and a set of attributes was taken in developing the abstract formal model concept (Wille and Ganter[1]). The notion of a formal concept was first developed for binary data matrix $R = (r_{ij}), i \in I, j \in J$, where all r_{ij} are either 1 or 0, which is the case we consider here. A formal concept (V, W), where $V \subseteq I, W \subseteq J$, corresponds to all $r_{ij} = 1$ for $i \in V, j \in W$ in such a way that adding elements to either V or W would break the equation at least at one pair (i, j). This notion is well justified in application to well developed "contexts" R, but seems somewhat rigid when applied to real world datasets. This is why researchers have been trying to relax the notion of formal concepts by admitting some zeros inside the box (V, W) and some unities outside it (see, for example, Pensa and Boullicaut[9]). In this respect, the box clustering approach proposed by Mirkin et al.[6] seems another form of relaxation of the notion of formal concept. Yet the box clustering algorithms proposed in Mirkin et al.[6] are not applicable to the binary data. Therefore, we put a threefold goal for this paper:

* Corresponding author

S.O. Kuznetsov et al. (Eds.): RSFDGrC 2011, LNAI 6743, pp. 248–256, 2011.

(1) To propose and explore a model and algorithm for biclustering boxes suitable for binary contexts;

(2) Extend it to triclustering of binary data involving three interrelated data objects;

(3) Apply both bi- and tri-clustering algorithms to real world datasets.

2 The Notions of Formal Concept and Box Clustering

A data matrix $R = (r_{ij})$, with $i \in I$ (objects) and $j \in J$ (attributes), such that either $r_{ij} = 1$ or $r_{ij} = 0$ is referred to as a conceptual context. A formal concept is a pair of sets (V, W), that is, a biset, such that $V \subseteq I$, $W \subseteq J$ and

$$r_{ij} = 1 \text{ for all } (i, j) \in V \times W \tag{1}$$

and neither V nor W can be increased without breaking the property (1). The cardinalities will be denoted by #V = n, #W = m.

The condition of all within-entries being non-zero can be too restrictive, especially with noisy data. There have been attempts at modifying both of these conditions by admitting a few zeros inside and most zeros outside (Pensa and Boulicaut[9], Rome and Haralick[11], Ignatov and Kuznetsov[4]).The data recovery clustering can be utilized to address this as well.

A set of box clusters $(\lambda_t, V_t, W_t), t = 1, \ldots, T$, forms a disjunctive box cluster model of data R if

$$r_{ij} = max_{t=1,\ldots,T}\lambda_t v_{it} w_{jt} + \lambda_0 + e_{ij} \tag{2}$$

where e_{ij} are sufficiently small, and $\lambda_0, 0 < \lambda_0 < 1$, plays the role of an intercept in linear data models. This model differs from those of additive bi-clustering since (2) involves the operation of maximization rather than summation. To fit (2) with a relatively small number of boxes, assume λ_0 to be constant and specified before the fitting of the model. Then the model in (2) can be rewritten by putting $r'_{ij} = r_{ij} - \lambda_0$ on the left, so that λ_0 becomes a similarity shift value rather than an intercept.

We apply here the one-by-one fitting strategy (Mirkin[7]) so that each box cluster (λ_t, V_t, W_t) in (1) is found as a most deviant from the "middle", that is, minimizing the residuals in a single cluster model (with a constant λ_0)

$$r'_{ij} = r_{ij} - \lambda_0 = \lambda v_i w_j + e_{ij} \tag{3}$$

with the least squares criterion. In this formulation, $v = (v_i)$ and $w = (w_j)$ are binary membership vectors of V and W, respectively, so that $v_i w_j = 1$ if and only if $(i, j) \in V \times W$ like it is in a formal concept.

Let us initially assume $\lambda_0 = 0$ so that $r'_{ij} = r_{ij}$. Box cluster (λ_t, V_t, W_t) minimizing the least squares criterion

$$L^2 = \Sigma_{ij}(r'_{ij} - \lambda v_i w_j)^2 \tag{4}$$

over real λ and binary v_i, w_j, must lead to optimal λ being equal to the within-box average:

$$\lambda = \Sigma_{i \in V, j \in W} r'_{ij}/nm \tag{5}$$

which is the proportion of ones within the box minus λ_0, and, assuming that the λ is optimal, criterion L^2 in (4) admits the following decomposition:

$$L^2 = \Sigma_{ij} r'^2_{ij} - \lambda^2 nm \tag{6}$$

thus implying the following criterion to maximize

$$g(V, W) = \lambda^2 nm \tag{7}$$

According to (6), this criterion expresses the contribution of the box (V, W) to the data scatter $\Sigma_{ij} r'^2_{ij}$ which is useful to see how closely the box follows the data. On the other hand, criterion (7) combines two contrasting criteria for a box to be optimal: (a) the largest area, (b) the largest proportion of within-box unities. If restricted to a within-box non-zero option, the criterion (7) would lead to the formal concepts of the largest sizes, nm, as the only maximizers.

3 Locally Optimal Box Cluster

As the optimal intensity of a box (V, W) is fully determined by the summary entries within the box formed, we identify the box cluster with just sets V and W, that is, biset (V, W). With no loss of generality assume that it is V' that differs from V, by adding or removing an entity $i^* \in I$, while $W' = W$:

$$Diff(i^*) = [z_i r^2(i^*, W) + 2z_i r(V, W) r(i^*, W) - r^2(V, W)/n]/(m(n + z)) \tag{8}$$

where z^*_i equals 1 if i^* is added to V and -1 if i^* is removed from V, and $r(V, W)$ denotes the sum of all R' entries within box $V \times W$ and $r(i^*, W)$ is the sum of all r'_{i*j} over $j \in W$. A symmetric expression holds for $Diff(j^*)$ at $j^* \in J$.

A local search algorithm can be drawn starting from every entity $i \in V$ (or $j \in W$):

Algorithm Bicluster Box (i)

0. Take $V = i$ and $W = \{j \mid r_{ij} = 1\}$ as the starting box.
1. Find $Diff(8)$ for all elements of I and J, take that D^* which is maximum.
2. If D^* is not positive, halt. Otherwise, perform the operation of adding/ removing for the corresponding entity and return to Step 1 with the updated box.

The resulting cluster box is provably rather contrast:

Statement 1
If box cluster (V, W) is found with $BBox()$ algorithm then, for any entity outside the box, its average similarity to it is less than the half of the within-box similarity λ; in contrast, for any entity belonging to the box, its average similarity to it is greater than or equal to the half of the within-box similarity λ.

Fitting model (2) can be done by applying algorithm $BBox()$ starting from each of the entities and retaining only different and most contributing solutions. Let us remind that the contribution of a box bicluster is but the value of criterion (7).

The subtracted λ_0 value can be used as a user-defined parameter to control, on average, the box cluster sizes. In our experiments we take λ to be the average value of R, that is, proportion of unities in the matrix.

4 Extending to Triclusters and P-Clusters

The model (2), as well as criteria (4) and (7) and algorithm $BBox$ are easily extended to the case when the data refer to relations between more than two sets. Specifically, one may suggest that there are p different entity sets, $I_1, I_2, \ldots, I_p$ such that relation R is p-ary, that is, corresponds to a subset of Cartesian product $\rho \subseteq I_1 \times I_2 \times \ldots \times I_p$. Then we would consider p-ary boxes. For the sake of simplicity, further on we consider only the case when $p = 3$, so that the three sets are denoted I, J, and K, whereas their subsets, by V, W, and U, respectively.

An extended form of model (2) is

$$r_{ijk} = max_{t=1T} \lambda_t v_{it} w_{jt} u_{kt} + \lambda_0 + e_{ijk} \quad (i \in I, j \in J, k \in K) \tag{9}$$

whereas criteria (4) and (7) lead to

$$L^2 = \Sigma_{ijk}(r'_{ijk} - \lambda v_i w_j u_k)^2 \tag{10}$$

and

$$g(V, W, U) = \lambda^2 nml \tag{11}$$

where n,m, and l are cardinalities of V, W, and U, respectively, and the optimal λ being the average value of all the R' three-way entries. The value of (11) shows contribution of the tricluster to the data scatter.

The value of difference $D(i^*) = g(V', W, U)g(V, W, U)$, where V' differs from V by the state of just one entity $i^* \in I$ so that i^* either belongs to V' if $i^* \notin V$ or does not, if $i^* \in V$, is expressed with a formula analogous to (8):

$$D(i^*) = [r^2(i^*, W, U) + 2z_{i^*} r(V, W, U) r(i^*, W, U) - z_{i^*} r^2(U, V, W)/n]/((n + z_{i^*})ml)$$

Here $z_{i^*} = 1$, if i^* is added to V and $z_{i^*} = -1$ otherwise, $r(V, W, U)$ is the sum of all the entries in R' over $(i, j, k) \in V \times W \times U$, and $r(i^*, V, W)$ is the sum of all the r'_{i^*jk} over $j \in W$ and $k \in U$. A symmetric expression holds for the changes in box (V, W, U) over $j^* \in W$ and $k^* \in U$. This leads to the following tricluster finding algorithm.

Algorithm Tricluster Box (i)

1.Take $V = \{i\}$, $W = \{j : r_{ijk} = 1 \text{ for some } k\}$ and $U = \{k \mid r_{ijk} = 1 \text{ for some } j\}$ as the starting box.

2. Find $D(i^*)$,$D(j^*)$ and $D(k^*)$ for all $i^* \in I$, $j^* \in J$, and $k^* \in K$ and J, take that of the values D which is maximum, denote it D^*.

3. If D^* is not positive, halt. Otherwise, perform the operation of adding/removing for the corresponding entity and return to Step 1 with the updated box.

A statement, similar to Statement 1, holds.

Statement 2
If box cluster $B = (V, W, U)$ is found with $TriclusterBox()$ algorithm then, for any entity outside the box, its average similarity to B is less than the half of the within-box similarity λ; in contrast, for any entity belonging to the box, its average similarity to B is greater than or equal to the half of the within-box similarity λ.

5 Experiments

We have run experiments with synthetic data sets, just to see that $BiclusterBox$ is competitive towards both generalized formal concepts algorithms and conventional bicluster algorithms. We also run experiments on three-way data, first, to see if our triclusters are any good, and second, to compare solutions found with tricluster and bicluster algorithms. That is possible if one considers a three-way dataset over $I \times J \times K$ as a two-way dataset over $I \times (J \times K)$, that is, over I and Cartesian product of the two other sets, $J \times K$. We describe here some of the experiments.

Experiment 1. Take a binary 30×15 data table $R0$ comprising three non-overlapping formal concepts from Pensa and Boulicaut[9]. All $R0's$ entries are zeros except for those within three boxes comprising, in respect, first 10 rows (from 1 to 10) and first 5 columns (from 1 to 5), second 10 rows (from 11 to 20) and second 5 columns (from 6 to 10), and third 10 rows (from 21 to 30) and third 5 columns (from 11 to 15), whose all entries are ones. Then this matrix is changed to a matrix R_p by randomly changing its every entry with the probability $p\%, p = 1, 2, \ldots, 40$. Algorithm $BBox(i)$ has been applied to each R_p, with its mean subtracted as λ_0, at each $i \in I$ and $j \in J$, and sets of differing results stored as B_p and D_p. To compare these results with the original concepts in $R0$, we utilized the extension of Jaccard coefficient described in Pensa and Boulicaut[9].

Specifically, given two bisets, (V, W) and (V', W'), ratio of the areas of the rectangles corresponding to their intersection and union is taken as the measure of similarity:
$S((V, W), (V', W')) = |V \cap V'||W \cap W'|/(|V \cup V'||W \cup W'|)$.
so that

$$\sigma(B, B') = \Sigma_i max_j S((V_i, W_i), (V'_j, W'_j))/|B| \quad (12)$$

The averaged results of runs of two versions of approximate $BBox$ algorithm through several rounds of generated matrices $R_p(p = 1, 2, , 40)$ are summarised in Figure 1.

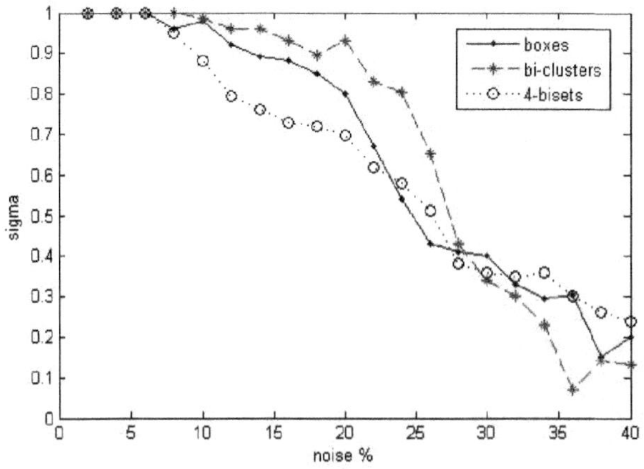

Fig. 1. Graphs of σ measure between the original three concepts and results of *Box* and *Dual* bicluster algorithms (Mirkin[8]) applied to the binary R_p matrix at different levels of random noise, $p = 1, 2, .., 40\%$. The third graph represents the σ values at 4-bisets by Pensa and Boullicaut[9].

Experiment 2. Pre-specified biclusters are set similarly here, yet the error is introduced via additive Gaussian noise so that the data become non-binary and general biclustering algorithms are applicable. We compare *BBox* with the best algorithm *Bimax* according to Prelic et al. [10] (see Figure 2). Here we refer to scoring functions measuring relevance and recovery as those utilized by Prelic et al. [10]. Given two sets of boxes on $I \times J$, B and B', consider expression

$$s_I(B, B') = \Sigma_{(v,w)\in B} max_{(v',w')\in B'} Jac(v, v')/|B| \tag{13}$$

where $Jac(v, v') = |v \cap v'|/|v \cup v'|$, the celebrated Jackard similarity index between sets v and v'. This is referred to as measure of recovery of B by B', and measure of relevance of B versus B'. *BBox* appears to be better than *Bimax* in recovery and worse than that in relevance.

Experiment 3. A binary file of a ternary relation between 250 movies, 738 keywords and 20 genres have been downloaded from Web-site:

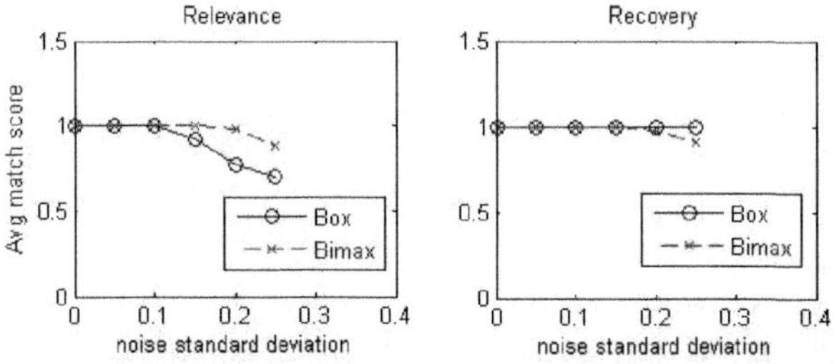

Fig. 2. Comparison of BBox and Bimax on the additive noise data

Table 1. Most contributing triclusters for 250 popular movies

Contrib.%	Movie	Keyword	Genre
28.5	'Star Wars:V-The Empire Strikes Back (1980)' 'Star Wars (1977)' 'Star Wars:VI - Return of the Jedi (1980)'	'Rebel' 'Princess' 'Empire' 'Death Star'	'Adventure' 'Action' 'Sci-Fi' 'Fantasy'
18.9	'12 Angry Men (1957)' 'Double Indemnity (1944)' 'Chinatown (1974)' 'The Big Sleep (1946)' 'Witness for the Prosecution (1957)' 'Dial M for Murder (1954)' 'Shadow of a Doubt (1943)'	'Murder' 'Trial' 'Widow' 'Marriage' 'Private' 'Detective' 'Blackmail' 'Letter'	'Crime' 'Drama' 'Thriller' 'Mystery' 'Film-Noir'
18.0	'The Return of the King (2003)' 'The Fellowship of the Ring (2001)' 'The Two Towers (2002)'	'Ring' 'Middle Earth'	'Adventure' 'Action' 'Fantasy'
18.0	'Terminator 2: Judgment Day(1991) ' 'The Terminator (1984)'	'Future' 'Cyborg'	'Thriller' 'Sci-Fi'

`http://www.imdb.com/chart/top` (accessed on 10 April 2009). In fact, the data involves more than 6500 keywords, but only those present at six or more of the movies have been used in our computations.

$TBox(i)$ algorithm found 84 triclusters containing more than one movie. Most contributing triclusters are in Table 1. Those who know of the movies can see that these are rather tight and meaningful indeed. We also took the data as a two-way table of movies over all possible pairs keyword-genre, to see if those results could be found without developing a novel algorithm. This gave a much smaller number of non-trivial biclusters, just 40 of them. Some of biclusters match the triclusters in Table 1 rather closely. Table 2 presents two of them, just to illustrate the differences. The most important difference is that biclusters are less expressive, in spite of more degrees of freedom: they can take any set of keyword-genre pairs whereas triclusters carry only Cartesian products. Take, for example, "Star Wars" biclusters it comprises just Cartesian product of set of keywords, Princess, Empire, of set of genres Adventure, Action, Sci-Fi, thus missing less trivial keywords "Rebel" and "Death Star" that have been picked up by the tricluster.

A similar observation can be made about tri- and bi-clusters containing "12 Angry Men" movie: the description of the biclusters, in Table 2, even if not a Cartesian product, seems rather dry and trivial in comparison to the description in Table 1.

Experiment 4. This experiment has been carried out with a $200 \times 50 \times 295$ data set C extracted from website: `http://www.automotive.com/index.html` (in November 2010; the dataset is available from the authors on request). This concerns 200 passenger car models along with their 50 possible features. Besides, each car model is associated with a set of car models that are said to be

Table 2. Comparable most contributing biclusters

Contrib.%	Movie	Keyword,Genre
18.0	'Star Wars:V-The Empire Strikes Back (1980)' 'Star Wars (1977)' 'Star Wars:VI - Return of the Jedi (1980)'	'Princess, Adventure' 'Princess, Action' 'Princess, Sci-Fi' 'Empire, Adventure' 'Empire, Action' 'Empire, Sci-Fi'
10.1	'12 Angry Men (1957)' 'To Kill a Mockingbird (1962)' 'Witness for the Prosecution (1957)"	'Murder, Drama' 'Trial, Crime' 'Trial, Drama' 'Trial, Mystery'

comparable to the model under consideration. The set of comparable cars appears to be somewhat greater than the car set, totalling to 295 car models. On comparing triclusters found at C with biclusters found at its two-way, $200 \times (50 \times 295)$, version, triclusters versus biclusters show trends similar to those observed at the movie data set as illustrated in Table 3 for a set of budget cars appeared at both of the approaches.

Table 3. A budget car tricluster (columns 1,2 and 3) versus its bicluster namesake (columns 1, 4)

Tricluster: V (Car)	Tricluster: W (Feature)	Tricluster: U (Comparable)
'Ford Fiesta Sdn' 'Hyundai Accent Sdn' 'Mazda 2 Hb'	'4-Door' 'Front Wheel Drive' '5 passengers' '5-speed Manual'	"Toyota Yaris sedan' 'Chevrolet Aveo 4-door sedan" 'Honda Fit'
Bicluster intent part for set V (Feature $\times$Comparable)		
'4-Door,Toyota Yaris sedan' '4-Door,Honda Fit' 'Front Wheel Drive,Toyota Yaris sedan' 'Front Wheel Drive,Honda Fit' '5 passengers,Toyota Yaris sedan' '5 passengers,Honda Fit'		

Another curiosity is that in all the most contributing triclusters (V, W, U), the set of comparable cars U does not necessarily cover the set V but is always biased towards more luxury cars than V, as can be seen in Table 3 in column 3 versus column 1.

6 Conclusion

The approximation approach proves flexible enough to develop effective methods for biclustering of binary data and extend them to p-way binary data.

Specifically, a viable algorithm for triclustering has been developed, for the first time in the literature, to our knowledge. In our experiments with two ternary contexts it has proved more effective than the corresponding biclustering procedures.

The approximation methods proposed in this paper can be viewed in two aspects: (a) finding one "best" p-cluster, or (b) filling in a disjunctive model of the p-cluster structure of the given p-ary context. The "p-Box" algorithms developed in the paper depend on a specific entry in the context, thus appear to be rather computationally intensive, and thus not competitive, in the aspect (a). In this regard, we tried to accelerate the repetitive process by applying the algorithm only to those entries that have not been processed in the previous iterations yet. Unfortunately, this leads to poor results and should be further elaborated. If, however, one concentrates on the aspect (b) – revealing the entire cluster structure in a disjunctive way, then we can safely claim that our approach leads to a solution to this problem. It is still time consuming and the future efforts should address this issue.

References

1. Ganter, B., Wille, R.: Formal Concept Analysis: Mathematical Foundations. Springer, Heidelberg (1999)
2. Hartigan, J.A.: Direct clustering of a data matrix. Journal of the American Statistical Association 67(337), 123–129 (1972)
3. Hartigan, J.A.: Clustering Algorithms. Wiley, Chichester (1975)
4. Ignatov, D., Kuznetsov, S.: Biclustering methods using lattices of closed subsets. In: Proceedings of 12th National Conference on Artificial Intelligence, Moscow, FML, vol. 1, pp. 175–182 (2010) (in Russian)
5. Madeira, S.C., Oliveira, A.L.: Biclustering algorithms for biological data analysis: A survey. IEEE/ACM Transactions on Computational Biology and Bioinformatics (TCBB) 1(1), 24–45 (2004)
6. Mirkin, B., Arabie, P., Hubert, L.: Additive two-mode clustering: the error-variance approach revisited. Journal of Classification 12, 243–263 (1995)
7. Mirkin, B.: Mathematical Classification and Clustering, p. 448. Kluwer, Dordrecht (1996)
8. Mirkin, B.: Two goals for biclustering: "Box" and "Dual" methods (2008) (unpublished manuscript)
9. Pensa, R.G., Boulicaut, J.-F.: Towards fault-tolerant formal concept analysis. In: Bandini, S., Manzoni, S. (eds.) AI*IA 2005. LNCS (LNAI), vol. 3673, pp. 212–223. Springer, Heidelberg (2005)
10. Prelic, A., Bleuler, S., Zimmermann, P., Wille, A., Bühlmann, P., Gruissem, W., Hennig, L., Thiele, L., Zitzler, E.: A Systematic Comparison and Evaluation of Biclustering Methods for Gene Expression Data. Bioinformatics 22(9), 1122–1129 (2006)
11. Rome, J.E., Haralick, R.M.: Towards a formal concept analysis approach to exploring communities on the World Wide Web. In: International Conference on Formal Concept Analysis, Lens, France (2005)

From Triconcepts to Triclusters

Dmitry I. Ignatov, Sergei O. Kuznetsov,
Ruslan A. Magizov, and Leonid E. Zhukov

National Research University Higher School of Economics, Moscow, Russia
dignatov@hse.ru

Abstract. A novel approach to triclustering of a three-way binary data is proposed. Tricluster is defined in terms of Triadic Formal Concept Analysis as a dense triset of a binary relation Y, describing relationship between objects, attributes and conditions. This definition is a relaxation of a triconcept notion and makes it possible to find all triclusters and triconcepts contained in triclusters of large datasets. This approach generalizes the similar study of concept-based biclustering.

Keywords: formal concept analysis, data mining, triclustering, three-way data, folksonomy.

1 Introduction

The term biclustering was coined by B.Mirkin in 1996 [15] and the appearance of triclustering and n-clustering was only a matter of time. The similar approach, called direct clustering, was proposed in early 70s by Hartigan [10]. In the Formal Concept Analysis method, introduced in 1982 by R. Wille [17,9], a particular kind of bicluster was used for analysis of binary data, notably formal concept. The Triadic (Formal) Concept Analysis (TCA) was introduced by Lehman and Wille [14] in 1995 as an extension of dyadic Formal Concept Analysis for the case of three-way binary data. The notions of formal concepts and triconcepts describe a useful pattern in binary data that is homogeneous and closed (maximal) in algebraic sense. Due to the rigid structure of formal concepts and computational complexity of processing algorithms (exponential w.r.t. size of input data), some relaxations of the formal concept notion were introduced for dyadic (relevant and dense bisets [3], concept factorization techniques [1], dense biclusters [11]) and triadic cases (triadic concept factors [2]). There are also exist several techniques for reduction of the number of formal concepts to only relevant ones, for example, iceberg lattices and stability indices mining. The need for scalable and efficient triclustering algorithms became clear with the growth of popularity and sizes of social resource tagging systems. The three-way data "user-tag-resource", so called folksonomy, is a core data structure in such systems. One of the known algorithms for mining folksonomies is the TRIAS algorithm [12]. There is also a promising approach to mine n-ary relational data [5,6]; its implementation (DataPeleer) is based on closed sets and outperforms another similar algorithm CubeMiner [13] for mining closed

S.O. Kuznetsov et al. (Eds.): RSFDGrC 2011, LNAI 6743, pp. 257–264, 2011.

trisets. Some researchers go further and actively apply closed trisets for mining complex attribute dependencies in three-way binary data, for example, triadic implications [8].

The paper is organized as follows. In the next section, we introduce main definitions of Triadic Concept Analysis. In the section 3 we define dense triclusters and describe their properties, present the algorithm and evaluate its complexity, discuss some heuristics, and also provide the reader with examples. Section 4 presents the description of the bibsonomy datasets and computer experiments with them. Section 5 concludes the paper.

2 Main Definitions

A triadic context $\mathbb{K} = (G, M, B, Y)$ consists of sets G (objects), M (attributes), and B (conditions), and ternary relation $Y \subseteq G \times M \times B$. An incidence $(g, m, b) \in Y$ shows that the object g has the attribute m under condition b.

For convenience, a triadic context is denoted by (X_1, X_2, X_3, Y). A triadic context $\mathbb{K} = (X_1, X_2, X_3, Y)$ gives rise to the following diadic contexts $\mathbb{K}^{(1)} = (X_1, X_2 \times X_3, Y^{(1)})$, $\mathbb{K}^{(2)} = (X_2, X_2 \times X_3, Y^{(2)})$, $\mathbb{K}^{(3)} = (X_3, X_2 \times X_3, Y^{(3)})$, where $gY^{(1)}(m, b) :\Leftrightarrow mY^{(1)}(g, b) :\Leftrightarrow bY^{(1)}(g, m) :\Leftrightarrow (g, m, b) \in Y$. The derivation operators (primes or concept-forming operators) induced by $\mathbb{K}^{(i)}$ are denoted by $(.)^{(i)}$. For each induced dyadic context we have two kinds of such derivation operators. That is, for $\{i, j, k\} = \{1, 2, 3\}$ with $j < k$ and for $Z \subseteq X_i$ and $W \subseteq X_j \times X_k$, the (i)-derivation operators are defined by:

$$Z \mapsto Z^{(i)} = \{(x_j, x_k) \in X_j \times X_k | x_i, x_j, x_k \text{ are related by Y for all } x_i \in Z\},$$

$$W \mapsto W^{(i)} = \{x_i \in X_i | x_i, x_j, x_k \text{ are related by Y for all } (x_j, x_k) \in W\}.$$

Formally, a triadic concept of a triadic context $\mathbb{K} = (X_1, X_2, X_3, Y)$ is a triple (A_1, A_2, A_3) of $A_1 \subseteq X_1, A_2 \subseteq X_2, A_3 \subseteq X_3$, such that for every $\{i, j, k\} = \{1, 2, 3\}$ with $j < k$ we have $A_i^{(i)} = (A_j \times A_k)$. For a triadic concept (A_1, A_2, A_3), the components A_1, A_2, and A_3 are called the extent, the intent, and the modus of (A_1, A_2, A_3). It is important to note that for interpretation of $\mathbb{K} = (X_1, X_2, X_3, Y)$ as a three-dimensional cross table, according to our definition, under suitable permutations of rows, columns, and layers of the cross table, the triadic concept (A_1, A_2, A_3) is interpreted as a maximal cuboid full of crosses. The set of all triadic concepts of $\mathbb{K} = (X_1, X_2, X_3, Y)$ is called the concept trilattice and is denoted by $\mathfrak{T}(X_1, X_2, X_3, Y)$.

3 Mining Dense Triclusters

3.1 Prime, Double Prime and Box Operators of 1-Sets

To simplify the notation, we denote by $(.)'$ all prime operators, as it is usually done in FCA. For our purposes consider a triadic context $\mathbb{K} = (G, M, B, Y)$ and

Table 1. Concept-forming operators of 1-sets

Prime operators of 1-sets	Their double prime counterparts
$m' = \{ (g,b) \mid (g,m,b) \in Y\}$	$m'' = \{ \tilde{m} \mid (g,b) \in m' \quad and \quad (g,\tilde{m},b) \in Y\}$
$g' = \{ (m,b) \mid (g,m,b) \in Y\}$	$g'' = \{ \tilde{g} \mid (m,b) \in g' \quad and \quad (\tilde{g},m,b) \in Y\}$
$b' = \{ (g,m) \mid (g,m,b) \in Y\}$	$b'' = \{ \tilde{b} \mid (g,m) \in b' \quad and \quad (g,m,\tilde{b}) \in Y\}$

introduce primes, double primes and box operators for particular elements of G, M, B, respectively. In what follows we write g' instead of $\{g\}'$ for 1-set $g \in G$ and similarly for $m \in M$ and $b \in B$: m' and b'.

We do not use double primes, but box operators that we introduce below:

$$\begin{aligned} g^{\square} &= \{ g_i \mid (g_i, b_i) \in m' \text{ or } (g_i, m_i) \in b'\} \\ m^{\square} &= \{ m_i \mid (m_i, b_i) \in g' \text{ or } (g_i, m_i) \in b'\} \\ b^{\square} &= \{ b_i \mid (g_i, b_i) \in m' \text{ or } (m_i, b_i) \in g'\} \end{aligned}$$

Let $\mathbb{K} = (G, M, B, Y)$ be a triadic context. For a triple $(g, m, b) \in Y$, the triple $T = (g^{\square}, m^{\square}, b^{\square})$ is called a tricluster.

The density of a tricluster (A, B, C) of a triadic context $\mathbb{K} = (G, M, B, Y)$ is given by the fraction of all triples of Y in the tricluster, that is $\rho(A, B, C) = |I \cap A \times B \times C|/|A||B||C|$.

The tricluster $T = (A, B, C)$ is called dense if its density is greater than a predefined minimal threshold, i.e. $\rho(T) \geq \rho_{min}$. For a given triadic context $\mathbb{K} = (G, M, B, Y)$ we denote by $\mathbf{T}(G, M, B, Y)$ the set of all its (dense) triclusters.

Property 1. For every triconcept (A, B, C) of a triadic context $\mathbb{K} = (G, M, B, Y)$ with nonempty sets A, B, and C we have $\rho(A, B, C) = 1$.

Property 2. For every triclucter (A, B, C) of a triadic context $\mathbb{K} = (G, M, B, Y)$ with nonempty sets A, B, and C we have $0 \leq \rho(A, B, C) \leq 1$.

Proposition 1. *Let $\mathbb{K} = (G, M, B, Y)$ be a triadic context and $\rho_{min} = 0$. For every $T_c = (A_c, B_c, C_c) \in \mathfrak{T}(G, M, B, Y)$ there exits a tricluster $T = (A, B, C) \in \mathbf{T}(G, M, B, Y)$ such that $A_c \subseteq A, B_c \subseteq B, C_c \subseteq C$.*

Example 1. Consider $3^3 = 27$ formal triconcepts, 24 with $\rho = 1$ and 3 void triconcepts with $\rho = 0$ (there are empty sets of either users, resources or tags). Although this data is small, we have 27 patterns to analyze (maximal number of triconcepts for the context size $3 \times 3 \times 3$); this is because the data is the power set triadic context. We can conclude that users u_1, u_2, and u_3 share almost the same sets of tags and resources. So, they are very similar in terms $(tag, resource)$ of shared pairs and it is convenient to reduce the number of patterns describing these data from 27 to 1. The tricluster $T =$

Table 2. A small example with Bibsonomy data

	t_1	t_2	t_3
u_1		×	×
u_2	×	×	×
u_3	×	×	×

r_1

	t_1	t_2	t_3
u_1	×	×	×
u_2	×		×
u_3	×	×	×

r_2

	t_1	t_2	t_3
u_1	×	×	×
u_2	×	×	×
u_3	×	×	

r_3

$(\{u_1, u_2, u_3\}, \{t_1, t_2, t_3\}, \{r_1, r_2, r_3\})$ with $\rho = 0.89$ is exactly such a reduced pattern, but its density is slightly less than 1. Each of triconcepts in $\mathfrak{T} = \{(\emptyset, \{t_1, t_2, t_3\}, \{r_1, r_2, r_3\}), (\{u_1\}, \{t_2, t_3\}, \{r_1, r_2, r_3\})...(\{u_1, u_2, u_3\}, \{t_1, t_2\}, \{r_3\})\}$ is contained, w.r.t. componentwise set inclusion, in T.

3.2 An Algorithm: TRICL

The idea is obvious: For all $(g, m, w) \in I$ with $\rho(g^\Box, m^\Box, w^\Box) \geq \rho_{\min}$ the algorithm stores $T = (g^\Box, m^\Box, w^\Box)$ in $\mathbf{T}$. In the pseudo-code of TRICL (Alg. 1) we provide computational details rather than simple algebraic description. It allows us to better evaluate the complexity of the main algorithm's steps and gives ideas on code implementation. The complexity of the first loop (steps 2-10) is $O(|I|)$. The main loop complexity (steps 15-19) is trickier: $O(|Y||\mathbf{T}|\log(|\mathbf{T}|)|G||M||B|)$ or, since we know that $|\mathbf{T}| \leq |Y|$, it is $O(|Y|^2||\log(|Y|)|G||M||B|)$. The factor $|G||M||B|$ appears due to the computation of a tricluster ρ density, this value is indeed hard to compute for large triclusters.

We propose a heuristic to compute $\rho(T)$, based on checking only some amount of randomly selected triples contained in the given tricluster T. For a tricluster $T = (A, B, C)$ we perform the density estimation $\hat{\rho}(T) = |P|/|N|$, where $P = \{(g, m, b) | (g, m, b) \in N \cap Y\}$, N is a set of $|N|$ randomly chosen elements of the tricluster. The parameter $|N|$ can be chosen relatively small, say $\frac{1}{10} \cdot |A||B||C|$.

4 Real Data and Experiments

In our experiments we analyzed the freely available data from the popular social bookmarking system bibsonomy [4]. We ran the TRICL algorithm on a part of the data consisting of all users, resources, tags and tag assignments to detect communities of users that have similar tagging behavior.

We used only tas file which is actually the list of tuples (tag assignments): who attached which tag to which resource/content.

- 1. user (number, no user names available)
- 2. tag
- 3. content_id (matches bookmark.content_id or bibtex.content_id)
- 4. content_type (1 = bookmark, 2 = bibtex)
- 5. date

For our purposes we need only fields 1, 2, and 3 of the tuple.

Algorithm 1. TRICL

```
Data: K = (G, M, B, Y) – formal context, ρ_min – density threshold
Result: T = {(A_k, B_k, C_k)|(A_k, B_k, C_k) – dense tricluster}
begin
    for (g, m, b) ∈ Y do
        if g not in PrimesObj then
            PrimesObj[g] = g'
        if m not in PrimesAttr then
            PrimesAttr[m] = m'
        if b not in PrimesCond then
            PrimesCond[b] = b'
        if g not in BoxesObj then
            BoxesObj[g] = g^□
        if m not in BoxesAttr then
            BoxesAttr[m] = m^□
        if b not in BoxeesCond then
            BoxesCond[b] = b^□
    for (g, m, b) ∈ Y do
        Tkey = hash((BoxesObj[g], BoxesAttr[m], BoxesCond[b]));
        if Tkey not in T then
            if ρ(BoxesObj[g], BoxesAttr[m], BoxesCond[b]) ≥ ρ_min then
                T[Tkey] = ((BoxesObj[g], BoxesAttr[m], BoxesCond[b]))
```

The resulting folksonomy (bibsonomy) consists of $|U| = 2\ 337$ users, $|T| = 67\ 464$ tags and $|R| = 28\ 920$ resources (bookmarks and bibtex's entries), that are linked by $|Y| = 816{,}197$ triples. We want to note that we have to deal here with a cuboid consisting of 4 559 624 602 560 cells.

We have also investigated the statistical distribution of the data before using TRICL algorithm. We calculated and plotted histogram for users and the number of (tag, document) assignment pairs, and similar histograms for tags and their number of (user, document) pairs, and for documents and their number of (user, tag) pairs. We found that the data follows Power Law distribution $p(x) = Cx^{-\alpha}$ with $\alpha = 3{,}6778$ and variance $\sigma = 0{,}0001$ in the case of documents vs number of (user, tag) assignments. For user and tag data we obtained $\alpha = 2{,}13$ and $\alpha = 1{,}8$ respectively. We computed α using ML estimator as described in [16] and verified the results by software mentioned in [7].

This introspection can afford us to use greedy approach to our data if we want to mine large and (relatively) dense triclusters, due to even not so big part of users have the most portion of (tag, user) assignments (similar conclusions for tags and documents distributions).

We measured the run-time performance of our implementation (in Python 2.7.1) on a Pentium Core Duo system with 2 GHz and 2 GB RAM. We used

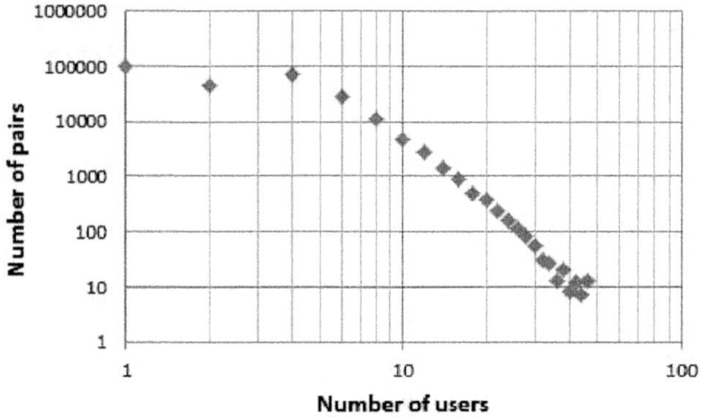

Fig. 1. Histogram of numbers of pairs (document, tag) for all triples of bibsonomy data

Table 3. Experimental results for k first triples of tas dataset with $\rho_{\min} = 0$

k, number of first triples	$\|U\|$	$\|T\|$	$\|R\|$	$\|\mathfrak{T}\|$	$\|\mathbf{T}\|$	Trias, s	TriclEx,s	TriclProb,s
100	1	47	52	57	1	0.2	0.2	0.2
1000	1	248	482	368	1	1	1	1
10000	1	444	5193	733	1	2	46,7	47
100000	59	5823	28920	22804	4462	3386	10311	976
200000	340	14982	61568	-	19053	> 24 h	> 24h	3417

Java implementation of Trias algorithm by R. Jäschke [12] to build all triconcepts of a certain context. The results of the experiments are presented in Table 3. The last two columns shows the mean execution time of Tricl with full and probabilistic density calculation strategies respectively.

Table 4. Density of triclusters distribution for 200 000 first triples of tas dataset with $\rho_{\min} = 0$

low bound of ρ	upper bound of ρ	number of triclusters
0	0,05	18617
0,05	0,1	195
0,1	0,2	112
0,2	0,3	40
0,3	0,4	20
0,4	0,5	10
0,5	0,6	8
0,6	0,7	1
0,7	0,8	1
0,8	0,9	0
0,9	1	49

In our experiments the $\hat{\rho}$ estimate has only 0.13 mean absolute error for a tricluster size $|N| = 1/10$, $\rho_{\min} = 0$, and 200 000 first triples of the bibsonomy data. The algorithm becomes drastically faster than Trias and TriclEx in the case of our probabilistic computational strategy.

Density distribution of triclusters for 200 000 first triples of bibsonomy dataset is given in the Table 4.

5 Conclusion

We proposed an FCA-based approach to triclustering. We showed that:

- The (dense) triclustering is a good alternative for TCA since the total number of triclusers in real data is significantly less than the number of triconcepts.
- The (dense) triclustering is able to cope with a large number of triconcepts. In the worst cases of tricontexts (or dense cuboids in them) only their main diagonal is empty and considers such cuboids as a whole tricluster. This is very relevant property for mining tricommunities in social bookmarking systems.
- The proposed algorithm has good scalability on real-world data especially when used with greedy covering approach and optimized version of the density calculation procedure.

We will continue our work on triclustering in several directions:

- Investigate mixing of several constraint-based approaches to triclustering (e.g., mining dense triclusters first and then frequent trisets in them).
- Search for better approaches to tricluster's density estimation.
- Develop a unified theoretical framework for triclustering based on closed sets.
- Take into account the nature of real-world data for optimization (data sparsity, distribution of values, etc.).

Acknowledgements. This work was partially supported by the Russian Foundation for Basic Research, project No. 08-07-92497-NTSNIL_a. We would like to thank our colleagues Sergei Obiedkov (NRU HSE), Mykola Pechenizsky (TU-Eindhoven) and Alena Pliskina (NRU HSE) for their suggestions and support.

References

1. Belohlavek, R., Vychodil, V.: Factor analysis of incidence data via novel decomposition of matrices. In: Ferré, S., Rudolph, S. (eds.) ICFCA 2009. LNCS, vol. 5548, pp. 83–97. Springer, Heidelberg (2009)
2. Belohlavek, R., Vychodil, V.: Factorizing three-way binary data with triadic formal concepts. In: Setchi, R., Jordanov, I., Howlett, R., Jain, L. (eds.) KES 2010. LNCS, vol. 6276, pp. 471–480. Springer, Heidelberg (2010)

3. Besson, J., Robardet, C., Boulicaut, J.F.: Mining a new fault-tolerant pattern type as an alternative to formal concept discovery. In: Schärfe, H., Hitzler, P., Øhrstrøm, P. (eds.) ICCS 2006. LNCS (LNAI), vol. 4068, pp. 144–157. Springer, Heidelberg (2006)
4. http://bibsonomy.org
5. Cerf, L., Besson, J., Robardet, C., Boulicaut, J.F.: Data peeler: Contraint-based closed pattern mining in n-ary relations. In: SDM, pp. 37–48. SIAM, Philadelphia (2008)
6. Cerf, L., Besson, J., Robardet, C., Boulicaut, J.F.: Closed patterns meet -ary relations. TKDD 3(1) (2009)
7. Clauset, A., Shalizi, C.R., Newman, M.E.J.: Power-law distributions in empirical data. SIAM Review 51(4), 661–703 (2009)
8. Ganter, B., Obiedkov, S.: Implications in triadic formal contexts. In: Wolff, K., Pfeiffer, H., Delugach, H. (eds.) ICCS 2004. LNCS (LNAI), vol. 3127, pp. 186–195. Springer, Heidelberg (2004)
9. Ganter, B., Wille, R.: Formal concept analysis: Mathematical foundations. Springer, Heidelberg (1999)
10. Hartigan, J.A.: Direct clustering of a data matrix. Journal of the American Statistical Association 67(337), 123–129 (1972)
11. Ignatov, D.I., Kaminskaya, A.Y., Kuznetsov, S.O., Magizov, R.A.: A concept-based biclustering algorithm. In: Proceedings of the Eight International conference on Intelligent Information Processing (IIP-8), pp. 140–143. MAKS Press (2010) (in russian)
12. Jäschke, R., Hotho, A., Schmitz, C., Ganter, B., Stumme, G.: Trias - an algorithm for mining iceberg tri-lattices. In: ICDM, pp. 907–911. IEEE Computer Society, Los Alamitos (2006)
13. Ji, L., Tan, K.L., Tung, A.K.H.: Mining frequent closed cubes in 3d datasets. In: Dayal, U., Whang, K.Y., Lomet, D.B., Alonso, G., Lohman, G.M., Kersten, M.L., Cha, S.K., Kim, Y.K. (eds.) VLDB, pp. 811–822. ACM, New York (2006)
14. Lehmann, F., Wille, R.: A triadic approach to formal concept analysis. In: Ellis, G., Levinson, R., Rich, W., Sowa, J. (eds.) ICCS 1995. LNCS, vol. 954, pp. 32–43. Springer, Heidelberg (1995)
15. Mirkin, B.: Mathematical Classification and Clustering. Kluwer, Dordrecht (1996)
16. Newman, M.E.J.: Power laws, pareto distributions and zipf's law. Contemporary Physics 46(5), 323–351 (2005)
17. Wille, R.: Restructuring lattice theory: an approach based on hierarchies of concepts. In: Rival, I. (ed.) Ordered Sets, Boston, pp. 445–470 (1982)

Learning Inverted Dirichlet Mixtures for Positive Data Clustering

Taoufik Bdiri and Nizar Bouguila

Concordia Institute for Information Systems Engineering
Concordia University, Montreal, Canada, Qc, H3G 2W1
t_bdiri@encs.concordia.ca, bouguila@ciise.concordia.ca

Abstract. In this paper, we propose a statistical model to cluster positive data. The proposed model adopts a mixture of inverted Dirichlet distributions and is learned using expectation-maximization (EM) for parameters estimation and the minimum message length criterion (MML) for model selection. Experimental results using both synthetic and real data are presented to show the advantages of the proposed model.

1 Introduction

Cluster analysis is one of the fundamental tools for exploring and analyzing the underlying structure of a given data set and has been applied in a variety of problems from different disciplines such as pattern recognition, biology, psychology, image processing, economy and medicine [1]. The main objective is to divide a given set of multidimensional vectors into homogeneous clusters. A lot of clustering algorithms can be found in the literature where finite mixtures are perhaps the most widely used and cited models [2]. Most of the conventional mixture models have been proposed for general data by considering, for instance, Gaussian distributions. In many applications, however, the generated data are not Gaussian. In previous works, we have shown that the Dirichlet distribution, for instance, can be an excellent alternative to the Gaussian for proportional data [3,4]. The goal of this work is to generalize it to the case of positive data. As it is to be expected, the chosen probability density function has to take into account this fact. We propose then to consider the inverted Dirichlet distribution. In particular, we propose an EM-based framework [2] for the estimation of the parameters of an inverted Dirichlet mixture model. One of the major difficulties associated with finite mixture models is that we have to determine automatically the number of mixture components. This challenging task has been the subject of extensive studies in the past (see, for instance, [4]) and is handled in our case by developing an MML-based criterion [4,5].

The paper is organized as follows: In Section 2, we briefly present the inverted Dirichlet mixture and an approach to learn this mixture model is proposed. Some examples are used and some simulation results are presented to demonstrate the effectiveness of the proposed model in Section 4. Finally, Section 5 concludes the paper.

S.O. Kuznetsov et al. (Eds.): RSFDGrC 2011, LNAI 6743, pp. 265–272, 2011.

2 The Inverted Dirichlet Mixture Model

If a D-dimensional positive vector $\boldsymbol{X} = (X_1, X_2, ..., X_D)$ follows an inverted Dirichlet distribution, the joint density function is given in [6]

$$p(\boldsymbol{X}|\boldsymbol{\alpha}) = \frac{\Gamma(|\boldsymbol{\alpha}|)}{\prod_{d=1}^{D+1}\Gamma(\alpha_d)} \prod_{d=1}^{D} X_d^{\alpha_d - 1}(1 + \sum_{d=1}^{D} X_d)^{-|\boldsymbol{\alpha}|} \tag{1}$$

where $X_d > 0, d = 1, 2, \ldots, D$, $\boldsymbol{\alpha} = (\alpha_1, \ldots, \alpha_{D+1})$ is the vector of parameters and $|\boldsymbol{\alpha}| = \sum_{d=1}^{D+1} \alpha_d$, $\alpha_d > 0, d = 1, 2, \ldots, D+1$. Let $\mathcal{X} = \{\boldsymbol{X_1}, \boldsymbol{X_2}, \ldots, \boldsymbol{X_N}\}$ be a data set of N D-dimensional positive vectors with a common, but unknown, probability density function $p(\boldsymbol{X}|\Theta)$. Generally $\mathcal{X}$ is composed of different, say M, clusters, thus $p(\boldsymbol{X}|\Theta)$ may be approximated with sufficient accuracy by a finite M-components mixture model:

$$p(\boldsymbol{X}|\Theta) = \sum_{j=1}^{M} p(\boldsymbol{X}|\boldsymbol{\alpha}_j)p_j \tag{2}$$

where p_j are the mixing proportions which are positive and sum to one, and $p(\boldsymbol{X}|\boldsymbol{\alpha}_j)$ is the inverted Dirichlet distribution. The symbol Θ refers to the entire set of parameters $\Theta = \{\boldsymbol{\alpha}_1, \boldsymbol{\alpha}_2, \ldots, \boldsymbol{\alpha}_M, p_1, p_2, \ldots, p_M\}$.

In the following, we first develop the maximum likelihood (ML) estimates of our mixture. Then, we develop an MML criterion and we give the complete learning algorithm. The ML estimate, associated with a sample of observations, is a choice of parameters which maximizes the probability density function of the sample. The log-likelihood is generally maximized, instead of the likelihood, within the EM framework where each $\boldsymbol{X}_n$ is supposed to have arisen from one of the M clusters. Thus, let $\mathcal{Z} = \{\boldsymbol{Z}_1, \ldots, \boldsymbol{Z}_N\}$ denote the missing group-indicator vectors where the jth element of $\boldsymbol{Z}_n$, Z_{nj}, is equal to one if $\boldsymbol{X}_n$ belongs to cluster j and zero, otherwise. The complete data in this case are $(\mathcal{X}, \mathcal{Z})$ and the associated complete-data log-likelihood is given by

$$\Phi_c(\mathcal{X}, \mathcal{Z}|\Theta) = \sum_{j=1}^{M}\sum_{n=1}^{N} Z_{nj}\Big(\log p_j + \log p(\boldsymbol{X}_n|\boldsymbol{\alpha}_j)\Big) \tag{3}$$

The EM algorithm proceeds iteratively in two steps, The expectation (E) step and the maximization (M) step. In the E-step, we compute the conditional expectation of $\Phi_c(\mathcal{X}, \mathcal{Z}|\Theta)$ which is reduced to the computation of the posterior probabilities (i.e. the probability that a vector $\boldsymbol{X}_n$ is assigned to a cluster j):

$$p(j|\boldsymbol{X}_n, \boldsymbol{\alpha}_j) = \frac{p_j p(\boldsymbol{X}_n|\boldsymbol{\alpha}_j)}{\sum_{j=1}^{M} p_j p(\boldsymbol{X}_n|\boldsymbol{\alpha}_j)} \tag{4}$$

Then, the conditional expectation of the complete-data log-likelihood given by

$$Q(\mathcal{X}, \Theta) = \sum_{j=1}^{M}\sum_{n=1}^{N} p(j|\boldsymbol{X}_n, \boldsymbol{\alpha}_n)\Big(\log p_j + \log p(\boldsymbol{X}_n|\boldsymbol{\alpha}_j)\Big) \tag{5}$$

is maximized in the M-step. To resolve this optimization problem, we must determine the solution to $\frac{\partial}{\partial \Theta} Q(\mathcal{X}, \Theta) = 0$. Calculating the derivative with respect to $\alpha_{jd}, d = 1, \ldots, D$, we obtain

$$\frac{\partial Q(\mathcal{X}, \Theta)}{\partial \alpha_{jd}} = \sum_{n=1}^{N} p(j|\boldsymbol{X}_n, \boldsymbol{\alpha}_j)\Big(\Psi(|\boldsymbol{\alpha}_j|) - \Psi(\alpha_{jd}) + \log\big(\frac{X_{nd}}{1 + \sum_{d=1}^{D} X_{nd}}\big)\Big) \quad (6)$$

where $\Psi(.)$ is the digamma function. The derivative with to respect α_{jD+1} is

$$\frac{\partial Q(\mathcal{X}, \Theta)}{\partial \alpha_{jD+1}} = \sum_{n=1}^{N} p(j|\boldsymbol{X}_n, \boldsymbol{\alpha}_j)\Big(\Psi(|\boldsymbol{\alpha}_j|) - \Psi(\alpha_{jD+1}) + \log\big(\frac{1}{1 + \sum_{d=1}^{D} X_{nd}}\big)\Big) \quad (7)$$

According to the previous two equations, it is clear that a closed-form solution to estimate $\boldsymbol{\alpha}_j$ does not exist. Thus, we will use an iterative approach namely the Newton-Raphson method expressed as

$$\hat{\boldsymbol{\alpha}}_j^{new} = \hat{\boldsymbol{\alpha}}_j^{old} - H_j^{-1} G_j \qquad j = 1, 2, \ldots, M \quad (8)$$

where H_j is the Hessian matrix associated with $Q(\mathcal{X}, \Theta)$ and G_j is the vector of first derivatives, $G_j = (\frac{\partial Q(\mathcal{X},\Theta)}{\partial \alpha_{j1}}, \ldots, \frac{\partial Q(\mathcal{X},\Theta)}{\partial \alpha_{jD+1}})^T$. To calculate the Hessian of $Q(\mathcal{X}, \Theta)$ we have to compute the second and mixed derivatives:

$$\frac{\partial^2 Q(\mathcal{X}, \Theta)}{\partial^2 \alpha_{jd}} = (\Psi'(|\boldsymbol{\alpha}_j|) - \Psi'(\alpha_{jd})) \sum_{n=1}^{N} p(j|\boldsymbol{X}_n, \boldsymbol{\alpha}_j) \qquad d = 1, \ldots, D+1 \quad (9)$$

$$\frac{\partial^2 Q(\mathcal{X}, \Theta)}{\partial \alpha_{jd_1} \partial \alpha_{jd_2}} = \Psi'(|\boldsymbol{\alpha}_j|) \sum_{n=1}^{N} p(j|\boldsymbol{X}_n, \boldsymbol{\alpha}_j) \quad d_1 \neq d_2 \quad d_1, d_2 = 1, \ldots, D+1 \quad (10)$$

where $\Psi'(.)$ is the trigamma function. Concerning the parameters p_j, it is straightforward to show that a closed-form solution does exist and is given by:

$$p_j = \frac{\sum_{n=1}^{N} p(j|\boldsymbol{X}_n, \boldsymbol{\alpha}_j)}{N} \quad (11)$$

In the following, we focus on the development of an MML criterion for model selection to have a certain trade-off between flexibility and model complexity. The message length for a mixture of distributions is given by [5]

$$MessLen \simeq -\log(h(\Theta)) - \log(p(\mathcal{X}|\Theta)) + \frac{1}{2}\log(|F(\Theta)|) + \frac{N_p}{2}(1 - \log(12)) \quad (12)$$

where $h(\Theta)$ is the prior probability, $p(\mathcal{X}|\Theta)$ is the likelihood, $F(\Theta)$ is the expected Fisher information matrix, $|F(\Theta)|$ is its determinant, and N_p is the number of free parameters which is equal to $M(D+2) - 1$. The selection of the number of clusters is carried out by finding the minimum with regards to Θ of the message length $MessLen$. The expected Fisher information matrix is generally approximated by complete-data Fisher information matrix [7], then:

$$|F(\Theta)| \simeq |F(p_1, \dots, p_M)| \prod_{j=1}^{M} |F(\boldsymbol{\alpha}_j)| \tag{13}$$

where $|F(p_1, \dots, p_M)|$ is given by [4]:

$$|F(p_1, \dots, p_M)| = \frac{N^{M-1}}{\prod_{j=1}^{M} p_j} \tag{14}$$

and $|F(\boldsymbol{\alpha}_j)|$ is the Fisher information with regards to $\boldsymbol{\alpha}_j$. For $F(\boldsymbol{\alpha}_j)$, let us consider the jth cluster of the mixture $\mathcal{X}_j = (\boldsymbol{X}_l, \dots, \boldsymbol{X}_{l+n_j-1})$, where $l \leq N$ and n_j is the number of elements in cluster j, with parameter $\boldsymbol{\alpha}_j$. Let $p(\mathcal{X}_j|\boldsymbol{\alpha}_j)$ be the log-likelihood function associated with $\mathcal{X}_j$, then

$$-\frac{\partial^2 \log(p(\mathcal{X}_j|\boldsymbol{\alpha}_j))}{\partial\alpha_{jd_1}\partial\alpha_{jd_2}} = -n_j\Psi'(|\boldsymbol{\alpha}_j|) \quad d_1, d_2 = 1, \dots, D+1 \quad d_1 \neq d_2 \tag{15}$$

$$-\frac{\partial^2 \log(p(\mathcal{X}_j|\boldsymbol{\alpha}_j))}{\partial^2\alpha_{jd}} = -n_j(\Psi'(|\boldsymbol{\alpha}_j|) - \Psi'(\alpha_{jd})) \quad d = 1, \dots, D+1 \tag{16}$$

$$|F(\boldsymbol{\alpha}_j)| = (1 - \Psi'(|\boldsymbol{\alpha}_j|) \sum_{d=1}^{D+1} \frac{1}{\Psi'(\alpha_{jd})})n_j^{D+1} \prod_{d=1}^{D+1} \Psi'(\alpha_{jd}) \tag{17}$$

By substituting the previous equation and Eq. 14 into Eq. 13, we obtain

$$F(\Theta) = \frac{N}{\prod_{j=1}^{M} p_j} \prod_{j=1}^{M} [(1 - \Psi'(|\boldsymbol{\alpha}_j|) \sum_{d=1}^{D+1} \frac{1}{\Psi'(\alpha_{jk})})n_j^{D+1} \prod_{d=1}^{D+1} \Psi'(\alpha_{jd})] \tag{18}$$

Regarding $h(\Theta)$, we make a common assumption in the case of finite mixture models by supposing that $\boldsymbol{\alpha}_j$ and the vector $(p_1, \dots, p_M)$ are independent:

$$h(\Theta) = h(p_1, \dots, p_M) \prod_{j=1}^{M} h(\boldsymbol{\alpha}_j) \tag{19}$$

We will now define the densities $h(\boldsymbol{\alpha}_j)$ and $h(p_1, \dots, p_M)$. We know that the vector $(p_1, \dots, p_M)$ is defined on the simplex $\{p_1, \dots, p_M | \sum_{j=1}^{M-1} p_j < 1\}$, then a natural choice, as a prior, for this vector is a symmetric Dirichlet Distribution with parameters set to one which gives a uniform prior [4]:

$$h(p_1, \dots, p_M) = (M-1)! \tag{20}$$

As for $h(\boldsymbol{\alpha}_j)$ and in the absence of other knowledge about the α_{jd}, we choose the following uniform prior, which has been found appropriate according to our experiments, over the range $[0, e^6|\hat{\boldsymbol{\alpha}}_j|/\hat{\alpha}_{jk}]$ where $\hat{\boldsymbol{\alpha}}_j$ is the estimated vector:

$$h(\alpha_{jk}) = \frac{e^{-6}\hat{\alpha}_{jk}}{|\hat{\boldsymbol{\alpha}}_j|} \tag{21}$$

By substituting Eqs. 21 and 20 in Eq. 19, we obtain the following:

$$\log(h(\Theta)) = \sum_{j=1}^{M-1} \log(j) - 6M(D+1) - (D+1)\sum_{j=1}^{M} \log(|\hat{\boldsymbol{\alpha}}_j|) + \sum_{j=1}^{M}\sum_{d=1}^{D+1} \log(\hat{\alpha}_{jd}) \quad (22)$$

Below we give the complete learning algorithm for each candidate value of M:

1. Input: N D-dimensional vectors $\boldsymbol{X}_n$ and a specified number of clusters M.
2. Initialization algorithm [1]
3. E-Step: Compute the posterior probabilities $p(j|\boldsymbol{X}_n, \boldsymbol{\alpha}_j)$ using Eq. 4.
4. M-Step:
   - Update the $\boldsymbol{\alpha}_j$ using Eq. 8, $j = 1, \ldots, M$.
   - Update p_j using Eq. 11, $j = 1, \ldots, M$.
5. If $p_j < \epsilon$, then discard component j and go to 3.
6. If the convergence test is passed terminate, else go to 3.
7. Calculate the associated criterion $MML(M)$ using Eq. 12.

3 Experimental Results

In this section, we first validate our algorithm using synthetic data and then using a real data set. The convergence test in our learning algorithm was based on the variation of the likelihood function and ϵ was set to 10^{-6}.

3.1 Synthetic Data

We tested our algorithm on several generated multi-dimensional data. In the following, we show one example of a two-dimensional data set that we have generated. We use $D = 2$ for ease of representation. Data were generated from three inverted Dirichlet densities (see figure 1) with different parameters as shown in table 1. A total of 100 samples for each of the two first densities and a total of 50 samples for the third distribution were taken. The message length values as a function of the number of clusters are presented in table 2, where we can see clearly that the MML criterion has found the exact number of clusters.

3.2 Real Data

Here we investigate the performance of our algorithm and compare the modeling capabilities of inverted Dirichlet and Gaussian mixtures using a well-known data set. The classification was performed using the Bayesian decision rule after the classes densities were estimated and the number of clusters was selected. The used data set, called Haberman dataset [9], contains cases from a study that was conducted between 1958 and 1970 at the University of Chicago's Billings Hospital on the survival of patients who had undergone surgery for breast cancer. The

[1] For the initialization we use the K-Means algorithm and the method of moments developed for the inverted Dirichlet as done for the Dirichlet in [8].

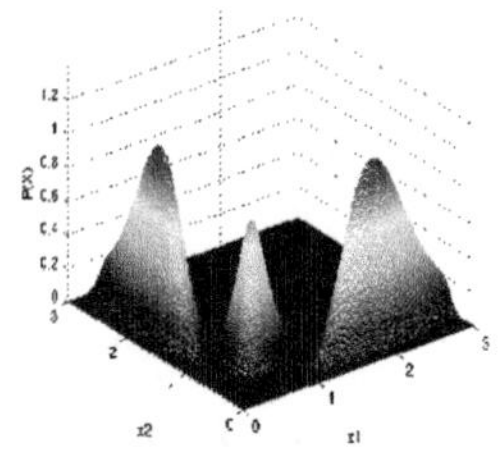

Fig. 1. A two-dimensional artificial mixture model with three components

Table 1. Real and estimated mixture parameters for the two-dimensional generated data set

Cluster 1	$p_1 = 0.4$	$\alpha_{11} = 15$	$\alpha_{21} = 65$	$\alpha_{31} = 30$
	$\hat{p}_1 = 0.4$	$\hat{\alpha}_{11} = 14.83$	$\hat{\alpha}_{21} = 64.38$	$\hat{\alpha}_{31} = 29.58$
Cluster 2	$p_2 = 0.4$	$\alpha_{12} = 65$	$\alpha_{22} = 15$	$\alpha_{32} = 30$
	$\hat{p}_2 = 0.4$	$\hat{\alpha}_{12} = 64.06$	$\hat{\alpha}_{22} = 14.90$	$\hat{\alpha}_{32} = 29.69$
Cluster 3	$p_3 = 0.2$	$\alpha_{13} = 30$	$\alpha_{23} = 34$	$\alpha_{33} = 35$
	$\hat{p}_3 = 0.2$	$\hat{\alpha}_{13} = 30.54$	$\hat{\alpha}_{23} = 34.32$	$\hat{\alpha}_{33} = 35.51$

Table 2. Message length values as a function of the number of clusters for the two-dimensional generated data set

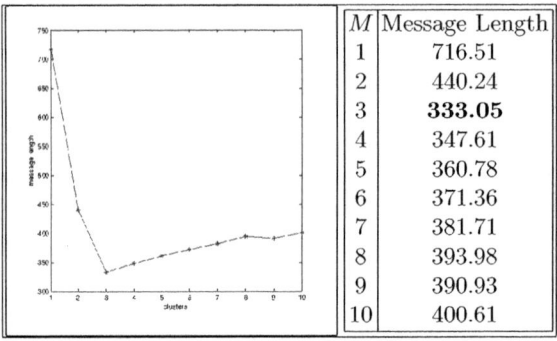

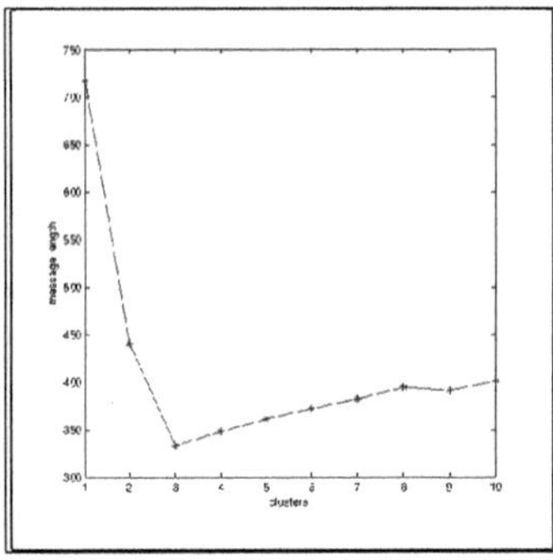

M	Message Length
1	716.51
2	440.24
3	**333.05**
4	347.61
5	360.78
6	371.36
7	381.71
8	393.98
9	390.93
10	400.61

dataset contains 306 instances, and four attributes including the class attribute. These attributes are: age of patient at time of operation, patient's year of operation, number of positive auxiliary nodes detected, survival status (1 = the patient survived 5 years or longer (S), 2 = the patient died within 5 year (D)). It has 225 instances from class 1, and 81 instances from class 2. By applying our algorithm to this dataset, the MML criterion has found that $M = 2$ leads us to the minimum message length. So we have two classes, which meets the specification of our dataset (see table 3). Using Gaussian mixture, however, we failed to obtain the exact number of clusters (i.e. $M = 3$ was wrongly favored) as

Table 3. The message length as a function of the number of clusters in the case of the Haberman dataset when using inverted Dirichlet mixture

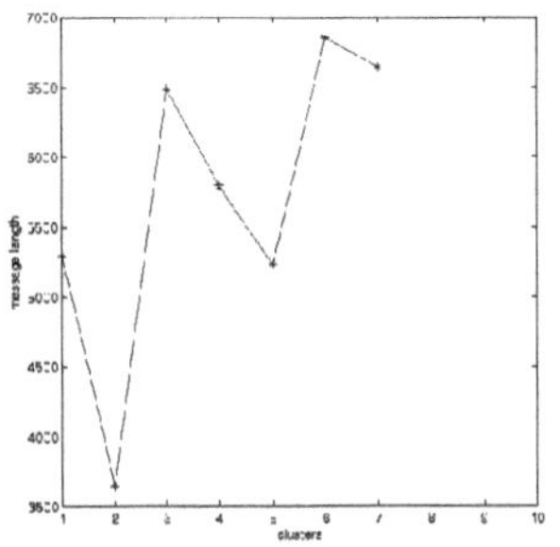

M	Message Length
1	5297.5
2	**3645.6**
3	6480.6
4	5791.3
5	5234.1
6	6851.5
7	6640.4
8	N/A
9	N/A
10	N/A

Table 4. The message length as a function of the number of cluster in the case of the Haberman dataset when using Gaussian mixture

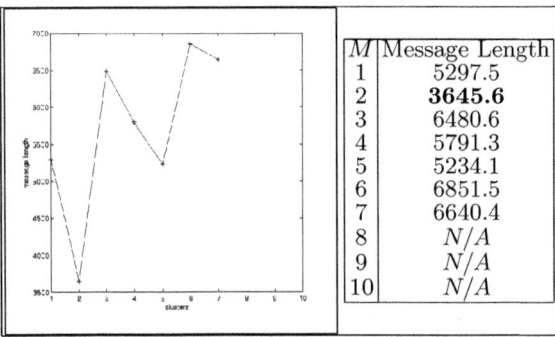

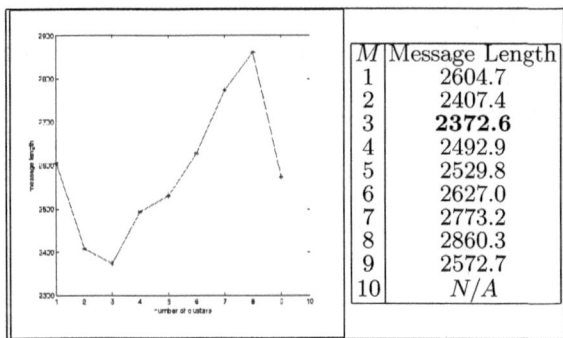

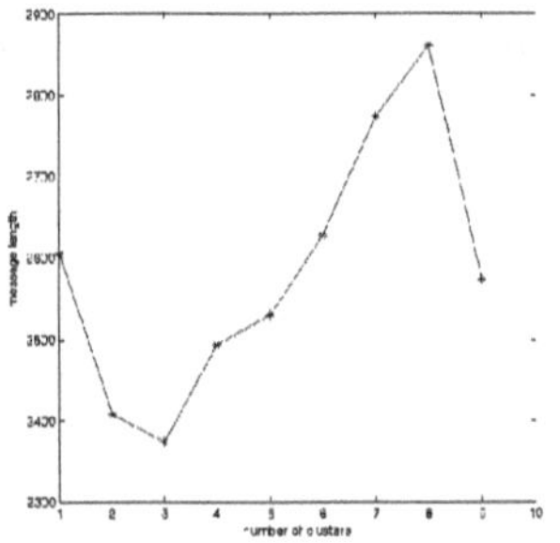

M	Message Length
1	2604.7
2	2407.4
3	**2372.6**
4	2492.9
5	2529.8
6	2627.0
7	2773.2
8	2860.3
9	2572.7
10	N/A

shown in table 4. Table 5 displays the confusion matrices for Haberman dataset classification when using both the inverted Dirichlet mixture and the Gaussian mixture which we have forced to consider $M = 2$. From this table we notice that the classification based on the inverted Dirichlet mixture is significantly more accurate and precise than the one based on the Gaussian mixture.

Table 5. Confusion matrices for Haberman dataset classification using both the inverted Dirichlet and the Gaussian mixture models

inverted Dirichlet

	S	**D**
S	195	30
D	51	30

Gaussian

	S	**D**
S	150	75
D	27	54

4 Conclusion

The main goal of this work was to find meaningful structure in a set of unlabeled non Gaussian positive vectors through inverted Dirichlet mixture-based modeling. A complete learning algorithm has been presented and applied to synthetic and real data.

Acknowledgment. The completion of this research was made possible thanks to the Natural Sciences and Engineering Research Council of Canada (NSERC).

References

1. Everitt, B.: Cluster Analysis. Arnold Publishers (2001)
2. McLachlan, G.J., Peel, D.: Finite Mixture Models. Wiley, New York (2000)
3. Bouguila, N., Ziou, D.: Using unsupervised learning of a finite Dirichlet mixture model to improve pattern recognition applications. Pattern Recognition Letters 26(12), 1916–1925 (2005)
4. Bouguila, N., Ziou, D.: Unsupervised Selection of a Finite Dirichlet Mixture Model: An MML-Based Approach. IEEE Transactions on Knowledge and Data Engineering 18(8), 993–1009 (2006)
5. Wallace, C.S.: Statistical and Inductive Inference by Minimum Message Length. Springer, Heidelberg (2005)
6. Tiao, G.G., Cuttman, I.: The Inverted Dirichlet Distribution with Applications. Journal of the American Statistical Association 60(311), 793–805 (1965)
7. Figueiredo, M.A.T., Jain, A.K.: Unsupervised Learning of Finite Mixture Models. IEEE Transactions on Pattern Analysis and Machine Intelligence 24(3), 4–37 (2002)
8. Bouguila, N., Ziou, D., Vaillancourt, J.: Unsupervised Learning of a Finite Mixture Model Based on the Dirichlet Distribution and its Application. IEEE Transactions on Image Processing 13(11), 1533–1543 (2004)
9. Murphy, P.M., Aha, D.W.: UCI Repository of Machine Learning Databases. Department of Information and Computer Science, University of California, Irvine (1998), `http://www.ics.ci.edu/mlearn/MLRepository.html`

Developing Additive Spectral Approach to Fuzzy Clustering

Boris Mirkin[1,2] and Susana Nascimento[3]

[1] Department of Computer Science, Birkbeck University of London, London, UK
[2] School of Applied Mathematics and Informatics, Higher School of Economics, Moscow, RF
[3] Department of Computer Science and Centre for Artificial Intelligence (CENTRIA), Faculdade de Ciências e Tecnologia, Universidade Nova de Lisboa, Caparica, Portugal

Abstract. An additive spectral method for fuzzy clustering is presented. The method operates on a clustering model which is an extension of the spectral decomposition of a square matrix. The computation proceeds by extracting clusters one by one, which allows us to draw several stopping rules to the procedure. We experimentally test the performance of our method and show its competitiveness.

In spite of the fact that many relational fuzzy clustering algorithms have been developed already [1,2,3,4,12], most of them are ad hoc and, moreover, they all involve manually specified parameters such as the number of clusters or threshold of similarity without providing any guidance for choosing them. We apply a model-based approach of additive clustering, combined with the spectral clustering approach, to develop a novel relational fuzzy clustering method that is both adequate and supplied with model-based parameters helping to choose the right number of clusters.

We assume the data in the format of what is called similarity or relational data, that is a matrix $W = (w_{tt'}), t, t' \in T$, of similarity indexes $w_{tt'}$, between objects t, t' from a set of objects T. We further assume that this similarity values are but manifested expressions of some hidden relational patterns which can be represented by fuzzy clusters. We propose to formalize a relational fuzzy cluster as represented by two items: (i) a membership vector $\mathbf{u} = (u_t)$, $t \in T$, such that $0 \leq u_t \leq 1$ for all $t \in T$, and (ii) an intensity $\mu > 0$ that expresses the extent of significance of the pattern corresponding to the cluster. With the introduction of the intensity, applied as a scaling factor to $\mathbf{u}$, it is the product $\mu\mathbf{u}$ that is a solution rather than its individual co-factors. Given a value of the product μu_t, it is impossible to tell which part of it is μ and which u_t. To resolve this, we follow a conventional scheme: let us constrain the scale of the membership vector $\mathbf{u}$ on a constant level, for example, by a condition such as $\sum_t u_t = 1$ or $\sum_t u_t^2 = 1$, then the remaining factor will define the value of μ. The latter normalization better suits the criterion implied by our fuzzy clustering method and, thus, is accepted further on.

S.O. Kuznetsov et al. (Eds.): RSFDGrC 2011, LNAI 6743, pp. 273–277, 2011.

To make the cluster structure in the similarity matrix sharper, we apply the spectral clustering approach to pre-process a raw similarity matrix W into A by using the so-called normalized Laplacian transformation as related to the popular clustering criterion of normalized cut [6]. This criterion relates to the minimum non-zero eigenvalue of the Laplacian matrix. To change this to the maximum eigenvalue, we further transform this to its pseudo-inverse matrix, which also increases the gaps between eigenvalues.

Our additive fuzzy clustering model follows that of [11,7,10] and involves K fuzzy clusters that reproduce the pseudo-inverted Laplacian similarities $a_{tt'}$ up to additive errors according to the following equations:

$$a_{tt'} = \sum_{k=1}^{K} \mu_k^2 u_{kt} u_{kt'} + e_{tt'}, \tag{1}$$

where $\mathbf{u}_k = (u_{kt})$ is the membership vector of cluster k, and μ_k its intensity.

The item $\mu_k^2 u_{kt} u_{kt'}$ is the product of $\mu_k u_{kt}$ and $\mu_k u_{kt'}$ expressing participation of t and t', respectively, in cluster k. This value adds up to the others to form the similarity $a_{tt'}$ between topics t and t'. The value μ_k^2 summarizes the contribution of the intensity and will be referred to as the cluster's weight.

To fit the model in (1), we apply the least-squares approach, thus minimizing the sum of all $e_{tt'}^2$. Within that, we attend to the one-by-one principal component analysis strategy for finding one cluster at a time by minimizing

$$E = \sum_{t,t' \in T} (b_{tt'} - \xi u_t u_{t'})^2 \tag{2}$$

with respect to unknown positive ξ weight (so that the intensity μ is the square root of ξ) and fuzzy membership vector $\mathbf{u} = (u_t)$, given similarity matrix $B = (b_{tt'})$.

At the first step, B is taken to be equal to A. After each step, the found cluster is subtracted from B, so that the residual similarity matrix for obtaining the next cluster is equal to $B - \mu^2 \mathbf{u}\mathbf{u}'$ where μ and $\mathbf{u}$ are the intensity and membership vector of the found cluster. In this way, A indeed is additively decomposed according to formula (1) and the number of clusters K can be determined in the process.

The optimal value of ξ at a given $\mathbf{u}$ is proven to be

$$\xi = \frac{\mathbf{u}' B \mathbf{u}}{(\mathbf{u}'\mathbf{u})^2} \tag{3}$$

which is obviously non-negative if B is semi-positive definite.

By putting this ξ in equation (2), we arrive at $E = S(B) - \xi^2 (\mathbf{u}'\mathbf{u})^2$, where $S(B) = \sum_{t,t' \in T} b_{tt'}^2$ is the similarity data scatter.

Let us denote the last item by

$$G(\mathbf{u}) = \xi^2 (\mathbf{u}'\mathbf{u})^2 = \left(\frac{\mathbf{u}' B \mathbf{u}}{\mathbf{u}'\mathbf{u}} \right)^2, \tag{4}$$

so that the similarity data scatter is the sum $S(B) = G(\mathbf{u}) + E$ of two parts, $G(\mathbf{u})$ explained by cluster $(\mu, \mathbf{u})$, and E, unexplained. Therefore, an optimal cluster is to maximize the explained part $G(\mathbf{u})$ in (4) or its square root

$$g(\mathbf{u}) = \xi \mathbf{u}'\mathbf{u} = \frac{\mathbf{u}'B\mathbf{u}}{\mathbf{u}'\mathbf{u}}, \tag{5}$$

which is the celebrated Rayleigh quotient: its maximum value is the maximum eigenvalue of matrix B, which is reached at its corresponding eigenvector, in the unconstrained problem.

This shows that the spectral clustering approach is appropriate for our problem. According to this approach, one should find the maximum eigenvalue λ and corresponding normed eigenvector z for B, $[\lambda, z] = \Lambda(B)$, and take its projection to the set of admissible fuzzy membership vectors.

Our clustering approach involves a number of model-based criteria for halting the process of sequential extraction of fuzzy clusters. The process stops if either is true:

1. The optimal value of ξ (3) for the spectral fuzzy cluster becomes negative.
2. The contribution of a single extracted cluster to the data scatter becomes too low, less than a pre-specified $\tau > 0$ value.
3. The residual data scatter becomes smaller than a pre-specified ϵ value, say less than 5% of the original similarity data scatter.

The described one-by-one Fuzzy ADDItive-Spectral cluster extraction method is referred to as FADDIS. We have experimentally compared FADDIS with other approaches, specifically, with those used at (a) ordinary graphs for revealing community structure, (b) affinity similarity data derived from feature based information, (c) small real-world benchmark dissimilarity datasets, and (d) genuine similarity data [8]. In this paper we describe one of the experiments - in comparing the performance of FADDIS with various versions of fuzzy c-means algorithm [1]. In this, we carry on the experiment described in [2]. This experiment concerns a two-dimensional data set, that we refer to as Bivariate4, comprising four clusters generated from bivariate spherical normal distributions with the same standard deviation 950 at centers (1000, 1000), (1000,4000), (4000, 1000), and (4000, 4000), respectively (see Fig. 1).

This data was analyzed in [2] by using the matrix D of Euclidean distances between the generated points. Five different fuzzy clustering methods have been compared, three of them relational, by Roubens [9], Windham [12] and NERFCFM [4], and two of the fuzzy c-means with different preliminary pre-processing options of the similarity data into the entity-to-feature format, FastMap and SMACOF [2]. Of these five different fuzzy clustering methods, by far the best results have been obtained with the fuzzy c-means method applied to a five-feature set extracted from D with FastMap method [2]. The adjusted Rand index [5] of the correspondence between the generated clusters and those the best is equal on average, of 10 trials, 0.67 (no standard deviation is reported in [2]).

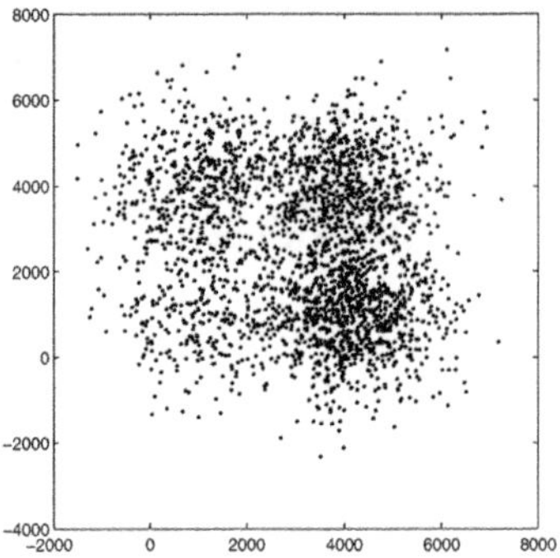

Fig. 1. Bivariate4: the data of four Gaussian bivariate clusters [2]

In our computations, five consecutive FADDIS clusters have been extracted for each of randomly generated ten Bivariate4 datasets. The algorithm halts at stop condition (2): 'cluster's contribution is too small'. Then the very first approximate cluster is discarded as reflecting just the general connectivity, and the remaining four are defuzzified into partitions so that every entity is assigned to its maximum membership class. The average values of the adjusted Rand index (ARI) in these experiments are 0.70 (0.03) at 500-and 1000-strong datasets, whereas ARI=0.73 (0.01) at 2500-strong generated datasets (the standard deviations are reported in the parentheses). This favorably compares with ARI value of 0.67 reported in [2] as the best achieved with fuzzy c-means.

Acknowledgments. This work has been supported by grant PTDC/EIA/69988/2006 from the Portuguese Foundation for Science & Technology. The support of the Laboratory for Analysis and Choice of Decisions at the National Research University Higher School of Economics, Moscow RF to BM is acknowledged.

References

1. Bezdek, J., Keller, J., Krishnapuram, R., Pal, T.: Fuzzy Models and Algorithms for Pattern Recognition and Image Processing. Kluwer Academic Publishers, Dordrecht (1999)
2. Brouwer, R.: A method of relational fuzzy clustering based on producing feature vectors using FastMap. Information Sciences 179, 3561–3582 (2009)
3. Davé, R., Sen, S.: Robust fuzzy clustering of relational data. IEEE Transactions on Fuzzy Systems 10, 713–727 (2002)
4. Hathaway, R.J., Bezdek, J.C.: NERF c-means: Non-Euclidean relational fuzzy clustering. Pattern Recognition 27, 429–437 (1994)
5. Hubert, L.J., Arabie, P.: Comparing partitions. Journal of Classification 2, 193–218 (1985)

6. von Luxburg, U.: A tutorial on spectral clustering. Statistics and Computing 17, 395–416 (2007)
7. Mirkin, B.: Additive clustering and qualitative factor analysis methods for similarity matrices. Journal of Classification 4(1), 7–31 (1987)
8. Mirkin, B., Nascimento, S.: Analysis of Community Structure, Affinity Data and Research Activities using Additive Fuzzy Spectral Clustering. Technical Report 6, School of Computer Science, Birkbeck University of London (2009)
9. Roubens, M.: Pattern classification problems and fuzzy sets. Fuzzy Sets and Systems 1, 239–253 (1978)
10. Sato, M., Sato, Y., Jain, L.C.: Fuzzy Clustering Models and Applications. Physica-Verlag, Heidelberg (1997)
11. Shepard, R.N., Arabie, P.: Additive clustering: representation of similarities as combinations of overlapping properties. Psychological Review 86, 87–123 (1979)
12. Windham, M.P.: Numerical classification of proximity data with assignment measures. Journal of Classification 2, 157–172 (1985)

Data-Driven Adaptive Selection of Rules Quality Measures for Improving the Rules Induction Algorithm

Marek Sikora[1,2] and Łukasz Wróbel[1]

[1] Silesian University of Technology, ul. Akademicka 16, 44-100 Gliwice, Poland
{Marek.Sikora,Lukasz.Wrobel}@polsl.pl
[2] Institute of Innovative Technologies EMAG, ul. Leopolda 31,
40-189 Katowice, Poland

Abstract. The proposition of adaptive selection of rule quality measures during rules induction is presented in the paper. In the applied algorithm the measures decide about a form of elementary conditions in a rule premise and monitor a pruning process. An influence of filtration algorithms on classification accuracy and a number of obtained rules is also presented. The analysis has been done on twenty one benchmark data sets.

Keywords: rules induction, rules quality measures, classification.

1 Introduction

Knowledge discovery in databases is a process of extraction of unknown, non-trivial and useful patterns from data. One of the most popular representations of such patterns, because of its simplicity, is a rule form (1).

$$\text{IF } a_1 \in V_{a_1} \text{and } \dots \text{ and } a_k \in V_{a_k} \text{THEN } d = V_d \tag{1}$$

Rules induction is made on the basis of training data set $\mathbf{DT} = (U, A \cup \{d\})$, where U is the finite set of objects characterized by the set of conditional attributes A and the decision attribute d. Each attribute $a \in A$ is treated as a function $a: U \to D_a$, where D_a is the range of the attribute a. The consequence of the assumed notation is that in the rule of the form (1) we have $\{a_1, \dots, a_k\} \subseteq A$, $V_{a_i} \subseteq D_{a_i}$, and $v_d \in D_d$. The expression $a \in V$ is called the conditional descriptor; a set of objects with equal values of decision attribute is the decision class (notation: $X_v = \{x \in U: d(x) = v\}$).

The algorithms building a coverage of training set are usually applied during the rule induction. The most popular algorithms are: RIPPER [4], CN2 [3], the AQ family [7] and the algorithms derived from the rough sets theory [5,11,13]. Rules are also employed for descriptive purposes (description based on rule induction, subgroup discovery [8,14]) because of their clearness. Rule induction algorithms exploiting a specific, domain or expert knowledge about an analyzed problem are also developed [12].

S.O. Kuznetsov et al. (Eds.): RSFDGrC 2011, LNAI 6743, pp. 278–285, 2011.

All of the algorithms mentioned above employ measures that decide either about form of a determined rule or about which of already determined rules may be removed or joined. These measures are called the rule quality measures and their main goal is such steering of induction and/or reduction processes that an output rule set contains rules with the best quality. A set composed of rules with good generalization (high classification accuracy) and description abilities (small number of output rules) is the rule set with high quality. Measures used in the rule growth process are usually different from measures employed in the pruning process. One can notice that at fixed induction algorithm, the kind of used measure influences results obtained by a classifier [6,10,11]. Results are here meant as the classification accuracy (average classification accuracy or the area under the ROC curve) and a number of induced rules. The majority of works concerning rules induction focus on finding one fixed method of induction that guarantees obtaining generally better classification results regardless of the kind of analyzed data set (in particular, the number of classes, examples distribution). In [6] an attempt at predicting the number of positive (p_{real}) and negative (n_{real}) examples covered by a rule on a data set irrespective of the training set was made. Thus, the matter is to define the real quality of the rule. Attempts at relating the numbers (p_{real} and n_{real}) with 9 parameters describing the rule on the training set by with the aid of the regression method or SVM were made. The parameters were, inter alia, rule precision, the true positive rate of rule, the false positive rate, the number of positive examples. Over 30 benchmark data sets have been evaluated in order to determine linear regression coefficients relating the mentioned 9 parameters with the values p_{val} and n_{val}. The quality measure defined in this way didn't prove better than other known measures (e.g. m-estimate).

A similar work is the earlier An's paper [1], in which attempts at defining conditions of recommendation for proper quality measure according to the analyzed data set characteristic were made. In the research, inter alia, the number of decision classes and distribution of examples among classes were taken into consideration. However, experimental justification that the obtained recommendations lead to better results didn't succeed. Both of the approaches are strongly dependent on the number and variety of data sets designed for a creation of the model recommending a quality measure.

In this paper, the other approach basing on adaptive measure selection during rules growing, pruning and filtering is presented. The appropriate measure selection is determined by the classifier quality gained by the measure on validation data sets independent of the testing set. The influence of one of the three rule filtration algorithms on the rule number and classification accuracy has been also researched. The number of conflicts during the classification and the number of conflicts resolved incorrectly were also monitored in order to verify whether the further improvement of the classifier quality is possible. We hope that the information will enable us to improve the classification accuracy through giving consideration in voting not only to the quality of voting rules, but also to information about a testing object neighbourhood.

The presented works are the continuation of the research contained in papers [10,11]. The mentioned paper by Jansen and Fürnkranz [6] had a strong influence on the paper, too. In the next sections, the process of adaptive measures selection in the rule induction process is discussed. Results of the experiments carried out on twenty one benchmark data sets are also described.

2 Rule Quality Measures

Values of most known rule quality measures [2,11] can be determined basing on analysis of a contingency table (or so-called PN space [2]), that allows to describe rules behavior with relation to the training set. Let p denote a set of positive examples covered by the rule r (P stands for all positive examples), and n denote a set of negative examples covered by the rule r (N stands for all negative examples). Then we can define the following table:

p	n	$p+n$
$P-p$	$N-n$	$P+N-p-n$
P	N	$P+N$

The basic measures defined for the rule are the accuracy (precision) $p/(p+n)$ and the coverage p/P. We aim at inducing rules characterized by maximal accuracy and maximal coverage.

The following known measures that evaluate the rule accuracy and the coverage simultaneously were used in tests described in this paper: laplace, m-estimate [2], g-measure [2], gain [11], Cohen [1], C2 [1], rule specificity and sensitivity, weighted relative accuracy [2]. The measures have been chosen due to high quality of results obtained by them [6,10,11]. Moreover, three additional measures have been tested. The first is the lift measure used for rules assessment by the See5 program. The second one is modified C2 measure. The C2 measure can be expressed by the product of two components. The accuracy occurs in one of the components, the rule coverage in the other. In the modified version of C2 the expression $(p-n)/(p+n)$ was used in place of accuracy. The expression is more restrictive than the precision and is applied in the RIPPER program [4] for rules pruning. The third measure, not used up till now, is p-value of a rule calculated for the exact Fisher test.

For the assessment of a statistical significance of a rule, we consider the following null hypothesis: *assignment of examples covered by the rule to the decision class indicated by the rule is equivalent to a random assignment of the examples to the class.* A p-value of the test is calculated by summing up probabilities obtained for all possible rules recognising as many examples as the analysed rule but characterized by higher precision. A p-value for the rule r which covers p positive and n negative examples can be computed using the formulas (2, 3).

$$p(r) = \frac{\binom{p+n}{p}\binom{P+N-p-n}{P-p}}{\binom{P+N}{P}} \tag{2}$$

$$p\text{-val}(r) = \sum_{k=0}^{\min\{P-p,n\}} p(p+k, n-k) \tag{3}$$

Values of the p-val measure belong to the interval $\langle 0, 1 \rangle$. The measure, like the others, is monotone with respect to p and n (provided that the second parameter value is fixed).

Each of the measures chosen for experiments evaluates a rule in a different manner. Some measures place stronger emphasis on the rule accuracy, another ones on its coverage. Some of them reflect examples distribution in the training set, and other ones not.

3 Data-Driven Selection of Rules Quality Measures in Rules Induction Algorithm

The process of creating a rule in the form (1) based on a certain data set consists in selection of conditional attributes that will create conditional descriptors and in establishing ranges of the descriptors (i.e. sets V_a).

The idea of rule induction algorithm presented below is a modification of the MODLEM algorithm [13]. Our version of the algorithm is named q-MODLEM. The rule premise generation process includes selection of conditional attributes, which create conditional descriptors, and determination of their ranges. The algorithm works in the following way: sorted in non-decreasing order values of each conditional attribute are one by one tested in order to find so called cut-off point g. The cut-off point is in the middle between two successive attribute a values (e.g. $v_a < g < w_a$) which separate positive examples from negative ones. The cut-off point g divides current range of values of attribute a into two ranges: $(-\infty, g\rangle$ and $\langle g, +\infty)$ and current set of training examples into two corresponding to these ranges subsets: U_1 and U_2. Let us assume that the set U_1 contains more positive examples than the set U_2, then the descriptor $(-\infty, g\rangle$ is added to already existing rule premise and the rule is evaluated by an established quality measure (algorithm's parameter). The cut-off point maximizing a value of the quality measure for the rule extended in this way is the best. Establishing an optimal descriptor causes its adding to the rule premise permanently. Then a process of searching a new cut-off point is initiated. Descriptors are added to the rule as long as adding the next one doesn't cause the increase of rule precision. After the grown phase, the pruning phase is initialized. The shortening algorithm uses the hill climbing strategy, which consists in removing elementary conditions as long as a rule keeps (or increases) its quality.

Independently on the method of rules induction, the output rules set can be reduced by the use of joining [9] and/or filtration [10,11] algorithms. The algorithms of joining and filtration of the rules set are briefly described below.

The next pruning procedure consists in rules filtration. Additionally, the filtration algorithm applies a tuning data set. During filtration, rules which are unnecessary on the ground of some criterion (for example, classification accuracy on the training and the tuning data set) are removed from the rules set.

The process of filtration is done by one of three following algorithms: Forward, Backwards, Coverage. The Forward algorithm, starting with one-rule (the best one – the higher quality) descriptions of decision classes, builds a classifier, and then in each iteration, successively adds a rule from the ranking list to the classifier if addition of this rule increases the quality of classifier. The process of rule addition stops if the filtered rules set has the same quality as the whole (unfiltered) rules set. The Backwards algorithm works in the opposite way – starting with the weakest rule it successively removes rules from the classifier. Forward and Backwards algorithms apply information about quality of single rules (verification on the training set) and their classification abilities (verification on the training and the tuning set). The third algorithm just builds the coverage of a training set of examples using the rules ranking established by the selected quality measure. Only rules that cover examples already covered by rules with higher quality are removed. A detailed description of the first version of the q-MODLEM algorithm can be found in [10].

In an adaptive selection of measures the induction algorithm receives a list of quality measures which can be used during cut-off points searching, growing, pruning and filtration of rules. The chosen measure is assumed to be applied consequently on each stage of rule induction and optimization. The selection of the best measure is made by cross-validation on the training set. The measure which obtains the best classification accuracy (or average classification accuracy – depending on what we want to optimize) is then used for rules induction based on the whole training set and applied for test examples classification. In the next section, the results of the experiments comparing the arbitrary and adaptive measure selection are presented.

4 Experiments

In the experiments described below twenty one benchmark data sets (balance-scale, breast-wisc, bupa, car, Australian credit, German credit, diabetes, ecoli, glass, heart Cleveland, heart statlog, ionosphere, iris, kdd-synthetic-control, lymphography, prnn-synthetic, segment, sonar, splice, wine, yeast) have been used. The 10-fold cross-validation was used as the testing methodology as well as for selection (done on training examples set) of the best measure. It means that various quality measures could be optimal for individual folds during one experiment. For measures m-estimate and g-measure values of parameters m and g were set to 22.4 and 2.

After rules induction, each of three filtration algorithms was activated. Forward and Backwards algorithms require the tuning examples set. In the described experiments, during rules induction, 15% of examples has been excluded from the training part and the set was treated as tuning set. Adding (the Forward algorithm) and removing (the Backwards algorithm) a rule occurred in the case when it caused a growth (the Forward algorithm) or did not cause a decrease (the Backwards algorithm) of the classification accuracy on any of training and tuning sets.

The results for all three algorithms are presented in the Table 1.

Table 1. Average results of 10x*CV* experiments

Quality	Filtration	Acc.	Std.	Avg. Acc.	Std.	Rules	Std.	Confl.	Neg. Confl.	p-Value
Auto	None	83.51	5.2	77.61	6.4	144.4	24.2	33	8	-
Auto	Coverage	83.09	5.0	76.36	6.1	50.5	10.0	30	8	*0.03809*
g-measure	None	82.86	5.6	75.62	6.3	169.4	7.8	27	8	**0.08881**
C2	None	82.56	5.1	76.09	6.1	148.4	8.1	28	8	**0.06525**
C2F	None	82.45	4.5	74.50	5.4	135.1	7.2	36	11	**0.12862**
m-estimate	Coverage	82.33	5.0	74.53	5.8	58.5	4.4	42	13	*0.00587*
Lift	None	82.23	5.7	79.43	6.7	185.9	7.3	24	9	*0.05553*
C2	Coverage	82.21	4.6	75.41	5.5	43.1	3.2	17	5	*0.01913*
m-estimate	None	82.18	5.0	74.35	5.7	101.7	7.6	45	13	*0.00030*
C2F	Coverage	82.06	4.7	73.89	5.6	52.3	3.6	27	8	**0.06864**
g-measure	Coverage	82.01	5.2	74.68	6.1	62.7	4.2	17	5	*0.00380*
Auto	Backwards	81.84	5.3	75.08	6.3	21.5	6.4	33	9	*0.00059*
m-estimate	Forward	81.80	5.0	75.40	6.1	29.1	6.9	21	7	*0.00036*
m-estimate	Backwards	81.68	5.5	75.32	6.5	27.1	6.7	25	7	*0.00093*
Auto	Forward	81.52	5.2	75.53	6.3	21.3	8.6	26	8	*0.00018*
Lift	Coverage	81.40	5.5	76.80	6.4	54.9	3.5	11	4	*0.00277*
Cohen	Backwards	81.18	6.1	75.09	6.8	17.3	4.4	48	14	*0.00036*
RIPPER	–	81.09	5.02	74.27	6.24	18.9	3.4	–	–	*0.00277*
C2	Forward	80.97	4.6	74.11	6.2	23.6	6.4	9	2	*0.00012*
C2F	Forward	80.90	4.9	73.19	6.0	25.5	6.7	14	4	*0.00036*
Cohen	Forward	80.86	5.7	76.14	6.6	10.7	2.4	36	11	*0.00016*
p-Val	Backwards	80.86	5.8	73.67	6.5	13.6	3.7	46	13	*0.00028*
Cohen	None	80.83	5.8	76.34	6.3	40.4	4.0	66	19	*0.00093*
C2	Backwards	80.81	4.9	74.42	5.9	20.2	5.2	9	2	*0.00005*
Gain	Forward	80.80	5.0	75.34	6.3	10.4	2.5	41	12	*0.00003*
Gain	Backwards	80.80	5.6	73.74	6.5	12.8	3.3	47	13	*0.00001*
C2F	Backwards	80.79	5.1	73.69	6.1	21.6	5.7	16	4	*0.00028*
Cohen	Coverage	80.71	5.8	76.21	6.4	26.0	2.6	65	19	*0.00059*
Wra	Backwards	80.43	5.2	73.83	5.7	14.6	3.7	57	16	*0.00004*
p-Val	Forward	80.41	5.7	74.26	6.5	10.7	3.0	38	11	*0.00002*
g-measure	Forward	80.20	5.1	73.27	6.7	33.8	7.2	7	2	*0.00003*
g-measure	Backwards	80.07	5.3	73.62	6.6	28.9	5.5	8	2	*0.00004*
Rss	Backwards	79.98	5.4	73.36	5.9	14.6	3.5	57	16	*0.00002*
Lift	Forward	79.79	5.2	74.02	6.3	31.6	7.7	5	2	*0.00002*
p-Val	None	79.66	5.7	73.00	6.4	39.0	4.8	70	23	*0.00119*
p-Val	Coverage	79.59	5.6	73.20	6.3	24.5	2.8	68	23	*0.00032*
Lift	Backwards	79.47	5.6	73.64	6.6	27.7	6.4	5	2	*0.00003*
Gain	None	79.35	5.2	72.92	5.9	34.6	5.0	71	23	*0.00054*
Wra	Forward	79.20	5.3	74.96	6.0	8.6	2.1	48	16	*0.00001*
Rss	Forward	79.18	5.3	74.88	6.0	8.6	2.0	47	16	*0.00010*
Gain	Coverage	79.16	5.3	73.00	6.3	22.8	2.9	70	23	*0.00003*
Rss	Coverage	76.58	5.7	71.79	6.4	19.3	2.3	79	28	*0.00010*
Wra	Coverage	76.46	5.7	71.68	6.4	19.4	2.2	79	28	*0.00000*
Rss	None	75.66	5.4	70.68	6.5	28.3	3.5	80	31	*0.00018*
Wra	None	75.61	5.5	70.62	6.5	28.3	3.5	81	31	*0.00016*

The classification algorithm uses the voting scheme; rule's voting strength is reflected by its quality. Unrecognized examples are assigned to majority decision class. The meaning of columns in Table 1 is the following: *Filtration* - denotes the applied filtration algorithm, *Acc.* - denotes the classification accuracy, *Avg.Acc.* - denotes the mean of decision classes accuracies, *Rules* - denotes the rules number, *Std* - denotes the standard deviation. The percentage of conflicts during classification and conflicts resolved incorrectly are also presented in the table. Results obtained by the RIPPER algorithm (the Rapid-Miner package implementation) are also given for comparative purposes. Additionally, p-value for the Wilcoxon test is given in the last column. The algorithm which uses automatic measure selection has been compared with each the other algorithms. The results in italics indicate p-value less than 0.05, and the results in bold – p-value less than 0.1. The algorithm with automatic quality measure selection gives statistically better results on the analyzed data sets, except one case, when the modified C2 measure was used.

It's worth noticing that improving classification accuracy in the automatic method doesn't happen at expense of average accuracy loss (which affects sensitivity and specificity of the classifier).

The analysis of the ranking presented in the Table 1 shows that higher classification accuracy is obtained at the cost of a bigger number of rules, which is confirmed by, inter alia, the [15] observations in relation to the Occam razor critique and rule number reduction. Filtration algorithms produce statistically worse results. The algorithm generating the coverage of a training set is here the best one. A considerable decrease of the rules number and decrease of the classification accuracy has been noticed for Forward and Backwards filtration algorithms. At the same time, a number of classification conflicts for the reduced rule set is relatively big, and a tidy part of them is settled incorrectly. It gives a chance to improve the accuracy of a classifier composed of a small number of rules. Authors want to obtain such improvement by taking the neighborhood of tested object covered by individual rules into consideration.

5 Conclusions

The method of automatic selection of quality measure that controls the process of growing and pruning of rules has been presented in the paper. The method enables us to obtain statistically better results than the algorithm which applies an arbitrary selected measure. The measure selection was optimized in order to obtain its highest classification accuracy. A similar optimization can be carried out for the average accuracy or for a complex measure which takes accuracy(and/or average accuracy) and complexity of a model into account.

Despite considerable reduction of the number of rules, filtration algorithms taking advantage of classification abilities of the filtered rule set has led to statistically worse accuracy. The underlying cause may be the method assumed, which assesses whether a rule should be added to the filtered rule set or not. There were no such decreases observed in the paper [11] in which the whole training set was used as the tuning set for the same algorithm. Further works will focus, inter alia, on the improvement of the voting algorithm by taking the neighborhood of tested object covered by individual rules into consideration, and by making use of information about a level of statistical significance of rule in the filtration (perhaps enabling the correction of so-called false discovery rate).

References

1. An, A., Cercone, N.: Rule quality measures for rule induction systems: description and evaluation. Computational Intelligence 17(3), 409–424 (2001)
2. Fürnkranz, J., Flach, P.A.: Roc'n' Rule Learning - Towards a Better understanding of covering Algorithms. Machine Learning 58, 39–77 (2005)
3. Clark, P., Niblett, T.: The CN2 Induction Algorithm. Machine Learning 3(4), 261–283 (1989)
4. Cohen, W.W.: Fast effective rule induction. In: Proc. of the 12th Int. Conference ICML 1995, pp. 115–123 (1995)
5. Grzymaa-Busse, J.W., Ziarko, W.: Data mining based on rough sets. In: Wang, J. (ed.) Data Mining Opportunities and Challenges, pp. 142–173. IGI Publishing, Hershey (2003)
6. Janssen, F., Fürnkranz, J.: On the quest for optimal rule learning heuristics. Machine Learning 78, 343–379 (2010)
7. Michalski, R.S., Mozetic, I., Hong, J., Lavrac, N.: The AQ15 inductive learning system: An overview and experiments. ISG Report No. 20. Department of Computer Sciences, University of Illinois at Urbana-Champaign (1986)
8. Mozina, M., Zabkar, J., Bratko, I.: Argument based machine learning. Artificial Intelligence 171, 922–937 (2007)
9. Sikora, M.: An algorithm for generalization of decision rules by joining. Foundation on Computing and Decision Sciences 30(3), 227–239 (2005)
10. Sikora, M.: Rule quality measures in creation and reduction of data rule models. In: Greco, S., Hata, Y., Hirano, S., Inuiguchi, M., Miyamoto, S., Nguyen, H.S., Słowiński, R. (eds.) RSCTC 2006. LNCS (LNAI), vol. 4259, pp. 716–725. Springer, Heidelberg (2006)
11. Sikora, M.: Decision rule-based data models using TRS and netTRS – methods and algorithms. In: Peters, J.F., Skowron, A. (eds.) Transactions on Rough Sets XI. LNCS, vol. 5946, pp. 130–160. Springer, Heidelberg (2010)
12. Sikora, M., Gruca, A.: Induction and selection of the most interesting Gene Ontology based multiattribute rules for descriptions of gene groups. Pattern Recognition Letters 32, 258–269 (2011)
13. Stefanowski, J.: Rough set based rule induction techniques for classification problems. In: Proc. 6th European Congress of Intelligent Techniques and Soft Computing, Achen, September 7-10, vol. 1, pp. 107–119 (1998)
14. Stefanowski, J., Vanderpooten, D.: Induction of Decision Rules in Classification and Discovery Oriented Perspectives. International Journal of Intelligent Systems 16, 13–27 (2001)
15. Webb, G.I.: Further experimental evidence against the utility of Occam's razor. Journal of Artificial Intelligence Research 4, 397–417 (1996)

Relationships between Depth and Number of Misclassifications for Decision Trees

Igor Chikalov, Shahid Hussain, and Mikhail Moshkov

Mathematical and Computer Sciences & Engineering Division
King Abdullah University of Science and Technology
Thuwal 23955-6900, Saudi Arabia
{igor.chikalov,shahid.hussain,mikhail.moshkov}@kaust.edu.sa

Abstract. This paper describes a new tool for the study of relationships between depth and number of misclassifications for decision trees. In addition to the algorithm the paper also presents the results of experiments with three datasets from UCI Machine Learning Repository [3].

Keywords: decision trees, depth, number of misclassifications.

1 Introduction

Decision trees are widely used as predictors, as a way of knowledge representation and as algorithms for problem solving. To have a more understandable decision tree we need to minimize the number of nodes in the tree. To have a faster decision tree we need to minimize the depth or average depth of the tree. And to have a more accurate decision tree we need to minimize the number of misclassifications.

We created a tool based on dynamic programming which allows us to optimize decision trees relative to the depth, the average depth, the number of nodes, and the number of misclassifications sequentially [1,2]. In this paper, we consider a new tool (an extension to our software) which allows us to study relationships between the depth and the number of misclassifications of a decision tree. We can find the minimum depth of a decision tree with at most n misclassifications as well as the minimum number of misclassifications among all decision trees with depth at most p. We consider the work of this tool on three decision tables from UCI ML Repository [3]: LYMPHOGRAPHY, BREAST-CANCER, and TIC-TAC-TOE.

2 Basic Notions

In this paper, we consider only decision tables with discrete attributes. These tables do not contain missing values and equal rows. Consider a *decision table* T depicted in Fig. 1. Here $f_1, \ldots, f_m$ are the names of columns (conditional attributes); $c_1, \ldots, c_N$ are nonnegative integers which can be interpreted as decisions (values of the decision attribute d); b_{ij} are nonnegative integers which are interpreted as values of conditional attributes (we assume that the rows

S.O. Kuznetsov et al. (Eds.): RSFDGrC 2011, LNAI 6743, pp. 286–292, 2011.

f_1 ... f_m	d
b_{11} ... b_{1m}	c_1
...	...
b_{N1} ... b_{Nm}	c_N

Fig. 1. Decision table

$(b_{11}, \ldots, b_{1m}), \ldots, (b_{N1}, \ldots, b_{Nm})$ are pairwise different). We denote by $E(T)$ the set of attributes (columns of the table T), each of which contains different values. For $f_i \in E(T)$ let $E(T, f_i)$ be the set of values from the column f_i.

Let $f_{i_1}, \ldots, f_{i_t} \in \{f_1, \ldots, f_m\}$ and $a_1, \ldots, a_t$ be nonnegative integers. We denote by $T(f_{i_1}, a_1) \ldots (f_{i_t}, a_t)$ the subtable of the table T, which consists of such and only such rows of T that at the intersection with columns $f_{i_1}, \ldots, f_{i_t}$ have numbers $a_1, \ldots, a_t$ respectively. Such nonempty tables (including the table T) will be called *separable subtables* of the table T. For a subtable Θ of the table T, we will denote by $R(\Theta)$ the number of unordered pairs of rows that are labeled with different decisions. A minimum decision value which is attached to the maximum number of rows in a nonempty subtable Θ will be called the *most common decision* for Θ.

A *decision tree* Γ *over* the table T is a finite directed tree with root in which each terminal node is labeled with a decision. Each nonterminal node is labeled with a conditional attribute, and for each nonterminal node the outgoing edges are labeled with pairwise different nonnegative integers. Let v be an arbitrary node of Γ. We now define a subtable $T(v)$ of the table T. If v is the root then $T(v) = T$. Let v be a node of Γ that is not the root, nodes in the path from the root to v be labeled with attributes $f_{i_1}, \ldots, f_{i_t}$, and edges in this path be labeled with values $a_1, \ldots, a_t$ respectively. Then $T(v) = T(f_{i_1}, a_1), \ldots, (f_{i_t}, a_t)$.

Let Γ be a decision tree over T. We will say that Γ is a *decision tree for* T if any node v of Γ satisfies the following conditions:

- If $R(T(v)) = 0$ then v is a terminal node labeled with the most common decision for $T(v)$;
- Otherwise, either v is a terminal node labeled with the most common decision for $T(v)$, or v is labeled with an attribute $f_i \in E(T(v))$ and if $E(T(v), f_i) = \{a_1, \ldots, a_t\}$, then t edges leave node v, and these edges are labeled with $a_1, \ldots, a_t$ respectively.

Let Γ be a decision tree for T. For any row r of T, there exists exactly one terminal node v of Γ such that r belongs to the table $T(v)$. Let v be labeled with the decision b. We will say about b as about the *result of the work of decision tree* Γ *on* r.

3 Representation of Sets of Decision Trees

Consider an algorithm for construction of a graph $\Delta(T)$, which represents the set of all decision trees for the table T. Nodes of this graph are some separable subtables of the table T. During each step we process one node and mark it with the symbol *. We start with the graph that consists of one node T and finish when all nodes of the graph are processed.

Let the algorithm has already performed p steps. We now describe the step number $(p+1)$. If all nodes are processed then the work of the algorithm is finished, and the resulting graph is $\Delta(T)$. Otherwise, choose a node (table) Θ that has not been processed yet. Let b be the most common decision for Θ. If $R(\Theta) = 0$, label the considered node with b, mark it with symbol * and proceed to the step number $(p+2)$. If $R(\Theta) > 0$, then for each $f_i \in E(\Theta)$ draw a bundle of edges from the node Θ (this bundle of edges will be called *f_i-bundle*). Let $E(\Theta, f_i) = \{a_1, \ldots, a_t\}$. Then draw t edges from Θ and label these edges with pairs $(f_i, a_1), \ldots, (f_i, a_t)$ respectively. These edges enter into nodes $\Theta(f_i, a_1), \ldots, \Theta(f_i, a_t)$. If some of the nodes $\Theta(f_i, a_1), \ldots, \Theta(f_i, a_t)$ are not present in the graph then add these nodes to the graph. Mark the node Θ with the symbol * and proceed to the step number $(p+2)$.

Fig. 2. Trivial decision tree

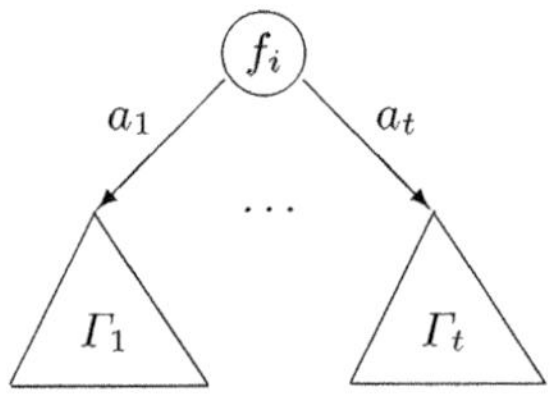

Fig. 3. Aggregated decision tree

Now for each node Θ of the graph $\Delta(T)$, we describe the set of decision trees corresponding to the node Θ. We will move from terminal nodes, which are labeled with numbers, to the node T. Let Θ be a node, which is labeled with a number b. Then the only trivial decision tree depicted in Fig. 2 corresponds to the node Θ.

Let Θ be a nonterminal node (table) then there is a number of bundles of edges starting in Θ. We consider an arbitrary bundle and describe the set of decision trees corresponding to this bundle. Let the considered bundle be an f_i-bundle where $f_i \in (\Theta)$ and $E(\Theta, f_i) = \{a_1, \ldots, a_t\}$. Let $\Gamma_1, \ldots, \Gamma_t$ be decision trees from sets corresponding to the nodes $\Theta(f_i, a_1), \ldots, \Theta(f_i, a_t)$. Then the decision tree depicted in Fig. 3 belongs to the set of decision trees, which correspond to this bundle. All such decision trees belong to the considered set, and this set does not contain any other decision trees. Then the set of decision trees corresponding to the node Θ coincides with the union of sets of decision trees corresponding to the bundles starting in Θ and the set containing one decision tree depicted in Fig. 2, where b is the most common decision for Θ. We denote by $D(\Theta)$ the set of decision trees corresponding to the node Θ.

The following proposition shows that the graph $\Delta(T)$ can represent all decision trees for the table T.

Proposition 1. *Let T be a decision table and Θ a node in the graph $\Delta(T)$. Then the set $D(\Theta)$ coincides with the set of all decision trees for the table Θ.*

4 Relationships between Depth and Number of Misclassifications

Let T be a decision table with N rows and m columns labeled with $f_1, \ldots, f_m$, and $D(T)$ be the set of all decision trees for T. Let $\Gamma \in D(T)$ then the *depth of* Γ, denoted as $h(\Gamma)$, is the maximum length of a path from the root to a terminal node of Γ and the *number of misclassifications for decision tree Γ for the table* T, denoted as $\mu(\Gamma)$, is the number of rows r in T for which the result of the work of decision tree Γ on r does not equal to the decision attached to the row r. It is clear that the minimum values of h and μ on $D(T)$ are equal to zero, an upper bound on the value of h on $D(T)$ is m, and an upper bound on the value of μ on $D(T)$ is N. We denote $B_h = \{0, 1, \ldots, m\}$ and $B_\mu = \{0, 1, \ldots, N\}$. We now define two functions $\mathcal{G}_T : B_h \to B_\mu$ and $\mathcal{F}_T : B_\mu \to B_h$ as follows:

$$\mathcal{G}_T(n) = \min\{\mu(\Gamma) : \Gamma \in D(T), h(\Gamma) \leq n\}$$

for any $n \in B_h$, and

$$\mathcal{F}_T(n) = \min\{h(\Gamma) : \Gamma \in D(T), \mu(\Gamma) \leq n\}$$

for any $n \in B_\mu$.

The function $\mathcal{G}_T$ can be represented by the tuple $(\mathcal{G}_T(0), \ldots, \mathcal{G}_T(m))$ of its values. The function $\mathcal{F}_T$ can also be represented similarly.

We now describe an algorithm which allows us to construct the function $\mathcal{G}_\Theta$ for any node (subtable) Θ from the graph $\Delta(T)$.We begin from terminal nodes and move to the node T.

Let Θ be a terminal node. It means that all rows of Θ are labeled with the same decision b and the decision tree Γ_b as depicted in Fig. 2 belongs to $D(\Theta)$. It is clear that $h(\Gamma_b) = 0$ and $\mu(\Gamma_b) = 0$ for the table Θ. Therefore $\mathcal{G}_\Theta(n) = 0$ for any $n \in B_h$.

Consider a node Θ, which is not a terminal node and a bundle of edges, which starts from this node. Let these edges be labeled with pairs $(f_i, a_1), \ldots, (f_i, a_t)$, and enter into nodes $\Theta(f_i, a_1), \ldots, \Theta(f_i, a_t)$, respectively, to which the functions $\mathcal{G}_{\Theta(f_i,a_1)}, \ldots, \mathcal{G}_{\Theta(f_i,a_t)}$ are already attached.

We correspond to this bundle (f_i-bundle) the function $\mathcal{G}_\Theta^{f_i}$, which for any $n \in B_h \setminus \{0\}$ is defined as follows:

$$\mathcal{G}_\Theta^{f_i}(n) = \min\{\mu(\Gamma) : \Gamma \in D(\Theta, f_i), h(\Gamma) \leq n\},$$

where $D(\Theta, f_i)$ is the set of decision trees for Θ corresponding to the considered bundle. In this set we have all trees from $D(\Theta)$ in which the root is labeled with f_i and only such trees. It is not difficult to show that for any $n \in B_h \setminus \{0\}$,

$$\mathcal{G}_\Theta^{f_i}(n) = \mathcal{G}_{\Theta(f_i,a_1)}(n-1) + \cdots + \mathcal{G}_{\Theta(f_i,a_t)}(n-1).$$

It is not difficult to prove that for any $n \in B_h \setminus \{0\}$,

$$\mathcal{G}_\Theta(n) = \min\{\mathcal{G}_\Theta^{f_i}(n) : f_i \in E(\Theta)\}.$$

We know that there is only one decision tree with depth zero in $D(\Theta)$. This is the tree Γ_b as depicted in Fig. 2, where b is the most common decision for Θ. For this tree, we have $h(\Gamma_b) = 0$ and $\mu(\Gamma_b)$ for Θ is equal to the number of rows in Θ which are labeled with decisions other than b. So,

$$\mathcal{G}_\Theta(0) = N(\Theta) - N(\Theta, b),$$

where $N(\Theta)$ is the number of rows in Θ and $N(\Theta, b)$ is the number of rows in Θ which are labeled with the decision b.

We can use the following proposition to construct the function $\mathcal{F}_T$.

Proposition 2. *For any* $n \in B_\mu$, $\mathcal{F}_T(n) = \min\{p \in B_h : \mathcal{G}_T(p) \leq n\}$.

Note that to find the value $\mathcal{F}_T(n)$ for $n \in B_\mu$ it is enough to make $O(\log|B_h|) = O(\log m)$ operations of comparison.

5 Experimental Results

We implemented the algorithm presented in this paper in DeepComputing (our software system for the study of decision trees, developed at KAUST) and performed several experiments on datasets (decision tables) acquired from UCI ML Repository [3]. In the following, we present the experimental results and show the plots depicting relationships between the number of misclassifications and the depth of decision trees.

5.1 Lymphography

Figure 4 contains two plots for the decision table LYMPHOGRAPHY (18 attributes and 148 rows). The first plot shows the relationship between the number of misclassifications and the depth (the minimum number of misclassifications among decision trees whose depth is at most the given value) and the second one shows the relationship between the depth and the number of misclassifications (the minimum depth among decision trees for which the number of misclassifications is at most the given value).

5.2 Breast-Cancer

Figure 5 contains two plots for the decision table BREAST-CANCER (9 attributes and 266 rows). The first plot shows the relationship between the number of misclassifications and the depth and the second one shows the relationship between the depth and the number of misclassifications.

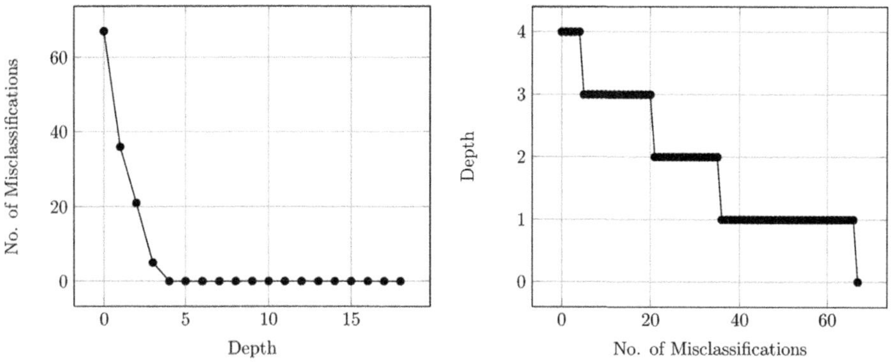

Fig. 4. Relationships between the depth and the number of misclassifications for LYMPHOGRAPHY

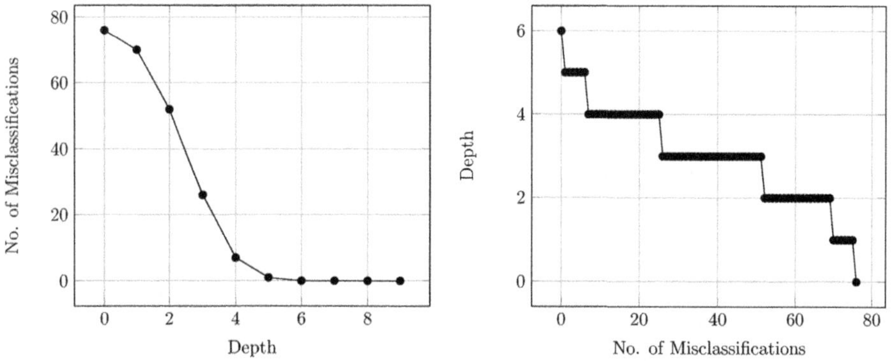

Fig. 5. Relationships between the depth and the number of misclassifications for BREAST-CANCER

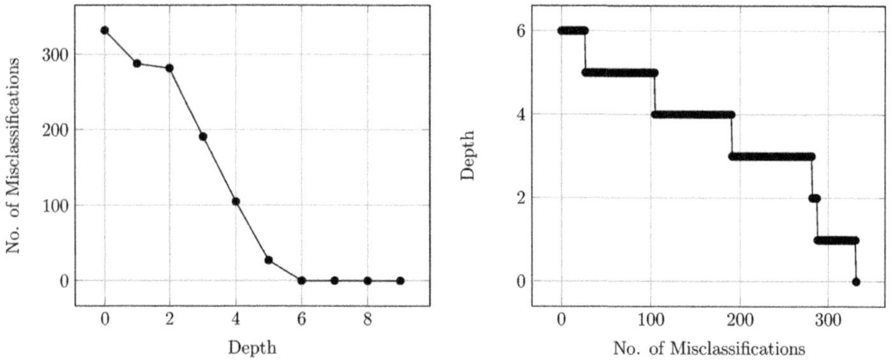

Fig. 6. Relationships between the depth and the number of misclassifications for TIC-TAC-TOE

5.3 Tic-Tac-Toe

Figure 6 contains two plots for the decision table TIC-TAC-TOE (9 attributes and 958 rows). The first plot shows the relationship between the number of misclassifications and the depth and the second one shows the relationship between the depth and the number of misclassifications.

6 Conclusions

The paper is devoted to the consideration of a tool for studying the relationships between the depth and the number of misclassifications for decision trees. The application of our software tool is illustrated by the experiments with three datasets from the UCI ML Repository [3]. Further studies will be connected with the extension of this tool to more complexity measures such as the average depth and the number of nodes of decision trees, and to inconsistent decision tables [4].

References

1. Alkhalid, A., Chikalov, I., Moshkov, M.: On algorithm for building of optimal α-decision trees. In: Szczuka, M., Kryszkiewicz, M., Ramanna, S., Jensen, R., Hu, Q. (eds.) RSCTC 2010. LNCS(LNAI), vol. 6086, pp. 438–445. Springer, Heidelberg (2010)
2. Chikalov, I., Moshkov, M., Zelentsova, M.: On optimization of decision trees. In: Peters, J.F., Skowron, A. (eds.) Transactions on Rough Sets IV. LNCS, vol. 3700, pp. 18–36. Springer, Heidelberg (2005)
3. Frank, A., Asuncion, A.: UCI Machine Learning Repository. University of California, School of Information and Computer Science, Irvine, CA (2010), `http://archive.ics.uci.edu/ml`
4. Pawlak, Z.: Rough sets – Theoretical Aspects of Reasoning About Data. Kluwer Academic Publishers, Dordrecht (1991)

Dynamic Successive Feed-Forward Neural Network for Learning Fuzzy Decision Tree

Manu Pratap Singh

Department of Computer Science, ICIS,
Dr. B. R. Ambedkar University,
Khandari Campus, Agra, Uttar Pradesh, India
manu_p_singh@hotmail.com

Abstract. Fuzzy decision trees have been substantiated to be a valuable tool and more efficient than neural networks for pattern recognition task due to some facts like computation in making decisions are simpler and important features can be selected automatically during the design process. Here we present a feed forward neural network which learns fuzzy decision trees during the descent along the branches for its classification. Every decision instances of decision tree are represented by a node in neural network. The neural network provides the degree of membership of each possible move to the fuzzy set $<< \mathit{good} \quad \mathit{move} >>$ corresponding to each decision instance. These fuzzy values constitute the core of the probability of selecting the move out of the set of the children of the current node. This results in a natural way for driving the sharp discrete-state process running along the decision tree by means of incremental methods on the continuous-valued parameters of the neural network. A simulation program in C has been deliberated and developed for analyzing the consequences. The effectiveness of the learning process is tested through experiments with three real-world classification problems.

Keywords: Decision tree, pattern classification, fuzzy system, artificial neural networks, fuzzy logic.

1 Introduction

It is well known that decision trees (DT) [1, 4] are the operational support of non-deterministic computations and also very efficient for pattern recognition task [3]. Actually, as long as pattern recognition is considered, a DT can be considered as more efficient than a neural network (NN). There are mainly two reasons, first the computations in making decisions are simpler – only one feature is used in each non-terminal node, and the only computation can be a very simple comparison (say, $x_i < a$). Second, important features can be selected automatically during the design process. In using an NN, since we do not know which feature is important, the only thing we can do is to use all features. However, the DTs do not have the adaptive or learning ability, and thus they cannot be used in changing environment. This problem can be avoided if we map a DT to neural network. Actually there is a very simple

S.O. Kuznetsov et al. (Eds.): RSFDGrC 2011, LNAI 6743, pp. 293–301, 2011.

mapping from DT to neural network. This mapping integrates the symbolic approach (DT) and sub-symbolic one (neural network). Specifically, this makes DTs adaptable, and at the same time, provides a systematic approach for structural learning of DTs.

In the stream of learning from examples to select the appropriate paths over a given decision tree, various architectural solutions and learning algorithms have been proposed [5-15]. Error back-propagation is the most used training algorithm, but other algorithms like the Widrow-Hoff learning rule are employed [14]. The use of backpropagation learning rule in neuro-fuzzy system to train fuzzy rules for the classification problem has been already consider in many real world applications [19].

Our learning approach follows simple backpropagation, with the following distinguishing features: (1) the same feed forward neural network is deputed to make decisions on each node. (2) The feed forward neural network is not a plain external classifier for the children of a given node, but is coupled with the decision tree in a tight way and (3) the training algorithm is especially devised to properly back propagate a particular error function along the branches of the decision tree .

The key thought of presented approach is to train a feed forward neural network to output the degree of membership [2] to the fuzzy set $<< \ good \ \ move \ >>$ in correspondence to an input move. Here each neuron is represented as a decision node capable of making decisions based on inputs. One important aspect of the proposed approach is any neuron can be replaced with a full feed forward neural network if a single neuron is not capable of making decisions. We have used a feed forward neural network in place of each neuron represented. The learning procedure we propose offers a set of operational options concerning the objective function to be minimized and the decision tree visiting methods. The output of a feed forward neural network plays a double role in incremental training methods: locally, it is the primer for subsequent states; globally, it is the input to the error function. This double role becomes extremely critical in the essentially discrete dynamics of our neural network, with the risk of getting the training process stuck in the meaningless fixed points. To avoid this drawback we give to the fuzzy values returned by the network the general meaning of conditional probability - after proper normalizations - of preferring one move among the available alternatives. There is one more issue pertaining related to generation of decision tree from real time information. For generation of decision trees, various algorithms have been proposed [16] and successfully worn.

The proposed learning procedure used for decision trees using simple backpropagation has been tested over well-known real time statistics i.e. IRIS [17], image segmentation [18], Postoperative Patient Data. The results obtained exhibit the inadequacy of backpropagation algorithm for classification. The adequate classification can be achieved using proposed procedure. The procedure envisages superior consequences in contrast with the feed forward neural network. The next section discusses the methodology and design of the problem. The experimental analysis and results have been shown in section 3. A brief discussion is presented in section 4. The Section 5 concludes this paper with a summary, the conclusions of this study.

2 Methodology and Simulation Design

The architecture of presented neural network system is extended from the multilayer feed forward neural network. At the lowest level, every decision instances of decision tree are represented by a node in neural network. The neural network whose inputs are the branches of the decision tree and whose outputs are the corresponding preference scores i.e. degree of membership. Since we presume that the branches of the decision tree do not look mutually independent to the neural network and we ask it sequentially node by node. The neural network system consists of the components of a conventional neural network system except that computation of degree of membership for each decision instance is performed by each neuron and the neural network's learning capacity is provided to enhance the system knowledge.

2.1 Simple Neural Network Architecture

The input-output stimuli's for a particular data set are trained with the neural network has three layers: one input layer, one output layer and a combination of hidden layer(s). Classification of fuzzy information can't be accomplished precisely with the help of conventional artificial neural network architecture.

In backpropagation training algorithm, an input pattern vector ***P*** having ***n*** features as $P_1, P_2, P_3, \ldots\ldots P_n$. Classification of these patterns will be in M classes having the output pattern respectively $C_1, C_2, C_3, \ldots\ldots, C_M$. In the backpropagation learning algorithm the change in weight vector is being done according to the calculated error in the network, after iterative training. The error and change in weights in the network can be calculated as,

$$\Delta w_{ho}(n+1) = -\eta \sum_{i=1}^{H} \frac{\partial E}{\partial w_{ho}} + \alpha \Delta w_{ho}(n) \tag{2.1.5}$$

$$\Delta w_{ih}(n+1) = -\eta \sum_{i=1}^{N} \frac{\partial E}{\partial w_{ih}} + \alpha \Delta w_{ih}(n) \tag{2.1.6}$$

$$E^p = \frac{1}{2} \sum_{m=1}^{M} \left(C_m^p - O_m^{p^o} \right)^2 \tag{2.1.7}$$

for $m = 1$ to M output pattern features and $p = 1$ to P presented input patterns and $\left(C_m^p - O_m^{p^o} \right)^2$ is the squared difference between the actual output value of output layer for pattern P and the target output value.

2.2 Representation of Decision Tree over Neural Network

A decision tree is constructed from a training set, which consists of objects. Each object is completely described by a set of attributes and a class label. Attributes can

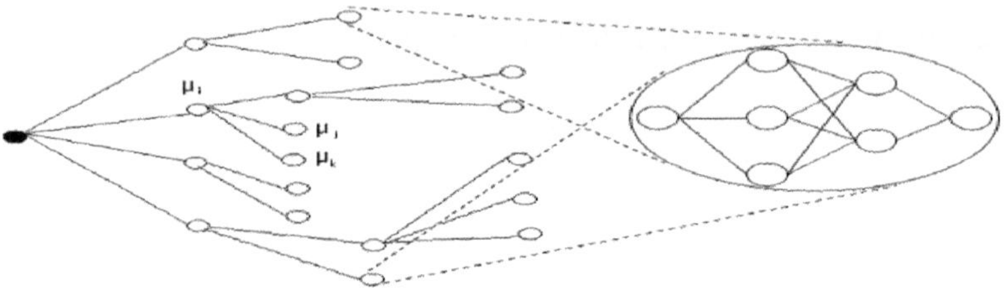

Fig. 1. Descent of the neural advisor for the decision tree. At each node the network outputs a value μ_{ab}. μ_{ab} is the degree of membership for deciding the output of each node. Node is represented by a neural network that learns a decision instance from a decision tree.

have ordered or unordered values. The concept underlying a data set is the true mapping between the attributes and class. A noise-free training set is one in which all the objects are generated using the underlying concept. A decision tree contains zero or more internal nodes and one or more leaf nodes. All internal nodes have two or more child nodes. All internal nodes contain splits, which test the value of an expression of the attributes. Arcs from an internal node t to its children are labeled with distinct outcomes of the test at t. each leaf node has a class label associated with it. The task of constructing a tree from the training set has been called tree induction, tree building and tree growing. Most existing tree induction systems proceed in a greedy top-down fashion. Starting with an empty tree and the entire training set, some variant of the following algorithm is applied until no more splits are possible.

If all the training examples at the current node t belong to category c, create a leaf node with the class c; otherwise, score each one of the set of possible splits S, using a goodness measure; choose the best split s* as the test at the current node; create as many child nodes as there are distinct outcomes of s*. Label edges between the parent and child nodes with outcomes of s*, and partition the training data using s* into the child nodes; a child node t is said to be pure if all the training samples at t belong to the same class. Repeat all above steps on all impure child nodes.

In our study, we adopted this representation of fuzzy decision tree as a list of 5-tuples. Each 5-tuples corresponds to a node. There are two kinds of nodes non-terminal & terminal node. Specifically a node is defined by

$$node = \{t, label, P, \mu, size\} \tag{2.2.1}$$

Here t is the node number. The node ($t = 0$) is called the root; label is the class label of a terminal node, and it is meaningful only for terminal nodes; P is pointer to the parent. For root it is NULL; μ is degree of membership for suggesting a decision for the next move.

Suppose we have ***p*** input-output stimuli, each having ***n*** features and each stimulus belongs to one of ***M*** classes. In fuzzy artificial neural network, the degree of membership for ith pattern (i = *1* to *P* patterns) with the jth class (j = *1* to *M* classes) can be generated as follows,

$$\mu_i^{C_j} = e^{-\frac{1}{2}\left[\frac{P_{ik} - c_k}{\sigma_k}\right]} \tag{2.2.2}$$

This method of generating degree of membership is taken from the standard Gaussian membership function (MF). A Gaussian MF is determined completely by **c** and **σ**; **c** represents the MFs center and **σ** determines the MFs width. Size is the size of the node when it is considered a sub-tree. The size of the root is the size of whole tree and of terminal node is 1.

We adopted backpropagation learning algorithm as the learning for our presented approach. Since the neural network has to learn to follow branching decision paths, the whole process must split into two stages; on the single tree node, the usual process feeds the inner dynamics of the network. Along the branches of the decision tree we have a process which obeys the syntactic rules of the decision process.

Now, in correspondence to each step of the second stage, the error to be minimized must depend in a non-linear way on the switching variables (network outputs at each layer) local to the current node of the decision tree. The method we used for computing the gradient of the error function is forward in both the stages. As discussed, the task of our neural network is to computer the degree of membership for each decision instance. The generated degree of membership will decide the next move. Indeed, among the possible modalities of carrying the decision process out, we followed the approach: at each decision node we select the moves that receive the highest score and continue from there our exploration of the decision tree. Intermediate strategies may plan to follow at each step a limited number of favorite next moves and to take a final decision from the collected paths.

Every node (i^{th} instance of decision tree) will generate degree of membership based on training data. This process will produce a vector **V (*p, m*)** of degree of membership corresponding to the relationship between various input-output stimuli's as follows;

$$V = \begin{matrix} \mu_1^{C_1} & \mu_1^{C_2} & \mu_1^{C_3} \ldots & \mu_1^{C_m} \\ \mu_2^{C_1} & \mu_2^{C_2} & \mu_2^{C_3} \ldots & \mu_2^{C_m} \\ \mu_i^{C_1} & \mu_i^{C_2} & \mu_i^{C_3} \ldots & \mu_i^{C_m} \\ \ldots & \ldots & \ldots\ \ldots & \ldots \\ \mu_p^{C_1} & \mu_p^{C_2} & \mu_p^{C_3} \ldots & \mu_p^{C_m} \end{matrix} \tag{2.2.3}$$

Here $\mu_i^{C_1}, \mu_i^{C_2}, \mu_i^{C_3} \ldots \mu_i^{C_m}$ represents the input pattern vector for training. For classification of this input pattern vector of degree of membership with the fuzzy-neural network system a target output corresponding each input pattern in the form of degree of membership may be defined as follows,

$$\mu_{\max}^{i} = \max\left(\mu_i^{C_1}, \mu_i^{C_2}, \mu_i^{C_3} \ldots \mu_i^{C_m}\right) \tag{2.2.4}$$

This vector of degree of membership will be used as input-output stimuli's for training to the fuzzy-neural network system in support of generating the appropriate classification using the backpropagation algorithm.

3 Experiments and Results

In order to consistently validate our method, we performed two experiments for three different sets of data i.e. IRIS, image segmentation and Postoperative Patient Data. First we are attempting to classify both the real-world data with the conventional artificial neural network, later the classification is carried out using the proposed approach of learning decision trees using neural network. First experiment was executed with the varying neural network architectures for generating the possible appropriate classification. Different combinations of hidden layers for artificial neural network have been used for investigating the adequacy of simple neural network. We have chosen three combinations of hidden layers i.e. one, two and three hidden layers. In the second approach, initially we used to generate the decision tree. These decision trees are now mapped to feed forward neural network. A decision instance is represented by a node in the feed forward neural network. A node can be a simple neuron or again a simple feed forward neural network depending upon the capability to produce degree of membership based on decision tree instance. Here we have used a simple feed forward neural network for a node. This node gives the degree of membership based on the inputs applied to decision trees. Tolerance of neural network has taken for error i.e. ($MAXE \leq 0.001$ or 0.1%). IRIS data contains four fuzzy input constraints to decide classification in three different classes named IRIS Setosa, IRIS Versicolor, and IRIS Virginica. Here we are considering only necessary 30 out of 150 rules for classification. Image segmentation contains 19 input constraints to decide classification in seven different classes named brickface, sky, foliage, cement, window, path and grass. Here we have all 210 training samples for learning of feed forward neural network based on backpropagation. Postoperative Patient data contains 8 input constraints to decide classification in three classes different named I (patient must be sent to Intensive Care Unit), S (patient prepared to go home) and A (patient is out of danger and ready to send to general hospital floor).

3.1 Feed-Forward Neural Network

Here we have performed the experiment for classification of IRIS, image segmentation and Postoperative Patient data with varying hidden layers. We have trained all the sample data for around 100 times. Figure 3 depicts the results. The results clearly show the difficulties of feed-forward neural network for learning and classifying various data.

3.2 Fuzzy Decision Tree over Feed Forward Neural Network

These decision trees are mapped to feed forward neural network as in figure-1. A decision instance is represented by a node in the feed forward neural network. A node can be a simple neuron or again a simple feed forward neural network depending

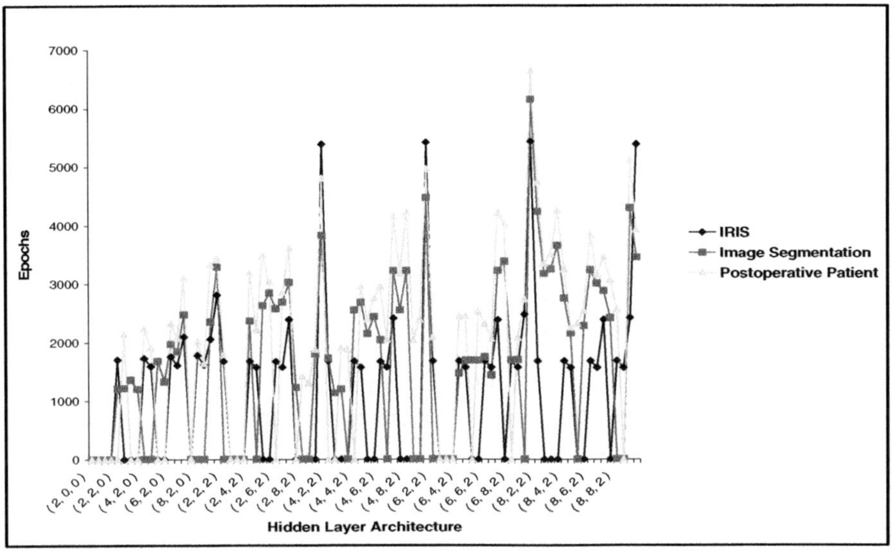

Fig. 2. Comparison of Classification of various data using Backpropagation algorithm

upon the capability to produce degree of membership based on decision tree instance and input parameters. We have used a simple feed forward neural network for a node. Here we have performed the experiment for classification of IRIS, image segmentation and Postoperative Patient data with varying hidden layers inside a node. We have trained all the sample data for around 100 times. Figure 2 depicts the result. The results clearly show the superiority of presented approach for learning and classifying various data over artificial neural network.

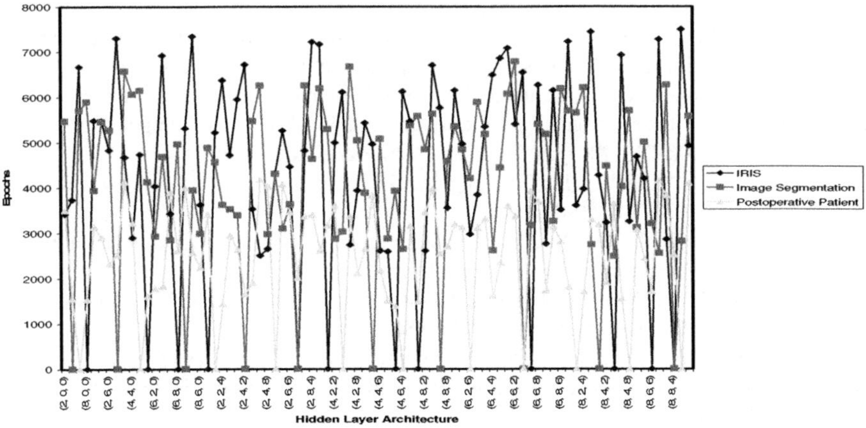

Fig. 3. Comparison of Classification of various data using presented approach

4 Conclusion

We mapped a decision tree generated from data related real world problems to neural network. The different decision instances are mapped to various nodes in neural network. We propose a higher order system, where a single neural network is the switcher of a decision tree. This single neural network consists of various nodes mapped to decision tree instances. We have used the backpropagation learning algorithm for generating the degree of membership based on parameters i.e. decision tree instances, input-output stimuli's, where the assessment of a discrete goal variable is driven by an incremental learning process.

The results demonstrated that, large significant differences exist between the performances of backpropagation feed forward neural network and presented approach for the classification problem of IRIS, image segmentation and Post-Operative Patient data in the terms of accuracy, convergence and epochs. These results recommend the adequacy of approach for classification. In first experiment i.e. using feed forward neural network, success percentage is quite lower for IRIS, image segmentation and Post-Operative Patient data.

References

1. Quinlan, J.R.: Induction of decision trees. Machine Learning 1, 81–106 (1986)
2. Zadeh, L.A.: Fuzzy Sets. Information and Control 8(3), 338–353 (1965)
3. Schalkoff, R.: Pattern Recognition: Statistical, Structural and Neural Appraoches. John Wiley & Sons, New Work (1992)
4. Olaru, C., Wehenkel, L.: A Complete Fuzzy Decision Tree Technique. Fuzzy Sets and Systems 138, 221–254 (2003)
5. Valiant, L.: A theory of the learnable. Communication of ACM 27, 1134–1142 (1984)
6. Kushilevitz, E., Mansour, Y.: Learning decision trees using the Fourier spectrum. Siam Journal of Computer Science 22(6), 1331–1348 (1993)
7. Hancock, T.: Learning 2m DNF and km decision trees. In: 4th COLT, pp. 199–308 (1991)
8. Bellare, M.: A technique for upper bounding the spectral norm with application to learning. In: 5th Annual Workshop on Computational Learning Theory, pp. 62–70 (1992)
9. Sakay, Y., Takimoto, E., Maruoka, A.: Proper learning algorithm for functions of k-terms under smooth distributions. In: Proc. of the 8th Workshop on Computational Learning Theory, pp. 206–213. Morgan Kaufmann, San Francisco (1995)
10. Erenfeucht, A., Haussler, D.: Learning decision trees from random examples. Inform. and Comp. 82(3), 231–246 (1989)
11. Hopfield, J., Tank, D.: Neural computations of decisions in optimization problems. Biological Cybernetics 52(3), 141–152 (1985)
12. Saylor, J., Stork, D.: Parallel analog neural networks for tree searching. In: Proc. Neural Networks for Computing, pp. 392–397 (1986)
13. Szczerbicki, E.: Decision trees and neural networks for reasoning and knowledge acquisition for autonomous agents. International Journal of Systems Science 27(2), 233–239 (1996)
14. Sethi, I.: Entropy nets: from decision trees to neural networks. Proceedings of the IEEE 78, 1605–1613 (1990)

15. Ivanova, I., Kubat, M.: Initialization of neural networks by means of decision trees. Knowledge-Based systems 8(6), 333–344 (1995)
16. Geurts, P., Wehenkel, L.: Investigation and reduction of discretization variance in decision tree induction. In: Lopez de Mantaras, R., Plaza, E. (eds.) ECML 2000. LNCS (LNAI), vol. 1810, pp. 162–170. Springer, Heidelberg (2000)
17. Anderson, E.: The Irises of the Gaspe peninsula, Bulletin America, IRIS Soc. (1935)
18. Budihardjo, A., Grzymala-Busse, J., Woolery, L.: Program LERS_LB 2.5 as a tool for knowledge acquisition in nursing. In: Proceedings of the 4th Int. Conference on Industrial & Engineering Applications of AI & Expert Systems, pp. 735–740 (1991)
19. Jain, M., Butey, P.K., Singh, M.P.: Classification of Fuzzy-Based Information using Improved backpropagation algorithm of Artificial Neural Networks. International Journal of Computational Intelligence Research 3(3), 265–273 (2007)

An Improvement for Fast-Flux Service Networks Detection Based on Data Mining Techniques

Ziniu Chen[1], Jian Wang[1], Yujian Zhou[2], and Chunping Li[1]

[1] Data Mining Group, School of Software, Tsinghua University, Beijing, China
czn09@mails.tsinghua.edu.cn
wj-217@163.com
cli@tsinghua.edu.cn
[2] MOST Information Center, Beijing, China
zhouyujian@most.cn

Abstract. Fast-flux is a kind of DNS technique used by botnets to hide the actual location of malicious servers. It is considered as an emerging threat for information security. In this paper, we propose an approach to detect the fast-flux service network (FFSN) using data mining techniques. Furthermore, we use the resampling technique to solve imbalanced classification problem with respect to FFSNs detection. Experiment results in the real datasets show that our approach improves the detective precision and effectiveness compared with existing researches.

1 Introduction

Botnets have been created to perform a wide range of illegal activities [1]. An emerging new use of botnets by cyber-criminals is a technique called Fast-Flux Service Networks (FFSNs). FFSNs are used in many illegal practices, including online pharmacy shops, money mule recruitment sites, phishing websites, illegal adult content, malicious browser exploit websites and the distribution of malware downloads, etc [2][5]. Generally speaking, FFSNs refer to rapidly changing the mapping between IP address and domain name. Each victim's request to visit the web server will thus reach one of the bots, and then the bot will proxy the request to the real server [9][12]. For this reason, it is hard to find out the mother-ship, which is the control unit behind FFSN. ICANN (The Internet Corporation for Assigned Names and Numbers) describes fast flux as 'rapid and repeated changes to host and/or name server resource records, which results in rapidly changing the IP address to which the domain name of an Internet host or name server resolves [3]. According to the Honeynet Project [5], FFSNs are categorized into two different types. One is single-flux and the other is double-flux. The single-flux only puts the IP address of the domain name in flux, while double-flux refers to dynamically and repeatedly changing the IP addresses of both the bots and their authoritative DNS [12]. The main difference between them is that the latter has an additional layer of protection.

S.O. Kuznetsov et al. (Eds.): RSFDGrC 2011, LNAI 6743, pp. 302–309, 2011.

However, FFSNs have some features which make it become possible to detect that kind of criminal behaviors [4]. The study of these malicious networks focused on characteristics of such networks. Alper Caglayan, et al [6] showed that such networks share common lifecycle characteristics, and form clusters based on size, growth and type of malicious behavior. Jose Nazario, et al [2] provided heuristics principles on the identification and qualification of domains. T. Holz, et al [7] presented the empirical study of fast-flux service networks and developed a metric to detect FFSNs. Three categories of features were extracted in [8]. Wu, et al [9] used data mining technique for the detection of fast-flux service networks based on four attributes.

Earlier works have their limitations mainly in the poor precision of detection of FFSNs class, therefore it is difficult to put them into practical use. In this paper, we propose a classification based approach to combine the resampling technique with the feature extraction from the collected datasets for the fast-flux detection. The experiment results show that our approach improves the detective precision and effectiveness significantly.

The organization of this paper is as follows. Section 2 gives a brief introduction on the characteristics of FFSNs and describes two additional extracted attributes as the fast flux features. Section 3 presents our approach and framework based on data classification technique for fast-flux detection. In Section 4, we show the experiment results and comparative study with other works. In Section 5, we have the concluding remarks and future work.

2 Characteristics of FFSN

Our detection strategy is based on the analysis of combination of following six distinguished attributes as fast flux features.

2.1 Num_asn

The number of unique ASNs (Autonomous System Number) for all *A records*, which keep the mapping between hostname and an IP address of the host. Normally, benign domains tend to return only *A records* from one particular AS (Autonomous System). On the contrary, FFSNs whose machines are scattered across different ISPs may return *A records* from different ASs.

2.2 Num_cname

The number of *cname*, where *cname* is an alias for one name to another. There is a tendency that FFSNs have less than 1 *cnames*.

2.3 Num_ns

The number of name server (NS). We can get it from NS records. The name server can also be hosted in FFSNs. Therefore, it often returns several NS records. The number of NS records returned by benign domains is usually very small.

2.4 Num_address

The number of different IP addresses. It is equal to the number of unique A records returned in all DNS lookups. In the Fast-Flux Service Networks, the mother-ship periodically updates these resource records to put new compromised machines and remove the fault ones. For this reason, compared with benign domains which commonly return only one to three A records, fast-flux domains often return more A records.

In addition, we compute two additional statistic values as the fast flux feature as follows, compared to existing approaches.

2.5 TTL (Time-To-Live)

It is a parameter that specifies the amount of seconds the response remains valid. Most of the flux-agents are end-user machines and consequently they will appear on-line and disappear very frequently [8]. To guarantee the high availability of the service offered through the fast-flux network, the set of active flux-agents has to be updated as soon as one of them changes its state. And then, the updating state must be rapidly transmitted across the Internet until the victims knows about the updating. So, the returned answers from DNS must change rapidly. That is the reason that most of FFSNs have lower TTL. However, there are also some benign domains especially Web portals with a low TTL. In spite of this, our experiment shows that TTL is a distinguishing feature attribute. The lower the time-to-live associated to the various DNS resource records of a domain, the higher the probability that the domain is malicious. We take the average time-to-live of *A records* as the value of TTL. The computational formula is defined as follows.

$$TTL = \frac{\sum_{1 \le i \le n_{SINGLE}} ttl_i}{n_{SINGLE}} \quad . \tag{1}$$

where the value ttl_i is the ttl value of ith A records, The value n_{SINGLE} is the number of IP addresses a single lookup returns.

2.6 Rate_flux

Due to the restrictions in establishing an FFSN, an attacker does not have directly control over the machines which run the FFSN. That is, flux-agents may go down at any time, so the attacker has to grasp a large number of compromised machines to make sure that when some flux-agents are halted, others can normally operate to offer illegal services. From these facts, we can infer that the total number of flux-agents in FFSNs could be more than currently available flux-agents, so we define $rate\_flux$ as following value to measure the diversification of IP addresses for a domain.

$$rate\_flux = \frac{n_{ALL}}{n_{SINGLE}} \quad . \tag{2}$$

Table 1. Comparison of five benign and malicious records using the selected features

	num_asn	*num_cname*	*num_ns*	*num_address*	*rate_flux*	*ttl*	*fast_flux*
Benign	1	3	4	1	1	15,635	N
	1	1	2	1	1	4,833	N
	1	1	2	1	1	1,252	N
	1	1	2	3	1	21,505	N
	1	1	2	3	1	65,703	N
Fast_flux	2	0	6	7	2.571	600	Y
	2	0	2	3	1.667	12	Y
	3	1	4	3	1.333	60	Y
	4	0	7	4	1	300	Y
	3	1	4	3	1.667	60	Y

Here we compute the value n_{ALL} as the number of unique IP addresses for a domain collected within a time period. A value $rate\_flux = 1$ means that the set of IP addresses remain constant within this period, which is common for benign domains. In contrast, $rate\_flux > 1$ indicates that the total number of IP addresses more than that of currently available IP addresses, which is a strong implication of FFSNs.

Table 1 shows the feature attribute values of both fast-flux networks and benign ones. It is observed that the $rate\_flux$ value of benign is always 1, whereas the value of fast-flux is almost always greater than 1. That is because fast-flux networks change IP addresses frequently. What's more, when it comes to *ttl*, the value of fast-flux is far more less than benign networks. There are also some differences of other feature attributes between fast-flux networks and benign ones.

3 The Approach

In this section, we present a three-steps approach for the fast-flux detection based on data classification technique, namely, generalize the patterns from the observed and collected real data of both the fast-flux networks and the benign ones to predict unknown URLs. First, we perform the DNS lookup with the tool dig [11] to collect the observed information from the given URLs. Then, the data is stored in database and we can extract discriminative features which reflect the differences between fast-flux and benign networks through the feature extractor. After that, feature vector is created. At last, these features are used to train a classifier to detect fast-flux networks. When the classifier is built, suspicious URLs were processed in the same way and feature vector are put into the classifier to test whether they are fast-flux networks or benign ones.

According to [9], the KNN method could perform better than other ones. On the other hand, due to imbalanced nature of the dataset, we here use the Random Forest(RF) classifier to make a comparative study. That is because its characteristics including bagging, ensembling and attribute raising that are widely used to overcome imbalanced data set problems and for classifier accuracy improvement [10].The classifier based on this method which overcome imbalanced data set problems performs better than generalized linear regression model.

4 Experiments and Analysis

The data sets of our experiment are from real data resource monitored by four enterprises and corresponding departments. All of them are divided into two domain lists, i.e., white list and black list. We use TP rate, FP rate, Precision, Recall and F-Measure as performance measures. Furthermore, we also use ROC Area [13] as performance metric.

The experiment is carried out in three parts. First, there are only four attributes including *num_asn*, *num_cname*,*num_ns* and *num_address*. The first part is a comparative study between these original attributes and combination of *rate_flux*. Meanwhile, considering the imbalanced class, resampling technique is used to solve the problem. Next, compared with the results of the five attributes we obtained, *TTL* is added to check whether the outcome could be improved significantly. Finally, we use the data from three departments as training data, the fourth department's data as testing data to further confirm that the feature attributes are discriminative during the real applications. In the following tables, the negative class (character N) is represented as benign networks while the positive class (character Y) represents fast-flux. The distribution of experimental instances are shown in Table 2.

Table 2. Positive and negative instances distribution

positive examples	1,697	6.34%
negative examples	24,806	93.66%

Considering the cost of predicting the fast-flux networks as benign ones is much higher than classifying the benign ones as the fast-flux networks, therefore we mainly focus on the classification performance of Y class. The results of the first part are shown in Table 3.

Here we use KNN classifier with parameter K = 1 and Random Forest (RF) classifier for the experimental comparison. With the Random Forest classifier, the TP rate of class Y is only 0.175, although the precision is 0.924. It means that most of the fast-flux networks are mistakenly classified. In contrast, using the KNN as classifier, the TP rate is raised to 0.467, but the precision is reduced to 0.249, so it is likewise an equally unsatisfactory result. Thus, we can conclude that the combination of these four attributes is not enough to classify the data into discriminating categories, and we therefore consider that new features need to be added.

When the *rate_flux* attribute is added, the improvement in TP rate of class Y is 0.167 using the KNN classifier and the value of precision is nearly doubled.

Table 3. Comparison between four attributes and five attributes with *rate_flux* +

		TP rate	FP rate	Precision	Recall	F-Measure	ROC Area	Class
Four	KNN	0.467	0.095	0.249	0.467	0.325	0.686	Y
Attributes	RF	0.175	0.001	0.924	0.175	0.294	0.878	Y
Rate_flux	KNN	0.634	0.046	0.485	0.634	0.55	0.794	Y
Added	RF	0.582	0.006	0.86	0.582	0.695	0.959	Y
Resample	KNN	0.741	0.031	0.621	0.741	0.676	0.955	Y
	RF	0.751	0.036	0.584	0.751	0.657	0.952	Y

On the other hand, the precision is still inadequate since it is less than 0.5. If the Random Forest classifier is used, the precision value reaches to 0.86. Meanwhile, the TP rate declines by 5.2% in comparison with KNN. It is not difficult to conclude that the *rate_flux* attribute is a useful one, although the outcomes are not very satisfactory. Then, the resample method is used to solve the problem of imbalanced classification. After several tests, the results are better when the resample rate is 1:4 or 1:5. On this occasion, TP rate is 0.751 most favorable, reflecting an increase of 11.7% compared with the best results of the preceding experiments. The low precision suggests that combination of five attributes is not a qualified discriminative feature. However, taking the improvements of performance into consideration, the *rate_flux* is a discriminative attribute in the classification.

In the second experiment, the attribute *TTL* is added. The results are shown in Table 4. From Table 4, a significant increase in both TP rate and precision value of class Y can be found in comparison with the results of five attributes, when *TTL* attribute is added. The highest TP rate is 0.942, with a precision value 0.955. What's more, TP rate and precision value are always above 0.92 whether you choose the KNN or the Random Forest classifier. Therefore, the *TTL* attribute is much more discriminative in the classification.

Table 4. Results of five attributes with *TTL* +

		TP rate	FP rate	Precision	Recall	F-Measure	ROC Area	Class
Six	KNN	0.942	0.003	0.955	0.942	0.948	0.999	Y
Attributes	(K=1)	0.997	0.058	0.996	0.997	0.997	0.999	N
With	RF	0.925	0.002	0.968	0.925	0.946	0.998	Y
TTL		0.998	0.075	0.995	0.998	0.996	0.998	N

Furthermore, the third experiment is made to simulate the detection of FF-SNs under real-world conditions and to make sure that the combination of six attributes is discriminative enough to classify them. In this part, there are 21,758 examples in the training set and 13,406 examples in the test set. Details of the data set are shown in Table 5.

The results of the third part are as shown in Table 6.

Table 5. The distribution of training set and test set

Training set	positive examples	1,551	7.12%
	negative examples	20,207	92.88%
Test set	positive examples	565	4.21%
	negative examples	12,841	95.79%

Table 6. Results of the third part of the experiment

		TP rate	FP rate	Precision	Recall	F-Measure	ROC Area	Class
Six Attributes	KNN (K=1)	0.935	0.002	0.95	0.935	0.942	0.999	Y
		0.998	0.065	0.997	0.998	0.997	0.999	N
	RF	0.904	0.001	0.966	0.904	0.934	0.998	Y
		0.999	0.096	0.996	0.999	0.997	0.998	N

The results illustrate that TTL is an important attribute in the classification and the six attributes we selected as feature attributes are more discriminative. Compared with the results in [9], almost each performance index is greatly improved. Taking run time into consideration, ca. 5 seconds is needed under Random Forest classifier,while it takes almost 106 seconds to get the result in KNN(K=1) classifier.Considering the data are collected under the real world conditions, we can conclude that the method will be applicable to the detection of fast-flux service networks.

At last, an additional experiment is made to determine whether TTL and $rate\_flux$ provide enough information for FFSNs detection.This time only TTL and $rate\_flux$ are chosen as feature attributes.Using the dataset of Table 2,TP rate of class Y is 0.818 and 0.797 in KNN(K=1) classifier and Random Forest classifier, respectively.If we use the dataset of Table 5 ,TP rate of class Y is 0.781 in KNN(K=1) classifier and 0.742 in Random Forest classifier.Apparently,the combination of TTL and $rate\_flux$ contains much information but not enough to detect FFSNs.

5 Conclusion

In this paper, the framework of detecting fast-flux service networks based on data mining technique is presented. Furthermore, we add two new attributes, i.e., fast-flux rate and TTL into the features of fast-flux. The results show that TP rate and Precision are significantly improved, which illustrates the combination of six attributes are more discriminative. To sum up, the features we selected provide an insight for implementing data mining techniques to detect fast-flux service networks and make it applicable. In the future work, we are going to find an effective way to classify legal fast-flux networks and the malicious ones.

Acknowledgments. This work was granted by Tsinghua National Laboratory for Information Science and Technology.

References

1. Castelluccia, C., Kaafar, M.A., et al.: Geolocalization of Proxied Services and its Application to Fast-Flux Hidden Servers. In: Proceedings of IMC 2009, November 4-6, pp. 184–189 (2009)
2. Jose, N., Thorsten, H.: As the Net Churns: Fast-Flux Botnet Observations. In: Proceedings of Int'l Conf, Malicious and Unwanted Software (Malware), pp. 24–31. IEEE Press, Los Alamitos (2008)
3. ICANN Security and Stability Advisory Committee.: SAC 025: SSA Advisory on Fast Flux Hosting and DNS (March 2008), http://www.icann.org/en/committees/security/ssac-documents.htm
4. Alper, C., Mike, T., et al.: Real-time detection of fast-flux service networks. In: Cybersecurity Applications & Technology Conference for Homeland Security, pp. 285–292 (2009)
5. The Honeynet Project.: Know Your Enemy: Fast-Flux Service Networks (2008), http://www.honeynet.org/papers/ff
6. Caglayan, A., Toothaker, M., et al.: Behavioral analysis of Fast Flux Service Networks. In: CSIIRW 2009, April 13-15 (2009)
7. Holz, T., Gorecki, C., et al.: Measuring and Detecting Fast-Flux Service Networks. In: Proceedings of the Network & Distributed System Security Symposium (2008)
8. Emanuele, P., Roberto, P., et al.: Fluxor: detecting and monitoring fast-flux service networks. In: Zamboni, D. (ed.) DIMVA 2008. LNCS, vol. 5137, pp. 186–206. Springer, Heidelberg (2008)
9. Wu, J., Zhang, L., et al.: A Comparative Study for Fast-Flux Service Networks Detection. In: Proceeding - 6th International Conference on Networked Computing and Advanced Information Management, NCM, pp. 346–350 (2010)
10. Sami, A., Yadegari, B., et al.: Malware Detection Based on Mining API Calls. In: Proceedings of SAC 2010, pp. 1020–1025 (2010)
11. Internet Software Consortium.: Dig: domain information groper (September 2007), http://www.isc.org/sw/bind/
12. Yu, S., Zhou, S., et al.: Fast-flux Attach Network Identification Based on Agent Lifespan. In: IEEE International Conference on Digital Object Identifier, Wireless Communications, Networking and Information Security (WCNIS), pp. 658–662 (2010)
13. Do, H.-H., Melnik, S., et al.: Comparision of schema matching evaluations. In: Proceedings of Web, Web-Services and Database Systems, pp. 221–337 (2002)

Online Learning Algorithm for Ensemble of Decision Rules

Igor Chikalov, Mikhail Moshkov, and Beata Zielosko*

Mathematical and Computer Sciences & Engineering Division
King Abdullah University of Science and Technology
Thuwal 23955-6900, Saudi Arabia
{igor.chikalov,mikhail.moshkov,beata.zielosko}@kaust.edu.sa

Abstract. We describe an online learning algorithm that builds a system of decision rules for a classification problem. Rules are constructed according to the minimum description length principle by a greedy algorithm or using the dynamic programming approach.

Keywords: decision rules, online learning, greedy algorithm, dynamic programming.

1 Introduction

Decision rules are widely used for representing knowledge extracted from large volumes of statistical or experimental data and to build classifiers that predict characteristics of new objects on the basis of information on existing objects [2].

Exact decision rules can be "overlearned", i.e., may depend on the "noise" present in the input data. Therefore, recent years particular attention has been devoted to the study of approximate decision rules [2,1].

Models with shorter descriptions are commonly believed to be more appropriate among the models with similar accuracy (minimum description length principle [3]). Following this principle, we are interested in building shortest decision rules with a given degree of accuracy.

In this paper, we consider online algorithm for construction of an ensemble of approximate decision rules. The decision rules built either by greedy algorithm [1] or by an algorithm based on dynamic programming [4]. We assume in the process of learning decision tables $T_1, T_2, \ldots, T_N$ appear consecutively. During learning, the algorithm is unable to store the tables itself, but can accumulate some information about the incoming data.

2 Decision Tables and α-Decision Rules

Decision table is a rectangular table T with n columns filled with nonnegative integers. Columns of the table are assigned attributes $f_1, \ldots, f_n$. The table rows

* This research was supported by the Russian Federal Program "Scientists and Educators in Russia of Innovations", contract number 02.740.11.5131.

S.O. Kuznetsov et al. (Eds.): RSFDGrC 2011, LNAI 6743, pp. 310–313, 2011.

are pairwise different, and each row r is labeled with a decision – a nonnegative integer $d(r)$. The rows are interpreted as tuples of attribute values.

A subtable of T is a table obtained from T by removing some rows with their assigned decisions. Let $j(1), \ldots, j(t) \in \{1, \ldots, n\}$ and $b_1, \ldots, b_t$ be nonnegative integers. Denote by $T(f_{j(1)}, b_1) \ldots (f_{j(t)}, b_t)$ the subtable of the table T, containing the rows from T, which at the intersection with the columns $f_{j(1)}, \ldots, f_{j(t)}$ have numbers $b_1, \ldots, b_t$ respectively.

Let $r = (a_1, \ldots, a_n)$ be a row of the table T. Denote by $U(T, r)$ the set of rows in T, which are labeled with decisions other than $d(r)$. Let $r' \in U(T, r)$. We say that an attribute f_i separates the row r from the row r' if at the intersection with the column f_i the rows r and r' contain different numbers. Let α be a real number, and $0 \leq \alpha < 1$. The expression $f_{i(1)} = a_{i(1)} \wedge \ldots \wedge f_{i(m)} = a_{i(m)} \rightarrow d(r)$ is called an α-decision rule for r and T if the attributes $f_{i(1)}, \ldots, f_{i(m)}$ separate from r at least $(1-\alpha)|U(T, r)|$ rows from $U(T, r)$ (or, equivalently, leave at most $\alpha|U(T, r)|$ rows from $U(T, r)$ unseparated). The number m is called the length of the decision rule.

3 Initial Classifier

Let $T_1, \ldots, T_N$ be a sequence of decision tables. For $i = 1, \ldots, N$, the decision table T_i contains n columns labeled with attributes $f_1, \ldots, f_n$. Table rows are pairwise different, and each row r is labeled with a decision – a number $d_i(r)$ from the set $\{1, ..., k\}$.

The online algorithm sequentially receives the tables $T_1, \ldots, T_N$. The algorithm processes one table T_i at a time and proceed with T_{i+1} only after processing of T_i is completed. Due to memory restrictions, only a limited information about each table can be stored. The online algorithm starts with constructing so-called initial classifier based on table T_1.

Let us split the table T_1 into two sub-tables T_c and T_p. The first subtable is used for construction of a system of α-decision rules and the second one for pruning of the system. Choose a real α, $0 \leq \alpha < 1$, and apply either the greedy algorithm [1] or the dynamic programming algorithm [4] to construct for each row r of the table T_c an α-decision rule for T_c and r. The greedy algorithm allows for an arbitrary α, but for the other algorithm, α values close to zero make the construction computationally expensive. Let the following α-decision rule was constructed for the row r: $f_{i(1)} = a_{i(1)} \wedge \ldots \wedge f_{i(m)} = a_{i(m)} \rightarrow d_1(r)$. Replace the right-hand side of this rule $d_1(r)$ with the tuple $(p_1, \ldots, p_k)$, where for $i = 1, \ldots, k$, p_i is the number of rows with the decision i in the table $T_c(f_{i(1)}, a_{i(1)}) \ldots (f_{i(m)}, a_{i(m)})$. Denote by rule$(T_c, r)$ the resulted rule

$$f_{i(1)} = a_{i(1)} \wedge \ldots \wedge f_{i(m)} = a_{i(m)} \rightarrow (p_1, \ldots, p_k). \tag{1}$$

We call (1) a generalized α-decision rule for T_c and r. Denote by $S_\alpha(T_c)$ the constructed system of rules $\{\text{rule}(T_c, r) : r \in T_c\}$. This system can be used as a classifier predicting decision for a new object O given by tuple of values $(b_1, \ldots, b_n)$ of attributes $f_1, \ldots, f_n$.

We say that the decision rule (1) is applicable to the object O, if $b_{i(1)} = a_{i(1)}, \ldots, b_{i(m)} = a_{i(m)}$. Let us find all the decision rules in $S_\alpha(T_c)$ applicable to the object O and sum up elementwise the tuples from the right-hand sides of these rules: $\sum(p_1, \ldots, p_k) = (\sum p_1, \ldots, \sum p_k)$. Denote the resulted tuple by $(P_1, \ldots, P_k)$. Then the predicted decision for the object O is the index of the maximum element in the tuple $(P_1, \ldots, P_k)$ (if the maximum is reached on several elements, take the first one).

If none of the rules of $S_\alpha(T_c)$ is applicable to the object O, then we take the most common decision in the table T_c as a decision for O.

Let us describe the operation of rule pruning. We assign to each rule rule(T_c, r) from $S_\alpha(T_c)$ (see (1)) a set of real numbers $A(T_c, r)$. For $j = 1, \ldots, m$, consider the rule

$$f_{i(1)} = a_{i(1)} \wedge \ldots \wedge f_{i(j)} = a_{i(j)} \rightarrow (p_1, \ldots, p_k). \tag{2}$$

Denote $\beta_j = 1 - |U(T_c, r, j)|/|U(T_c, r)|$, where $|U(T_c, r, j)|$ is the number of rows in $U(T_c, r)$ that are separated from the row r by the attributes $f_{i(1)}, \ldots, f_{i(j)}$. Then $A(T_c, r) = \{\beta_1, \ldots, \beta_m\}$. Denote $A = \cup A(T_c, r)$ where the union is taken on all rows r of the table T_c.

Let $\beta \in A$. For each rule rule(T_c, r) (see (1)) from $S_\alpha(T_c)$ find the minimum $j \in \{1, \ldots, m\}$, for which the rule (2) is a generalized β-decision rule for T_c and r, i.e., $1 - |U(T_c, r, j)|/|U(T_c, r)| \leq \beta$. Denote the resulting rule by rule$_\beta(Tc, r)$ and denote by $S_\beta(T_c)$ the constructed system of rules $\{\text{rule}_\beta(T_c, r) : r \in T_c\}$.

For each $\beta \in A$, apply the classifier $S_\beta(T_c)$ to all rows of the table T_p. Denote by $ER_\beta(T_p)$ the number of rows from T_p, for which the predicted value of the decision differ from the actual decision. Chose $\gamma \in A$ that minimizes $ER_\gamma(T_p)$. Denote $S_1 = S_\gamma(T_c)$. We call S_1 the initial classifier for the table T_1.

4 Online Algorithm

The classifier S_1 continues to learn on the rows of tables $T_2, T_3, \ldots$. Let r be one of these rows and $f_{i(1)} = a_{i(1)} \wedge \ldots \wedge f_{i(j)} = a_{i(j)} \rightarrow (p_1, \ldots, p_k)$ an arbitrary rule from S_1 that is applicable to r. Then processing of the row r changes the right-hand side of the considered rule: the value p_l is increased by one, where l is the decision of the row r. If no rule in S_1 is applicable to the row r, then the row r with its decision l is added to the table D, which was empty initially.

We assume that some threshold Δ is defined, such that once the number of rows in D exceeds Δ, then finished processing the current table T_i, we construct the initial classifier for the table D and denote it S_2. It may happen that D contains identical rows. Then from each group of identical rows we leave one, labeled with the most common decision. Once the classifier S_2 is built, all rows are removed from the table D.

The initial classifiers S_1 and S_2 now form a new classifier, which continues to learn on the rows of tables $T_{i+1}, T_{i+2}, \ldots$. Let r be one of these rows. If S_1 contains rules applicable to r, then the right-hand sides of these rules are changed as described at the beginning of this section. Rules from S_2 do not change in this case. Otherwise, if none of the rules from S_1 is applicable to the row r, then start working with the classifier S_2. If S_2 contains rules applicable to r, then the right-hand sides of these rules are updated as described at the beginning of this section. Otherwise, the row r with its decision is added to the table D. If the number of rows in D exceeds the threshold Δ, then finished processing the current table T_q, leave in the table D only pairwise different rows, construct the initial classifier S_3 for the table D, remove all rows from D, etc.

Let at some moment initial classifiers $S_1, S_2, \ldots, S_t$ have been built. This sequence of classifiers continues to learn on the rows of not yet processed tables $T_e, T_{e+1}, \ldots$. Let r be one of these rows. The algorithm looks for the minimum $i \in \{1, \ldots, t\}$, for which S_i contains rules applicable to r. Right-hand sides of these rules are updated as described at the beginning of this section. Rules of the systems $S_1, \ldots, S_{i-1}, S_{i+1}, \ldots, S_t$ remain unchanged. If the systems $S_1, S_2, \ldots, S_t$ contain no rules applicable to r, then the row r with its decision is added to the table D, etc.

The process of learning of the sequence of elementary classifiers $S_1, S_2, \ldots, S_t$ can be interrupted at any time, and the sequence of modified initial classifiers $S_1, S_2, \ldots, S_t$ (with modified right-hand sides of decision rules) can be used for predicting the decision for a new object O, given by the values $(b_1, \ldots, b_n)$ of the attributes $(f_1, \ldots, f_n)$ for this object. Let us find the minimum $i \in \{1, \ldots, t\}$, for which S_i contains rules applicable to O. Then find in S_i all decision rules applicable to O and sum up their right-hand side tuples elementwise: $\sum(p_1, \ldots, p_k) = (\sum p_1, \ldots, \sum p_k)$. Denote by $(P_1, \ldots, P_k)$ the resulting tuple. Then the predicted decision for the object O is the index of the maximum element in the tuple $(P_1, \ldots, P_k)$ (if the maximum is reached on several elements, take the first one).

If none of the rules in the systems $S_1, S_2, \ldots, S_t$ are applicable to the object O, then take the most common decision assigned to the rows of already processed tables $T_1, \ldots, T_N$ as decision for the object O.

5 Conclusions

We described a new online algorithm for construction of an ensemble of approximate decision rules. We are planning to compare it with the existing algorithms on a representative set of test examples.

References

1. Moshkov, M., Piliszczuk, M., Zielosko, B.: Partial Covers, Reducts and Decision Rules in Rough Sets: Theory and Applications. Studies in Computational Intelligence, vol. 145. Springer, Heidelberg (2008)
2. Pawlak, Z.: Rough Sets—Theoretical Aspects of Reasoning about Data. Kluwer Academic Publishers, Dordrecht (1991)
3. Rissanen, J.: Modeling by shortest data description. Automatica 14, 465–471 (1978)
4. Zielosko, B., Moshkov, M., Chikalov, I.: Optimization of decision rules based on methods of dynamic programming. Bulletin of Nizhny Novgorod State University (to appear) (in Russian)

Automatic Image Annotation Based on Low-Level Features and Classification of the Statistical Classes*

Andrey Bronevich and Alexandra Melnichenko

Mathematics Department, Technological Institute of Southern Federal University, Taganrog, Nekrasovskij Street 44, Taganrog 347928, Russia
brone@mail.ru, alexandramelnichenko@gmail.com

Abstract. This work is devoted to the problem of automatic image annotation. This problem consists in assigning words of a natural language to an arbitrary image by analyzing textural characteristics (low-level features) of images without any other additional information. It can help to extract intellectual information from images and to organize searching procedures in a huge image base according to a textual query. We propose the general annotation scheme based on the statistical classes and their classification. This scheme consists in the following. First we derive the low-level features of images that can be presented by histograms. After that we represent these histograms by statistical classes and compute secondary features based on introduced inclusion measures of statistical classes. The automatic annotation is produced by aggregating secondary features using linear decision functions.

Keywords: automatic image annotation, image retrieval, low-level features, statistical classes, inclusion measures.

1 Introduction

Nowadays there are many visual data bases accessible in Internet, but it is hard to find such information because the usual textual query cannot be processed properly because, in general, images do not have additional description, that can be matched with textual representation. To solve this problem, it is necessary to provide effective methods for automatic image annotation enabling to obtain image annotations consisting of words and describing the image content.

There are many methods proposed for this problem which differ in some aspects such as image features representation and type of classifier used [1,2]. In the recent investigations two main features representations are used equally often: *regional representation*, when each image region is described separately with own feature vector, and *global representation*, where whole image is described by the one vector of low-level features. While regional representation could be more

* This work is supported by RFBR, projects #10-07-00135, #10-07-00478, #11-07-00591.

S.O. Kuznetsov et al. (Eds.): RSFDGrC 2011, LNAI 6743, pp. 314–321, 2011.

discriminative, the accurate image segmentation, required for it, is hard problem. As regards annotation model, in the early years various probabilistic methods dominate here (such as Bayesian classifiers and particularly Cross-Media Relevance Models [3,4])along with some machine learning algorithms [2,5]).Recent years Theory of Rough Sets proposes promising approaches for building classifiers and dicision rules for solving this sort of classification tasks [13].

The automatic image annotation can be considered as a classification problem, in which we should choose words from a given vocabulary that describe the image relevantly. In this investigation we don't consider the semantic textual descriptions and our output result should be the list of words ordered by their significance or relevance to a given image. Let us notice that the annotations based on textural features of images can reflect general image characteristics, that represented by words like "day", "night", "sea", "tree", "city", that describe the image in the whole. And we apply for this low-level features based on evaluation of gradient, colors distribution and various textural characteristics. As a rule, low-level image features can be some numerical characteristics or samples that are often represented by histograms. This way of representation can be used for describing colors distributions, the distribution of gradient directions and so on. These characteristics should be stable to scene illumination and scaling. Because it is hard to find an explicit connection between low-level features and words, it is reliable to use methods from pattern recognition theory: according to the problem statement we have a learning sample of annotated images and we have to build decision functions allowing us to classify an arbitrary image using words of a given vocabulary. The main characteristics of this pattern recognition problem are the following:

1. A huge number of classes to which a given image can belong (the number of classes is equal to the cardinality of a chosen vocabulary).
2. Classes are not disjoint in general.
3. It is impossible precisely to define boundaries between classes.
4. As a rule, low-level features can be represented as independent samples of a random variable that characterizes the image.
5. A very high dimension of feature space in which the classification problem should be solved.

These characteristics of the classification problem can be easily derived by analyzing its nature. For example, a city landscape can include houses, trees, and some times a part of the picture can include sea outlook. If a picture contains a palm, it is not possible to judge whether the photo was made within the house or outside. If we classify images using words "morning", "day", "evening", and "night", then it is hard to define exact boundaries between classes "day" and "evening", "night" and "morning". Trying to increase classification quality, we should increase the number of the used low-level features, and this also forced the increasing of the feature space dimension. These characteristics lead to the following classification scheme, depicted on Fig. 1.

According to this scheme, images should be processed first for extracting low-level features, then the secondary features are derived, and the annotation is

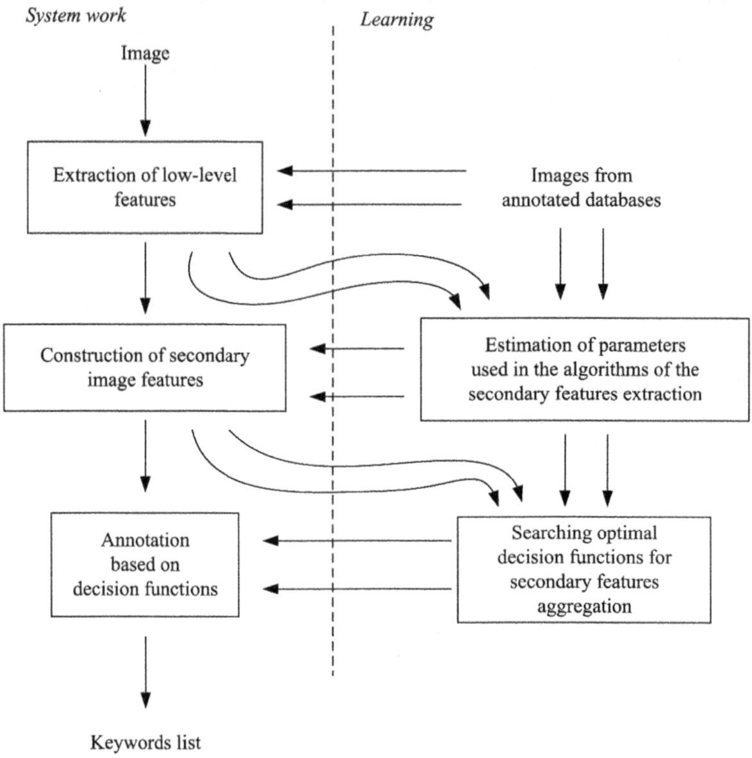

Fig. 1. General scheme of the automatic image annotation procedure

constructed by aggregating secondary features. Sometimes the secondary features extraction can be understood as a problem of decreasing feature space dimension.

2 Low-Level Image Features

Low-level feature extraction is a very important stage of automatic annotation algorithm because it provides the base for building the image representation. Low-level features producing image annotation should meet humans perception of image similarity and satisfy certain conditions allowing to consider these features as patterns for classification. The main of such requirements are:

- invariance with respect to image scaling and to lighting conditions of image capturing;
- small correlation of different features;
- the dimension of obtained patterns should be adequate to the size of keywords used.

We briefly describe here four low-level feature types which are the most appropriate in terms of required properties. We have successfully used these features for the construction of automatic image annotation system.

Histogram of Oriented Gradient (HoG). allows to determine appearance of the local objects and to recognize their shape [5]. HoG-descriptors calculation is performed using rectangular grid. The grid consists of cells small spatial regions which are combined to the larger intersected areas called "blocks". For an image the matrices of gradient magnitudes in horizontal and vertical directions are calculated for each color channel in RGB color space. As the orientation and magnitude for each pixel the corresponded values are taken from the color channel with the maximal magnitude. For each cell one calculates histograms of gradient orientations and joins histograms which compose a block. Histograms belonging to the one block are normalized to achieve invariance to the local illumination changes.

Measure of the background homogeneity. can be useful, for example, to distinguish such scene types as landscape, portrait or macro [6]. Image background is relatively large areas of connected pixels with the similar color characteristics. We calculate the measure of background homogeneity using Shannon entropy of every color channel as described in [6]. To obtain more informative image representation, we calculate the measure of background homogeneity for particular rectangles from some regular grid.

Color histograms. evaluate an image property that is very important for human visual perception — the color distribution. The key issue here is the choice of the appropriate color space. We use for this purpose *CIELab* color space, which has been designed as the space with linear color changes with respect to the human perception. To achieve illumination invariance we discard L ("lightness") component of the (L, a, b) pixel component and build two-dimensional histogram from two remaining chrominance color components.

Texture image features. Texture determines surfaces features which are helpful for objects recognition. One of the most informative texture features is one introduced in [7], which includes such characteristics as coarseness, contrast and directionality of the texture. The coarseness characterizes the size of the structural units forming the texture. The texture contrast value indicates how much gray levels vary within an image and in what degree their distribution is biased to white or black. Feature of texture directionality is calculated in the each pixel from the magnitudes of histogram peaks. For detailed description of calculation procedure see [7].

3 Statistical Classes Classification

3.1 Notion of Statistical Class

Here we use the **notion of statistical class** [8], that is introduced for the finite case as follows. Let $X = \{x_1, ..., x_n\}$ be a finite universal set and let $\mathcal{U} = 2^X$ be the powerset of X. Assume also that the space X is equipped with an additive measure V, called the volume measure. Then any statistical class F can be defined by a probability measure P on $\mathcal{U}$, that has to be absolutely continuous w.r.t. the volume measure V. Because we assume that all low-level features can be described by histograms, we postulate that any such feature is a histogram, which can be considered as an evaluation of a probability distribution. Secondary features are computed by using inclusion measures of statistical classes. Let us remind that the absolute continuity for a finite case means that $x_i \in X$ and $P(\{x_i\}) > 0$ implies that $V(\{x_i\}) > 0$. Therefore, it is possible to define a probability density by formula

$$h(x) = \begin{cases} P(\{x\})/V(\{x\}) & \text{if V}(\{\text{x}\}) > 0, \\ 0, & \text{otherwise.} \end{cases}$$

Using the probability density we can compute the probability of any event $A \in \mathcal{U}$ by $P(A) = \sum_{x \in A} h(x)V(\{x\})$. The last sum can be considered as an integral sum for Lebesgue integral, therefore, we can write: $P(A) = \int_A h(x)dV$. It is clear that the density function can be considered as another way for defining the statistical class. In real applications, the volume measure has to be chosen such that it can discriminate statistical classes in the best way. If we have no sufficient prior information, we can assume $V(\{x\}) = c > 0$ for all $x \in X$, in particular, $c = 1$ or $c = 1/n$. Obviously, in the last case, V is a probability measure on $\mathcal{U}$.

Theoretically the **inclusion relation** of statistical classes is introduced with the so-called minimal events. In this paper we drop this theoretical construction (see for details [8]). For practical applications it is sufficient to know of how this relation is defined by using membership functions. Given a statistical class F, defined by a probability measure P_F with a density $h_F(x)$. Then functions

$$\underline{\mu}_F(x) = \sum_{y \in X | h_F(y) < h_F(x)} P_F(\{y\}) \text{ and } \bar{\mu}_F(x) = \sum_{y \in X | h_F(y) \leq h_F(x)} P_F(\{y\})$$

are called a lower and an upper membership functions of the statistical class F respectively. By definition, the statistical class F_1 is included to the statistical class F_2, i.e. $F_1 \subseteq F_2$, if $\underline{\mu}_{F_1}(x) \leq \underline{\mu}_{F_2}(x)$ and $\bar{\mu}_{F_1}(x) \leq \bar{\mu}_{F_2}(x)$ for all $x \in X$. It is possible to prove that membership functions define each statistical class uniquely. In the next, we consider set-theoretical operations on statistical classes, which are produced with the help of min and max operations:

1. $\underline{\mu}_{F_1 \cap F_2}(x) = \min\left(\underline{\mu}_{F_1}(x), \underline{\mu}_{F_2}(x)\right)$, $\bar{\mu}_{F_1 \cap F_2}(x) = \min\left(\bar{\mu}_{F_1}(x), \bar{\mu}_{F_2}(x)\right)$ are membership functions of the statistical class $F_1 \cap F_2$;
2. $\underline{\mu}_{F_1 \cup F_2}(x) = \max\left(\underline{\mu}_{F_1}(x), \underline{\mu}_{F_2}(x)\right)$, $\bar{\mu}_{F_1 \cup F_2}(x) = \max\left(\bar{\mu}_{F_1}(x), \bar{\mu}_{F_2}(x)\right)$ are membership functions of the statistical class $F_1 \cup F_2$.
   It is possible that a statistical class $F_1 \cap F_2$ or $F_1 \cup F_2$ can not be generated by a probability measure. The sense of this can be explained while considering classification problems.

An **inclusion measure of statistical classes** $\mu(F_1 \subseteq F_2)$ is introduced for evaluating an inclusion degree of statistical class F_1 into statistical class F_2. By definition, $\mu(F_1 \subseteq F_2) \in [0, 1]$ and $\mu(F_1 \subseteq F_2) = 1$ iff $F_1 \subseteq F_2$. This functional can be introduced in various ways but we don't describe all of them, see [8,9,10] for details. Here we use the inclusion measure axiomatically defined in [9]. Let us introduce an auxiliary function

$$\psi(F_1 \subseteq F_2) = 0.5 \int_X \left(\underline{\mu}_{F_2}(x) + \bar{\mu}_{F_2}(x)\right) dP_1 = 0.5 \sum_{x \in X} \left(\underline{\mu}_{F_2}(x) + \bar{\mu}_{F_2}(x)\right) P_1(x).$$

Then an inclusion measure is defined as

$$\mu(F_1 \subseteq F_2) = \psi(F_1, F_1 \cap F_2) + 1 - \psi(F_2, F_1 \cup F_2).$$

Let us notice that in the last formula it is necessary to compute membership functions of statistical classes $F_1 \cap F_2$ and $F_1 \cup F_2$.

3.2 Secondary Features Construction Based on Inclusion Measures

Let we have a set $\{S_1, S_2, ..., S_n\}$ of etalon statistical classes. Then the classification of any statistical class F consists in computing the following classifying vector: $(\mu(F \subseteq S_1), ..., \mu(F \subseteq S_n))$.

Let us show how to get secondary features using inclusion measures. Since secondary features can be represented by means of histograms, we can assume that any low-level feature is a histogram being an estimate of probability distribution. For example, features derived with the help of the gradient directions are the set of histograms that correspond to different positions of the scanning window. Features of the texture coarseness and background homogeneity can be represented as a histogram if we calculate these features for different positions of the scanning window.

Let us consider how to build etalon classes. Assume that images are annotated such that a keyword corresponds to the part or the whole image. For example, if an image is annotated by a word "building", then the annotation procedure would be more precise if to compute the statistical characteristics for the part of the image, where the building is situated. Assume that we are going to extract secondary features for buildings. For this aim, we should first construct etalon classes corresponding to the word "building". The simplest way to do so is to compute the histogram for the all images from the learning sample that are annotated by the word "building". Let us assume that the etalon class S has been constructed that corresponds to the word w ("building") and a to given low-level feature b. Then for image classification, we has to compute histograms, corresponding to b. Let these histograms are statistical classes $F_1, ..., F_l$. Then the secondary feature is $p(w|b) = \max\{\mu(F_1 \subseteq S), ..., \mu(F_l \subseteq S)\}$. Notice that in the last formula the maximum is used, because we choose in this case the part of the image that is the most relevant to the keyword w. Notice also that the inclusion measure has a probabilistic interpretation: it is a mean value of conditional probability of minimal events that correspond to the etalon class S provided that we observe the statistical class F. Hence, the greater value $p(w|b)$ is, the greater probability is that the image is annotated by the keyword w.

3.3 The Aggregation of Secondary Features

Assume that on this stage we have a vocabulary $W = \{w_1, ..., w_m\}$ and a set of features $B = \{b_1, ..., b_n\}$. The secondary features are presented by $p(w_i|b_j)$. The next procedure is to construct aggregation functions $\varphi_i : [0,1]^n \to [0,1]$, allowing us to compute the global features: $p(w_i) = \varphi_i(p(w_i|b_1), ..., p(w_i|b_n))$, $i = 1, ..., n$, which have to give us the global evaluation that the keyword w_i is relevant to the analyzed image.

Let $\varphi : [0,1]^n \to [0,1]$ be a aggregation function. Then it has the following properties [11]: 1) $\varphi(\mathbf{0}) = 0$ and $\varphi(\mathbf{1}) = 1$, where $\mathbf{0} = (0, ..., 0)$ $\mathbf{1} = (1, ..., 1)$; 2) $\varphi(\mathbf{x}) \leq \varphi(\mathbf{y})$ for $\mathbf{x} \leq \mathbf{y}$, where $\mathbf{x} = (x_1, ..., x_n)$, $\mathbf{y} = (y_1, ..., y_n)$, and $\mathbf{x} \leq \mathbf{y}$ if $x_i \leq y_i$ for all $i \in \{1, ..., n\}$.

If we assume that the features are independent, then it is rational to use linear aggregation functions of the type $\varphi(\mathbf{x}) = \sum_{i=1}^{n} a_i x_i$, where $a_i \geq 0$, $i = 1, ..., n$, and $\sum_{i=1}^{n} a_i = 1$. Let we annotate images by the rule: an image is annotated by a keyword w_i if $p(w_i) > \varepsilon_i$. In this scheme parameters of aggregation functions φ_i and non-negative numbers ε_i have to be estimated using the learning sample. Suppose that the learning sample consists of N images. In this case any image with a number $k \in \{1, ..., N\}$ is described by a vector of secondary features $\mathbf{p}_k = (p_k(w_i|b_1), ..., p_k(w_i|b_n))$, that characterizes the relevance of the keyword w_i. Assume further that we code with a number $\delta_k \in \{-1, 1\}$ the information whether or not the image with the number k is annotated by the keyword w_i, assuming that $\delta_k = 1$ if w_i is in the image annotation, and $\delta_k = -1$, otherwise. Then we have a learning problem of searching a vector $\mathbf{a} = (a_1, ..., a_n)^T$ and a threshold value ε so that the number of false classifications would be minimal. In other words, the number of true inequalities $\delta_k(\mathbf{p}_k\mathbf{a} - \varepsilon) > 0$ $k = 1, ..., N$, would be maximal. Such optimization problem of finding a linear classifier is classical in pattern recognition theory and can be solved by any well-known algorithm, in particular, perceptron algorithm [12].

4 Conclusion

In this paper a problem of automatic image annotation is considered and a general scheme for this problem is presented based on low-level image features extraction. The key properties of low-level features are discussed and several feature types with desired properties are briefly described. The further annotation procedure is based on extracting secondary features from the low-level features and on classifying the obtained patterns. For this purpose, the notion of statistical class and the inclusion measure of statistical classes are introduced. In our problem, we propose to use statistical classes for representing probability distributions of low-level features. A scheme of classifying statistical classes into etalon classes, which correspond to keywords, is given. The generation of annotations is produced by the aggregation of secondary features using linear decision functions constructed by the learning procedure based on perceptron algorithm. The presented annotation scheme is implemented practically and has shown its effectiveness provided by the proposed algorithms.

References

1. Tsai, C., Hung, C.: Automatically Annotating Images with Keywords: A Review of Image Annotation Systems. Recent Patents on Computer Science 1, 55–68 (2008)
2. Hanbury, A.: A Survey of Methods for Image Annotation. Journal of Visual Languages & Computing 19(5), 617–627 (2008)
3. Duygulu, P., Barnard, K., de Freitas, J.F.G., Forsyth, D.: Object Recognition as Machine Translation: Learning a Lexicon for a Fixed Image Vocabulary. In: Heyden, A., Sparr, G., Nielsen, M., Johansen, P. (eds.) ECCV 2002. LNCS, vol. 2353, pp. 97–112. Springer, Heidelberg (2002)
4. Jeon, J., Lavrenko, V., Manmatha, R.: Automatic Image Annotation and Retrieval Using Cross-Media Relevance Models. In: Proc. of the ACM SIGIR Conference, vol. 1, pp. 119–126 (2003)
5. Dalal, N., Triggs, B., Schmid, C.: Human detection using oriented histograms of flow and appearance. In: Leonardis, A., Bischof, H., Pinz, A. (eds.) ECCV 2006. LNCS, vol. 3952, pp. 428–441. Springer, Heidelberg (2006)
6. Abramov, S.K., Lukin, V.V., Ponomarenko, N.N.: Entropy Based Background Measure Calculation for Images Searching and Sorting in the Large Collections. Electronics and Computer Systems 2(21), 24–28 (2007)
7. Tamura, H., Mori, S., Yamawaki, T.: Texture Features Corresponding to Visual Perception. IEEE Trans. On Sys. Man, and Cyb. 8(6), 460–473 (1978)
8. Bronevich, A.G., Karkishchenko, A.N.: Statistical Classes and Fuzzy Set Theoretical Classification of Possibility Distributions. In: Bertoluzza, C., Gil, M.A., Ralescu, D.A. (eds.) Statistical Modeling, Analysis and Management of Fuzzy Data, pp. 173–198. Physica-Verl., Heidelberg (2002)
9. Bronevich, A.G., Karkishchenko, A.N.: Application of Possibility Theory for Ranking Probability Distributions. In: Proc. of the European Congress on Intelligent Techniques and Soft Computing, pp. 310–314 (1997)
10. Bronevich, A.G., Karkishchenko, A.N.: Fuzzy Classification of Probability Distributions. In: Proc. of the Fourth European Congress on Intelligent Techniques and Soft Computing, vol. 1, pp. 120–124 (1996)
11. Grabisch, M., Pap, E., Mesiar, R., Marichal, J.-L.: Aggregation Functions. Cambridge University Press, Cambridge (2009)
12. Tsypkin, Y.Z.: Adaptation and Learning in Automatic Systems. Academic Press, Inc., Orlando (1971)
13. Skowron, A., Swiniarski, R.W.: Information Granulation and Pattern Recognition. Rough-Neural Computing. In: Techniques for Computing with Words, pp. 599–636. Springer, Heidelberg (2004)

Machine Learning Methods in Character Recognition

Lev Itskovich[1] and Sergei Kuznetsov[2]

[1] ABBYY, Moscow, Russia,
Moscow Institute of Physics and Technology, Moscow, Russia,
Lev_I@abbyy.com
[2] Higher School of Economics, Moscow, Russia,
Moscow Institute of Physics and Technology, Moscow, Russia,
skuznetsov@hse.ru

Abstract. In this paper we consider applications of well-known numerical classifiers to the problem of character recognition (optical character recognition, OCR). We discuss the requirements which these classifiers should meet to solve this problem. Various modifications of well-known algorithms are proposed. Recognition rates of these classifiers are compared on real character datasets.

Keywords: OCR, numerical classifiers, recognition rate, naive Bayes, nearest neighbor, decision tree, concept lattice.

1 Introduction

Numerical classifiers based on various machine learning methods [1] are often used in character recognition software. These methods can automatically build classification rules based on numerical descriptions of known objects (training samples).

The aim of this paper is to determine, which learning approaches can solve best the character recognition problem. Learned classifiers are tested as a part of recognition schema, which is used in ABBYY OCR Technologies. Starting from structure of this recognition schema, we formulate requirements for numerical classifiers used in this schema. Experimental results obtained with classifiers used in ABBYY OCR Technologies are compared with results of various "classical" algorithms.

2 ABBYY OCR Technologies Recognition Schema

Structure of ABBYY OCR Technologies recognition schema [2] is shown in Fig. 1.

Three numerical classifiers on the left part of Fig. 1 are build on different sets of numerical attributes. In classification process they operate in a sequence. If the recognition confidence of at least one of them is enough, the classification process stops.

S.O. Kuznetsov et al. (Eds.): RSFDGrC 2011, LNAI 6743, pp. 322–329, 2011.

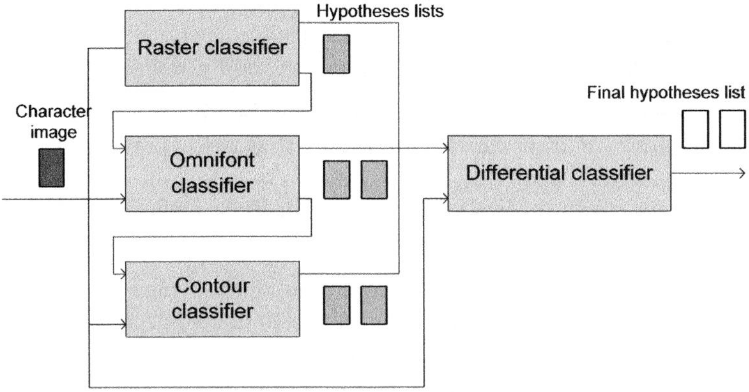

Fig. 1. ABBYY OCR Technologies recognition schema

Raster classifier operates directly with black-and-white character image without numerical attributes calculation, so it's description is beyond the scope of this paper. Omnifont and contour classifiers are numerical and are based on following decision rule. First, training samples of each class are clustered separately. The next stage of classification is organized using nearest neighbor classification, where Mahalanobis distance [2] is used as a measure of distance between test set objects and training set clusters.

In this schema the aim of numerical classifiers is only to build list of hypothesis (possible results of classification) arranged by their confidence. The next stage of classification process is differential classifier. It rearranges hypothesis list using modified bubble sort, where each two hypothesis (classes) are compared by their individual linear classifier.

3 Requirements Imposed on Classifiers by Recognition Schema

Starting from considered recognition schema, we will formulate requirements for any numerical classifier that can be potentially used in this schema.

High recognition rate. The most evident requirement for such classification algorithm is its high recognition rate. Experimental results are listed in section 5, but the approximate lower boundary is 95% right classification. If recognition rate of some classifier is considerably lower than 95%, such classifier is not suitable for practical symbol recognition.

Hypothesis generation. Recognition schema as a whole must be able to generate not only single classification results, but several classification variants (hypothesis). This feature allows us to correct recognition results using linguistic information (dictionary) when all symbols are already recognized. Thus numerical classifier is responsible for hypothesis generation inside the recognition schema.

Therefore additional restriction is imposed on previous requirement. The classifier has to generate first right hypothesis in 95% cases, but moreover in the rest 5% cases right hypothesis should appear in the hypothesis list.

Hypothesis confidence estimation. Classifier must be able to estimate confidence of each hypothesis (in other words, probability that the hypothesis is right). This property of algorithm allows us to use cut-offs to improve classification speed: if one of the classifiers recognized symbol with high confidence, then it is not necessary to launch next classifiers.

High speed. First of all, it is necessary to emphasize that this requirement implies high recognition speed of the algorithm on particular training set (not asymptotic complexity of the algorithm). This particular speed obviously depends on algorithm realization. Therefore here we formulate only qualitative requirement: learning stage of classification should include the greater of algorithm complexity. Results of learning stage ought to be saved in compact data, which can further be easily and rapidly used on the classification stage. Such data is called *samples.*

4 Modifications of Classical Algorithms

Nearest neighbor [3] and *Naive Bayes* [3] classifiers can be easily improved in order to generate hypotheses and to calculate their confidence.

Nearest neighbor algorithm. We can simply consider k nearest neighbors instead of single one. An obvious way to determine hypothesis confidence is to use distance function $d(s, s_0)$ (distance between object s and object/cluster s_0 which represent particular hypothesis).

The best complexity is shown by the modification of the kNN algorithm which involves training set clusterization. In that case after stage of learning we need to store only cluster centers (instead of all attribute values for all objects of training set). Thus, we reduce complexity of classification stage. Notice that this modification of kNN algorithm is used in ABBYY OCR Technologies [2] so it is treated as a starting point for comparison.

Naive Bayes classifier. Modification of this algorithm which generates hypotheses and calculates their confidence is also obvious. We can use probability value to select best hypotheses.

The values of probabilities can be calculated in advance at the stage of learning. Therefore this algorithm is considered to be rather fast, but it needs large amount of precomputed data.

5 Modifications of Decision Tree Classifier

Modifications of *decision tree classifier* [1] and *concept lattice classifier* [4] that generate hypotheses is not as evident as previous modifications. Here we describe two approaches to this problem.

Post-fuzzification. Using classical algorithm C4.5 [5] without any changes to build decision tree, we modify classification stage according to method described in [6]. Contrary to deterministic choice between two successors T_1 and T_2 of node T, we consider "fuzzy" classification step, moving to the node T_j ($j = \overline{1,2}$) with probability p_j. Integral probability of the classification path is the multiplication of elementary step probabilities. Hypothesis list contains classification results obtained with the most probable paths. To optimize classification we propose well-known beam search method [7]. Probability p_j can be defined as a piecewise function which is constant inside the interval of the attribute a_i values of objects from the node T_j and exponentially decreases outside the interval.

Random forest. Another method which we propose for decision tree hypothesis generation is random forest [8] (voting of several decision trees). Random forest is widely used in order to improve classifier recognition rate (we show in section 6 that this effect really takes place). But we also involve random forest as a decision tree hypothesis generation method.

Each of random forest decision trees is build using randomly generated subsets of initial training set and attribute set. Classical random forest generates single classification result (class which was returned by most of decision trees). We treat classes returned by all trees as hypotheses list, and we define the confidence of hypothesis as number of trees which returned such classification result.

Notice that this approach can be also applied to combine results of concept lattice classifiers.

Combination of these methods. Considered methods can be easily combined. Each tree from the random forest can generate several hypotheses itself. We can define many methods to combine confidences obtained from different trees and make the final list of hypotheses. For instance, we can use sum of each class confidences or their maximum.

6 Experimentation

Experiments were carried out with the training set (9000 symbols, 73 numerical attributes) and test set (300000 symbols) used in ABBYY OCR Technologies.

6.1 Decision Tree Construction Method

There exist various classical methods for decision tree construction. One can perform input data discretization (using either, one or another cutting criteria) and then launch C4.5 algorithm for discrete attributes. On the other hand, modification of C4.5 algorithm for continuous input data can be used. Perhaps, it is better to use concept lattice instead of decision tree. That is the reason why the first experiment aimed at the comparison of classification accuracies of the approaches (Table 1).

Therefore, decision tree built directly on numerical data shows the best result. Concept lattice classifier seems to be appropriate only for little training sets [4].

Table 1. Recognition rates of tree-like classifiers

Algorithm	Cutting criteria	Recognition rate (%)
decision tree (discretization)	entropy	87.63
decision tree (discretization)	Hotelling coefficient	87.98
decision tree (no discretization)	**entropy**	**89.94**
decision tree (no discretization)	Gini index	88.32
concept lattice		89.60

Probably the reason is that it is difficult to find a priori logical relationships between attributes in case when training set contains thousands of objects.

Notice that all classical decision tree algorithms do not reach required 95% recognition rate, but further we deal with their modifications discussed above, which demonstrate higher recognition rate.

6.2 Decision Tree Hypotheses Generation

Let us compare methods for decision tree hypotheses generation using best decision tree algorithm from previous experiment. Recognition rates of these algorithms are shown in Fig. 2. For post-fuzzification algorithm, N maximal number of tree paths maintained by beam search algorithm. For random forest, N is a number of voting decision trees.

Random forest demonstrates the best hypotheses generation, post-fuzzification results are worse and almost independent from hypotheses count.

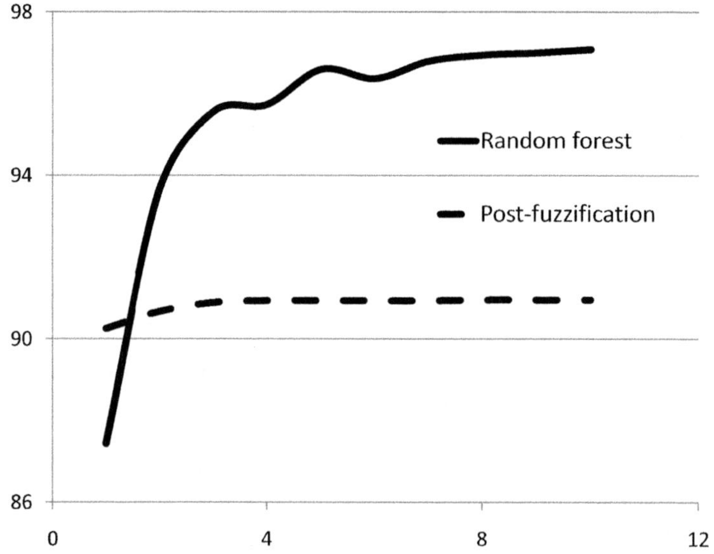

Fig. 2. Decision tree hypotheses generation comparison

6.3 Recognition Rate Comparison

To compare recognition rates of considered algorithms, we measure percent of cases, when right hypothesis appears among N first ones generated by algorithm. This measure is defensible, because differential classifier which operates after numerical classifier will possibly choose right hypothesis from any position in the hypotheses list.

Experimental results are shown in Fig. 3.

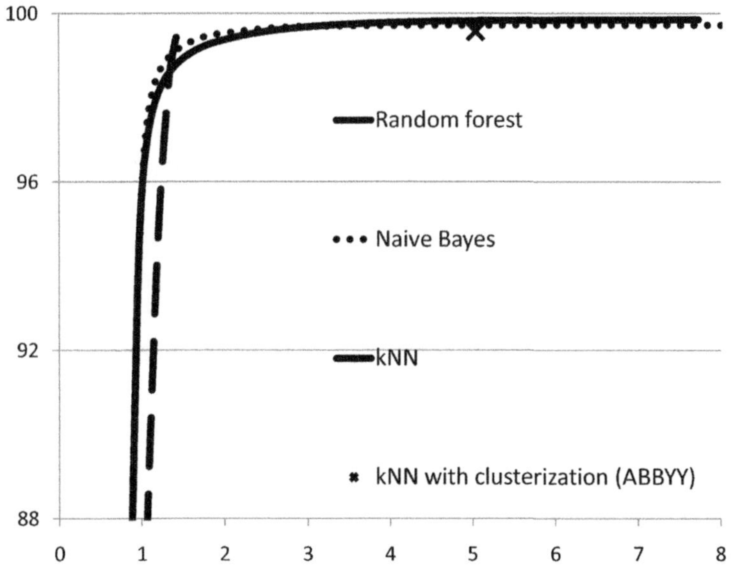

Fig. 3. Classifiers recognition rate comparison

Therefore, the best results for all values of N are shown by random forest classifier. For high lengths of hypothesis list, high recognition rate is also demonstrated by naive Bayes classifier.

6.4 Ability to Organize Cut-Offs Efficiently

Trying to estimate ability to organize cut-offs of various classifiers quantitatively, we encounter following problem. An obvious way to organize cut-offs is to specify the boundary value of confidence and to reject hypotheses with the confidence less than this boundary value. But every classification algorithm uses its own method to define confidence, so such confidences are incomparable. Our proposal is to avoid direct comparison of confidences, considering following parametric dependence. To characterize the strength of cut-off, we consider the average length of hypotheses list $N(p)$. To characterize recognition rate, we consider the percent of cases when right hypothesis appears in hypotheses list.

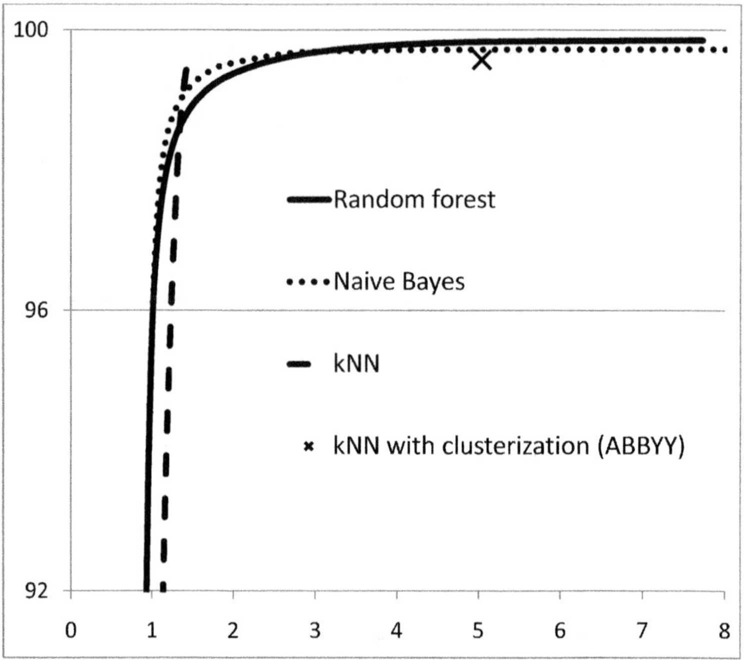

Fig. 4. Cut-offs organization comparison

Thus, for short hypotheses list, the best results are demonstrated by naive Bayes algorithm. But for long hypotheses list best recognition rate is shown by random forest classifier.

7 Conclusion

Based on our experiments, we can propose modifications of naive Bayes classifier, nearest neighbor classifier (with clusterization) and random forest classifier, which generate first hypothesis with more than 95% precision and a list of 8 hypotheses with 99% precision. The best recognition rate is reached by the random forest algorithm (more than 97%). This algorithm also demonstrates best ability to organize cut-offs on high average hypotheses list lengths.

References

1. Mitchell, T. M.: Machine Learning. McGraw-Hill, New York (1997)
2. Tereschenko, V.V.: Development and realization of new principles of handprint documents automatic recognition for computer systems. Moscow (1999) [in Russian]
3. Hastie, T., Tibshirani, R., Friedman, J.: The elements of statistical learning. Springer-Verlag, New York (2001)

4. Itskovich, L.A., Kolotienko, S.S., Kuznetsov, S.O.: Character recognition using concept lattices: realization and comparison with the other approaches. Proceedings of 51th MIPT scientific conference, Moscow (2008) [in Russian]
5. Quinlan, J.R.: C4.5: Programs for Machine Learning, Morgan Kaufmann, Los Altos, CA (1993)
6. Nguyen, H.S., Nguyen S.H.: Fast split selection method and its application in decision tree construction from large databases. International Journal of Hybrid Intelligent Systems 2, 149–160 (2005)
7. Russell, S., Norvig, P.: Artificial Intelligence: A Modern Approach. Prentice Hall (2010)
8. Breiman, L.: Random forests. Machine Learning 45/1, 5–32 (2001)

A Liouville-Based Approach for Discrete Data Categorization

Nizar Bouguila

Concordia Institute for Information Systems Engineering
Concordia University, Montreal, Canada, Qc, H3G 2W1
bouguila@ciise.concordia.ca

Abstract. In this paper, we describe a learning approach based on the smoothing of multinomial estimates using Beta-Liouville distributions. Like the Dirichlet, the Beta-Liouville is conjugate to the multinomial. It has, however, an important advantage which is its more general covariance matrix. Empirical results indicate that the proposed approach outperforms previous smoothing techniques based mainly on the Dirichlet distribution.

1 Introduction

The phenomenal growth of internet has resulted in the availability of huge amount of data composed of natural language texts, images and videos. A highly desirable objective is the automatic organization and modeling of this content. Several approaches have been proposed in the past. A common important step in all these approaches is the transformation of these data into feature vectors representations that can be used by learning algorithms. In many applications, these feature vectors are discrete and describe the frequency of features (ex. frequency of words in a given text or visual words in a given image) [10].

Various statistical techniques have emerged to meet the needs of scientific workers dealing with discrete data. Finite mixture models are among the most widely used techniques [4]. An important problem in this case is the choice of an appropriate probability density function to model the data. Several studies have shown that the widely used Gaussian distribution, based on asymptotic normality assumption, is inappropriate for discrete data [14]. The multinomial is then generally used as an alternative. This assumption has, however, several drawbacks especially in the case of rare features since it is based directly on the counts. The most widely used approach to overcome this problem is to use the Dirichlet as a prior to the multinomial to smooth the multinomial parameters estimates [11,8]. But, even this well-accepted technique has its own drawbacks. Indeed, although some success has been reported, there has also been criticism point out that this approach may not behave well when the covariance structure of the parameters is not negative.

The goal of this paper is to present another alternative based on the Liouville family of distributions from which we extract the Beta-Liouville. Like the Dirichlet, the Beta-Liouville is conjugate to the multinomial, yet it has a more general

S.O. Kuznetsov et al. (Eds.): RSFDGrC 2011, LNAI 6743, pp. 330–337, 2011.

covariance structure which makes it more useful in real-life applications. This fact is shown through an application involving automatic objects categorization.

The rest of this paper is organized as follows. In Section 2 we present our smoothing model. Experimental results are presented in Section 3. Finally, we give our conclusion in Section 4.

2 The Smoothing Model

2.1 Background

Given a set of N frequency (or count) vectors $\mathcal{X} = \{\boldsymbol{X}_1, \ldots, \boldsymbol{X}_N\}$ representing N textual (or visual) documents where $\boldsymbol{X}_n = (X_{n1}, \ldots, X_{nV})$, X_{nv} denotes the frequency of feature (e.g. word, visual word, etc) w_v in document n among the set of features (e.g. vocabulary) $\mathcal{V} =< w_1, \ldots, w_V >$. V denotes the total number of features (e.g. total number of words in the vocabulary). A given vector $\boldsymbol{X} \in \mathcal{X}$ is generally considered to have a multinomial distribution with parameters $\boldsymbol{\pi} = (\pi_1, \ldots, \pi_{V-1})$:

$$p(\boldsymbol{X}|\boldsymbol{\pi}) \propto \prod_{v=1}^{V} \pi_v^{X_v} \tag{1}$$

where $\pi_v > 0$ denotes the probability of observing the particular v^{th} feature w_v in the document represented by $\boldsymbol{X}$, and $\pi_V = 1 - \sum_{v=1}^{V-1} \pi_v$.

Consider the task of estimating the parameters over the set $\mathcal{X}$. Using only the frequencies, we obtain the following:

$$\hat{\pi}_v = \frac{X_v}{\sum_{v=1}^{V} X_v} \qquad v = 1, \ldots, V \tag{2}$$

Many studies, however, have shown that this estimator is "poor" especially in the case of large sparse data where the number of features is high. In this case the frequencies can be small and then the observed proportions will tend to zero [3]. The usual approach to tackle this problem is to smooth the estimates by using the Dirichlet as a prior to the multinomial (i.e. suppose that the multinomial parameters are random variables which follow a Dirichlet distribution). As a basis for such choice is the fact that the Dirichlet is conjugate to the multinomial which gives us the following smoothed estimates [10]:

$$\hat{\pi}_v = \frac{X_v + \alpha_v}{\sum_{v=1}^{V}(X_v + \alpha_v)} \tag{3}$$

where $(\alpha_1, \ldots, \alpha_V)$ is the vector of hyperparameters (i.e. the parameters of the Dirichlet taken as a prior). Smoothing approaches based on Dirichlet priors have several main weaknesses. Indeed, in spite of its flexibility and the fact that it is conjugate to the multinomial, the Dirichlet has a very restrictive negative covariance matrix which violates generally experimental observations in practical situations [6,2]. Then, it is necessary to postulate a plausible prior for the multinomial parameters.

2.2 The Model

If a vector $\boldsymbol{\pi} = (\pi_1, \ldots, \pi_{V-1})$ has a $(V-1)$-variate Beta-Liouville distribution with positive parameters $\theta = (\alpha_1, \ldots, \alpha_{V-1}, \alpha, \beta)$, then [15]:

$$p(\boldsymbol{\pi}|\theta) = \frac{\Gamma(\sum_{v=1}^{V-1} \alpha_v)\Gamma(\alpha+\beta)}{\Gamma(\alpha)\Gamma(\beta)} \prod_{v=1}^{V-1} \frac{\pi_v^{\alpha_v - 1}}{\Gamma(\alpha_v)} (\sum_{v=1}^{V-1} \pi_v)^{\alpha - \sum_{v=1}^{V-1} \alpha_v} (1 - \sum_{v=1}^{V-1} \pi_v)^{\beta - 1} \tag{4}$$

It is worth pointing out that the Beta-Liouville is reduced to the Dirichlet when $\alpha = \sum_{v=1}^{V-1} \alpha_v$ and $\beta = \alpha_V$. Let us assume that $\boldsymbol{\pi}$ follows a finite Beta-Liouville mixture [5]:

$$p(\boldsymbol{\pi}|\Theta) = \sum_{j=1}^{M} p_j p(\boldsymbol{\pi}|\theta_j) \tag{5}$$

where $p(\boldsymbol{\pi}|\theta_j)$ is a Beta-Liouville distribution with parameters θ_j, $\{p_j\}$ is the set of mixing parameters which are positive and sum to one, and $\Theta = \{\{p_j\}, \{\theta_j\}\}$. Having this mixture as a prior, the joint distribution of $\boldsymbol{X}$ and $\boldsymbol{\pi}$ is

$$p(\boldsymbol{X}, \boldsymbol{\pi}|\Theta) \propto \sum_{j=1}^{M} p_j \Bigg[\frac{\Gamma(\sum_{v=1}^{V-1} \alpha_{jv})\Gamma(\alpha_j+\beta_j)}{\Gamma(\alpha_j)\Gamma(\beta_j)} \prod_{v=1}^{V-1} \frac{\pi_v^{\alpha_{jv}+X_v-1}}{\Gamma(\alpha_{jv})} \tag{6}$$
$$\times (\sum_{v=1}^{V-1} \pi_v)^{\alpha_j - \sum_{v=1}^{V-1} \alpha_{jv}} (1 - \sum_{v=1}^{V-1} \pi_v)^{\beta_j + X_V - 1} \Bigg]$$

Then, it is easy to show that the marginal is

$$p(\boldsymbol{X}|\Theta) \propto \sum_{j=1}^{M} p_j \frac{\Gamma(\sum_{v=1}^{V-1} \alpha_{jv})\Gamma(\alpha_j+\beta_j)}{\Gamma(\alpha_j)\Gamma(\beta_j)\prod_{v=1}^{V-1}\Gamma(\alpha_{jv})} \frac{\Gamma(\alpha'_j)\Gamma(\beta'_j)\prod_{v=1}^{V-1}\Gamma(\alpha'_{jv})}{\Gamma(\sum_{v=1}^{V-1}\alpha'_{jv})\Gamma(\alpha'_j+\beta'_j)} \tag{7}$$

where $\alpha'_{jv} = \alpha_{jv} + X_v$, $\alpha'_j = \alpha_j + \sum_{v=1}^{V-1} X_v$ and $\beta'_j = \beta_j + X_V$. Having the joint and marginal distributions in hand, we can show that π_v can be estimated as follows:

$$\hat{\pi}_v = \sum_{j=1}^{M} p(j|\boldsymbol{X}) \frac{\alpha'_j}{\alpha'_j + \beta'_j} \frac{\alpha'_{jv}}{\sum_{v=1}^{V-1} \alpha'_{jv}} \qquad v = 1, \ldots, V-1 \tag{8}$$

$$\hat{\pi}_V = 1 - \sum_{v=1}^{V-1} \hat{\pi}_v \tag{9}$$

where

$$p(j|\boldsymbol{X}) = \frac{p_j \frac{\Gamma(\sum_{v=1}^{V-1} \alpha_{jv})\Gamma(\alpha_j+\beta_j)}{\Gamma(\alpha_j)\Gamma(\beta_j)\prod_{v=1}^{V-1}\Gamma(\alpha_{jv})} \frac{\Gamma(\alpha'_j)\Gamma(\beta'_j)\prod_{v=1}^{V-1}\Gamma(\alpha'_{jv})}{\Gamma(\sum_{v=1}^{V-1}\alpha'_{jv})\Gamma(\alpha'_j+\beta'_j)}}{\sum_{j=1}^{M} p_j \frac{\Gamma(\sum_{v=1}^{V-1} \alpha_{jv})\Gamma(\alpha_j+\beta_j)}{\Gamma(\alpha_j)\Gamma(\beta_j)\prod_{v=1}^{V-1}\Gamma(\alpha_{jv})} \frac{\Gamma(\alpha'_j)\Gamma(\beta'_j)\prod_{v=1}^{V-1}\Gamma(\alpha'_{jv})}{\Gamma(\sum_{v=1}^{V-1}\alpha'_{jv})\Gamma(\alpha'_j+\beta'_j)}} \tag{10}$$

and can be viewed as the posterior probability that the vector $\boldsymbol{X}$ will be assigned to cluster j when the marginal distribution $p(\boldsymbol{X}|\Theta)$ in Eq. 7 is taken as the parent distribution to model the data. Note that when $M = 1$, Eq. 8 is reduced to

$$\hat{\pi}_v = \frac{\alpha'}{\alpha' + \beta'} \frac{\alpha'_v}{\sum_{v=1}^{V-1} \alpha'_v} \tag{11}$$

Finally, it is noteworthy that Eq. 11 is itself reduced to Eq. 3 if we take $\alpha = \sum_{v=1}^{V-1} \alpha_v$ and $\beta = \alpha_V$.

2.3 Model Learning

According to Eq. 8 the smoothing of the multinomial parameters requires the estimation of $p(j|\boldsymbol{X})$, α_j, β_j and α_{jv}. Traditionally, the estimation of finite mixture models has been based on the maximum likelihood approach:

$$\max_{\Theta} \left\{ p(\mathcal{X}|\Theta) = \prod_{n=1}^{N} p(\boldsymbol{X}_i|\Theta) \right\} \tag{12}$$

In some situations, however, maximizing the likelihood is not straightforward or appropriate. In our case, for instance, the maximization of the likelihood leads to the following estimate for the p_j parameters:

$$p_j = \frac{1}{N} \sum_{n=1}^{N} p(j|\boldsymbol{X}_N) \tag{13}$$

However, a closed-form solution does not exist for the $\theta_j = (\alpha_j, \beta_j, \{\alpha_{jv}\})$ parameters. Thus, we use a Newton-Raphson approach, based on the first and second derivatives of the loglikelihood function, to estimate these parameters:

$$\theta_j^{new} = \theta_j^{old} - \left(\frac{\partial^2 \log p(\mathcal{X}|\Theta)}{\partial^2 \theta_j}\right)^{-1} \frac{\partial \log p(\mathcal{X}|\Theta)}{\partial \theta_j} \tag{14}$$

Note that one needs to have a criterion to allow a trade-off between goodness of fit and the complexity of the smoothing mixture model. Here, we use the MDL criterion [1] given by [25]

$$MDL(M) = -\log(p(\mathcal{X}|\Theta)) + \frac{1}{2} N_p \log(N) \tag{15}$$

where $N_p = M(D+3) - 1$ is the number of free parameters in the mixture model. Concerning the initialization, we use of the spherical K-means [13], rather than the well-known K-means with Euclidean distance. This choice is justified by the fact that count data lack a Euclidean structure since they are represented in terms of multinomial models. The spherical K-means is applied in conjunction

[1] One may use other selection criteria (see, [9], for instance, for discussions about other selection criteria).

with the method of moments [10] based on the first and second moments of the Beta-Liouville distribution. Having the initialization algorithm and the MDL criterion in hand, the complete smoothing parameters learning algorithm can be summarized as the following:

Algorithm
For each candidate value of $M \in [M_{min}, M_{max}]$:

1. Apply the initialization algorithm.
2. E-Step: Compute the *posterior* probabilities $p(j|\boldsymbol{X}_n)$ using Eq. 10.
3. M-Step:
   (a) Update the p_j using Eq. 13.
   (b) Update the θ_j using Eq. 14.
4. Calculate the associated criterion MDL(M) using Eq. 15.
5. Select the optimal model M^* such that: $M^* = \arg\max_M MDL(M)$

3 Experimental Results

We are now ready to illustrate how to apply the learning approach developed in this paper. We consider in particular the problem of objects categorization in images. Indeed, an increasingly overwhelming quantities of images are generated everyday. A crucial problem is the analysis, modeling and categorization of these images [8,1,7]. The main goal of this section is to compare our approach to previous smoothing techniques. The majority of these techniques can be viewed actually as special cases of the Dirichlet-based smoothing such as the one proposed in [18,19] which suggests adding a $\frac{1}{2}$ count to every frequency (Jeffreys smoothing). In an earlier work, the suggestion was to add a count of one to every frequency [17] (Laplace smoothing). The same suggestions can be found in [16]. The authors in [20] have increased the counts by $\frac{1}{V}$, where V is the dimensionality of the vector (Perks smoothing).

Objects categorization involves two main phases. First, feature extraction which maps each image to a vector in high-dimensional space. Second, the clustering of the resulted vectors. Several approaches and techniques have been proposed in the past. In particular, an interesting approach based on image patches, extracted at points of interest, has been proposed in [12]. This approach that we will consider here can be summarized as follows. First, up to 1000 square image patches are taken as image features and are extracted around interest points obtained using the approach described in [22]. Moreover, 300 patches are added from a uniform grid of 15×20 cells that is projected onto the image. The main goal of these added patches is to take into account the homogeneity of objects. Having the patches in hand, a PCA dimensionality reduction is applied by keeping only 40 coefficients. The resulting data are then clustered with a Linde-Buzo-Gray algorithm [21] by considering the Euclidean distance. Thus, each image patch is assigned to a cluster which allows to represent each image by a histogram of cluster frequencies (i.e. each entry in the histogram is created by counting how many patches belong to its associated cluster). As each image is now represented by a vector of counts, we can obviously assume that it is generated by a multinomial distribution which parameters can be estimated using our developed algorithm. In the following experiments we set the number of clusters to 512 (i.e. we use 512-dimensional count vectors to represent the images) and the results are averaged over 10 runs of the algorithm.

Two image databases are selected to evaluate our approach and are the Columbia Object libraries (COIL-20 and COIL-100). COIL-20 contains 1440 images of 20 objects (72 images per object) [23]. Each object is represented in the database by 72 images obtained by the rotation of the object through 360° in 5° steps. COIL-100 complete the COIL-20 with additional 80 objects (72 images per object) and consists then of 7200 images [24]. Figure 1 shows some of the 20 objects in the COIL-20 and figure 2 shows examples of images from the additional 80 objects. Both databases have been divided into disjuncts sets of 50% training and 50% test images.

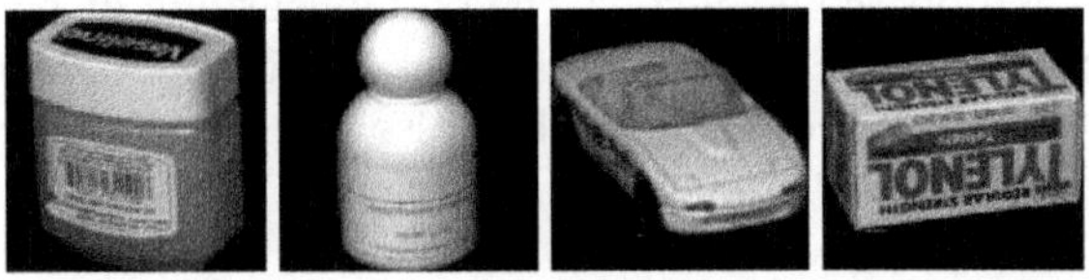

Fig. 1. Examples of images from the COIL-20 data set

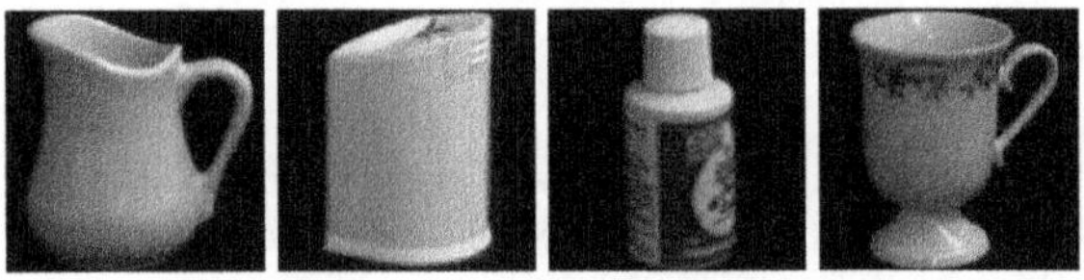

Fig. 2. Examples of images from the COIL-100 data set

Table 1 shows the recognition rates for the COIL-20 and COIL-100 databases using the multinomial with several smoothing techniques.

Table 1. Recognition rates (%) for the COIL-20 and COIL-100 databases using different smoothing methods

Method	COIL-20	COIL-100
Laplace	83.11 ± 0.32	81.07 ± 0.66
Jefferys	82.82 ± 0.54	80.95 ± 0.59
Perks	82.79 ± 0.86	80.67 ± 0.54
Dirichlet	84.26 ± 0.65	83.28 ± 0.71
Beta-Liouville	87.09 ± 0.63	86.22 ± 0.64

According to the categorization results, we can see clearly that Beta-Liouville smoothing performs better than the other approaches which can be explained by the flexibility of this model and by the fact that the other smoothing techniques are actually just special cases.

4 Conclusion

A new smoothing technique for multinomial parameters estimation has been proposed in this paper. The proposed approach is based on the Beta-Liouville distribution which general covariance nature together with its conjugacy to the multinomial make it an attractive alternative to the Dirichlet. The proposed model is illustrated by an application which involves objects categorization. According to this application, we find that our smoothing approach performs better than other previous approaches. The model is capable of two forms of extension of practical importance. The first one could involve the introduction of feature selection to automatically detect the most important features for a given application. The second one could involve the online learning of parameters to take into account the dynamic nature of databases. Finally, it is noteworthy that the proposed framework could be applied also for other problems such as text segmentation and natural language processing.

Acknowledgment. The completion of this research was made possible thanks to the Natural Sciences and Engineering Research Council of Canada (NSERC).

References

1. Bouguila, N.: Spatial Color Image Databases Summarization. In: IEEE International Conference on Acoustics, Speech, and Signal Processing (ICASSP), Honolulu, HI, USA, vol. 1, pp. 953–956 (2007)
2. Bouguila, N.: Clustering of count data using generalized Dirichlet Multinomial distributions. IEEE Trans. Knowledge and Data Engineering 20(4), 462–474 (2008)
3. Bouguila, N.: A Model-Based Approach for Discrete Data Clustering and Feature Weighting Using MAP and Stochastic Complexity. IEEE Transactions on Knowledge and Data Engineering 21(12), 1649–1664 (2009)
4. Bouguila, N.: Count Data Modeling and Classification Using Finite Mixtures of Distributions. IEEE Transactions on Neural Networks 22(2), 186–198 (2011)
5. Bouguila, N.: Bayesian Hybrid Generative Discriminative Learning Based on Finite Liouville Mixture Model. Pattern Recognition 44(6), 1183–1200 (2011)
6. Bouguila, N., ElGuebaly, W.: On Discrete Data Clustering. In: Washio, T., Suzuki, E., Ting, K.M., Inokuchi, A. (eds.) PAKDD 2008. LNCS (LNAI), vol. 5012, pp. 503–510. Springer, Heidelberg (2008)
7. Bouguila, N., ElGuebaly, W.: A Generative Model for Spatial Color Image Databases Categorization. In: Proc. of the IEEE International Conference on Acoustics, Speech, and Signal Processing (ICASSP), Las Vegas, Nevada, USA, pp. 821–824 (2008)
8. Bouguila, N., Ziou, D.: Improving Content Based Image Retrieval Systems Using Finite Multinomial Dirichlet Mixture. In: Proc. of the IEEE Workshop on Machine Learning for Signal Processing (MLSP 2004), Sao Luis, Brazil, pp. 23–32 (2004)

9. Bouguila, N., Ziou, D.: Unsupervised Selection of a Finite Dirichlet Mixture Model: An MML-Based Approach. IEEE Transactions on Knowledge and Data Engineering 18(8), 993–1009 (2006)
10. Bouguila, N., Ziou, D.: Unsupervised Learning of a Finite Discrete Mixture: Applications to Texture Modeling and Image Databases Summarization. Journal of Visual Communication and Image Representation 18(4), 295–309 (2007)
11. Bouguila, N., Ziou, D., Vaillancourt, J.: Novel Mixtures Based on the Dirichlet Distribution: Application to Data and Image Classification. In: Perner, P., Rosenfeld, A. (eds.) MLDM 2003. LNCS(LNAI), vol. 2734, pp. 172–181. Springer, Heidelberg (2003)
12. Deselaers, T., Keysers, D., Ney, H.: Discriminative Training for Object Recognition Using Image Patches. In: Proc. of the IEEE Conference on Computer Vision and Pattern Recognition (CVPR), pp. 157–162 (2005)
13. Dhillon, I.S., Modha, D.S.: Concept Decompositions for Large Sparse Text Data Using Clustering. Machine Learning 42(1-2), 143–175 (2001)
14. Dunning, T.: Accurate Methods for the Statistics of Surprise and Coincidence. Computational Linguistics 19(1), 61–74 (1993)
15. Fang, K.T., Kotz, S., Ng, K.W.: Symmetric Multivariate and Related Distributions. Chapman and Hall, New York (1990)
16. Gart, J.J., Zweifel, J.R.: On the Bias of Various Estimators of the Logit and its Variance with Application to Quantal Biossay. Biometrika 54(1/2), 181–187 (1967)
17. Goodman, L.A.: Interactions in Multidimensional Contingency Tables. The Annals of Mathematical Statistics 35(2), 632–646 (1964)
18. Goodman, L.A.: The Multivariate Analysis of Qualitative Data: Interactions among Multiple Classifications. Journal of the American Statistical Association 65(329), 226–256 (1970)
19. Goodman, L.A.: The Analysis of Multidimensional Contingency Tables: Stepwise Procedures and Direct Estimation Methods for Building Models for Multiple Classifications. Technometrics 13(1), 33–61 (1971)
20. Grizzle, J.E., Starmer, C.F., Koch, G.G.: Analysis of Categorical Data by Linear Models. Biometrics 25(3), 489–504 (1969)
21. Linde, Y., Buzo, A., Gray, R.M.: An Algorithm for Vector Quantization Design. IEEE Tranactions on Communications 28, 84–95 (1980)
22. Loupias, E., Sebe, N., Bres, S., Jolion, J.: Wavelet-Based Salient Points for Image Retrieval. In: Proc. of the IEEE International Conference on Image Processing (ICIP), pp. 518–521 (2000)
23. Nene, S.A., Nayar, S.K., Murase, H.: Columbia Object Image Library (COIL-20). Technical Report CUCS-005-96, Columbia University (1996)
24. Nene, S.A., Nayar, S.K., Murase, H.: Columbia Object Image Library (COIL-100). Technical Report CUCS-006-96, Columbia University (1996)
25. Rissanen, J.: Modeling by Shortest Data Description. Automatica 14, 465–471 (1978)

Image Recognition with a Large Database Using Method of Directed Enumeration Alternatives Modification

Andrey V. Savchenko

National Research University – Higher School of Economics, N. Novgorod, Russia
avsavchenko@hse.ru

Abstract. A new modification of the method of directed alternatives' enumeration using the Kullback–Leibler discrimination information is proposed for half-tone image recognition.Results of an experimental study in the problem of face images recognition with a large database are presented. It is shown that the proposed modification is characterized by increased speed of image recognition (5-10 times vs exhaustive search).

1 Introduction

Processing large image databases [1] is a well-known challenging problem [2]. Traditional image recognition [3] methods based on exhaustive search [4] cannot be implemented in real-time applications. Thus method of directed enumeration of alternatives (MDEA) has been proposed [5]. The practical capabilities of our method are limited because of distances matrix containing the distances between given alternatives from the database. This matrix could be too huge to be stored in the RAM. In this paper we propose novel MDEA modification to decrease recognition complexity using the most valuable part of this matrix.

The rest of the paper is organized as follows. Section 2 introduces new MDEA modification to reduce the amount of necessary memory. In Section 3, we present the experimental results in application to faces recognition problem. Concluding comments are presented in Section 4.

2 MDEA Modification

In this paper, we use a histogram-based method [6], which applies the minimum information discrimination criterion [7]. Let a set of R half-tone images $X_r = \|x^r_{uv}\|$, $(u = \overline{1,U},\ v = \overline{1,V}, r = \overline{1,R})$ be specified. Here U and V are the image height and width, $x^r_{uv} \in \{0, 1, \ldots, x_{\max}\}$ is the intensity of an image point with coordinates (u, v); and x_{max} is the maximum intensity. It is required to assign a new input image $X = \|x_{uv}\|$ to one of the R classes.

According to approach [8], we consider a random variable - color of image X_r. Its distribution $H_r = \left[h^r_1, h^r_2, \ldots, h^r_{x_{\max}}\right]$ is known as "color histogram" [4]. Then color histogram H is defined for the input image X. It is required to verify

S.O. Kuznetsov et al. (Eds.): RSFDGrC 2011, LNAI 6743, pp. 338–341, 2011.

R hypotheses on the distribution H_r. The optimal decision in Bayesian terms is equivalent [7] to the minimum discrimination information criterion

$$\rho_{KL}\left(X/X_r\right) = \sum_{x=1}^{x_{\max}} h_x \ln\left(h_x/h_x^r\right) \to \min . \tag{1}$$

Statistic $\rho_{KL}(X/X_r)$ defines the Kullback–Leibler information discrimination [7]. Based on metric properties of discrimination (1), we first transform criterion (1) to a simplified form, suitable for practical implementation:

$$\rho_{KL}\left(X/X_r\right) < \rho_0 = const \tag{2}$$

Here ρ_0 is the threshold for the admissible discrimination on the images from one class due to their known variability.

Following the general computation scheme (1), (2), we reduce the image recognition problem to a check of the first N variants $X_1,\ldots,X_N$ from the database. Let us arrange these images in decreasing order of their discriminations (1). As a result, we have an ordered sequence of template images $\{X_{i_1}, X_{i_2} \ldots X_{i_N}\}$. This procedure is used to obtain the first local optimum X_{i_N}. In the second step, for the image X_{i_N} from the matrix $\mathrm{P} = \|\rho_{ij}\|$ of values $\rho_{ij}{=}\rho_{KL}\ (X_i/X_j)$, we find the set of $M < R$ images $X^{(M)} = \left\{X_{i_{N+1}}, \ldots X_{i_{N+M}}\right\}$:

$$\left(\forall X_i \notin X^{(M)}\right)\left(\forall X_j \in X^{(M)}\right) \Delta\rho\left(X_i\right) \geq \Delta\rho\left(X_j\right) \tag{3}$$

where

$$\Delta\rho(X_j) = \left|\rho_{KL}\left(X_j/X_i\right) - \rho_{KL}\left(X/X_{i_N}\right)\right| \tag{4}$$

We add one more $(M+1)$-th element $X_{i_{N+M+1}}$ that did not fall in the control sample in the previous computation step. As a result, for the analysis we obtain the second sample $\left\{X_{i_1},, \ldots, X_{i_N}, \ldots X_{i_{N+M+1}}\right\}$. Next, all computations of the first step are repeated cyclically until, in some step, an element X^* satisfies condition (2).

Generally, there may be a considerable gain in the total number of checks (1) compared to the database size R. It's explained by the fact that probability p of desired image X^* containing in$X^{(M)}$, usually exceeds the probability of belonging X^* to M alternatives for random choice

$$p = P\left\{X^* \in X^{(M)}\right\} >> p_0 = M/R \tag{5}$$

Actually, the probability p (5) should depend also on the distance between X and X_{i_N}. We could assume that image X_{i_N} contains valuable information to obtain X^* if it's closer (or further) to object X, than the majority of other images from database. To show this fact, we measure the dependence of p from $\rho_{KL}\left(X/X_{i_N}\right)$ for large faces database [9]. The 6000 photographs of 400 different people were selected as templates $R = 900$ of the most different images using clusterization [4].

Parameter M was fixed to 64. Dependence of probability p on the discrimination $\rho_{KL}\left(X/X_{i_N}\right)$ is shown at Fig.1. Based on it we suppose that though

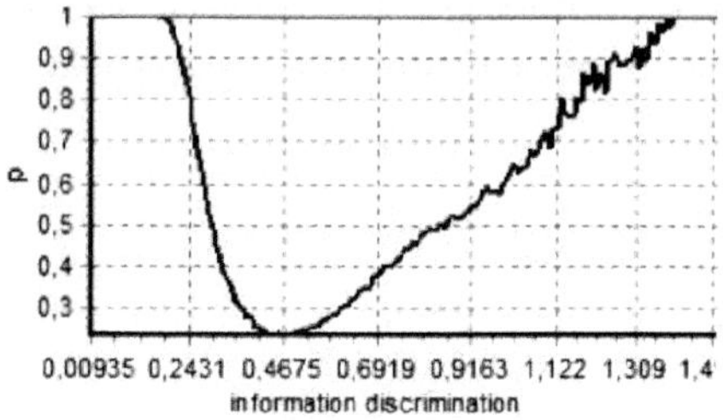

Fig. 1. Dependence of Probability p on $\rho_{KL}(X/X_{i_N})$

minimum probability p (0.25 based on this graph) is quite greater than random search success probabilityp_0=*M*/*R*=*64*/*900*=*0.07*, the most valuable (in terms of further computations decrease) distances are concentrated in the Fig.1 "corners"

Thus we propose to store not all matrix P, but only the most T lower and the most T higher distances to reference images for each image from database. Here T=*const*<*R*/*2* is parameter of proposed modification. The ratio $2T/R$ determines the decrease of memory usage. This approach causes modification of rule (4) to select set $X^{(M)}$. If stored part of P isn't enough to determine $X^{(M)}$ based on (4),(5) we just miss this step and select randomly one of image from database until procedure (4),(5) could be applied.

3 Experimental Results

The face recognition procedure was tested for large database [8]. The photos were preliminary processed to detect faces using OpenCV library. Then detected faces were divided into 16 (4x4) parts for information discrimination computation. Each part was normalized [4]. Such fragmentation [10] is used to take into account heterogeneous illumination of images. The discrimination between images was calculated as a sum of discriminations (1) between these parts.

In the first case 1000 test images and the Kullback–Leibler discrimination (1) were used with the original MDEA and the following method parameters were chosen: $N = 9$ and $M = 32$, $\rho_0 = 0.19$. Using the MDEA (2)–(6), we obtained an average number of discrimination (1) calculation equal to 11.1% of R. In this case, condition (2) was not satisfied for any template from the given database for 6.9% of the test images; therefore, all R alternatives were checked. The recognition accuracy is 98.1%.

In the second case, proposed modification was used and the parameter $T = 32$ was chosen. This approach shows practically the same result as for the previous experiment. With a probability of 90%, the number of template images to check does not exceed 15% of R. The error probability increases a bit (to 2.3%) with 12.8% of average number of checks. The increase of latter factor to less than 2% is appropriate to the most applications as we achieved memory economy in $2\times32/900\times100\%\approx7\%$. I.e. the proposed modification needs $900\times2\times32\times(4+8)/1024=675$ Kb additional RAM (in comparison with 9.27 Mb

RAM for original implementation). The memory to store whole database (counting for 1 byte per pixel) is approximately 8Mb.

At the end, MDEA was used with conventional l_1 metric to compare pixels, instead of information discrimination (1). The error probability increases to 4.5%, and the average number of distance calculation was 21% of R. And again, proposed modification achieves the same accuracy with 23.5% distance calculations and even 4Mb RAM as l_1 is a symmetric metric.

4 Conclusion

The problem of increasing the computation speed has attracted considerable interest of experts in both the theory and practice of pattern recognition. Despite a huge number of approaches, most of the algorithms compare an input image with each template image, and unavoidably cannot be implemented in real-time mode for large databases. For solution of that problem MDEA [5] may be used to reduce the computational complexity by 5-10 times. The efficiency of this method depends on the matrix of distances between given alternatives from the database. Storage of the whole matrix demands twice more RAM in comparison with the amount of memory needed for image database storage This paper showed that proposed modification overcomes this limitation both for information discrimination [7] and conventional l_1 criterions [4].

References

1. Jia, Z., Amselang, L., Gros, P.: Content-based image retrieval from a large image database. Pattern Recognition 11(5), 1479–1495 (2008)
2. Russ, J.: The Image Processing Handbook, 5th edn. CRC Press, Boca Raton (2007)
3. Rui, Y., Huang, T., Chang, S.F.: Image retrieval: current techniques, promising directions and open issues. Visual Communication and Image Representation 10, 39–62 (1999)
4. Theodoridis, S., Koutroumbas, C.: Pattern Recognition, 4th edn. Elsevier, Amsterdam (2009)
5. Savchenko, A.V.: Method of directed enumeration of alternatives in the problem of automatic recognition of half-tone images. Optoelectronics, Instrumentation and Data Processing 45(3), 83–91 (2009)
6. Santini, S.: Exploratory Image Databases: Content-Based Retrieval. Academic Press, London (2001)
7. Kullback, S.: Information Theory and Statistics. Dover Pub., New York (1978)
8. Savchenko, A.V.: Image retrieval using minimum information discrimination criterion. In: The Proc. of IASTED ACIT-CDA, Novosibirsk, pp. 345–349 (2010)
9. Essex Faces database, `http://cswww.essex.ac.uk/mv/allfaces/index.html`
10. Pedrycz, W., Kreinovich, V., Skowron, A. (eds.): Handbook of Granular Computing. Wiley, Chichester (2008)

Comparators for Compound Object Identification

Łukasz Sosnowski[1,2] and Dominik Ślęzak[3,4]

[1] Systems Research Institute, Polish Academy of Sciences
ul. Newelska 6, 01-447 Warsaw, Poland
[2] Dituel Sp. z o.o.
ul. Ostrobramska 101 lok. 206, 04-041 Warsaw, Poland
[3] Institute of Mathematics, University of Warsaw
ul. Banacha 2, 02-097 Warsaw, Poland
[4] Infobright Inc.
ul. Krzywickiego 34 lok. 219, 02-078 Warsaw, Poland
l.sosnowski@dituel.pl, slezak@infobright.com

Abstract. We discuss theoretical foundations and practical implementation of the compound object identification methodology based on information granules, fuzzy relations, and the architecture of comparators. We report its application in the commercial project aimed at visualization of the Polish Self-Government Elections in 2010, where one of the main technical challenges was to be able to identify the administrative areas by basing on their imprecise images.

Keywords: Comparators, Fuzzy Sets, Image Analysis, Granulation.

1 Introduction

Generally, the paper deals with a task of image-related spatial object identification based on available finite set of already known objects (see e.g. [2]).

The process of identification can be designed in many ways. One may measure some properties, e.g., the lengths of a quadrangle's edges, and identify it as a square if the lengths are positive and equal to each other. This simple example illustrates the following two stages: obtaining information about the object on the basis of various kinds of techniques and, secondly, interpreting the results of the first stage (see e.g. [1]). The second phase does not refer to the object but only to its measured values, which are further subject to transformation and logical interpretation. Measures respond to some questions formulated as logical sentences that lead through a set of rules towards the final identification.

Another methodology is to demonstrate the identity to an already known reference object [3]. For the above example, it would mean finding the identical quadrangle known as a square and consequently deciding that the investigated object is a square as well. The objects' identity may be examined relatively to the given scopes. Various scopes can be analyzed using various features extracted analogously to the above-discussed rule-based approach, although they are now used for the object comparisons rather than for the logical interpretation.

S.O. Kuznetsov et al. (Eds.): RSFDGrC 2011, LNAI 6743, pp. 342–349, 2011.

Regardless of whether objects are compared to each other or matched against some rules, it is impossible to expect such comparisons/matchings to provide the exact outcomes. Actually, in such areas as compound object identification or recognition, there is a need to work with similarities at the level of both whole objects and their particular components (see e.g. [6]). In our previous research (see [8] for further references), we investigated various applications of compound object comparators based on similarities modeled using fuzzy relations [4].

A comparator is a logical structure responsible for the process of comparing objects by means of predefined features. Its performance depends on the choice of reference objects and the search method. It can also refer to the repository of forbidden objects and features. If object b is not forbidden for object a, we compute fuzzy membership of (a, b) to a fuzzy similarity relation. Definition of membership can be adjusted to reflect a general similarity within a given object class or, e.g., similarity of some specific aspects of objects. Membership can be represented as a function $\mu : R \times R \to [0, 1]$. The degree of similarity can be further treated as an input to an activation function $f : [0, 1] \to \{0, 1\}$, with a threshold adjusted according to the expert knowledge or, e.g., some heuristic optimization process. As a result, for each input object, we can get: a) no reference objects (because of forbidden features or not exceeding the activation threshold for μ), b) exactly one reference object, or c) multiple reference objects [8].

In this paper, we report a usage of object identification methodology in the project aimed at visualization of the results of the Polish Self-Government Elections in 2010.[1] The task included color-based presentation of attendance in the administrative areas of Poland, such as provinces, counties, and communes. There were three sources of input data: 1. Attendance information for every commune; 2. Contour of every commune; 3. The map of Poland divided onto communes. Attendance results and contours were labeled with the communes' administrative codes. However, because of the project limitations, those codes were not present at the map of Poland. Thus, there was a need to identify communes on the map. As manual identification was out of the question and the quality of images extracted from the map did not allow for exact matching (different scales, resolution, etc.), the implementation was based on the above-discussed comparators, where the input commune images constituted the repository of reference objects and the images extracted from the map of Poland were treated as the objects to be identified. We refer to Figure 1 for illustration.

The paper is organized as follows: Section 2 shows how to extract the reference images from the map of Poland. Section 3 explains how to granulate images in order to approximate their features. In particular, we show how to express the coverage of granules by the investigated areas, as well as how to find extreme points in each of granules and encode directions of lines connecting those points as a string. Section 4 discusses how to introduce a fuzzy relation over granulated representations of images. Section 5 presents an illustrative example and final results. Section 6 describes some of our future research directions.

[1] wybory2010.pkw.gov.pl/att/1/eng/000000.html

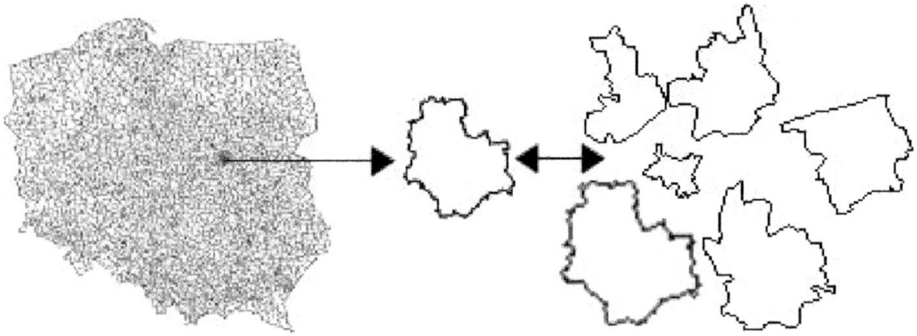

Fig. 1. The administrative area identification schema. The arrows correspond to particular stages of the process outlined in Sections 2-4.

2 Data Segmentation

The first phase of our application aims at extracting the images of particular administrative areas of Poland. Usually, they are connected. Otherwise, further steps involve comparisons of the connected subareas in order to identify the whole areas. There are also cases of areas contained in another areas. Surely, all such situations need to be detected and appropriately processed.

At the beginning, we need a map of the administrative divisions of Poland. We can assume that the map has two colors: RGB(255,255,255) for the inside area and RGB(0,0,0) for the borders. The first step is to eliminate the area outside the borders of Poland. We use the *flood fill*[2] method to paint the external area with RGB(125,125,125). Then we compute the standard histogram [7]. Its score for brightness 255 indicates how many pixels are not painted.

The next step is repeated until there are no RGB(255,255,255) pixels left. We start from the left side of the map and choose a fixed number of coordinates of RGB(255,255,255) pixels. For each chosen pixel, we use *flood fill* to repaint the corresponding (sub)area to RGB(50,50,50). Then we read the newly colored pixels, add the single-pixel border, and save the new image with the coordinates of the corresponding pixel in its filename. The image stores the smallest polygon that can cover the given area. The polygon's size is $[0, w] \times [0, h]$, where w and h denote the area's width and height. The area is in RGB(250,250,250), the border is in RGB(0,0,0), and the rest of polygon is in RGB(255,255,255).

After extraction of each new image file, we repaint RGB(50,50,50) pixels on the main map to RGB(125,125,125) and partially recompute the histogram.[3] The termination condition – no RGB(255,255,255) pixels left on the map – means that there are no more areas to extract. This finishes the segmentation phase and defines the input to the next steps, as illustrated in Figure 2.

[2] en.wikipedia.org/wiki/Flood_fill

[3] The map's fragment to be recomputed is decided using an additional algorithm.

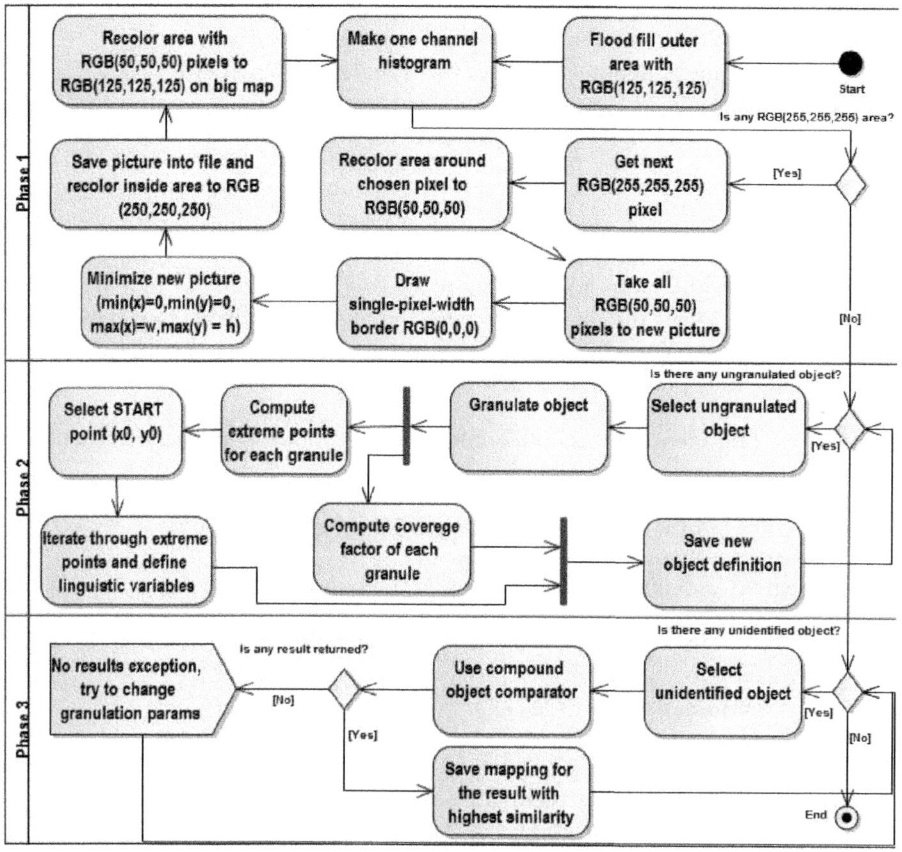

Fig. 2. Activity diagram of the described algorithm. We refer to [8] for more details about the third phase, at the level of arbitrary compound objects.

3 Image Granulation

Once the areas are isolated, we need a layer that describes them in a way convenient for imprecise comparisons. This section outlines how to construct such a layer by means of information granules [6]. In Section 4, we show how to use the obtained descriptions to conduct the identification process.

Resolution of every image is parameterized by integers m and n, such that $0 < m < w$ and $0 < n < h$. This means dividing the image onto $n \times m$ granules. Parameters m and n can be chosen based on the expert knowledge or tuned experimentally. They have a significant impact on the quality of the process. If m and n are too high, the algorithm may not find a sufficiently good solution. If they are too low, we can get many equivalent solutions. Resolution may vary for each of images and, actually, it can be recomputed dynamically if the images are stored appropriately (see Section 6; cf. [8]). However, when comparing two images, we should set up the same m and n for both of them.

For a given image, denote the set of its granules by $G = \{g_1, \ldots, g_{n\times m}\}$. Each granule corresponds to a rectangular subsurface of the image's polygon. Using granules, we can approximate various aspects of images, such as size, proportions, or shape. In this paper, we focus on the following features:

Coverage. The idea is to compute the degrees of granules' overlap with the area represented by the given image. We can easily do it by computing each granule's histogram and reading its score for brightness 250. For image a and its granule $g_i \in G$, let us divide this score by the number of pixels in g_i and denote the result as cov_i^a. We will use such coefficients in the next section in order to define similarities between the pairs of images.

Contour. The idea is to choose some extreme points and connect them. For every $g_i \in G$, extreme points look as follows:

$$(x^i_{\leftarrow}, y^i_{\leftarrow}), \quad (x^i_{\rightarrow}, y^i_{\rightarrow}), \quad (x^i_{\uparrow}, y^i_{\uparrow}), \quad (x^i_{\downarrow}, y^i_{\downarrow}) \tag{1}$$

with coordinates defined over C_i, which is the contour of g_i:

$$\begin{array}{llll} x^i_{\leftarrow} &= \min\{x : (x,y) \in C_i\} & y^i_{\leftarrow} &= \max\{y : (x,y) \in C_i, x = x^i_{\leftarrow}\} \\ x^i_{\rightarrow} &= \max\{x : (x,y) \in C_i\} & y^i_{\rightarrow} &= \max\{y : (x,y) \in C_i, x = x^i_{\rightarrow}\} \\ y^i_{\uparrow} &= \min\{y : (x,y) \in C_i\} & x^i_{\uparrow} &= \max\{x : (x,y) \in C_i, y = y^i_{\uparrow}\} \\ y^i_{\downarrow} &= \max\{y : (x,y) \in C_i\} & x^i_{\downarrow} &= \max\{x : (x,y) \in C_i, y = y^i_{\downarrow}\} \end{array} \tag{2}$$

The next step is to draw straight lines between the above points and describe them by some linguistic variables. For each image and its related set of granules G, consider (x_0, y_0) such that:

$$\begin{array}{ll} x_0 &= \min\{x : (x,y) \in \bigcup_i g_i\} \\ y_0 &= \max\{y : (x,y) \in \bigcup_i g_i, x = x_0\} \end{array} \tag{3}$$

Starting from (x_0, y_0), we can express directions of lines leading to each next extreme point[4] by means of variables such as *right*, *up*, *left*, *down*, *right-up*, *right-down*, etc. Going further, we can label each image with a string that is concatenation of abbreviations of particular directions. For instance, we can use two-letter codes, e.g.: *RR* for *right* and *RU* for *right-up*.

We may also use various types of so called modifiers to describe directions more or less precisely. The lengths of strings depend on the choice of m and n, as well as the applied modifiers. As already mentioned, the choice of granulation's resolution should be the same for each pair of compared objects. However, modifiers responsible for the precision of directions may differ.

In summary, the output of this phase takes the form of the set of images' descriptions, computed under specified granulation parameters.

---

[4] We choose each next extreme point clockwise, basing on the 8-point neighborhood, remembering the recently visited points in order to backtrack if necessary.

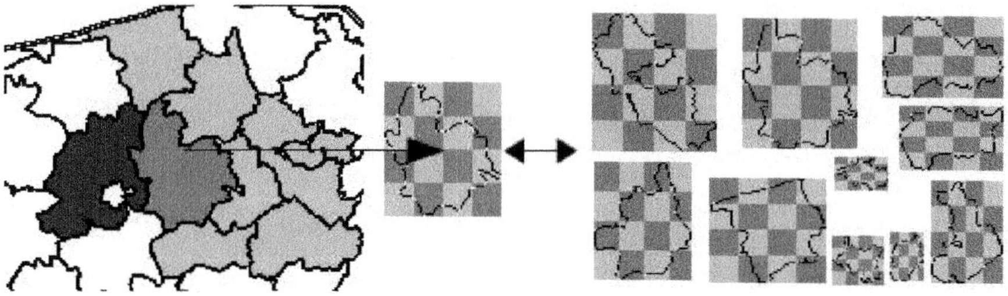

Fig. 3. The Wejherowski County (grey). One of the areas is selected for identification (dark-grey). Its granulation is compared with granulations of the reference objects (right part). We can also see an example of the area (dark) identified as including a smaller area (white) at the first phase of the algorithm.

4 Area Identification

Finally, we need to find the most similar reference object(s). Similarities should be handled differently for different representations. For linguistic descriptions of contours described in the previous section, we follow the experience of the first author with some other applications requiring string comparisons (see [8] for references) and define the following membership function:

$$\mu_{contour}(a,b) = 1 - DL(a,b) \,/\, \max(n(a), n(b)) \tag{4}$$

where $DL(a,b)$ is the Levenshtein distance[5] between linguistic descriptions of objects a, b, and $n(a), n(b)$ denote the lengths of these descriptions. One may surely consider also other measures [5]. Our application is designed in a way that enables to replace the formulas for μ easily.

With regards to the granules' coverage, we may consider the following:

$$\mu_{coverage}(a,b) = 1 - \sum_i^{n\times m} \left|cov_i^a - cov_i^b\right| \,/\, n \times m \tag{5}$$

For the purposes of this paper, we use the following aggregated similarity:

$$\mu(a,b) = \tfrac{1}{2}\left(\mu_{contour}(a,b) + \mu_{coverage}(a,b)\right) \tag{6}$$

Comparing to [8], we did not consider any forbidden features that may block comparisons of specific objects even prior to computation of memberships. On the other hand, we carefully tuned the activation threshold for μ (see Section 1). Also, we implemented an additional procedure for the following cases:

1. If some reference object was not chosen for any of investigated objects, then use it for the most similar unidentified object even if its degree of similarity is not greater than the activation threshold.
2. If some reference object was chosen for many investigated objects, then use it for the most similar of them and re-identify the remaining ones excluding the already used reference objects.

[5] en.wikipedia.org/wiki/Levenshtein_distance

Table 1. The values of $\mu_{contour}$ and $\mu_{coverage}$ for 9 communes in the Wejherowski County. The numbers in brackets denote whether correct reference objects are the most similar, 2nd most similar, etc., to particular objects to be identified.

Area	Contour	Coverage
1	0.670 (2)	0.974 (1)
2	0.735 (1)	0.953 (1)
3	0.596 (4)	0.849 (3)
4	0.632 (2)	0.972 (1)
5	0.761 (1)	0.904 (1)
6	0.676 (1)	0.970 (1)
7	0.660 (1)	0.936 (1)
8	0.628 (1)	0.888 (1)
9	0.573 (4)	0.944 (1)

5 Example and Results

For illustration purposes, let us consider a sub-map of one of the counties in Poland – the Wejherowski County (Figure 3). As a result of the first phase of the algorithm (Section 2), we obtain 9 image files to be identified.

The first image has width $w = 61$ and height $h = 69$. For parameters $m = n = 4$, after rounding 61/4 to 15 and 69/4 to 17, we obtain granules $g_1 = \{(x, y) : x \in [0, 15), y \in [0, 17)\}$, $g_2 = \{(x, y) : x \in [15, 29), y \in [0, 17)\}$, etc.

Let us now take a look at how the contour's description is built. For the analyzed image, not all extreme points are distinguished. For g_1 we obtain only two of them: $(7, 16)$ and $(14, 2)$. For g_2 we have all four: $(15, 1)$, $(17, 0)$, $(29, 15)$, and $(28, 16)$. This shows that the corresponding strings of directions can vary in length and the formulas for $\mu_{contour}$ need to take it into account.

Table 1 presents the results for all 9 communes in the Wejherowski County. In this case, our algorithm was 100% accurate, although the numbers reported in brackets might suggest otherwise. Out of two components of function (6), $\mu_{coverage}$ looks better. However, our tests show that using $\mu_{coverage}$ itself would provide worse results. It seems that $\mu_{coverage}$ plays the leading role but $\mu_{contour}$ contributes additionally in situations when comparator based only on $\mu_{coverage}$ would provide multiple reference objects or no reference objects at all. With this respect, modifications of (6) are on our future research roadmap.

The presented implementation enabled us to identify 338 out of 380 administrative areas of Poland. All those 338 areas were identified correctly. In order to accomplish the project related to visualization of the Polish Self-Government Elections in 2010, the remaining 42 areas were analyzed manually.

It is important to add that without an extra verification based on two rules outlined in the end of the previous section the number of unidentified areas would increase by 10. One of such cases is actually the 3rd item in Table 1. Indeed, its correct identification was possible only because other communes in the Wejherowski County were matched with high enough confidence.

6 Conclusions

We discussed the identification mechanism based on comparing information obtained by granulating and aggregating compound objects. As a case study, we presented practical implementation of our approach for the purposes of visualization software. The proposed approach is a continuation of our previous research related to comparators (see [8] for further references). It can be applied to different types of objects (not necessarily images). Surely, for other types of objects some other values will be aggregated. However, the scheme of handling the resulting granules and their aggregated descriptions remains the same.

Out of many technical details that were skipped in this paper, let us mention about data management. In our approach, compound objects are stored in the Infobright's RDBMS,[6] which is optimized with respect to large volumes of data (therefore, the objects can be represented in an extremely detailed way; e.g., images can be stored at the level of particular pixels) and analytical types of SQL statements (therefore, e.g., image granulation and comparison operations can be quickly executed with various resolution settings; [8]). Ultimately, our goal is to establish the database environment for flexible identification and recognition of various types of compound objects (images, texts, sequences, processes).

Acknowledgment. The second author was supported by the National Centre for Research and Development (NCBiR) under Grant No. SP/I/1/77065/10 by the strategic scientific research and experimental development program: "Interdisciplinary System for Interactive Scientific and Scientific-Technical Information".

References

1. Chellappa, R., Wilson, C.L., Sirohey, S.: Human and Machine Recognition of Faces: A Survey. Proceedings of the IEEE 83(5), 705–741 (1995)
2. Deb, S. (ed.) Multimedia Systems and Content-Based Image Retrieval. IGI Global (2004)
3. Duin, R.P.W., Pękalska, E.: The Dissimilarity Representation for Pattern Recognition: A Tutorial. Technical Report no. 2009_10 (2009)
4. Kacprzyk, J.: Multistage Fuzzy Control: A Model-Based Approach to Fuzzy Control and Decision Making. Wiley, Chichester (1997)
5. Navarro, G.: A guided tour to approximate string matching. ACM Comput. Surv. 33(1), 31–88 (2001)
6. Pedrycz, W., Kreinovich, V., Skowron, A. (eds.): Handbook of Granular Computing. Wiley, Chichester (2008)
7. Russ, J.: The Image Processing Handbook, 5th edn. CRC Press, Boca Raton (2007)
8. Ślęzak, D., Sosnowski, Ł.: SQL-Based Compound Object Comparators: A Case Study of Images Stored in ICE. In: Kim, T.-h., Kim, H.-K., Khan, M.K., Kiumi, A., Fang, W.-c., Ślęzak, D. (eds.) ASEA 2010. CCIS, vol. 117, pp. 303–316. Springer, Heidelberg (2010)

[6] www.infobright.org

Measuring Implicit Attitudes in Human-Computer Interactions

Andrey Kiselev[1], Niyaz Abdikeev[2], and Toyoaki Nishida[1]

[1] Dept. of Intelligence Science and Technology, Graduate School of Informatics, Kyoto University; Yoshida-Honmachi, Sakyo-ku, Kyoto 606-8501 Japan
andrewak@ii.ist.i.kyoto-u.ac.jp, nishida@i.kyoto-u.ac.jp
[2] Plekhanov Russian Academy of Economics; 6, Stremyanny lane, Moscow 116998 Russia
nabd@rea.ru

Abstract. This paper presents the ongoing project which attempts to solve the problem of measuring users' satisfaction by utilizing methods of discovering users' implicit attitudes. In the initial stage, authors attempted to use the Implicit Association Test (IAT) in order to discover users' implicit attitudes towards a virtual character. The conventional IAT procedure and scoring algorithm were used in order to find possible lacks of original method. Results of the initial experiment are shown in the paper along with method modification proposal and preliminary verification experiment.

Keywords: Embodied Conversational Agent, Human-Computer Interaction, Implicit Association Test.

1 Introduction

Measuring users' satisfaction is the integral part of the Human-Computer Interaction (HCI) field [1]. At the same time measuring user satisfaction with any computer-based electronic appliance is a very challenging task. Being often interconnected with and referred to users' attitudes, methods of discovering humans' satisfaction signs have been developed very intensively.

In our work we assume that although humans can adapt to any kind of interface, the most effective one is given to us by nature. It is natural face-to-face communication. Providing computers with the abilities to communicate with us in a shape of Embodied Conversational Agents (ECA) [2], we can dramatically increase the efficiency of human-computer collaborative performance.

Nevertheless we believe that introducing state-of-art technologies in graphics, speech processing and dialog management is not the only condition of success in this task. A number of issues related to social and cultural aspects of communications among humans and computers should be considered. Gathering feedback from humans about their attitudes towards agents is used in a number of research projects as a measure of team effectiveness. This is supported by the assumption that the more natural and pleasant interaction is with an agent, the more effective a team performance on a collaborative task. In order to discover humans' attitudes a number

S.O. Kuznetsov et al. (Eds.): RSFDGrC 2011, LNAI 6743, pp. 350–357, 2011.

of report based methods such as surveys and interviews are used. All of these methods have unquestionable advantage such as simplicity and low cost, however they all tend to rely upon a humans' awareness, honesty, cultural aspects.

Measuring subjects' performance in different kinds of collaborative tasks is another method for investigating effects of ECAs. Normally a combination of statistically significant surveys with collaborative task performance measurement gives high validity, however other metrics are often needed in order to verify results.

The Implicit Association Test [3] is a powerful psychological tool which has been already used for more than a decade in psychology. Further developments of the procedure and scoring algorithms of the IAT are ongoing, aiming to solve a number of different issues of the original IAT [4]. One of the best known modifications of the test is the Go/No-go Association Task [5], which aims to eliminate the need to bring a pair of categories into comparison. There are also some other modifications of the test, such as a Brief IAT [6] and Single Attribute IAT [7], which aim to solve known issues, simplify and improve the original test.

This goal of our research is to investigate a modification of the IAT which will extend the application domain of the conventional test to the possibility of using it with unfamiliar information and for indirect attitude measurements. Particularly, we use the conventional IAT in order to evaluate humans' attitudes towards slightly different kinds of presenting information by the same social actor. However our application method differs in principle from the original test application, we used a conventional test without any modification in order to discover it's possible drawbacks and find solutions. In this paper we propose modifications in the procedure and the scoring algorithm of the original IAT along with preliminary verification data.

The paper is organized as follows. Section 2 shows key differences between conventional IAT and applying a test to assess different types of presenting information. Section 3 contains a description and results of a conducted experiment along with our attempts to modify the original test and preliminary verification data. Known issues and future work directions are discussed in Section 4. The paper is concluded in Section 5.

2 Hypothesis

The Implicit Association Test is a very powerful psychological tool which can be used in order to discover a subjects' implicit preferences towards different categories. Particularly, IAT can be used to measure attitudes towards different kinds of objects and concepts, stereotypes, self-esteem and self-identity. The test requires a subject's rapid (in fact, as fast as possible) categorization of stimuli which appear on a screen. The problem is that the test will give a reliable result if and only if subjects make a reasonable number of mistakes, trying to keep a balance between rapid categorization without thinking and spontaneous key pressing. In order to achieve reliable results, subject are required to be a fluent English readers (if a test is conducted in English) and be aware of the topic of the test.

Some very well-known tests allows us to measure attitudes towards, for example, flowers and insects, different races, and food preferences. If flowers and insects are assessed, the stimuli are names or images of particular flowers and insects.

In our case we want to compare two methods of presenting information. We just want to know whether a virtual agent with rich animation can attract subjects more than just audio-presentation by the same voice as used in the case of agent. Noticeably, in this case we are talking about the same social actor. We do not compare different agents. We would like to go deeper and compare interfaces of the same agent.

In order to achieve this goal we use two slightly different information blocks, one of which is presented by a full-featured agent and the second by voice only. This causes the principal difference between our application and original usage of the IAT: in our case subjects are not familiar with these information blocks. These are not things which the subject uses every day. Thus, as opposed to the original IAT, in our test subject are expected to make not only misprints, but also mistakes. We believe that results given by the conventional test are not reliable because correct answers are always shown to subjects during a test and this can cause a learning-while-testing side-effect, which can tamper results.

Thus, our experiment has two goals. The first one is to discover whether the body of an agent really make sense for humans. The second is to find whether the learning-while-testing effect really exists.

As a result of this experiment, we expect to see at least a weak preference for one of the methods of presenting information from most of subjects. No preference will mean that our test is not well designed and it can not "catch" the difference in two presentation styles. We also expect to see a learning-while-testing side effect, which is caused by the procedure of the conventional test and which should make results of the test less reliable, because we obviously should eliminate any learning during the test.

3 Presenter Agent Experiment

The objective of the experiment is to investigate by using conventional IAT, whether the body of the ECA has an effect on subjects' attitudes towards two different methods of information mediation: full-featured ECA-based presentation and vocal presentation.

The experiment consists of two stages. In the first stage subjects were asked to learn two different stories from the presenter agent. Both stories are biographies of two famous Russian writers and both stories are unfamiliar to subjects. The key difference between the two stories is the presence of the ECA on the screen. One story was presented by the female agent with a synthesized female voice and a rich set of non-verbal cues, while the other story was presented by the same female voice only. Both stories were accompanied with the same number of illustrations which were used later as categories and items in the IAT. Both stories are approximately the same size (200 and 226 words) and difficulty of memorization. The order of stories and method of presenting (which of two stories is presented by ECA and which by voice only) were chosen individually for each subject. Before the experiment subjects were told that the test will evaluate their attitudes towards interactions with agents. Subjects were not told that they should memorize information with will be presented. A screen-shot of the full-featured ECA-based presentation is shown in Fig. 1. As opposed to the previous method, in vocal presentation the agent does not appear on the screen, but the entire environment remains absolutely the same.

Fig. 1. Screen-shot of the full-featured ECA-based presentation. ECA (female) gives an oral presentation which is accompanied by images (presentation pane on the right wall of the room).

According to [8] people tend to unconsciously interpret the same voice as the same social actor. By choosing the same female voice for both stories we eliminated the necessity of comparing distinct social actors. Instead, two methods of information mediating from the same social actor were assessed by the IAT.

In the second stage, subjects' attitudes towards information mediation methods were assessed by the IAT. Thus, two key differences between our experiment and the conventional IAT are (as we previously mentioned in Section 2):

a) attempt to utilize indirect method of measurement with IAT (we assess methods of information mediation by assessing information blocks which were presented to subjects);

b) attempt to use the IAT on items which subjects are not familiar with (information blocks had been learned at most half-an-hour before the IAT).

3.1 Results

In total 10 subjects have participated in the experiment. Eight of them are students and two are administrative staff of Kyoto University, eight male and two female, all Asians, and all can listen and read in English fluently. Their results are shown in Figure 2.

The horizontal axis of this graph represents the number of correct answers and the vertical axis represents the D measure. The positive value of the D measure shows subject's preference towards full-featured ECA-based style of presentation and vice versa, the negative value shows preference towards voice-only presentation. All subjects had an error rate of less than 25% during the test. It should be noted that in this figure we do not distinguish between mistakes and misprints as well as between misprints in testing and reference categories. This means that the real number of meaningful mistakes might be less than shown on the graph.

Thus, subjects #8, and #2 do not show any significant preference for any kind of presentation. Subjects #3, #9, and #10 show slight preference for ECA-based presentations, however subjects #1, and #7 show slight preference for voice-only

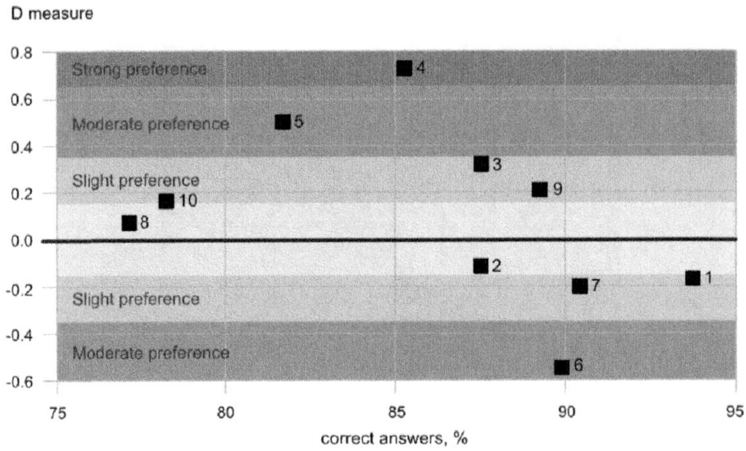

Fig. 2. Results of the IAT. X axis: number of correct answers in %; Y axis: result of the test (D measure)

presentations. Subject #5, and #6 have moderate preference for ECA-based and voice-only presentations respectively. Finally, subject #4 shows a strong preference for the ECA-based presentation. Altogether, five subjects show significant preference for ECA-based presentations, three subjects show preference for voice-only presentations and only two subjects do not show any significant preferences. This conforms to the first part of our hypothesis.

An important fact is that the three subjects who show significant preference for one of presentation styles, reported that they could memorize correct answers during the test. Thus, they confirmed that they experienced the learning-while-testing effect. Some other subjects also experienced the same effect, however they did not report clearly about it. According to Fig. 3 subjects #2, #5, #6, #9, and #10 gave more correct answers in blocks 6 and 7 than in blocks 3 and 4. Please note, that for Fig. 3 and Fig. 4 we calculated only meaningful mistakes and misprints, eliminating misprints in reference categories. The total number of answers is 32. And as we can see, none of the subjects made zero mistakes. The best result was given by subject #6 in the blocks 6 and 7 – 29 correct answers.

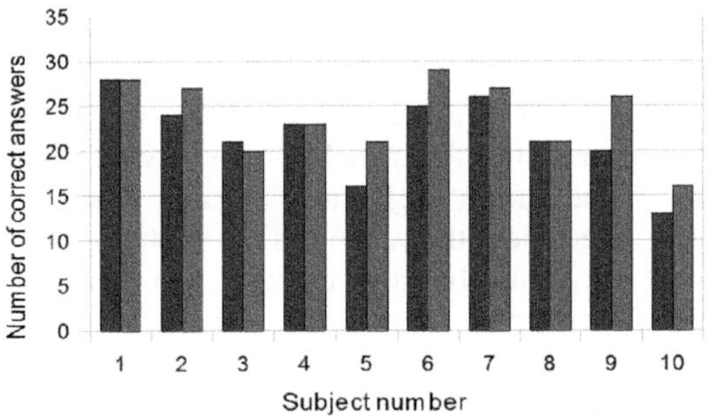

Fig. 3. Number of mistakes made in compared categories. Dark gray column – number of correct answers in block 3 + block 4; light gray column – block 6 + block 7.

Within ten minutes after the experiment subjects were asked to take the test once again. D-measures of the second test were not used and do not appear in Fig. 2, however we tried to find and analyze changes in the numbers of mistakes. These results are shown in Fig. 4.

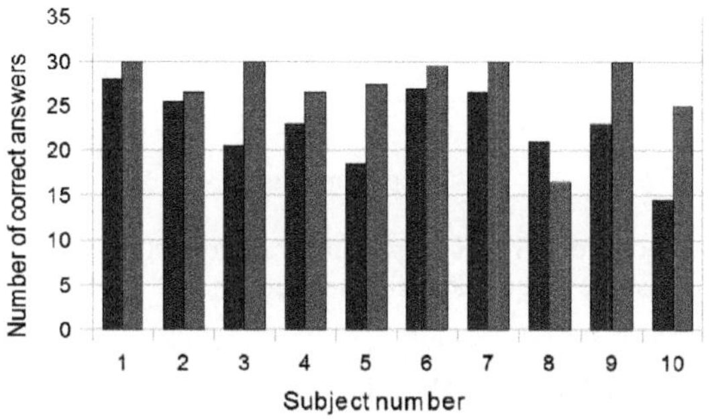

Fig. 4. Number of mistakes made in compared categories. Dark gray column – number of correct answers in the first test (mean of all pairing blocks); light gray column – second test (mean of all pairing blocks).

As we can see from the graph, subject #8 made quite a lot of mistakes in the second test, but all other subjects gave significantly more correct answers. In our opinion this means that subject learned correct answers during the first test and this confirms the second part of hypothesis.

3.2 Revised IAT Procedure and Scoring Algorithm

According to the aforesaid we propose to modify the procedure and scoring algorithm of the original IAT in order to address described issues.

The only difference in the procedure of the proposed test from the original one is that mistakes are not emphasized to subjects. We tried to not modify the essentials of scoring algorithm in order to be able to compare results of both tests. The difference is that before scoring we entirely remove each stimuli for which the number of mistakes exceeds 35% of their total numbers of presentations. In addition a further processing of wrong answers (such as giving penalties as in the original coring) was eliminated.

The modified test was preliminary verified by using a "flowers-insects" test scenario. This test is designed to measure subjects implicit preference towards flowers or insects. Results of the experiment are presented in Fig. 5.

In total 7 subjects have participated in the preliminary experiment. All are Europeans, aged from 26 to 37, students and lecturers of different schools of Kyoto University. The English ability is ranged from intermediate to fluent.

Subjects were asked to first pass a conventional test and then a modified test without any time lag. Results of the modified test are presented on the graph. D-measure of the modified test shows that all subjects show strong preference towards flowers (from 0.63 up to 1.25). This correlates with results of the conventional IAT.

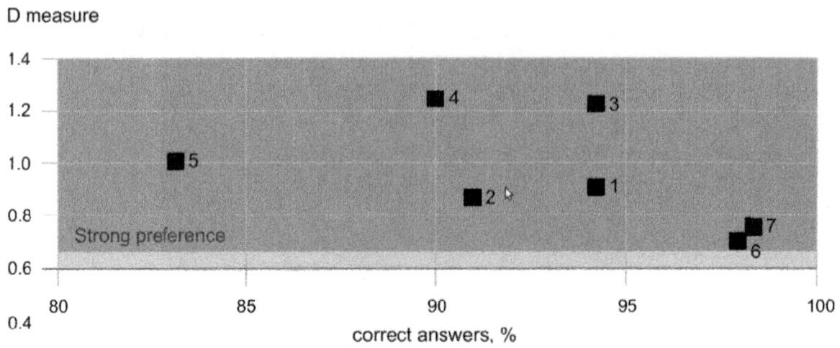

Fig. 5. Results of the modified IAT. X axis: number of correct answers in %; Y axis: result of the test (D measure). Positive value stands for the preference towards flowers.

4 Conclusions and Future Work

Taking into account the difference between conventional usage of an IAT and using an IAT for evaluating ECAs, several key issues can be defined and addressed in future work.

The one of conceptual issues is related to the fact that a conventional IAT deals with well known concepts while in the case of evaluating ECAs users deal with just-learned information, and this can cause mistakes in addition to misprints which are normal for the conventional IAT. In the conducted experiment subjects had only one chance to memorize information. Before the experiment they were not told that they should memorize information presented during the experiment, so they were expected to make mistakes.

On the other hand, during the conventional IAT wrong answers are always shown. Bearing in mind that in conventional IAT mistakes are not supposed to happen (misprints only, because subject deal with very familiar concepts only) this approach is very reasonable. However, for unfamiliar concepts, which we use in the experiment, it may cause a learning-while-testing side effect since each item is shown several times during the experiment.

Our proposed solution for the described problem includes several steps. The first is to not to emphasize wrong answers during experiment. Essentially, this will minimize the learning-while-testing effect, but at the same time can distort final results. We propose to eliminate all stimuli for which the total number of mistakes exceeds a fixed number. The proposed test is preliminary verified by experiment, however further verification is needed.

5 Summary

The goal of this work is to evaluate the potential possibility of using the Implicit Association Test where subjects' awareness of comparison concepts is less than in the case of conventional IAT, and to figure out possible issues related to this specific application. This paper presents results of the initial experiment where we used the conventional IAT procedure and scoring algorithms without any modifications along with our proposal for modifying the test. The data collected during the experiment shows how significant the difference between conventional usage of IAT and the proposed method is and which key issues should be addressed in future work. We showed our initial experiment which confirms our hypothesis about the effect of agent presence on the screen during presentation and procedural drawbacks of the conventional IAT in our particular circumstances. We made an analysis of D-measures and numbers of mistakes for each participant. We conducted a preliminary experiment with the modified test and outlined the directions of future research.

References

1. Sears, A., Jacko, J.A. (eds.): Human-Computer Interaction Handbook, 2nd edn. CRC Press, Boca Raton (2007) ISBN 0-8058-5870-9
2. Cassell, J., Sullivan, J., Prevost, S., Churchilll, E.F. (eds.): Embodied Conversational Agents. MIT Press, Cambridge (2000)
3. Greenwald, A.G., McGhee, D.E., Schwartz, J.K.L.: Measuring individual differences in implicit cognition: The Implicit Association Test. Journal of Personality and Social Psychology 74, 1464–1480 (1998)
4. Greenwald, A.G., Nosek, B.A., Banaji, M.R.: Understanding and Using the Implicit Association Test: I. An Improved Scoring Algorithm. Journal of Personality and Social Psychology 85, 197–216 (2003)
5. Nosek, B.A., Banaji, M.R.: The go/no–go association task. Social Cognition 19, 625–664 (2001)
6. Sriram, N., Greenwald, A.G.: The Brief Implicit Association Test. Experimental Psychology 56, 283–294 (2009)
7. Penke, L., Eichstaedt, J., Asendorpf, J.B.: Single-Attribute Implicit Association Tests (SA-IAT) for the Assessment of Unipolar Constructs. Experimental Psychology 53(4), 283–291 (2006)
8. Nass, C., Steuer, J.S., Tauber, E.: Computers are social actors. In: Proceeding of the Computer-Human Interaction (CHI 1994) Conference, pp. 72–78 (1994)

Visualization of Semantic Network Fragments Using Multistripe Layout

Alexey Lakhno and Andrey Chepovskiy

Higher School of Economics,
Data Analysis and Artificial Intelligence Department,
Pokrovskiy boulevard 11, 109028 Moscow, Russia
alakhno@gmail.com, achepovskiy@hse.ru

Abstract. Semantic network is an information model of knowledge domain. Objects and their relations are specified with an attributed graph. Multistripe layout is suitable for visualization of relations incident to the selected set of objects. The method provides a compact drawing that is guaranteed to avoid link crossings and label overlaps for objects and relations of corresponding subnetwork. In this paper we describe a common scheme of the multistripe layout approach and propose the way of visualization of semantic network fragments. These fragments may contain additional relations and objects in comparison with subnetworks considered earlier.

Keywords: semantic networks, relations visualization, multistripe layout, attributed graph drawing, link crossings, label overlaps.

1 Introduction

Semantic networks provide a natural representation of information about relations between objects. Formally semantic network can be considered like an attributed graph that contains labels on vertices and edges. The vertices of this graph correspond to the objects of knowledge domain, while the edges can be treated as the relations between them. The labels on vertices and edges specify the descriptions for corresponding objects and relations.

Multistripe layout, proposed in [1], is a method for drawing subnetworks induced by the set of relations incident to the selected objects. This method can be used for visualization of selected objects' direct relations. Multistripe layout provides regular and easy to follow drawings that can be used for visual analysis and report creation. Multistripe layout guarantees no link crossings and label overlaps. However the structure of concerned subnetworks is quite limited. There can be only selected objects and the objects directly adjacent to them (secondary objects). All other objects are ignored by the algorithm. Relations between the secondary objects are also out of scope. In this paper we propose an extension of the multistripe layout method that handles the limitations stated above.

Graph drawing covers a wide range of problems concerned with the visualization of networks and related combinatorial structures. A solid survey of this

S.O. Kuznetsov et al. (Eds.): RSFDGrC 2011, LNAI 6743, pp. 358–364, 2011.

area can be found in [2,3]. Multistripe layout combines several ideas from different graph drawing approaches. In a visibility representation, originally proposed in [4], each vertex is mapped to a horizontal segment and each edge to a vertical segment. This idea is used for visualization of the selected objects and their relations. The secondary objects are represented with rectangles bounding their labels. For visualization of relations multistripe layout uses polyline drawing convention — each edge is drawn as a polygonal chain. Edge labels are also represented with their bounding rectangles.

The rest of the paper is organized in the following way. Section 2 provides a formal description of subnetworks that can be visualized with the multistripe layout method and its extension. In Sect. 3 we describe a basic idea of the multistripe layout and its construction procedure. Section 4 presents the idea of layout extension. Finally, we summarize and conclude our work in Sect. 5.

2 The Object of Visualization

Multistripe layout method deals with the visualization of subnetworks induced with a set of relations incident to the selected objects. We assume that we are given

- A (possibly directed) graph $G_0 = \langle V_0, E_0 \rangle$, where V_0 is a set of vertices and E_0 is a set of edges. There are no selfloops in G, but it may contain multiple edges.
- Vertex and edge labels specified with the dimensions of bounding rectangles: $w(v)$, $h(v)$ for $v \in V_0$ and $w(e)$, $h(e)$ for $e \in E_0$, where w is the width and h is the height of rectangle.
- The selected vertices set $V' \subseteq V_0$ corresponding to the selected objects set.

The object of multistripe layout visualization is a subnetwork specified with a subgraph $G = \langle V, E \rangle$ of the graph G_0 where:

$$E = \{e \in E_0 | \text{ the edge } e \text{ is incident to some vertex } u \in V'\} \ ; \quad (1)$$

$$V = \{v \in V_0 | \text{ the vertex } v \text{ is incident to some edge } e \in E\} \ . \quad (2)$$

The graph G contains the selected vertices from V' and the vertices directly adjacent to them. Let's call the vertices from $V \setminus V'$ as *secondary* ones. There are no edges between the secondary vertices in G as each edge $e \in E$ is incident to some vertex $u \in V'$. So each edge of the graph G connects either a pair u_1, u_2 of selected vertices from V' or a selected vertex $u \in V'$ and a secondary vertex $v \in V \setminus V'$.

The extension, proposed in this paper, allows to use the multistripe layout method for visualization of network fragments of more general type. These fragments may incorporate the vertices, which are not directly adjacent to the selected vertices from V' but are connected to them through a chain of edges. Denote the set of additional vertices as V_{add}. Besides there can be a number of additional edges $e_{\text{add}} = (v_1, v_2)$ where $v_1, v_2 \in (V \setminus V') \cup V_{\text{add}}$. Let E_{add} be the set of such edges. So the extended subnetwork is specified with a graph $G_{\text{ext}} = \langle V_{\text{ext}}, E_{\text{ext}} \rangle$ where $V_{\text{ext}} = V \cup V_{\text{add}}$ and $E_{\text{ext}} = E \cup E_{\text{add}}$.

3 Multistripe Layout

Let's illustrate the idea of multistripe layout with a network fragment, which contains two selected vertices (Fig. 1). The selected vertices are represented with horizontal segments. The space between the segments is divided into three stripes: stripe A is used for layout of the secondary vertex labels, stripes B' and B'' are used for layout of the edge labels.

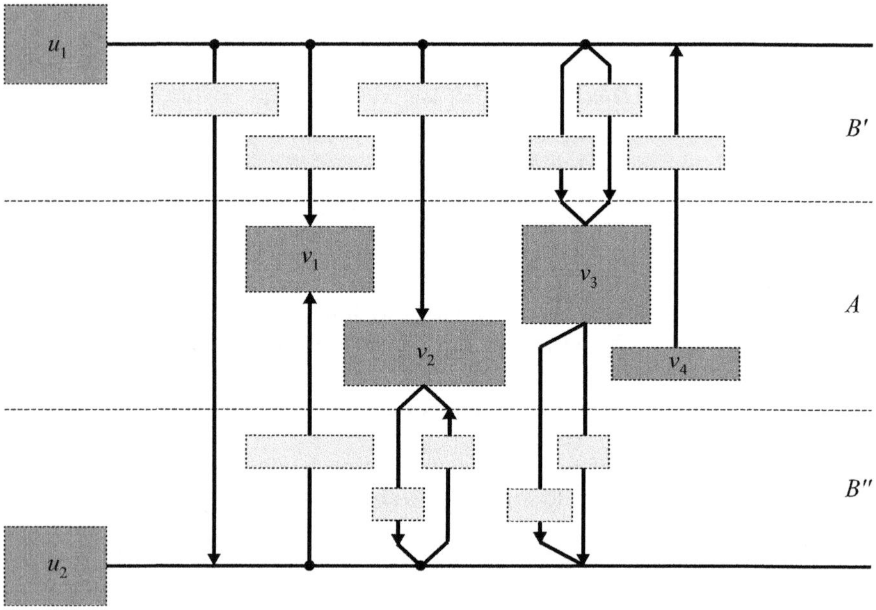

Fig. 1. Multistripe layout fragment: u_1, u_2 — selected vertices; v_1, v_2, v_3, v_4 — secondary vertices. *Dark shaded rectangles* correspond to vertex labels, *light shaded rectangles* correspond to edge labels. A, B' and B'' — layout stripes.

In a general case, if the selected set V' contains n vertices, multistripe layout uses $n+1$ stripes for the secondary vertex labels and $2n$ stripes for the edge labels (Fig. 2). The algorithm of multistripe layout construction consists of six steps:

1. Fix a relative order of the selected vertices $u_1, \ldots, u_n \in V'$.
2. Choose an addition order for the edges connecting selected vertices.

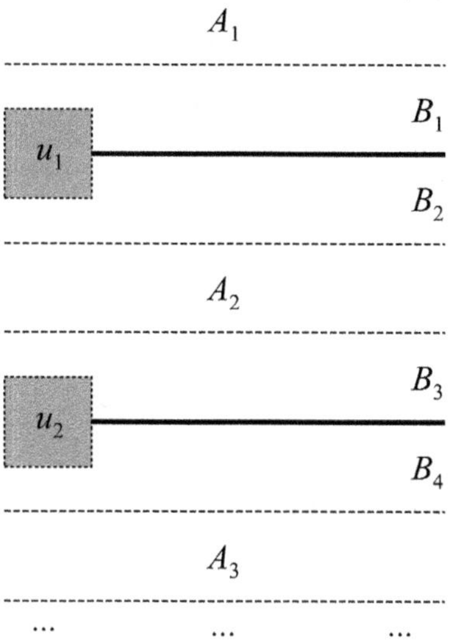

Fig. 2. Layout stripes: $A_1, A_2, \ldots$ — the stripes for layout of the secondary vertex labels; $B_1, B_2, \ldots$ — the stripes for layout of the edge labels

3. For every secondary vertex $v \in V \setminus V'$ define some layout stripe A_i.
4. Choose an addition order for the secondary vertices $v_1, \ldots, v_m \in V \setminus V'$.
5. Perform the layout of the edges that connect selected vertices. The edges are added to the drawing one by one according to the order defined in Step 2. The layout of each edge is performed in such a way to avoid label overlaps with the edges added earlier.
6. Perform the layout of the secondary vertices and the edges adjacent to them. The vertices are added one by one according to the order defined in Step 4. The layout procedure of each vertex performs the layout of adjacent edges.

The detailed description of the algorithm can be found in [1]. Here we shall focus on Step 5 and Step 6 as their understanding is essential for the proposed layout extension. For each edge e considered in Step 5 denote the set of stripes crossed by e as $C(e)$. The position of e is defined by the state of the stripes from $C(e)$. So for each of the stripes we keep the profile that describes the border between the busy part and the free part of the stripe (Fig. 3). After the addition of edge e all profiles from $C(e)$ are updated. Similarly for each secondary vertex v considered in Step 6 let $C(v)$ be the set of stripes that are crossed by the edges incident to v or used for v label placement. The profiles of the stripes from $C(v)$ are used for proper layout of v that is done in the following way:

- calculate the limitations on the placement of v and its adjacent edges;
- compare the limitations and perform the coordinated layout;
- update the profiles from $C(v)$ according to performed layout changes.

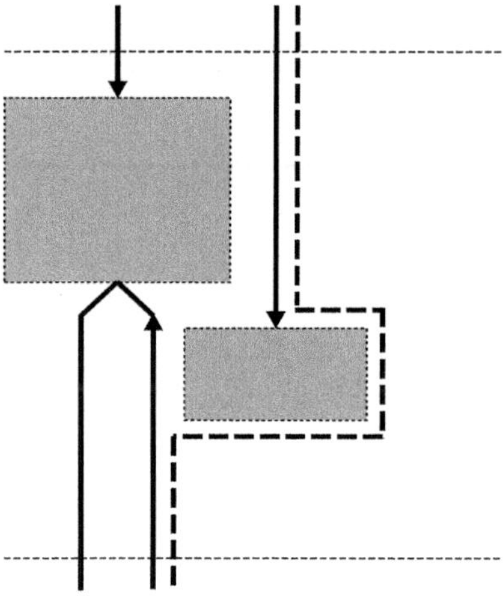

Fig. 3. Stripe profile. The *dotted line* separates busy and free parts of the stripe.

4 Layout Extension

The original multistripe layout method can be used for visualization of relations incident to the selected objects (Sect. 2). The corresponding subnetwork is specified with a graph $G = \langle V, E \rangle$ where V contains the selected vertices V' and the secondary ones $V \setminus V'$. There are two main ideas behind the visualization of $G_{\text{ext}} = \langle V \cup V_{\text{add}}, E \cup E_{\text{add}} \rangle$ using multistripe layout. The first one is the incorporation of V_{add} into the general mulistripe layout scheme temporarily connecting the vertices from V_{add} to the selected vertices. Vertex $v_{\text{add}} \in V_{\text{add}}$ should be connected to selected vertex $u \in V'$ if and only if they are connected with a chain of edges that does not pass through the other selected vertices. So according to the definition of V_{add} each vertex $v_{\text{add}} \in V_{\text{add}}$ will be adjacent to some selected vertex $u \in V'$ and can be treated as a secondary vertex. The second idea is the consideration of additional edges from E_{add} in Step 4 during the secondary vertices ordering. Connected vertices should be placed as close as possible. If the secondary vertices v_1 and v_2 connected with an edge $e_{\text{add}} \in E_{\text{add}}$ are placed to the same stripe A_i this idea allows to perform the automatic layout of e_{add} and its label in A_i. This perfectly works if there are one or two selected objects. However it can be used in a general case if there are no edges between the secondary vertices placed in different stripes.

The extension was implemented as a layout plugin for i2 Analyst's Notebook analytical system [5]. This software is designed for security investigations, risk management and fraud detection in business, law enforcement and counter terrorism activity support. We considered a problem of visualization of mobile contacts network. The objects of this network correspond to subscribers while the edges correspond to calls and messages. The analysis of such networks is actively used in police investigations for detection of criminal groups [6]. The proposed extension allows to perform automatic layout of complementary objects and relations on the schemes and thus visualize additional information (Fig. 4). It seems that multistripes layout method and its extension may also appear to be useful for social networks visulization. If users are treated as the objects of corresponding semantic network then multistripe layout method can provide drawings of the acquaintance circles of selected sets of users.

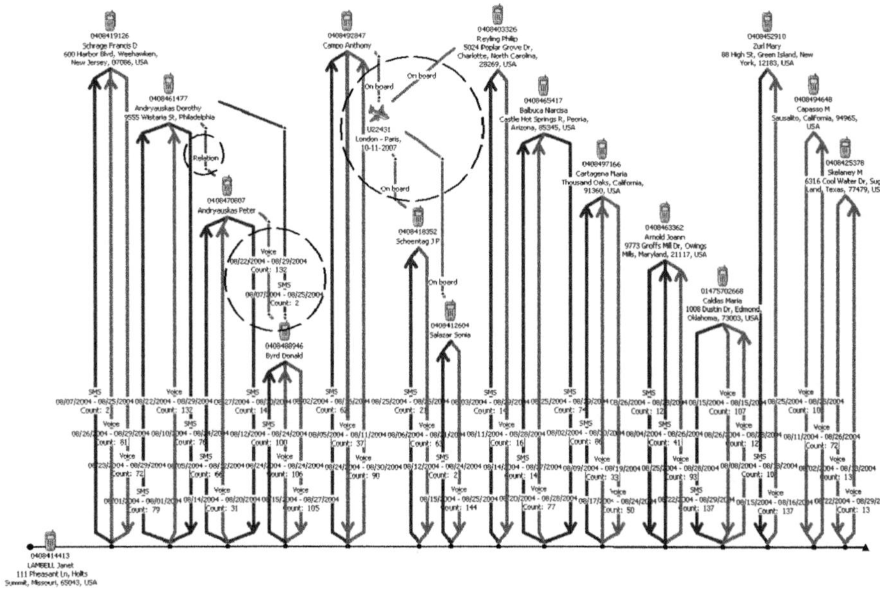

Fig. 4. The layout of mobile contacts network fragment. Additional objects and relations are marked with *circles.*

5 Conclusion

Multistripe layout is a method of visualization of relations incident to the selected set of objects. In this paper we presented the way to extend the applicability of multistripe layout to the network fragments of more general type. These fragments may contain additional vertices, which are not directly adjacent to the selected set of objects, and the edges that connect secondary vertices. The approbation of proposed method was performed on the base of i2 Analyst's Notebook. The method perfectly works if there are one or two selected objects. However it can be used for bigger selected sets on the assumption of some restrictions on the structure of relations.

References

1. Lakhno, A.P., Chepovskiy, A.M., Chernobay, V.B.: Visualization of Selected Objects Relations in a Semantic Network. Applied Informatics 6(30), 24–30 (2010)
2. Di Battista, G., Eades, P., Tamassia, R., Tollis, I.G.: Graph Drawing: Algorithms for the Visualization of Graphs. Prentice Hall, New Jersey (1999)

3. Kaufmann, M., Wagner, D. (eds.): Drawing Graphs: Methods and Models. Springer, London (2001)
4. Di Battista, G., Tamassia, R.: Algorithms for Plane Representations of Acyclic Digraphs. Theoret. Comput. Sci. 61, 175–198 (1988)
5. i2 Analyst's Notebook, http://www.i2group.com/us/products--services/analysis-product-line/analysts-notebook
6. Xu, J., Chen, H.: Criminal Network Analysis and Visualization: a Data Mining Perspective. Communications of the ACM 48(6), 101–107 (2005)

Pawlak Collaboration Graph and Its Properties

Zbigniew Suraj, Piotr Grochowalski, and Łukasz Lew

Chair of Computer Science, University of Rzeszów, Poland
{zsuraj,piotrg,llew}@univ.rzeszow.pl

1 Introduction

Nowadays, special kind of information gaining popularity is the one coming from social networks. In the paper we study basic statistical and graph-theoretical properties of the collaboration graph, which is an example of a large social network. To build such graph we use the data collected in the Rough Set Database System [9]. The collaboration graph contains data, among others, on Z. Pawlak, his co-authors, their co-authors, et cetera. In principle, the main idea presented in the paper is similar to the one of Erdos number [3], enriched with some concepts and techniques from social network analysis [1]. Analyzing our data we discover hidden patterns of collaboration among members of the rough set community [6],[8] which can be interesting for this community and others. Our data also provides fairly large, appealing real-life graphs on which one can test graph algorithms, in the spirit of [4].

Professor Zdzisław Pawlak (1926-2006) is one of the most known Polish computer scientists. He is a creator of the rough set theory [5] and a promoter of collaboration within the rough set community. This was the major inspiration for introducing the Pawlak number and the Pawlak collaboration graph.

The paper is organized as follows. Section 2 provides a definition of the Pawlak collaboration graph. In Section 3, we describe basic analysis results of the Pawlak collaboration graph. Section 4 includes concluding remarks and further work considerations.

2 Pawlak Collaboration Graph

In order to reveal a social phenomenon of collaboration in rough set research, we defined the collaboration graph in the paper [6]. In the considered graph the vertices represent all researchers (rough set paper authors [9] in particular), whereas the edges represent collaboration relations between two given authors. Two vertices of the graph are joined with an edge, if the two authors have had a joint research paper published, with or without other co-authors. A simple edge fixed between two authors in the graph means one or more co-publications. The structure of the collaboration graph together with its basic properties have been presented in [6]. In order to characterize more precisely existing collaboration between the rough set community members we define a subgraph of the graph with a distinguished vertex corresponding to Pawlak.

S.O. Kuznetsov et al. (Eds.): RSFDGrC 2011, LNAI 6743, pp. 365–368, 2011.

Table 1. The evolution of the Pawlak graph over time

	$n_P = 0,1$		$n_P = 2$		$n_P = 3$		$n_P = 4$		$n_P = 5$		$n_P = 6$		$n_P = 7$		Graph G	
Year	$\|V_1\|$	$\|E_1\|$	$\|V_2\|$	$\|E_2\|$	$\|V_3\|$	$\|E_3\|$	$\|V_4\|$	$\|E_4\|$	$\|V_5\|$	$\|E_5\|$	$\|V_6\|$	$\|E_6\|$	$\|V_7\|$	$\|E_7\|$	$\|V\|$	$\|E\|$
2006	23	411	251	518	198	219	130	149	74	64	16	5	4	5	724	1566
2007	23	424	261	559	220	237	134	157	82	67	16	5	4	8	776	1680
2008	23	433	266	595	242	319	180	440	169	160	37	10	6	3	923	2161
2009	23	439	271	630	260	393	210	453	192	206	51	33	12	3	1019	2382
2010	23	439	271	631	269	393	210	453	192	206	51	33	12	3	1019	2383

Before introducing such graph definition, we need the one of the Pawlak number. *The Pawlak number* n_P of an author is defined as follows: Pawlak himself has $n_P = 0$; people who have written a joint paper with Pawlak have $n_P = 1$; and their co-authors, with the Pawlak number not defined yet, have $n_P = 2$; etc. Pawlak numbers can be interpreted as vertex distances (the number of edges in a shortest path joining two given vertices) from Pawlak vertex.

The experiments showed that the number of people signified with the Pawlak number from 0 to 7, according to the RSDS data, is: 1, 22, 271, 260, 210, 192, 51, 12, respectively. Thus, the median of Pawlak numbers is 3; the mean is 3.47, and the standard deviation - 1.32. In our case the standard deviation is low which indicates that the data points tend to be very close to the mean. This in turns most authors (about 68 percent, assuming normal distribution) have the Pawlak number from the interval [2.15,4.79], considering one standard deviation. When it comes to two standard deviations almost all the authors (approximately 95 percent) obtain the Pawlak number falling into [0.83,6.11].

A graph $G = (V, E)$, where V is a set of vertices representing known authors in our RSDS database with $n_P \leq 7$ and E is a set of edges connecting two authors, if they wrote a joint paper, and at least one of them has $n_P \in \{0, 1, \ldots, 6\}$. The graph G is called *the Pawlak collaboration graph* (*the Pawlak graph* in short). Currently, the data on collaboration among authors with $n_P = 8$ is not available in our database, yet.

3 Basic Analysis of Pawlak Collaboration Graph

We can turn now to the issue of collaboration in rough set research. Firstly, we provide basic statistics of the Pawlak graph G, then more advanced graph-theoretical analysis of its properties. Table 1 shows the evolution of the Pawlak graph over time. It is clear that the graph's size grows significantly in time. However, the size of subgraphs related to particular Pawlak numbers decreases with the vary numbers' increase (omitting Pawlak numbers 0 and 1).

As Table 2 indicates, the average degree (average number of co-authors collaborating with an author) fluctuates between 21.59 for the Pawlak number 1 and 2.42 for the Pawlak number 7 with distinctive decreasing trend. A similar tendency can be observed in the case of the maximum degrees.

Table 2. Basic statistics on degrees in the Pawlak graph

	$n_P \in \{0,1\}$	$n_P = 1$	$n_P = 2$	$n_P = 3$	$n_P = 4$	$n_P = 5$	$n_P = 6$	$n_P = 7$
Minimum	2	2	1	1	1	1	1	2
Median	24.5	4.5	13.0	9.5	16.5	6.5	3.5	2.5
Average degree	21.61	21.59	5.35	3.54	5.12	3.42	2.73	2.42
Maximum	63	63	37	30	55	21	6	3

If we remove Pawlak himself and his connections from the graph G we get so called *the truncated Pawlak collaboration graph* G'.

The data used in this article covers the period from 1981 to 2010. The latest, 2010 edition, of the graph G contains 1019 vertices and 2383 edges, and the graph G' has 1018 vertices and 2361 edges. There are 1294 vertices outside G, which for this analysis purpose will be ignored as they do not collaborate with so called Pawlak research group. Other graph-theoretical properties of G' provide further insight into the rough set researchers' interconnections. There are 3 connected components in G'. The largest component contains 996 authors and two remaining ones are small (2 and 20 authors). Next, we concentrate on the largest component of G'. The diameter (maximum distance between two vertices) of the largest component is 12 and the radius (minimum eccentricity of a vertex, with an eccentricity defined as the maximum distance from that vertex to any other) is 6. For any fixed vertex u in the largest component, we can enquire about the shape of the distance distribution from u to the other 995 vertices in this component. The distance from u to v is certainly the Pawlak number of v, when u is Pawlak. It would be interesting to determine the shape of the distance distribution from a given u to other vertices in the largest component of G', and compare the outcome with the results presented in [2].

As a final measure of collaboration, we use the concepts of a k-core and the collaborativeness defined below.

Let $G = (V, E)$ be a graph, $W \subseteq V$, and let $v \in V$. A maximal subgraph $H_k = (W, E|W)$ induced by the set W is called a *k-core* iff $\forall v \in W : \deg_{H_k}(v) \geq k$ [1]. The core of maximum order is called the *main* core.

In the experiments as a measure of author's *collaborativeness* [1] we use the quantity $coll(v) = \frac{core(v)}{\overline{core}(v)}$, where $core(v)$ is the largest value k for v such that it belongs to a k-core, and $\overline{core}(v)$ is the average core number of all co-authors for v such that $\overline{core}(v) = 0$, if $N(v) = \emptyset$ otherwise $N(v) = \frac{1}{|N(v)|}\sum_{u \in N(v)} core(u)$, where $N(v) = \{u \in V : (v, u) \in E\}$ called *neighborhood* of vertex v. We assume that $coll(v) = 0$, if $core(v) = 0$. This parameter measures the openness of the author v towards external authors.

In G' the main core consists of 21 vertices (total number of authors), and its order is 20. The average number of all co-authors in G' is 26.6, and the average of their collaborativeness is 1.275. For all the authors from the main core of G' the minimal value of the parameter *coll* is 1.0, and the maximal one - 2.231.

4 Conclusions and Future Work

The analysis' results of the Pawlak graph using the authors' own software have been presented in the paper. They provide hidden patterns of collaboration among members of the rough set community. Additional restrictions on co-authors have been set for the sake of other interpretation of obtained results and more rigorous analysis. In the approach we have computed the characteristics of the Pawlak graph in which two authors are linked in the graph, if they have written a joint paper whether, or not, other authors were involved. It is interesting to define the Pawlak collaboration graph in such a way that we put an edge between two vertices, if the authors have a joint paper, with no other co-authors. It is clear that this new definition of the Pawlak graph is more restrictive than previous one. It provides a wonderful opportunity for further study on publishing patterns among rough set researchers. This exemplifies the problems we would like to investigate by applying the approach presented in the paper. Moreover, following papers will be devoted to some additional techniques for analysis of large social networks and their parts' visualisations, in the case of the Pawlak graph (cf. [1]).

Last but not least, seeing the following statement: '*My Pawlak number is...*' on home pages of the rough set researchers or people interested in that field, would be a great pleasure. Authors of this paper collected the related data and made them available at the URL: http://rsds.univ.rzeszow.pl (Pawlak numbers)

Acknowledgment. We wish to thank the anonymous referees for constructive remarks and useful suggestions to improve the presentation of the paper.

References

1. Batagelj, V., Mrvar, A.: Some Analyses of Erdös Collaboration Graph. Social Networks 22(2), 173–186 (2000)
2. Grossman, J.W.: Patterns of Collaboration in Mathematical Research. SIAM News 35(9) (2002)
3. Grossman, J.W.: The Erdös Number Project (1996), `http://www.oakland.edu/grossman/erdoshp.html`
4. Knuth, D.: The Stanford GraphBase. Addison-Wesley, Reading (1993)
5. Pawlak, Z.: Rough Sets - Theoretical Aspects of Reasoning About Data. Kluwer, Dordrecht (1991)
6. Suraj, Z., Grochowalski, P.: Patterns of Collaborations in Rough Set Research. Studies in Fuzziness and Soft Computing, vol. 224, pp. 79–92. Springer, Heidelberg (2008)
7. Suraj, Z., Grochowalski, P.: The Rough Set Database System. In: Peters, J.F., Skowron, A. (eds.) Transactions on Rough Sets VIII. LNCS, vol. 5084, pp. 307–331. Springer, Heidelberg (2008)
8. Suraj, Z., Grochowalski, P.: Some Comparative Analyses of Data in the RSDS System. In: Yu, J., Greco, S., Lingras, P., Wang, G., Skowron, A. (eds.) RSKT 2010. LNCS, vol. 6401, pp. 8–15. Springer, Heidelberg (2010)
9. Website of the RSDS system, `http://rsds.univ.rzeszow.pl`

Author Index

Batch number: 09490872

Printed by Printforce, the Netherlands